MODELLING OF BIOMOLECULAR STRUCTURES AND MECHANISMS

THE JERUSALEM SYMPOSIA ON
QUANTUM CHEMISTRY AND BIOCHEMISTRY

Published by the Israel Academy of Sciences and Humanities,
distributed by Academic Press (N.Y.)

1. *The Physicochemical Aspects of Carcinogenesis* (October 1968)
2. *Quantum Aspects of Heterocyclic Compounds in Chemistry and Biochemistry* (April 1969)
3. *Aromaticity, Pseudo-Aromaticity, Antiaromaticity* (April 1970)
4. *The Purines: Theory and Experiment* (April 1971)
5. *The Conformation of Biological Molecules and Polymers* (April 1972)

Published by the Israel Academy of Sciences and Humanities,
distributed by D. Reidel Publishing Company (Dordrecht, Boston, Lancaster, and Tokyo)

6. *Chemical and Biochemical Reactivity* (April 1973)

Published and distributed by D. Reidel Publishing Company
(Dordrecht, Boston, Lancaster, and Tokyo)

7. *Molecular and Quantum Pharmacology* (March/April 1974)
8. *Environmental Effects on Molecular Structure and Properties* (April 1975)
9. *Metal-Ligand Interactions in Organic Chemistry and Biochemistry* (April 1976)
10. *Excited States in Organic Chemistry and Biochemistry* (March 1977)
11. *Nuclear Magnetic Resonance Spectroscopy in Molecular Biology* (April 1978)
12. *Catalysis in Chemistry and Biochemistry Theory and Experiment* (April 1979)
13. *Carcinogenesis: Fundamental Mechanisms and Environmental Effects* (April/May 1980)
14. *Intermolecular Forces* (April 1981)
15. *Intermolecular Dynamics* (Maart/April 1982)
16. *Nucleic Acids: The Vectors of Life* (May 1983)
17. *Dynamics on Surfaces* (April/May 1984)
18. *Interrelationship Among Aging, Cancer and Differentiation* (April/May 1985)
19. *Tunneling* (May 1986)
20. *Large Finite Systems* (May 1987)

Published and distributed by Kluwer Academic Publishers
(Dordrecht, Boston, London)

21. *Transport through Membranes: Carriers, Channels and Pumps* (May 1988)
22. *Perspectives in Photosynthesis* (May 1989)
23. *Molecular Basis of Specificity in Nucleic Acid-Drug Interaction* (May 1990)
24. *Mode Selective Chemistry* (May 1991)
25. *Membrane Proteins: Structures, Interactions and Models* (May 1992)
26. *Reaction Dynamics in Clusters and Condensed Phases* (May 1993)
27. *Modelling of Biomolecular Structures and Mechanisms* (May 1994)

VOLUME 27

MODELLING OF BIOMOLECULAR STRUCTURES AND MECHANISMS

PROCEEDINGS OF THE TWENTY-SEVENTH JERUSALEM SYMPOSIUM ON QUANTUM CHEMISTRY AND BIOCHEMISTRY HELD IN JERUSALEM, ISRAEL, MAY 23–26, 1994

Edited by

A. PULLMAN
Institut de Biologie Physico-Chimique
(Fondation Edmond de Rothschild), Paris, France

J. JORTNER
The Israel Academy of Sciences and Humanities,
Jerusalem, Israel

and

B. PULLMAN
Institut de Biologie Physico-Chimique
(Fondation Edmond de Rothschild), Paris, France

KLUWER ACADEMIC PUBLISHERS

DORDRECHT / BOSTON / LONDON

Library of Congress Cataloging-in-Publication Data

```
Jerusalem Symposium on Quantum Chemistry and Biochemistry (27th : 1994
   : Jerusalem, Israel)
     Modelling of biomolecular structures and mechanisms : proceedings
of the Twenty-seventh Jerusalem Symposium on Quantum Chemistry and
Biochemistry held in Jerusalem, Israel, May 23-26, 1994 / edited by
A. Pullman, J. Jortner, and B. Pullman.
       p.   cm. -- (The Jerusalem symposia on quantum chemistry and
biochemistry ; v. 27)
     Includes bibliographical references.
     ISBN 0-7923-3102-8
     1. Biomolecules--Structure--Computer simulation--Congresses.
2. Biomolecules--Structure--Mathematical models--Congresses.
3. Structure-activity relationships (Biochemistry)--Congresses.
I. Pullman, Alberte.  II. Jortner, Joshua.  III. Pullman, Bernard,
1919-   .  IV. Title.  V. Series.
QP517.M3J47  1994
574.19'28--dc20                                        94-32858
```

ISBN 0-7923-3102-8

Published by Kluwer Academic Publishers,
P.O. Box 17, 3300 AA Dordrecht, The Netherlands

Kluwer Academic Publishers incorporates
the publishing programs of
D. Reidel, Martinus Nijhoff, Dr W. Junk and MTP Press.

Sold and distributed in the U.S.A. and Canada
by Kluwer Academic Publishers,
101 Philip Drive, Norwell, MA 02061, U.S.A.

In all other countries, sold and distributed
by Kluwer Academic Publishers Group,
P.O. Box 322, 3300 AH Dordrecht, The Netherlands

Printed on acid-free paper

Printed in the Netherlands

Table of Contents

*Contributions indicated with an asterisk will also appear in the journal *Molecular Engineering*, Volume 5, Nos. 1–3.

Preface

The 27th Jerusalem Symposium was devoted to the theme of the modelling of biomolecular structures and mechanisms, a theme which has grown recently very much in importance and in audience. The main topics on which the Symposium focused were: nucleic acids and their interactions, proteins and their interactions, membranes and their interactions, enzymatic processes and pharmacological and medical aspects. The meeting gathered a number of the foremost world experts on these subjects and made possible a thorough evaluation of the status of our present knowledge in these fields.

Held under the auspices of the Israel Academy of Sciences and Humanities and the Hebrew University of Jerusalem, the Twenty Seventh Jerusalem Symposium was sponsored by the Institut de Biologie Physico-Chimique (Fondation Edmond de Rothschild) of Paris. We wish to express our deep thanks to Baron Edmond de Rothschild for his continuous and generous support, which makes him a true partner in this important endeavour. We would also like to express our gratitude to the Administrative Staff of the Israel Academy and, in particular, to Mrs. Avigail Hyam for the efhciency and excellency of the local arrangements.

Alberte PULLMAN
Joshua JORTNER
Bernard PULLMAN

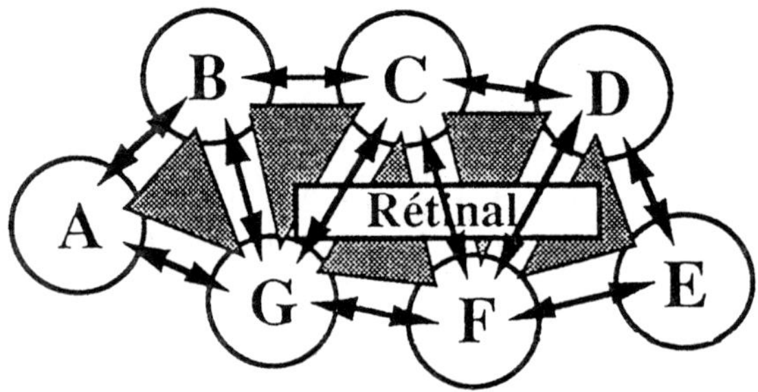

Fig. 1. Schematic view of the seven helices of bacteriorhodopsin viewed from the cytoplasmic face. The topology of the bundle leads to a decomposition of the bundle into 11 helix pairs (arrows), and five helix triplets (shaded triangles).

helices together, followed by a quasi-Newton energy minimisation including all side chains as well as helical rotation and axial displacement variables.

Tests on the Determination of the Rotational Orientation of α-Helix Bundles

The hierarchical approach that has been described involves several approximations whose effects have to be evaluated.

A preliminary study was performed on myohemerythrin [13], a four-helix-bundle, soluble protein, for which an X-ray structure exists at 1.7Å resolution. This study was followed by an investigation of bacteriorhodopsin (BR) [3], a much more complex protein involving seven helices surrounding a chromophore (retinal). The current model of BR, based on high resolution electron microscopy, defines the arrangement and axis positions of its seven transmembrane helices [14]. Helix rotational orientations have been studied by electron microscopy [14] and neutron diffraction [15,16]. The position of the retinal prosthetic group has also been investigated (reviewed in [14]). Because atomic resolution data are not yet available, approximate side-chain positions are known only for the largest residues. The structure of the extramembrane regions is not known in detail.

For the small four helix bundle of myohemerythrin, only the pair level was used during the rotational search. For the seven helices of BR, due to the topology of the bundle, we have considered an approximation composed of 11 pairs and five triplets of helices (Figure 1). These studies already allow us to draw two conclusions concerning our search strategy:

Firstly, it seems that a 20° step size for rotational scanning is sufficient since relaxation of the full bundle led individual helices to rotate by no more than 15°. It may be added that, during the full bundle refinement, helix displacements along the helical axis were limited to roughly 1 Å and thus searching axial displacement would require a step size of the order of 1.5–2 Å.

Secondly, it seems that a reasonable number of pseudo-optimal solutions kept at the pair level stage is of the order of several thousand (typically, we kept 5000 solutions for the 7 helices of BR). This is a very small number compared to the

TABLE III

Best solutions obtained for the rotational positioning of the helices with the bacteriorhodopsin bundle. Only the solutions presenting a stable packing are presented. Energies are in kcal/mol, axial rotations (r) in degrees and axial translations (t) in Å. *set* Corresponds to the origin of the solution: * indicated the D helix was not displaced along its axis, while other conformations began with this helix displaced by 2.5 Å. *Epck.* is the difference between the energy of the bundle and the sum of the energies of the isolated helices *Eint.* is the sum of the helix interaction energies

set	Epck.	Eint.	rA	rB	rC	rD	rE	rF	rG	tA	tB	tC	tD	tE	tF	tG
	-160	-388	4	-1	-4	-13	4	-7	-1	0.5	0.7	0.1	-2.3	0.6	-0.1	0.2
	-123	-380	4	-1	-2	-16	14	-5	0	0.8	0.7	0.3	-2.5	-0.3	-0.1	0.2
	-121	-386	2	-3	-3	-5	-10	-11	-2	0.8	0.7	0.1	-1.7	0.8	-0.6	0.1
	-84	-385	1	0	-4	-5	-15	-7	-1	0.3	0.7	0.1	-1.8	0.8	-0.3	0.2
	-81	-376	0	0	-4	-2	-16	-8	-2	0.2	0.7	0.2	-1.7	0.8	-0.3	0.2
*	-79	-353	6	0	-1	17	2	-4	0	0.7	0.4	-0.4	-0.2	-0.2	-0.3	-0.1
	-76	-385	1	0	-9	-17	1	-6	0	0.4	0.8	0.4	-2.5	0.4	-0.1	0.3
	-74	-380	0	1	-8	-12	-1	-4	0	0.4	0.8	0.3	-2.5	0.2	0.1	0.4
*	-73	-342	4	3	0	21	2	-8	0	0.6	0.5	-0.5	-0.1	-0.4	-0.5	0.0
	-69	-358	2	6	-5	-13	7	-10	-1	0.2	0.6	0.2	-2.2	0.5	0.1	0.2
	-54	-354	1	-3	-6	-14	9	-3	-1	0.8	0.8	0.5	-2.4	0.4	-0.1	0.0
	-52	-360	2	6	-6	-14	-7	-11	-1	0.3	0.8	-0.1	-2.3	0.9	-0.4	0.3
*	-34	-320	6	1	3	17	6	-3	0	0.7	0.4	-0.1	-0.3	-0.4	-0.3	-0.1
	-24	-344	0	1	-4	0	-18	-126	1	0.5	1.0	-0.1	-2.6	-0.0	0.1	0.6
*	-20	-322	3	2	3	16	6	-3	0	0.2	0.5	-0.0	-0.4	-0.3	-0.2	0.0
	-18	-314	2	4	-3	24	99	-4	-1	0.3	0.8	-0.0	-1.9	0.2	-0.1	0.3
	-11	-338	5	1	-7	22	105	-120	4	0.4	0.5	-0.0	-2.5	0.2	0.6	0.4
	-2	-341	1	2	-9	-18	7	-1	-1	0.2	0.6	0.7	-2.3	0.4	0.0	0.2
*	-1	-341	4	2	0	18	111	-119	2	0.0	0.5	-0.6	-0.1	-0.3	0.4	0.0
	0	-332	2	-2	-14	-8	-22	-119	2	0.5	0.8	0.1	-3.8	0.1	1.1	0.7

total number of possible orientations. It is also found that the solutions retained at the triplet level are mainly composed mixtures of the few best solutions or minor deviations from these (generally by only a single 20° step). These deviations can easily be removed during the refinement of the full bundle.

Finally, in both cases studied so far, the rotational orientations found show that our method is indeed able to find rotational orientations in good agreement with experimental data. In the case of the myohemerythrin, correct helix orientations were obtained if a simple correction was made for hydration effects not included in our force field. This consisted of using the amphipathic profiles of each helix to partially restrict the rotational orientations searched (A, D ± 90°, C ± 60° D no restriction). This study also allowed us to show that the terminal residues of the helices seemed to have little influence on their optimal packing. In the case of the BR, good orientations were found for all seven helices forming the bundle (the D helix was however restrained to ±90° rotations due to its lateral amphipathicity), with no more than a few degrees of rotation with respect to the best experimental model available, from high resolution electron microscopy (see Table III). Our calculations also supported the suggestion of the authors of the experimental study that the D helix should be displaced along its axis by roughly 2 Å toward its N terminus. If this same displacement is made in the experimental conformation, the Cα positions differ from those of our best model by an overall RMS of only 0.37 Å. For the side chains, the overall RMS is 2.06 Å, with a value of 1.33 Å for internal residues. It should however be recalled that the electron microscopy data used to generate the experimental structure was not detailed enough to place any but the largest aromatic side chains with any degree of certainty. Several major changes in the side chain conformation involving aromatic residues on the periphery of the bundle are seen. These changes may be related to the lack of any explicit model for the lipids surrounding BR, but these changes seem to be of little importance in determining the optimal packing of the bundle.

Toward Building Models Starting from Electron Crystallography Density Maps ?

Our results for the rotational orientations of helices are still dependant on experimental information: The helix axes used were those of the experimental models, and their relative longitudinal positioning was preserved for all but one helix of BR. Transversal displacement was not considered. In addition, the backbone geometry was kept in its experimental conformation, that is in some case far from ideal helical conformation, since some helices (like the C helix of BR) exhibit important curvature. Although our approach is thus still far from *ab initio* structural prediction, we have begun to examine the possibility of building models starting from low resolution electron microscopy maps, such as that recently obtained for rhodopsin [17]. Two points are currently under investigation.

The first concerns the geometry of the helix backbones. A major question concerns the factors determining the distorsions of some helices (notably those containing proline residues) from ideal α-helical geometry. If such curvature is mainly determined by the presence of surrounding helices, this would imply the inclusion of backbone degrees of freedom at each level of the methodology. On the contrary, curvature mainly governed by the helix amino acid sequence could be easily solved by simply studying the best backbone geometry prior to studies of the helix bundle. We have thus explored the possibility of reproducing the curvature of isolated helices containing prolines, starting from ideal α-helical backbone geometries. Our preliminary results, issued from bending maps for each helix, show that the optimal conformations exhibit a bending close to that observed within the experimental model.

The second point under investigation is a test of the sensitivity of our results concerning the rotational orientation of the helices when taking into account experimental uncertainties in the axes positions. This study is expected to help defining of a more general strategy which would allow helix repositioning during rotational searches.

Conclusion

The approach that has been developed to predict the packing of helices within transmembrane bundles is based on the assumption that helices are prefolded within the membrane before packing takes place, and that inter-helix packing is mainly directed by helix side chains. We also assume that the extra membrane loop regions connecting the helices do not play a crucial role in packing. The strategy described has enabled us to treat very complex problems with only reasonable computational expense. The results obtained for rotational orientation of the bundles of myohemerythrin and BR, in good agreement with experimental data, would appear to justify our hypotheses, and suggest that this technique may now be applied to a variety of new problems.

Possible applications include the refinement of high resolution electron microscopy data and also modelling of site directed mutagenesis with the aim of locating those residues most important for determining the bundle conformation.

References

1. J. L. Popot, C. de Vitry, and A. Atteia: 'Folding and assembly of integral membrane proteins. An introduction', in White (Ed.), *Membrane Protein Structure: Experimental Approaches*, Oxford University Press, Oxford, in press (1994).
2. J. L. Popot and C. de Vitry: *Ann. Rev. Biophys. Chem.* 369 (1990).
3. P. Tuffery, C. Etchebest, J. L. Popot, and R. Lavery: *J. Mol. Biol.* **236**, 1105 (1994).
4. J. M. Baldwin: *Embo J.* **12**, 1693 (1993).
5. D. Donelly, J. P. Overington, S. V. Ruffle, J. H. Nugent and T. L. Blundell: *Protein Sci.* **2**, 55 (1993).
6. W. R. Taylor, D. T. Jones, and M. Green: *Proteins* **18**, 281 (1994).
7. J. L. Popot and D. Engelman: *Biochemistry*, **29**, 4031 (1990).
8. B. J. Borman and D. M. Engelman: *Ann. Rev. Biophys. Biomol. Struct.* **21**, 223 (1992).
9. J. L. Popot: *Curr. Opin. Struct. Biol.* **3**, 532 (1993).
10. P. Tuffery, C. Etchebest, S. Hazout, and R. Lavery: *J. Biomol. Struct. Dynam.* **8**, 1267 (1991).
11. P. Tuffery, C. Etchebest, S. Hazout, and R. Lavery: *J. Comp. Chem.* **14**, 790 (1993).
12. H. Schrouber, F. Eisenhaber, and P. Argos: *J. Mol. Biol.* **230**, 592 (1993).
13. P. Tuffery and R. Lavery: *Proteins* **14**, 413 (1993).
14. R. Henderson, J. M. Baldwin, T. A. Ceska, F. Zemlin, E. Beckmann, and K. H. Downing: *J. Mol. Biol.* **213**, 899 (1990).
15. J. L. Popot, D. M. Engelman, O. Gurel, and G. Zaccaï: J. Mol. Biol. **210**, 829 (1989).
16. F. A. Samatey, G. Zaccaï, D. M. Engelman, C. Etchebest, and J. L. Popot: *J. Mol. Biol.* **236**, 1093 (1994).
17. G. F. X. Schertler, C. Villa, and R. Henderson: *Nature.* **362**, 770 (1993).

Binding Sites of Acetylcholine in the Aromatic Gorge Leading to the Active Site of Acetylcholinesterase

ALBERTE PULLMAN
Institut de Biologie Physico-Chimique, Fondation Edmond de Rothschild, 13, rue Pierre et Marie Curie, 75005 Paris, France.

Abstract. Theoretical calculations performed on the interactions of acetylcholine with the 'aromatic gorge' of acetylcholinesterase indicate the existence of a number of local minima for the substrate. These minima are clustered in four regions of increasing interactions from top to bottom of the gorge, culminating in the region of the 'active site'. The results allow the delineation of the role of the different aminoacids lining the walls, emphasizing, in particular, that of Trp 279 and Trp 84 while smaller interactions involve tyrosines 70, 121, 130, 334 and Phe 330. The influence of D72 is stressed, as well as the orientating role of A 201 and the strong driving influence of E199.

Key words. Acetylcholine, binding site, acetylcholinestesrase, molecular modelling

Introduction

Acetylcholinesterase (ACHE) catalyses with considerable efficiency the hydrolysis of the ester linkage of the neurotransmitter acetylcholine:

$$CH_3COOCH_2CH_2N^+(CH_3)_3$$

into choline and acetic acid, thereby terminating the transmission of nerve impulse at cholinergic synapses [1].

The active site, at which the reaction takes place, was described very early [2,3] as comprising two subsites: the 'esteratic site', where the ester linkage of the substrate is broken after intermediary binding of its carbonyl carbon atom to an 'activated' serine residue [4], and the 'anionic site', assumed to interact with the trimethyammonium head. In the esteratic site, the identification [5] of the active serine as serine 200 of the sequence (the numbering of the residues utilized is that of the Torpedo Californica enzyme) was confirmed by site-directed mutagenesis [6] which attributed a functional importance also to His 440 and Glu 199. The resolution, in 1992 [7], of the X-ray structure of the enzyme of Torpedo Californica showed the relative spatial disposition of these and other residues surrounding the active serine and indicated, in particular, that His 440 is inserted within hydrogen-bonding distance between Ser 200 and another glutamate, Glu 327, "in a planar array reminiscent of that formed by the better-known catalytic triad of the serine proteases".

In regard to the anionic subsite, its molecular composition, initially assumed to involve one or more anionic residues [8] was disputed for many years, but a role in it for uncharged lipophilic residues was emphasized early [9], until the presence of Trp 84 in this area was detected (see references in [7]) and its probable involvement as a 'putative' component of the anionic site was explicitly envisioned [10]. A considerable progress was then made possible by the spatial localization

A. Pullman et al. (eds.), Modelling of Biomolecular Structures and Mechanisms, 11–23.
© 1995 *Kluwer Academic Publishers. Printed in the Netherlands.*

of this tryptophan in the X-ray structure: this localization and that of the residues surrounding serine 200 led to the proposal [7] of a model of the binding of the substrate at the active site in which, by manual docking, the choline head of the molecule could be placed in van der Waals contact with Trp 84, while the carbonyl carbon atom was appropriately oriented for making a tetrahedral adduct with $O\gamma$ of the active serine, a disposition assumed to occur as an intermediary step in the early phase of the enzymic reaction [1]. We shall come back to this structure later.

Another important feature revealed by the X-ray structure is the location of the active site near the bottom of a 20 Å-long cavity baptized the 'active-site gorge', the inner walls of which are lined with fourteen aromatic residues. It was suggested [7,11,12] that this unusual aromatic lining could serve "to help the substrate slide down to the active site along a series of low-affinity sites involving aromatic-quarternary ammonium interactions". The existence of such 'cation-π-interactions' and their possible importance in various biological phenomena have been under-lined in a number of papers [13–22].

With a view to clarifying the nature and characteristics of the interactions invoked, preliminary theoretical calculations have been performed [23] of the intrinsic affinities of the individual aminoacid side-chains for the tetramethylam-monium ion (TMA). The calculations showed that the aromatic residues do indeed have a distinct affinity for TMA, with a definite tendency to place this ligand above or below the aromatic plane, a disposition determined by the dispersion (van der Waals London) component of the interaction energy, the largest value occuring in the case of the most extended π-electron systems. In the case of the aromatic polar side-chains tryptophan and tyrosine, the dispersion attraction is supplemented by a smaller, but appreciable, electrostatic attractive component which reinforces the preeminence of these residues over phenylalanine in the total interaction. These results provide a quantitative basis for the concept that, when a few aromatic side chains are appropriately disposed in a protein, they can combine their individual intrinsic attractions in different ways, so as to form local 'binding sites' of diverse stabilities for quarternary organic ions.

In order to characterize more precisely such sites, if any exist, in the active site gorge of ACHE, we have deemed it interesting to perform further theoretical calculations on the interactions, in the gorge proper, of the substrate of the enzyme. Aside from the question of their possible role in the sliding of the molecule down towards the reaction site, the characterization of local attractive sites for acetylcholine may help to throw light on the binding of known or potential inhibitors, in conjunction with the results of other techniques such as photoaffinity labelling and/or crystallographic experiments [12,24,25].

Standpoint and Methodology

The hypothesis enunciated above raises a few preliminary questions:

(1) Are there really such sites along the path of the substrate inside the aromatic gorge of acetyicholinesterase? If such sites exist, what are their characteristics (energy, structure, role of the aromatics and/or polar residues)?

(2) Is the 'active site' a stable site? Is it, in fact, the most favorable site for the substrate?

(3) Ancilliary questions concern the role of the residues flanking the active serine, in particular Glu 199, Ala 201, and also His 440 and Glu 327.

To address these questions we have adopted the simplest possible procedure, in the spirit of our preliminary study [23]. First, a molecular system was built comprising the fourteen aromatic residues observed to be on the inner walls of the gorge, namely five tyrosines (70, 121, 334, 130, 442 from top to bottom), five tryptophans (279, 233, 432, 84, 114 from top to bottom) and four phenylalanines (290, 288, 331, 330), supplemented by aspartate 72, also on the inner wall roughly halfway down, and by aspartate 285 and glutamate 173 which 'guard' the top entrance to the gorge. To these residues were added the aminoacids observed at, or close to, the esteratic subsite, namely the sequence Glu 199, Ser 200, Ala 201, Gly 202, supplemented by His 440 and Glu 327. To the X-ray coordinates of these residues, kindly provided by J. Sussman, we added those of the missing hydrogens as obtained by the program Polypep developed in our laboratory [26]. The interactions of acetylcholine with this system were then computed using the program Ligand [27] developed for energy optimization of complexes of flexible molecules in interaction, utilizing the Flex force field [28,29]. The energy of the system comprising the 23 residues enumerated above and the acetylcholine molecule was optimized, starting from a large number of initial positions and conformations inside the gorge, in each case letting the ligand optimize its conformation (defined by eight dihedral angles) and keeping its position with respect to the gorge frozen. No explicit water molecules were included but a sigmoidal distance-dependent dielectric constant, based on a proposal by Hingerty [30], was adopted, using the recent reformulation [31] and the reparametrization elaborated by Lavery *et al.* [32].

A remark must be made concerning the methodology: clearly, the procedure adopted (energy optimization) is by no means destined to find either the best possible position of the substrate ready to undergo the enzymatic reaction, or the trajectory followed to the active site; but it is considered quite capable of detecting the existence of binding positions, or rather 'sites of favorable interactions', inside the gorge and determining their relative strengths and the role, in them, of the aromatic and/or of the charged residues lining the walls. Furthermore, by examining the evolution of the energies of a sufficient number of local minima from the top to the bottom of the gorge, one may hope to see whether they follow a definite pattern, possibly compatible with the concept of 'aromatic guidance'. A subsidiary question to be considered is that of the disposition of the ligand at the active site itself.

In order to span as thoroughly as possible the space contained within the inner walls of the gorge, a large number of starting points for the optimizations were chosen systematically, putting the acetylcholine molecule in different orientations and conformations, placing its pivot successively at regular intervals, from top to bottom of lines parallel to the central axis of the gorge (see Figure 1). The stability of the results was carefully tested by restarts and reoptimizations.

We used the facilities of the ligand programme to characterize a complex by its

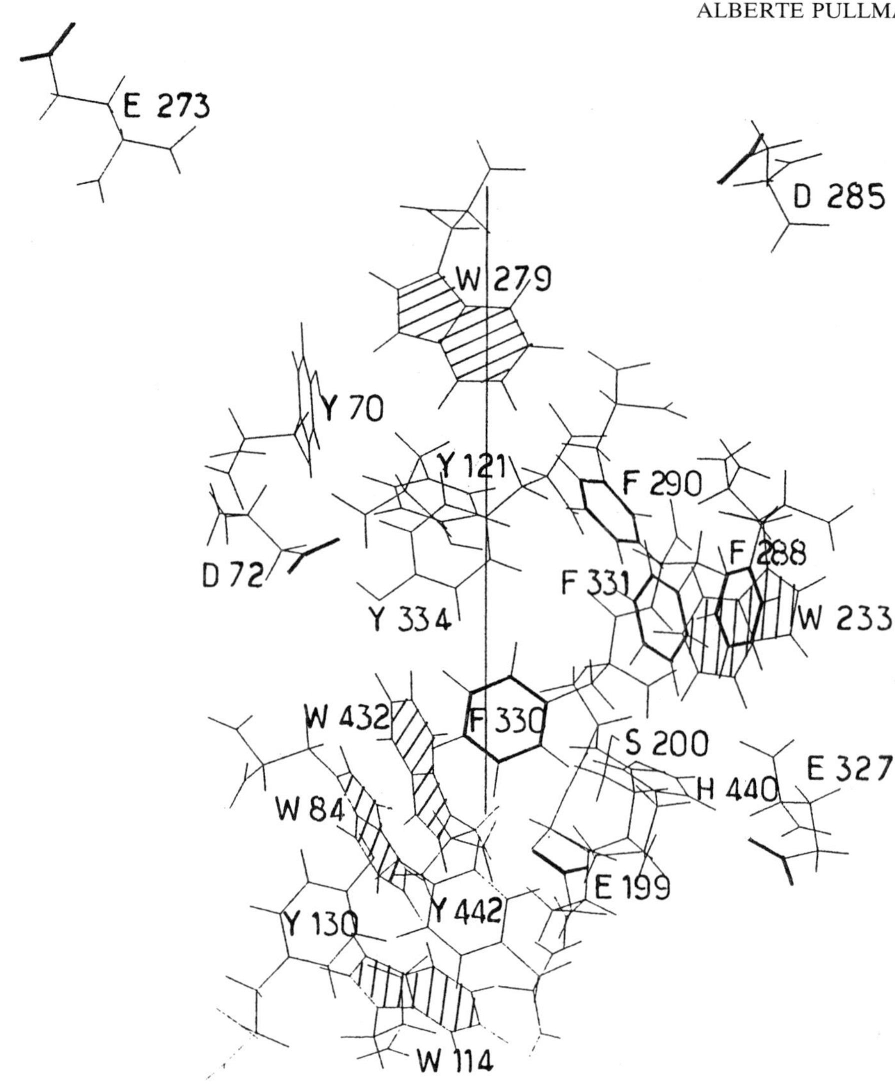

Fig. 1. The disposition of the 23 aminoacids in the system representing the gorge. The plane of the figure contains the axis of the gorge (defined manually) and the Cα carbon of Asp 285. To facilitate the visualization, the contours of the phenyl rings of phenylalanines are in thick lines, and the indole rings of the tryptophans are hatched. (D = Asp, E = Glu, F = Phe, H = His, S = Ser, Y = Tyr, W = Trp).

stability, S, expressed as the difference between its optimized energy, E, and the sum, C, of the energies of the two isolated partners (namely the empty gorge and the isolated optimized ligand). S can be written:

$$S = E - C = I + \delta e$$

where I is the total interaction of the substrate with the gorge in the minimum considered and δe is the deformation energy of the molecule from its optimal isolated conformation to its conformation in the complex, all quantities being obtained in the calculations. Since the gorge is kept frozen, C is a constant.

Results and Discussion [33]

Numerous distinct local minima were found. They can be characterized energywise either by the global stability of the complex, S, or by the total interaction energy, I, of the substrate with the gorge in the minimum. The two quantities run essentially parallel to each other although a few cases were found where, in order to achieve a better interaction, the ligand spends a larger amount of energy on its deformation (*vide infra*).

A scrutiny of the spatial location of the minima in the gorge indicates that they are distributed in three zones of increasing interaction from top to bottom, namely:

zone I for I = −23.1 to −32. 1 kcal/mole.
zone II for I = −45 to −54 kcal/mole
zone III for I = −63.7 to −69.2 kcal/mole

To this must be added a fourth category, which we classify separately as zone IIIbis, where I reaches −70 to −71.5 kcal/mole.

Examples of the typical distribution, in each zone, of the individual interaction energies of the substrate with the aminoacids of the gorge are given in Figures 2 and 3.

A characteristic feature common to all minima in zone I (see Figure 2a) is the appreciable interaction of the substrate with tryptophan 279, while the rest of the interaction is shared in different ways between the two tyrosines 70 and 121 and also Asp 72, with an occasional participation of phenylalanines 290 and 331. The 'sharing' of the substrate molecule by W279, Y70, Y121 and D72 in minimum A of Figure 2 is illustrated in Figure 4. A particular complex of zone I deserves a special mention, namely minimum C in Figure 2, in which the substrate lies between Trp 279 and one of the 'guards' of the entrance, Asp 285 (see stereoview, Figure 5), a disposition which suggests a possible role for this aspartate in the binding, as yet unexplored, of blockers of appropriate structure or of the substrate itself. It is tempting to speculate that minimum C corresponds to the position involved in the binding of the substrate to the 'peripheral site' when the molecule itself acts as inhibitor at high concentration [34].

In the minima of zone II (see Figure 2b), the interactions of the substrate with Trp 279 and Tyr 70 have been lost and replaced by increased contributions of Tyr 121 and Tyr 334, the largest interaction occuring, however, with Asp 72 which dominates this zone (see stereoview Figure 6, for one structural example). Here again, the phenylalanines are little involved (or not at all) in the complexations.

In zone III (see Figure 3), most of the interaction with the residues situated in the upper part of the gorge has been lost and replaced by contributions from those involved in the active site, among which prevails strongly the interaction with the sequence 199–202, itself dominated by the attraction of Glu 199, as seen in Table I. The other appreciable contributions in this zone are due to Trp 84 and His 440, with values of 6.7 to 9 kcal/mole, and to Glu 327 (about 3 kcal/mole), and

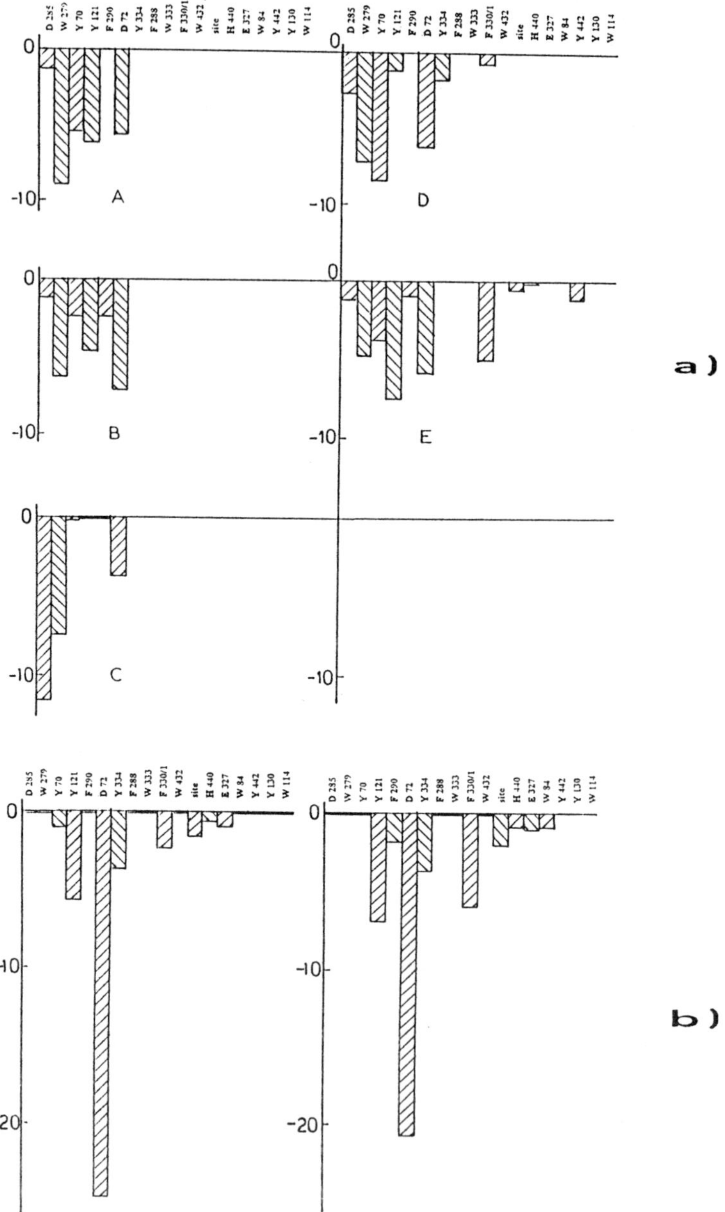

Fig. 2.   Individual interaction energies of the substrate acetylcholine with the aminoacids from top to bottom of the gorge: (a) in various minima A, B, C, D, E of Zone I. (b) for typical minima of zone II (For technical reasons, F 330–F 331 (F 330/1) and the sequence E 199–G 202 (site) are given here as single entities). The scale on the ordinate is in kcal/mole.

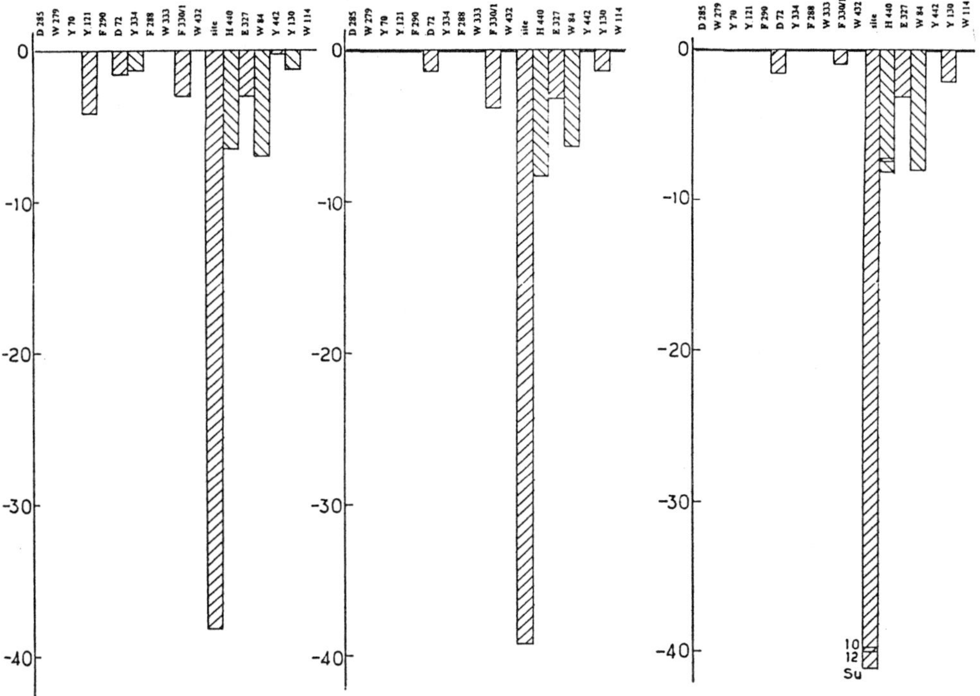

Fig. 3.   Same as Figure 2 for two typical minima of zone III (left) and the minima of zone IIIbis (right).

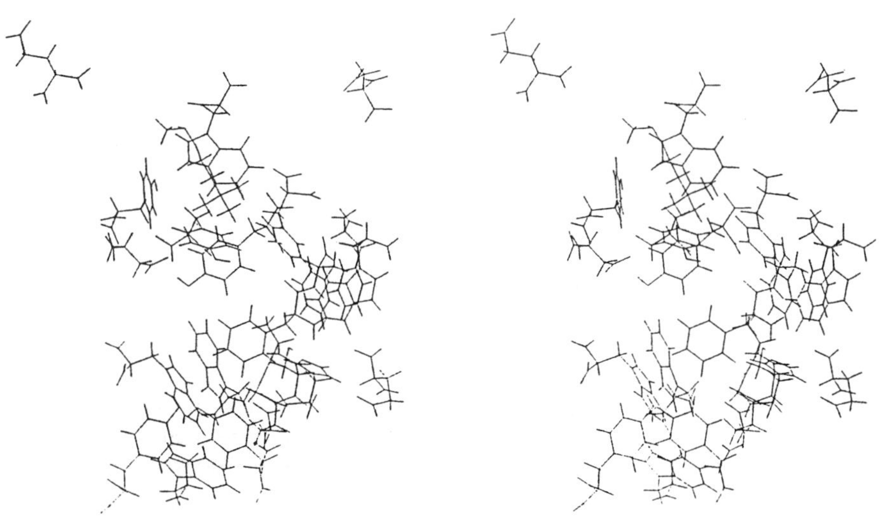

Fig. 4.   Stereoview of minimum A in zone I.

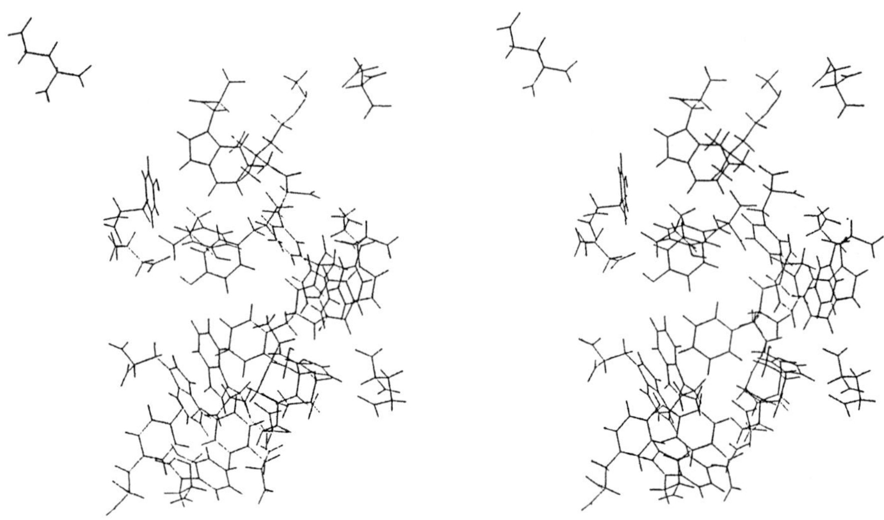

Fig. 5.   Stereoview of minimum C in zone I.

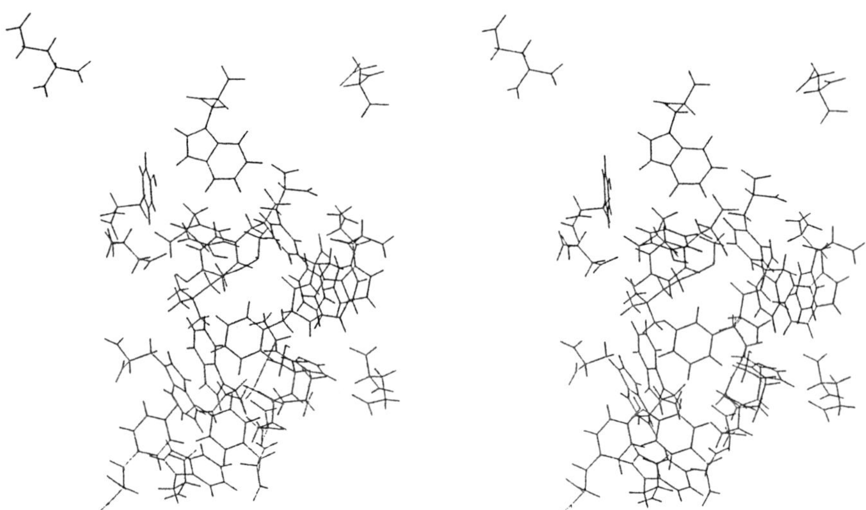

Fig. 6.   Stereoview of a minimum in zone II.

occasionally to a residual interaction with the pair Phe 330–331 still in contact
with the tail of the molecule (example in the stereoview Figure 7). When this
contact is lost, the substrate can lean towards the esteratic subsite, forming various
complexes of similar energies and can reach the minima of zone IIIbis which
deserve a special comment: in this category, we have found only 3 minima, labelled
10, 12 and Su, in Figures 3 and 8. Their individual interactions have globally the
same characteristics as those described in zone III, with a still larger dominance

TABLE I

Interactions in different minima of zones III and IIIbis. The names are internal conventions, I is the total interaction of the substrate with the system; the following lines give the global interaction with the sequence E 199-Gly 202 (called 'site' in Figures 2–5), then the individual participation in it, of the residues 199, 200 and 201 (that of G202, negligible, is omitted). The last line gives, for comparison, the individual interactions involving W84 (energies in kcal/mole)

Name	Y6	Y8	Y9	Y10	Y12	Su
I	−67.1	−65.6	−67.1	−70.0	−70.4	−71.4
sequence 199–202	−38.2	−39.3	−39.5	−45.7	−46.0	−47.1
E199	−40.5	−41.7	−42.2	−41.3	−42.4	−43.7
S200	1.3	1.3	1.8	−0.7	−0.4	0.5
A201	0.8	0.7	0.6	−3.0	−2.6	−2.4
W84	−7.1	−6.6	−7.5	−8.0	−7.4	−7.4

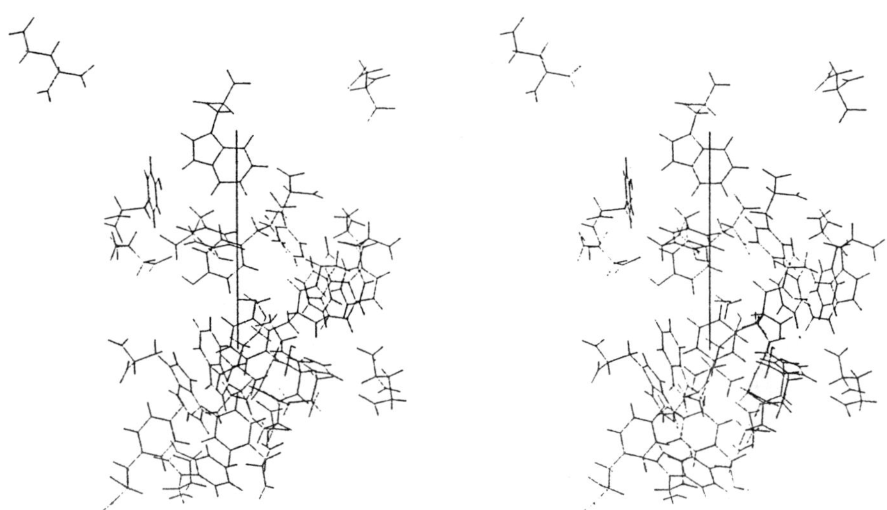

Fig. 7.   Stereoview of the minima in zone III.

of the contribution of the sequence 199–202. But the distribution of this interaction among the constituent aminoacids indicates (Table I), aside from the major role of Glu 199, a novel feature, namely a favorable interaction with Ala 201, absent in the other minima where the corresponding energy term is repulsive. An analysis of the underlying structural features of this extra binding energy shows that its source is the existence of a distorted hydrogen bond between the carbonyl oxygen O7 of the substrate and the amide hydrogen of Ala 201. This constitutes an interesting confirmation of a hypothesis emitted on the basis of the disposition of this residue in the X-ray structure [7]. (The hydrogen bonds obtained in the computations have distances O7----H 2.26, 2.35 and 2.20 Å in the minima 10, 12 and Su respectively, with NH----O angles of 144, 145 and 142 degrees, such depar-

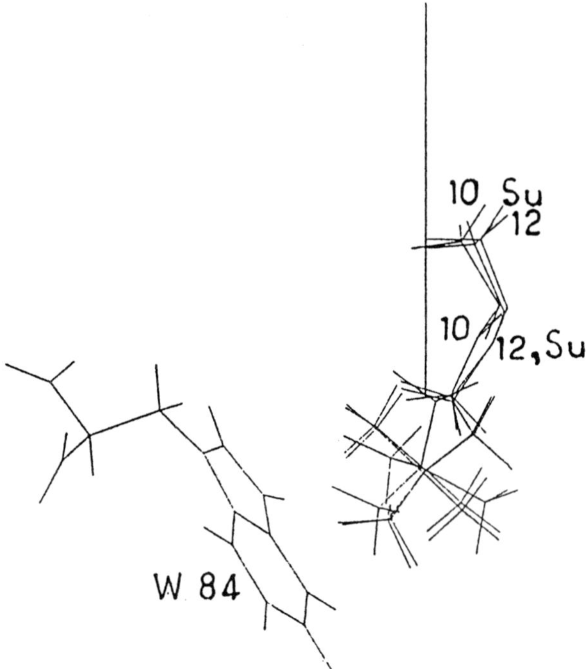

Fig. 8.   Comparison of the three minima in zone IIIbis.

tures from ideal lengths and linearity being largely within the limits of the significance of the calculations). It is in zone IIIbis that the largest deformation energy of the ligand, mentioned earlier, is observed, a feature possibly related to a strain due to the formation of the O7----Ala hydrogen bond. Note that the hydrogen bonds observed were obtained in the gorge kept frozen; their strength could easily be enhanced by small displacements in the sequence 199–202 itself or in its surroundings in the protein. The relatively small structural differences between the three minima of zone IIIbis can be visualized in Figure 8.

Two remarks may be added, concerning the individual interactions in zone III and IIIbis: (a) the active serine contributes very little to the energy even in the best complexes; (b) the energy contribution of Glu 327 to the interaction is fairly constant and relatively small in spite of the anionic character of the residue, obviously too far and oriented unfavorably, contrary to Glu 199.

One precision is in order concerning the best complex at the active site (called Su in what precedes): it should not be identified with the model structure of the acetylcholine molecule given in the X-ray coordinates of the Protein Data Bank, which was obtained [7], as recalled in the Introduction, by manual docking, so as to form a tetrahedral adduct with O7 of Ser 200 at the carbonyl carbon of the substrate, using a strained conformer of the ester bond. When we calculated the energy of this structure, we found it strongly repulsive, due to the strain and to the close approach needed to achieve the tetrahedron. Energy minimization relieves the strain at the ester bond and suppresses the short contacts imposed by the

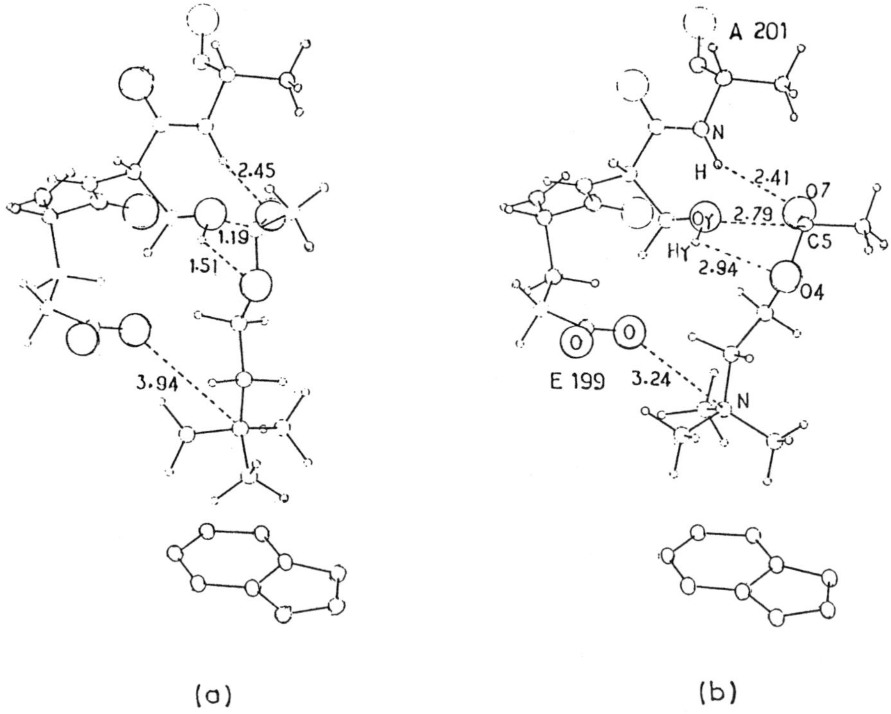

Fig. 9.   The substrate acetylcholine (a) in the complex constructed manually in [7]; (b) in the complex optimized from the model.

model (see Figure 9). The passage from the optimal disposition, corresponding to the best interaction at the site, to a tetrahedral intermediate like that of the model requires obviously an energy supply, the study of which is beyond the scope of our investigation.

Conclusion

If we go back to the three questions formulated in our standpoint, above, we can now give the following answers:

(1) There are, indeed, a number of local minima for the substrate in the aromatic gorge of the enzyme. They cluster in four separate categories according to their interaction energies which increase (discontinuously) from the top zone to that comprising the active site. In each minimum, the molecule interacts with a few aminoacids in a characteristic fashion: the first zone emphasizes in particular the importance of Trp 279, of two tyrosines, and occasionally of Asp 285; the second zone, lower, is dominated by the strong attraction of Asp 72 with smaller contributions from the neighbour aromatics; in the third region, the interaction, strongly dominated by the attraction of Glu 199,

involves also variable contributions from Trp 84, His 440, Glu 327, Tyr 130, Phe 330/331, Ser 200 and Ala 201 according to the disposition and conformation of the substrate, the best interaction occuring when the molecule achieves an orientation where its carbonyl oxygen forms a hydrogen bond with the amide hydrogen of Ala 201.

(2) The region defined as the active site contains a minimum which corresponds to the best interaction energy among all the minima found.

(3) Concerning the residues flanking the activated serine: the role of Ala 201 appears as facilitating the disposition of the substrate for a better interaction with the site; the energy results for His 440 show that its interaction with the substrate is appreciable; Glu 327 has apparently a role rather structural than energetic, but Glu 199, in contrast, is clearly a strong driving force for leading the substrate to the site.

As concerns the role of the aromatics, they intervene in all the minima detected, with a particular significance of the attraction of the two tryptophans 279 and 84, a result in agreement with the present concepts on the probable importance of these residues in the enzyme (identification of Trp 279 as part of the peripheral site, role of Trp 84 at the anionic site, role of both residues in the binding of decamethonium [24]). The other tryptophans of the gorge do not participate in the minima found, being apparently out of reach of the substrate, as seems to be the case also for Tyr 442 and Phe 288. As a rule, the phenylalanines intervene less than the tyrosines, clearly owing to their intrinsic smaller attraction [23] for the molecule, the role played occasionally by Phe 330 seeming due essentially to its central location. Aside from the role of the aromatics, that of the negative residues belonging to the gorge proper or to its entrance have been assessed: in none of the optimizations starting in the top part of the gorge has the substrate shown any tendency to back-up towards Glu 173, probably owing essentially to the unfavorable orientation of its carboxylate group, but the attraction of Asp 285 was manifest: in particular, we have put into evidence the possibility of its cooperation with Trp 279 in a secondary binding site in the early part of the gorge. On the other hand, we have stressed the role of Asp 72 in the minima of the second zone and that of Glu 199 in zones III and IIIbis. As stated in the introduction, the passage from one minimum to another or from one zone to the other is beyond the reach of our methodology. A search for barriers and minimal paths would in fact be illusory without taking into account the rest or at least a large part of the protein.

References

1. D. M. Quinn: *Chem. Rev.* **87**, 955–979 (1987).
2. I. B. Wilson and F. Bergmann: *J. Biol. Chem.* **185**, 479–489 (1950).
3. I. B. Wilson, F. Bergmann, and D. Nachmanson: *J. Biol. Chem.* **186**, 683–692 (1950).
4. H. C. Froede and I. B. Wilson: *J. Biol. Chem.* **259**, 11010–11013 (1984).
5. K. McPhee-Quigley, P. Taylor, and S. Taylor: *J. Biol. Chem.* **260**, 12185–12189 (1985).
6. G. Gibney, S. Camp, M. Dionne, K. McPhee-Quigley, and P. Taylor: *Proc. Natl. Acad. Sci. USA* **87**, 7546–7550 (1990).

7. J. L. Sussman, M. Harel, F. Frolow, C. Oefner, A. Goldman, L. Toker, and I. Silman: *Science* **253**, 872–879 (1991).

8. H. Noble, T. L. Rosenberry, and E. Neumann: *Biochemistry* **19**, 3705–3711 (1980).

9. S. G. Cohen, H. Salih, M. Solomon, S. Howard, S.B. Christi, and J.B. Cohen: *Biochim. Biophysica Acta* **997**, 167–175 (1989).

10. C. Weise, H. J. Kreienkamp, R. Raba, A. Pedak, A. Aaviksaar, and F. Hucho: *Embo J.* **9**, 3885–3888 (1990).

11. J. L. Sussman and I. Silman: *Current Opinions in Structural Biology* **2**, 721–729 (1992).

12. J. L. Sussman, M. Harel, and I. Silman: in A. Pullman *et al.* (Ed.), *Membrane Proteins, Structures, Interactions and Models*, KIuwer Academic Publishers, Dordrecht pp. 161–175 (1992).

13. D. A. Dougherty and D. A. Stauffer: *Science* **250**, 1558–1560 (1990).

14. M. Dhaenens, L. Lacombe, J. M. Lehn, and J. P. Vigneron: *J. Chem Soc. Chem. Commun.* 1097–1099 (1984).

15. F. B. Hasan, S. G. Cohen, and J. B. Cohen: *J. Biol. Chem.* **255**, 3898–3904 (1980).

16. J. L. Galzi, F. Revah, A. Bessis, and J. P. Changeux: *Ann. Rev. Pharmacology* **31**, 37–72 (1991).

17. Y. G. Satow, H. Cohen, E. A. Padlan, and D. R. Davies: *J. Mol. Biol.* **190**, 593–604 (1986).

18. G. F. Tomaselli, J. T. McLaughlin, M. E. Jurman, E. Hawrot, and G. Yellen: *Biophys. J.* **60**, 721–727 (1991).

19. S. Pierce, P. Preston-Hurlburt, and E. Hawrot: *Proc. R. Soc. Lond. Biol.* **241**, 207–213 (1990).

20. J. L. Galzi, D. Bertrand, A. Devillers-Thiery, F. Revah, S. Bertrand, and J. P. Changeux: *FEBS Lett.* **294**, 198–202 (1991).

21. R. McKinnon: *Curr. Opin. Neurobiol.* **1**, 14–19 (1991).

22. M. L. Verdank, G. J. Boks, H. Kooijman, J. A. Kanlersand, and J. Kroon: *Computer-Aided Molecular Design* **7**, 173–182 (1993).

23. A. Pullman and X. W. Hui: in A. Pullman *et al.* (Ed.), *Membrane Proteins. Structures. Interactions and Models*, Kluwer Academic Publishers, Dordrecht, pp. 229–232 (1992).

24. M. Harel, I. Schalk, L. Ehret-Sabatier, F. Bouet, M. Goeldner, C. Hirth, P. Axelsen, I. Silman, and J.L. Sussman: *Proc. Natl. Acad. Sci. USA* **90**, 9031–9035 (1993).

25. M. Harel, J. L. Sussman, E. Krejci, S. Bon, P. Chanal, J. Massoulié, and I. Silman: *Proc. Natl. Acad. Sci. USA* **89**, 10827–10831 (1992).

26. *Polypep*, C. Etchebest and R. Lavery: Laboratoire de Biochimie Théorique du CNRS. Institut de Biologie Physico-Chimique, Paris, France (1989)

27. R. Lavery: *Manual to Ligand*, Laboratoire de Biochimie Théorique du CNRS, Institut de Biologie Physico-Chimique, Paris, France (1990).

28. R. Lavery, H. Sklenar, K. Zakrzewska, and B. Pullman: *J. Biomol. Struct. Dynam.* **3**, 989–1014 (1986).

29. R. Lavery, I. Parker, and J. Kendrick: *J. Biomol. Struct. Dynam.* **4**, 443–461 (1986).

30. B. Hingerty, R. H. Richie, T. L. Ferrel, and J. E. Turner: *Biopolymers* **24**, 427–439 (1985).

31. B. Hartmann, B. Malfoy, and R. Lavery: *J. Mol. Biol.* **207**, 433–444 (1985).

32. R. Lavery and K. Zakrzewska: in D. Beveridge, and R. Lavery (Eds.) *Theoretical Chemistry and Molecular Biophysics. Vol. 1 DNA*, Adenine Press, pp. 173–190 (1990).

33. A preliminary summary of some of the results of this study was given in A. Pullman and X. Hui: *Biophysical J.* **66**, (2, part 2) A 345 (1994).

34. Z. Radic, E. Reiner, and P. Taylor: *Molecular Pharmacology* **39**, 98–104 (1991).

Binding of Cations and Protons in the Active Site of Acetylcholinesterase

STANISLAW T. WLODEK, JAN ANTOSIEWICZ,[1] ANDREW McCAMMON, and MICHAEL K. GILSON[2]
Department of Chemistry, University of Houston, Houston, TX 77204–5641, U.S.A.

Abstract. The active site of acetylcholinesterase contains a number of ionizable residues. The pK_as of the catalytic histidine, His 440, is believed to be 6.3, based upon enzyme kinetic studies. However, the pK_as of the other residues have not been measured. Here, we describe calculations of the pK_a of the ionizable groups in this enzyme. Interestingly, the initial calculations predict a pK_a of 9.3 for His 440. The deviation of 3 pK_a units from the measured pK_a is traceable to the influence of Glu 199 and Glu 443 upon His 440. We argue that the deviation does not represent a failure of the computational method. Rather it points to the need for an adjustment in the model of the protein. The adjustment we suggest involves a monovalent cation bound in the active site, near Glu 199 and His 440. Including such a cation in the calculations brings the computed pK_a of His 440 into agreement with the measured value. Furthermore, the idea that a bound cation substantially reduces tbe pK_a of His 440 leads to satisfying explanations of a number of otherwise puzzling experimental data.

Key words. Acylation, deacylation, acetylcholinesterase, titration, pK_a, inhibitors, lipase.

Introduction

Acetylcholinesterase (AChE) is a critical enzyme in many organisms, including humans. It serves primarily to clear neuronal and neuromuscular synapses of the neurotransmitter acetylcholine (ACh), through hydrolysis of the ester bond in the substrate. AChE has attracted continued interest from researchers over several decades, partially because of its physiologic, pharmacologic, and toxicologic significance, and partially because of its ready availability in the electric organs of certain eels and rays.

A combination of experimental approaches has identified the key catalytic groups in the enzyme as Ser 200, His 440, and Glu 327 [1–4]. (The numbering is appropriate to *Torpedo californica* AChE (TcAChE).) These residues are arrayed much like the classic catalytic triad of the serine proteases, with the histidine presumably poised to accept the hydroxyl hydrogen of the serine as the serine oxygen attacks the ester carbon of the substrate [5]. Therefore, catalysis cannot proceed if the histidine is ionized, and the inactivation of AChE as the pH is lowered through 6.3 presumably results from the protonation of His 440 [6–8]. Note that this pK_a is for the 'free' enzyme in the absence of substrate [7].

Despite the resemblance between the catalytic triad of TcAChE and the serine proteases, the active site of TcAChE differs in several interesting ways. These include the depth of its active site 'gorge' [5], and the presence of two glutamic

[1] On leave from Dept. of Biophysics, University of Warsaw, 02-089 Warsaw, Poland.
[2] Author for correspondence. Present address: Center for Advanced Research in Biotechnology, 9600 Gudelsky Drive, Rockville, MD 20850.

A. Pullman et al. (eds.), Modelling of Biomolecular Structures and Mechanisms, 25–37.

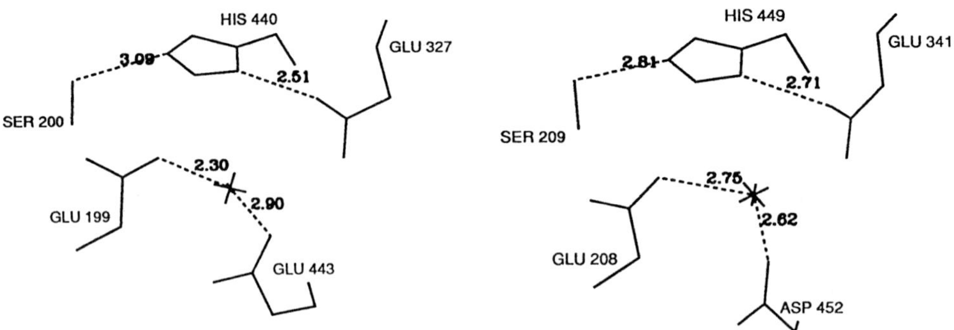

Fig. 1.   (A) Ionizable groups and solvent (star) in the active site of TcAChE. (B) Ionizable groups and solvent in the active site of *Candida rugosa* lipase. Distances in Ångstroms.

acid residues, Glu 199 and Glu 443, which lie near the catalytic triad and the probable choline-binding site (see Figure 1a). The carboxylic oxygens of Glu 199 form part of the active site surface, but Glu 443 is buried within the protein. An internal solvent molecule, water 43 in the Protein Data Bank [9] crystal structure 1ACE [5], lies within hydrogen-bonding distance of the carboxylic groups of both Glu 199 and Glu 443 (Figure 1a), and also of the $O\gamma$ atom of Ser 126 (not shown). This bound solvent, its liganding groups, and its location relative to the catalytic triad, are conserved within a family of enzymes to which AChE belongs [10]. The strikingly similar active site of another family member, *Candida rugosa* lipase [11,12], is diagrammed in Figure 1b. The lipase of *Geotrichum candidum* also displays this arrangement of functional groups and solvent [13]. The conservation of the triad of carboxylic-solvent-carboxylic within this family of enzymes, and its location relative to the catalytic triad, suggests a functional role. It has been proposed that Glu 199 and Glu 443 facilitate hydrolysis through organization of water [10], or that the electrostatic field they generate helps to expel the anionic reaction products from the active site [14]. Experimentally, single-site mutations of Glu 199 and Glu 443 in AChE reduce, but do not abolish, enzyme activity [4,15–17]. It will be difficult to understand the roles of the various ionizable groups in the active site of AChE without information on their ionization states as a function of pH. However, the only group whose pK_a is known with any certainty is His 440, the catalytic histidine (see above).

Here we present the results of a computational study of the protonation equilibria of the ionizable groups in the active site of TcAChE. The methodology, recently developed, has been shown to be predictive in comparisons with a large number of experimental data [18]. We find that calculations on the unmodified crystal structures of TcAChE predict a pK_a for His 440 which is well above the expected value of about 6.3. Because the discrepancy is larger than any error so far observed in tests of the methodology against measured pK_as, the apparent error probably results from some inadequacy in the structural model. Further analysis suggests that at least part of the discrepancy can be explained by binding of a monovalent cation in the active site. This model is particularly attractive, because the picture of competition between cation-binding and protonation of His 440 proves to fit remarkably well with a number of experimental data, as described below.

Methods

The theoretical model, computational method, and atomic parameters used in the titration calculations are detailed elsewhere [18,19]. Briefly, the program UHBD [20] is used to solve the linearized Poisson–Boltzmann equation in calculations of the intrinsic pK_a [21] of each ionizable group, as well as of the electrostatic interaction energies among the ionizable groups. The 'cluster' method [19] is used to solve the multiple-site titration problem, and thus to compute the average charge of each ionizable group over a range of pH values. The reported pK_a of a group is the pH at which it is half ionized. Note that although we present pK_as for only a few ionizable groups in TcAChE, the calculations include all.

The solvent dielectric constant is set to 78 and the protein dielectric constant to 20, which has been shown [18] to yield much better agreement with measured pK_as than the more customary value of 4 [22–25]. The ionic strength of the solvent is set, somewhat arbitrarily, to 85 mM. We have previously shown that computed pK_as are rather insensitive to ionic strength above about 50 mM [18]. Accordingly, when a sample calculation here is repeated for 150 mM (physiologic) ionic strength, the pK_as change trivially. (Compare lines 1 and 2 of Table II.) The neutral tautomers of the histidines in TcAChE are assigned as described previously [26].

The AChE calculations are based upon several structural models of TcAChE. Most are derived from the original Protein Data Bank entry, 1ACE [5]. Because this structure was subsequently found to contain the bis-quaternary inhibitor decamethonium in the active site [27], it is referred to here as the decamethonium structure. Two calculations are based upon the separately solved structure of TcAChE complexed with the drug 9-amino-1,2,3,4-tetrahydroacridine (tacrine). In accord with our usual procedure, all solvent molecules and ligands are stripped out before the calculations are initiated [18].

However, as noted above and detailed below, calculations based upon these two protein structures yield pK_as for His 440 which are outside the limits of accuracy of the computational methodology. This suggests that these two structural models are inadequate. We therefore consider the influence of a cation bound in the active site through the inclusion of: (1) a model-built ACh molecule placed in a plausible 'pre-catalytic' conformation; (2) tacrine, placed in accordance with the crystal structure obtained in the presence of that inhibitor; and (3) an artificial spherical, monovalent cation of radius 2 Å placed at the same site as that assigned to the choline nitrogen of ACh. The treatment of these ligands is detailed below.

The ligands ACh and tacrine are treated as follows. Their structures are optimized through *ab initio* quantum mechanical calculations *in vacuo*, at the 6-31G** level, using the program Gaussian 92 [28]. Atomic charges are assigned by means of electrostatic potential fitting, using the program CHELPG [29]. In the case of tacrine, the optimized structure is rotated and translated to fit the position and orientation provided in the crystal structure. In the case of ACh, for which no crystallographic coordinates are available, the optimized structure is placed in the

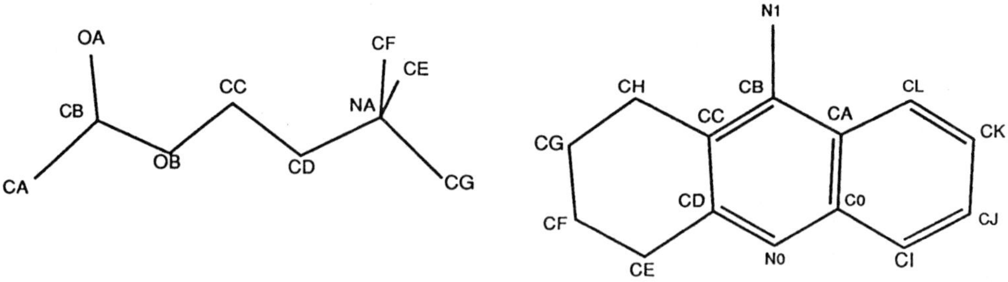

Fig. 2.   (A) Acetylcholine. (B) Tacrine. Atom names correspond to those in Table I.

active site with the choline moiety near the carboxylic group of Glu 199 and *en face* to the indole ring of Trp 84, and with the carbonyl carbon of the ester group near the side chain oxygen of Ser 200. With protein coordinates fixed, the system is energy-minimized for 1000 steepest-descent steps using the CFF91 force field [30] in the program DISCOVER [31]. Because this procedure alters the internal structure of the ACh molecule, the original 6–31**-optimized structure of ACh is then rotated and translated to fit the position of ACh generated by the energy minimization. In the end, the nitrogen atom of ACh lies near the location of the ring nitrogen of tacrine as solved crystallographically. (See coordinates in Table I.)

Both the ring and the amino nitrogens of tacrine are basic. They are treated as ionizable sites, with initial pK_as of 9.8 [32] and 4.0 (based upon the low pK_a of aniline), respectively. (Resonance stabilization causes the ring nitrogen to be the more basic.) In the electrostatic calculations, the atomic radii of ACh and tacrine are set to 0.5σ for similar atom types within the OPLS parameter set [33], where α is the usual nonbonded parameter. Figure 2 and Table I provide the atomic charges, radii and coordinates used in the electrostatic calculations. Note that the neutral form of tacrine is given; its ionizations are modeled as the addition of a $+1\ e$ charge to the amino or the ring nitrogen.

Results

UNMODIFIED STRUCTURES

Lines 1–3 of Table II present computed pK_as for TcAChE in the absence of any ligands. The first line is for the decamethonium structure (1ACE) and the third for the structure solved with bound tacrine. The most striking prediction is the pK_a of 9.3 for His 440. As noted in the Introduction, experimental data, and the example of chymotrypsin [34, 35] indicate that the true pK_a of His 440 is about 6.3. Furthermore, the putative catalytic mechanism of AChE, which involves attack by the hydroxyl of Ser 200 on the ester carbon of ACh and simultaneous transfer of the hydroxyl proton of Ser 200 to His 440, requires that His 440 be neutral at the start of catalysis. If this scheme is correct, the fact that the enzyme functions at pH 7 indicates that the pK_a of His 440 cannot be 9.3.

On the other hand, it is unlikely that the computational method used here is

TABLE Ia

Atomic parameters and coordinates for ACh. Hydrogens are named by their carbon atom. See Methods Section for details.

Atom	Charge (e)	Radius (Å)	X	Y	Z
CA	−0.458	1.900	4.587	63.137	63.910
CB	0.949	1.875	4.441	64.405	63.125
CC	0.442	1.953	4.754	66.714	63.169
CD	−0.216	1.953	5.285	67.707	64.197
CE	−0.247	1.980	3.918	69.642	63.456
CF	−0.247	1.980	6.176	69.353	62.567
CG	−0.331	1.980	5.850	69.962	64.899
OA	−0.615	1.480	4.022	64.518	62.027
OB	−0.598	1.500	4.856	65.480	63.830
NA	0.121	1.625	5.299	69.153	63.763
HA1	0.142	0.000	4.244	62.307	63.311
HA2	0.145	0.000	5.626	62.994	64.186
HA3	0.145	0.000	4.006	63.202	64.823
HC1	0.011	0.000	3.724	66.904	62.902
HC2	0.011	0.000	5.342	66.698	62.264
HD1	0.152	0.000	6.302	67.447	64.457
HD2	0.152	0.000	4.683	67.654	65.094
HE1	0.159	0.000	3.280	69.455	64.307
HE2	0.145	0.000	3.535	69.131	62.586
HE3	0.155	0.000	3.963	70.702	63.258
HF1	0.159	0.000	7.158	68.958	62.780
HF2	0.145	0.000	5.752	68.847	61.714
HF3	0.155	0.000	6.244	70.410	62.360
HG1	0.180	0.000	5.870	71.002	64.611
HG2	0.172	0.000	6.850	69.622	65.120
HG3	0.172	0.000	5.217	69.831	65.763

simply in error by $9.3 - 6.3 = 3$ pK_a units for His 440. The basis for this statement is illustrated in Figure 3, which presents a histogram of errors between measured pK_as and those computed with the present method, for about 60 ionizable residues in several proteins. (These data have been presented previously in a different format [18].) The largest absolute error between the computed and measured pK_as is 2.1 pK_a units, and the root-mean-square error is 0.9 pK_a units. The error of 3 pK_a units for His 440 of TcAChE is larger than any error found previously for the present method, so it makes sense to seek a source of discrepancy in the structural model of the protein, rather than in the method. (For comparison, Figure 3b presents a histogram of errors for the 'null' model [18], which assumes that proteins do not shift pK_as at all. The errors are considerably greater than those for the present computational method.)

Before proceeding to a possible explanation of the discrepancy, we compare the results for TcAChE with the results of similar calculations for chymotrypsin. Chymotrypsin is a serine protease whose catalytic triad is similar to that of TcAChE. Titration calculations for chymotrypsin, carried out with an identical methodology, yield a pK_a of about 7 for the catalytic histidine, His 57, and a pK_a of about 2 for the catalytic carboxylic acid, Asp 102 [18]. These results agree well

TABLE Ib

Atomic parameters and coordinates for tacrine. Hydrogens are named by their carbon atom. See Methods Section for details.

Atom	Charge (e)	Radius (Å)	X	Y	Z
NO	−0.760	1.625	6.169	70.122	65.555
CO	0.550	1.775	5.740	70.481	66.779
CA	−0.230	1.775	4.440	70.376	67.186
CB	0.470	1.750	3.470	69.848	66.236
CC	−0.480	1.875	3.923	69.476	64.991
CD	0.510	1.875	5.306	69.635	64.709
N1	−0.860	1.625	2.141	69.709	66.581
CE	−0.070	1.953	5.868	69.231	63.363
CF	0.080	1.953	4.811	69.110	62.269
CG	−0.010	1.953	3.654	68.251	62.767
CH	0.270	1.953	2.960	68.912	63.958
CI	−0.270	1.775	6.699	71.000	67.690
CJ	−0.050	1.775	6.333	71.390	68.936
CK	−0.190	1.775	4.983	71.285	69.350
CL	−0.070	1.775	4.047	70.792	68.498
H11	0.360	1.200	1.499	69.640	65.826
H12	0.360	1.200	1.805	70.305	67.300
HE1	0.030	0.000	6.639	69.944	63.098
HE2	0.040	0.000	6.368	68.272	63.489
HF1	−0.030	0.000	4.442	70.097	61.995
HF2	−0.010	0.000	5.254	68.680	61.375
HG1	−0.010	0.000	4.036	67.276	63.063
HG2	−0.020	0.000	2.928	68.078	61.977
HH1	−0.030	0.000	2.303	68.186	64.430
HH2	−0.050	0.000	2.318	69.715	63.588
HI	0.140	1.210	7.715	71.068	67.350
HJ	0.110	1.210	7.067	71.781	69.619
HK	0.120	1.210	4.701	71.596	70.340
HL	0.100	1.210	3.030	70.723	68.836

TABLE II

Computed pK_as of ionizable groups in active site of TcAChE. Protein structures are either that solved in the presence of bound decamethonium ('Decameth': Protein Data Bank accession code 1ACE) or that solved in the presence of bound tacrine ('Tacr' [27]). The results in line 2 are for 150 mM ionic strength; all others are for 85 mM. See Methods section for details of the models.

Line	Xtal Struct.	State	His 440	Glu 327	Glu 199	Glu 443
1	Decameth	No ligand	9.3	−1.7	1.7	5.8
2	Decameth	No ligand, 150 mM	9.1	−1.2	2.0	5.8
3	Tacr	No ligand	9.2	−1.1	0.86	6.1
4	Decameth	Glu 199 & Glu 443 neutral	6.1	−1.8	−	−
5	Decameth	Glu 199 neutral	7.3	−1.7	−	1.8
6	Decameth	Glu 199 neutral	8.5	−1.7	1.0	−
7	Decameth	With ACh	6.9	−1.9	−2.1	5.1
8	Tacr	With tacrine	6.3	−1.2	−2.6	5.4
9	Decameth	With bound small cation	7.4	−2.0	−1.5	5.2

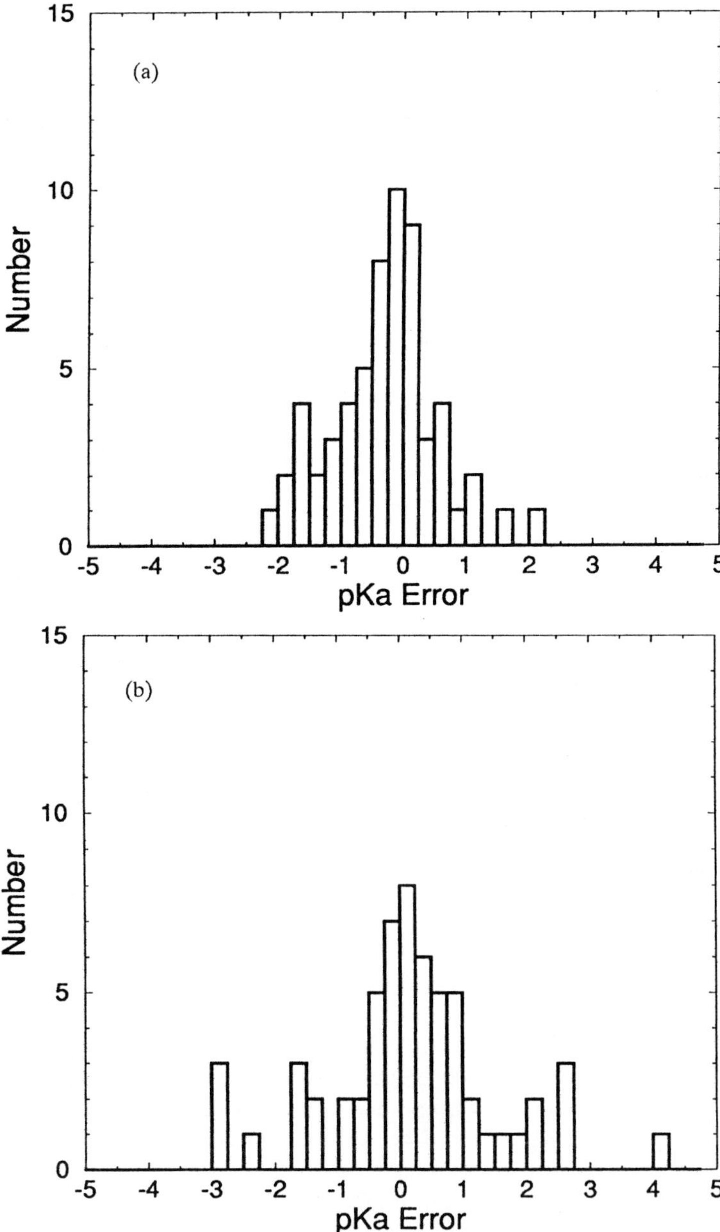

Fig. 3. (A) Histogram of errors of 60 pK_as computed by the present method, relative to experiment. (B) Errors for same set of pK_as predicted by the 'null' model of zero pK_a shifts.

with the experimental data [34,35]. Why, then, are such different results obtained for TcAChE? One possible explanation is that the electrostatic potential generated by Glu 199 and Glu 443 in TcAChE makes His 440 in TcAChE appear more basic than His 57 in chymotrypsin. Recall that chymotrypsin contains no homologs to these two carboxylic residues. (In fact, no ionizable side chain other than Asp 102 lies within $\sim$10 Å of His 57.) We test this idea by repeating the titration calculation on the decamethonium structure of TcAChE, but with Glu 199 and Glu 443 constrained to remain neutral. (This is accomplished by artificially raising their initial pK_as to 20.) The result, presented in line 4 of Table II, is that neutralization of these groups brings the pK_a of His 440 down to a very reasonable value of 6.1. Glu 199 produces most of the effect: neutralizing this residue alone lowers the pK_a of His 440 to 7.3 (Table II, line 5). In contrast, neutralizing Glu 443 alone lowers the pK_a of His 440 only to 8.5 (Table II, line 6).

In comparing chymotrypsin and TcAChE, it makes sense to ask whether the catalytic histidine of the latter is more highly desolvated than that of the former. If so, the increased basicity of the histidine in TcAChE due to Glu 199 and Glu 443 might be compensated by an increased desolvation penalty for ionizing the histidine. Clearly, our computations do not yield this prediction. Moreover, graphical examination of the histidines and their surroundings show that the catalytic histidine in TcAChE is at least as accessible to solvent as is that of chymotrypsin.

That Glu 199 is predicted to influence the pK_a His 440 so strongly makes sense, because one of its carboxylate oxygens lies only 4 Å from Nϵ2 of His 440. These considerations lead one to wonder what compensates for the influence of Glu 199 and Glu 443 on His 440, and permits it to have a pK_a of about 6.3. The rest of this paper considers one possible explanation.

A CATION IN THE LIGAND BINDING SITE

Acetylcholinesterase is a cation-binding enzyme. Although it has been argued that its ability to hydrolyse neutral substrates implies that it does not actually possess a cation-specific binding site [36], it seems more likely that when neutral substrates bind to AChE, they bind in a somewhat different fashion, as argued by Quinn [37].

If the active site of AChE does have a high affinity for cations, it may hold a solvent cation even in the absence of substrate or inhibitor. If this cation shifts the pK_a of His 440 significantly, it might account for at least part of the discrepancy between our computed pK_a of His 440, and the actual pK_a of about 6.3 for the free enzyme. A cationic substrate would presumably displace this solvent cation when it binds, but a neutral substrate might not. Thus, a cation would be present under essentially all conditions, fixing the pK_a of His 440 at the appropriate value. In this view, the negative electrostatic potential generated by Glu 199 and Glu 443, which is responsible for pushing up the pK_a of His 440 to 9.3, might also be responsible for binding a cation which would bring the pK_a of His 440 back down to about 6.3.

We test this idea by recalculating the pK_as of TcAChE in the presence of three monovalent, cationic ligands: ACh, tacrine, and a small, spherical, cation (see Methods). As shown in lines 7, 8 and 9, of Table II, including these ligands in the calculation does indeed shift the computed pK_a of His 440 to values of 6.3–7.4, which agree with the correct value of 6.3 to within the demonstrated error of the computational method. It therefore appears possible that a cation in the choline binding site compensates for the upward pK_a shift of His 440 produced by Glu 199 and Glu 443.

Discussion

The calculations described here strongly suggest that the influence of Glu 199 and Glu 443 upon the pK_a of His 440 must be compensated in some way, given experimental data suggesting a pK_a of about 6.3 for His 440. In retrospect, the same impression may be reached without calculations, by comparison of the active sites of chymotrypsin and AChE. We find, furthermore, that artificial placement of a monovalent cation near Glu 199 brings the computed pK_a of His 440 to within the error of the model of the correct value. We are not aware of experimental data which fully prove or disprove the hypothesis that a bound cation substantially compensates for the influence of Glu 199 and Glu 443 upon the pK_a of His 440. However, several experiments in the literature do strongly support the more limited conclusion that when the active site of AChE binds a cation, the proton affinity of His 440 diminishes significantly. We now discuss these experiments in light of the present model.

INTERPRETATION OF KINETIC DATA

As preparation for this discussion, we note that the present model predicts that the influence of cations in solution upon the acylation step of catalysis will differ from their influence upon the deacylation step, at least in the case of a cationic substrate such as ACh. The basis for this assertion is as follows. When ACh binds to the active site, its cationic choline moiety compensates for the effect of Glu 199 upon the pK_a of His 440. Dissolved cations, such as sodium or potassium, are expected to diminish the affinity of the enzyme for ACh, by competing for the binding site, but should not be needed to lower the pK_a of His 440 and permit catalysis. In contrast, because the choline moiety may have exited the active site when deacylation occurs, binding of a cation from solution may be needed to lower the pK_a of His 440 and permit catalysis. However, this state of affairs will be pH-dependent: at higher pH, His 440 will be neutral whether or not a cation is bound. These considerations lead to the following predictions:

1. At low pH, the deacylation rate should be increased by cations. This effect should diminish with increasing pH, as cations become unnecessary to neutralize His 440.
2. The observed enzyme activity should increase with cation concentration under conditions where deacylation is rate-limiting.
3. Deacylation should become the rate-limiting chemical step at low pH and low cation concentration.
4. The enzyme-substrate dissociation constant (the Michaelis constant, K_m) should increase with increasing cation concentration.

These predictions imply that, under conditions where deacylation is at least partly rate-limiting, increasing ionic strength (and therefore cation concentration) will increase both K_m and k_{cat}. This is precisely what has been observed in at least two studies [38,39]. Even more striking, Krupka has observed that at low pH, for substrates rate-limited by deacylation, the hydrolysis rate is enhanced by tetramethylammonium, a cation presumed to bind at the same site as the choline of ACh. This rate-enhancement is not observed for substrates rate-limited by acylation [8]. Indeed, Krupka hypothesized, as we have here, that these effects result from a reduction in the pK_a of His 440 by bound tetramethylammonium. (It is also worth mentioning that choline itself accelerates deacylation [40].)

The present model may also offer an explanation for an otherwise puzzling data set. The effects of gallamine, a cationic inhibitor of AChE, were examined by the use of carbamylating agents [4] These create carbamylated enzyme intermediates which resemble the usual acyl intermediates, but are more long-lived. Their long half-lives facilitate study of the second step of catalysis. Kitz and coworkers found that gallamine accelerates the decarbamylation reaction in much the same way that choline and trimethylammonium accelerate deacylation [40]. This acceleration was found to increase with decreasing pH. This is consistent with the idea that binding of a cation is required to neutralize His 440 at low pH, but is less important at high pH. (This explanation requires the reasonable assumption that the cation binding site has greater affinity for gallamine than for the solvent cations, primarily Mg^{2+} in this study.)

Remarkably, it was also found that the degree to which gallamine inhibits the hydrolysis of AChE declines with decreasing pH. The authors argue that this result implies the existence of two gallamine binding sites. One, responsible for acceleration of decarbamylation, retains its affinity for gallamine at low pH. The other, responsible for inhibition of ACh hydrolysis, loses its affinity for gallamine at low pH. The conclusion was that gallamine accelerates deacylation by binding to an allosteric site distinct from the substrate-binding site.

It would appear, however, that the model developed in the present paper can provide a simpler explanation for these data. The inhibitory action of gallamine upon hydrolysis presumably results from its competition for the substrate binding site. This process is 'upstream' from the deacylation step. As a consequence, gallamine's competition for the substrate binding site may well fail to slow the overall hydrolysis rate under conditions where it is the deacylation step which is rate-limiting. If, as we have suggested, the deacylation step tends to become rate-limiting at low pH, it follows that the inhibitory effect of gallamine should decrease with decreasing pH. This is exactly what is observed.

We conclude this section by noting that, although the present arguments appear to be reasonable, they still must be elaborated in a more complete kinetic scheme. This will make it possible to determine whether experimental data can indeed be explained by a model employing realistic binding- and rate-constants.

MUTATION OF Glu 199

A puzzling result obtained from mutation studies of TcAChE is that substitution of Glu 199 by a neutral residue yields an enzyme whose activity shows the same pH-dependence as the wild type [4]. At first glance, this result seems to provide a strong argument against the notion that Glu 199 is positioned to have a strong influence upon the pK_a of His 440. In fact, our initial calculations on TcAChE suggest that neutralization of Glu 199 will lower the pK_a of His 440 by 2 pK_a units (see Table II, lines 1 and 5).

However, closer examination suggests that the present model, involving binding of a cation in the active site, may account for the experimental observation quite well. If binding of the cation depends upon the electrostatic potential generated by Glu 199, then removal of the negative charge of this residue is associated with removal of the positively charged cation too. Therefore the net charge change in the vicinity of His 440 is zero, and the effect upon the pK_a of His 440 is small.

This idea is supported by the present calculations. The computed pK_as of His 440 for the models with a bound cation (Table II, lines 7–9) range between 6.3 and 7.4. The computed pK_a of His 440 in the absence of a bound cation, and with Glu 199 set neutral, is 7.3 (Table II, line 5). Thus, if neutralization of Glu 199 is associated with loss of a bound cation, the pK_a of His 440 is expected to change little, in agreement with experiment.

LIPASES

As noted above, the active sites of several lipases possess catalytic triads extremely similar to that of TcAChE. They also possess carboxylic residues corresponding to Glu 199 and Glu 443 in AChE, and it seems probable that their catalytic histidines also must be neutral for hydrolysis to occur. If so, one again wonders what compensates for the influence of the carboxylic acids upon the pK_as of the catalytic histidines. Unfortunately, much less is known about the pH-dependence and ionic strength-dependence of the activity of the lipases, so we are unable to determine whether cation-binding might be explanatory for the lipases as it seems to be for AChE. That the substrates of these lipases are usually electrically neutral glycerol esters means that these enzymes are not known to bind cations. On the other hand, they possess at least as many negatively charged residues in their active sites. Furthermore, like TcAChE [5], the lipase active sites possess a number of aromatic residues, which offer possible stabilizing cation-aromatic interactions [41]. Thus it is reasonable to hypothesize that a cation lies in the active site near the catalytic histidine when the enzyme is functioning. Although examination of the crystallographic structures does not reveal particularly promising cation-binding sites in these structures, such sites might nonetheless form during the process of interfacial activation [42].

ROLE OF Glu 199 AND Glu 443

We have not attempted to explain why the carboxylic-solvent-carboxylic triad is so well conserved in this family of enzymes. Perhaps it positions one or more water molecules involved in hydrolysis, as previously proposed [10]; or perhaps the electrostatic field it generates facilitates the chemical reaction by some other mechanism. It might also accelerate the dissociation of the carboxylic acid products of catalysis, through electrostatic repulsion [14]. The present model suggests that the specific arrangement of ionizable groups in the active site of TcAChE causes the catalytic histidine, and therefore the entire catalytic process, to have an effective pK_a which is very sensitive to the concentration of cations. We have proposed that this sensitivity is responsible for the experimentally observed rise in k_{cat} with increasing ionic strength at physiologic pH. Berman and coworkers point out that this rise in k_{cat} with ionic strength partially compensates for the simultaneous rise in K_m with increasing ionic strength. They suggest that this compensation serves to stabilize the overall activity of the enzyme in the face of the large variations in ionic strength which are supposed to occur in the synapses where the enzyme resides [39]. Perhaps, then, the active site of AChE has evolved to hydrolyse a cationic ligand at a fairly consistent rate despite fluctuations in ionic strength. On the other hand, this concept does not seem applicable to the lipases, with their neutral substrates. Finally, perhaps it will be impossible to ascribe any special function to this substructure; it may simply represent one way of forming a stable, functional active site.

Acknowledgements

We are grateful to Dr. Joel Sussman for kindly providing us with the coordinates of the tacrine-TcAChE complex, and to Drs. Palmer Taylor and Hillary S. R. Gilson for helpful discussions. This work was supported in part by the N.I.H., the Robert A. Welch Foundation, and the N.S.F. Supercomputer Centers Metacenter Program. M. K. G. is a Howard Hughes Physician Research Fellow.

References

1. K. MacPhee Quigley, P. Taylor, and S. Taylor: *J. Biol. Chem.* **260**, 12185 (1985).
2. M. K. Gentry and B. P. Doctor: in *Cholinesterases: Structure. Function, Mechanism, Genetics and Cell Biology*, J. Massoulie, F. Bacou, E. Barnard, A. Chatonnet, B. P. Doctor and D. M. Quinn (Eds.), American Chemical Society, pp. 394–398 (1991).
3. B. P. Doctor *et al.*: in *Computer.Assisted Modeling of Receptor-Ligand Interactions, Theoretical Aspects and Applications to Drug Design*, R. Rein and A. Golombek (Eds.), Vol. 289, Liss. pp. 305–316 (1989).
4. G. Gibney, S. Camp, M. Dionne, K. MacPhee-Quigley, and P. Taylor: *Proc. Natl. Acad. Sci. U.S.A.* **87**, 7546 (1990).
5. J. L. Sussman, M. Harel, F. Frolow, C. Oefner, A. Goldman, L. Toker, and I. Silman: *Science* **253**, 872 (1991).
6. T. L. Rosenberry: *Adv. Enzymol. Relat. Areas Mol. Biol.* **43**, 103 (1975).
7. R. M. Krupka: *Biochemistry* **5**, 1983 (1966).
8. R. M. Krupka: *Biochemistry* **5**, 1988 (1966).
9. F. C. Bernstein, T. F. Koetzle, T. F. Williams, G. J. B. Meyer, Jr., M. D. Brice, J. R. Rodgers, O. Kennard, T. Shimanouchi, and M. Tasumi: *J. Mol. Biol.* **112**, 535 (1977).
10. M. Cygler, J. D. Schrag, J. L. Sussman, M. Harel, I. Silman, M. K. Gentry, and B. P. Doctor: *Protein Science* **2**, 366 (1993).
11. P. Grochulski, Y. Li, J. D. Schrag, F. Bouthillier, P. Smith, D. Harrison, B. Rubin, and M. Cygler: *J. Biol. Chem.* **268**, 12843 (1993).
12. P. Grochulski, Y. Li, J. D. Schrag, and M. Cygler: *Prot. Sci.* **3**, 82 (1994).
13. J. D. Schrag and M. Cygler: *J. Mol. Biol.* **230**, 575 (1993).
14. P. Grochulski, F. Bouthillier, R. J. Kaslauskas, A. N. Serreqi, J. D. Schrag, E. Ziomek, and M. Cygler: *Biochemistry* **349**, 3495 (1994).

15. Z. Radíc, G. Gibney, S. Kawamoto, K. MacPhee-Quigley, C. Bongiorno, and P. Taylor: *Biochemistry* **31**, 9760 (1992).
16. A. Ordentlich, C. Kronman, D. Barak, D. Stein, N. Ariel, D. Marcus, B. Velan, and A. Shafferman: *FEBS Lett.* **334**, 215 (1993).
17. A. Shafferman, B. Velan, A. Ordentlich, C. Kronman, H. Grosfeld, M. Leitner, Y. Flashner, S. Cohen, D. Barak, and N. Ariel: *EMBO J.* **11**, 3561 (1992).
18. J. Antosiewicz, J. A. McCammon, and M. K. Gilson: *J. Mol. Biol.* **238**, 415 (1994).
19. M. K. Gilson: *Proteins: Struct. Func. Gen.* **15**, 266 (1993).
20. M. E. Davis, J. D. Madura, B. A. Luty, and J. A. McCammon: *Comput. Phys. Commun.* **62**, 187 (1991).
21. C. Tanford and J. G. Kirkwood: *J. Am. Chem. Soc.* **79**, 5333 (1957).
22. D. Bashford and M. Karplus: *Biochemistry* **9**, 327 (1990).
23. D. Bashford and K. Gerwert: *J. Mol. Biol.* **224**, 473 (1992).
24. A. S. Yang, M. R. Gunner, R. Sampogna, K. Sharp, and B. Honig: *Proteins: Struct. Func. Gen.*, **15**, 252 (1993).
25. A.-S. Yang and B. Horng: *J. Mol. Biol.* **231**, 459 (1993).
26. J. Antosiewicz, M. K. Gilson,.and J. A. McCammon: *Israeli. J. Chem.* **34**, 151 (1994).
27. M. Harel, I. Schalk, L. Ehret-Sabatier, F. Bonet, M. Goeldner, C. Hirth, P. Axelsen, I. Silman, and J. L. Sussman: *Proc. Natl. Acad. Sci. U.S.A.* **90**, 9031 (1993).
28. M. J. Frisch, G. W. Trucks, M. Head-Gordon, P. M. W. Gill, M. W. Wong, J. B. Foresman, B. G. Johnson, H. B. Schlegel, M. A. Robb, E. S. Replogle, R. Gomperts, J. L. Andres, K. Raghavachari, J. S. Binkley, C. Gonzalez, R. L. Martin, D. J. Fox, D. J. Defrees, J. Baker, J. J. P. Stewart, and J. A. Pople: Gaussian, Inc., Pittsburgh, PA.
29. C. M. Breneman and K. B. Wiberg: *J. Comput. Chem.* **11**, 361 (1990).
30. U. Dinur and A. T. Hagler: in *Review of Computational Chemistry*, K. B. Lipkowitz and D. Boyd (Eds.), VCH, New York, Vol. 2, Ch. 4 (1991).
31. Biosym Technologies, San Diego, CA.
32. E. Pop, M. E. Brewster, J. J. Kaminski, and N. Bodor: *Int. J. Quant. Chem.* **35**, 315 (1989).
33. W. L. Jorgensen and J. Tirado-Rives: *J. Am. Chem. Soc.* **110**, 1657 (1988).
34. M. L. Bender, G. E. Clement, F. J. Kezdy, and H. d'A. Heck: *J. Am. Chem. Soc.* **86**, 3680 (1964).
35. A. R. Fersht and J. Sperling: *J. Mol. Biol.* **74**, 137 (1973).
36. F. B. Hasan, S. G. Cohen, and J. B. Cohen: *J Biol. Chem.* **255**, 3898 (1980).
37. D. M. Quinn: *Chem. Rev.* **87**, 955 (1987).
38. J.-P. Changeux: *Mol. Pharmacol.* **2**, 369 (1966).
39. H. A. Berman, K. Leonard, and M. W. Nowak: 'Function of the peripheral anionic site of acetylcholinesterase', in *Cholinesterases: Structure, Function, Mechanism, Genetics and Cell Biology*, J. Massoulie, F. Bacou, E. Barnard, A. Chatonnet, B. P. Doctor and D. M. Quinn, Eds., American Chemical Society, pp. 229–234 (1991).
40. R. J. Kitz, L. M. Braswell, and S. Ginsberg: *Mol. Pharmacol.* **6**, 108 (1969).
41. D. A. Dougherty, and D. A. Stauffer: *Science* **250**, 1558 (1990).
42. P. Desnuelle: in *The Enzymes*, 3rd Ed., P. D. Boyer, Ed., Academic Press, New York, Vol 7, pp. 575–616 (1972).

Structure Modeling of the Acetylcholine Receptor Channel and Related Ligand Gated Channels

EBERHARD VON KITZING
Max-Planck-Institut für Medizinische Forschung, P.O. Box 10 38 20, D-69028 Heidelberg, Germany.

Abstract. In this model-building study a model for the pore of the acetylcholine receptor channel is proposed. The pore is formed by five α-helices of the M2 segment where three rings of hydrophilic side chains point into the channel lumen. This model is in agreement with most experimental data like photolabeling, drug affinity studies, single channel conductivity measurements and cryo electron microscopy known about this channel.

This study predicts a strong coupling of the motion of the ions in the channel to that of the charged and highly hydrophilic amino acid side chains at the channel wall. Due to the negative net-charge in the pore more than a single cation may occupy the pore region. The resulting strong local electric fields make the commonly used constant field approximation obsolete for this type of ion channel.

Key words: Acetylcholine receptor channel, model building, energy profile, ion conduction mechanism, constant field approximation, channel pore, multi ion channel, coupled motion.

1. Introduction

A flux of ions through cell membranes is utilized for intracellular communication and was probably established early during biological evolution [1,2]. Lipid membranes surrounding biological cells provide an unfavorable medium for alkali and earthalkali metal ions [3,4]. They represent a high energy barrier for ionic fluxes. Therefore, ion channels may be considered as special enzymes [5] mediating the flux of ions through membranes of biological cells. In typical cases at $100\,\mathrm{mV}$ about 10^8 ions pass a channel each second [6]. Many channels are rather specific with respect to the transported ionic species [7].

These 'enzymes' are regulated in various ways from within the cell and from outside [8]. The most obvious type of regulation is that the ability for ions to pass the channel may be switched on and off. This opening and closing of the channel may be induced by extra- or intracellular transmitter substances (ligand gated channels [8]), G-proteins or electric potential difference across the membrane (voltage gated channels [9]).

A typical function of voltage gated channels is to drive the electric signal along axons. Ligand gated channels are used to transmit the nerve signal from one nerve cell to the next; this action may be excitatory or inhibitory. Muscle fibers are typically excited by means of the acetylcholine receptor channel (see Figure 1). After electric excitation the synapse secretes acetylcholine into the synaptic cleft. This transmitter substance activates acetylcholine receptor channels in the post-synaptic membrane resulting in the influx of ions into the muscle cell which in turn induces the contraction of the muscle fiber. The neuro-transmitter acetylcholine either diffuses away or it is degraded by acetylcholine esterase.

In this article a model for the pore of the acetylcholine receptor channel is presented. Due to the high homology of the amino acid sequence of the acetylcho-

A. Pullman et al. (eds.), Modelling of Biomolecular Structures and Mechanisms, 39–57.
© 1995 *Kluwer Academic Publishers. Printed in the Netherlands.*

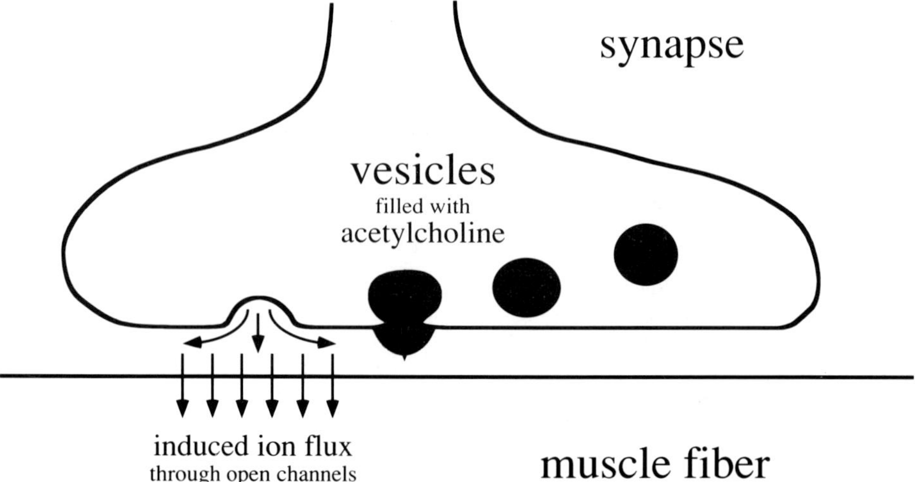

Fig. 1. The synapses which induce the contraction of muscle fibers contain vesicles filled with the neuro-transmitter substance acetylcholine. Upon excitation the synapse releases acetylcholine into the synaptic cleft which in turn opens acetylcholine receptor channels in the post-synaptic membrane. The ionic current through these open channels initiates the contraction of the muscle fiber. The neuro-transmitter either diffuses away or is degraded by acetylcholine esterase.

line receptor channel with serotonine [10], glycine and GABA receptor channels [11], the results of this study has also implications for the ligand gated ion channels of those families.

2. Experiments Revealing Structural Information about the Acetylcholine Receptor Channel

To understand the function of enzymes requires a detailed knowledge of their three dimensional structure. Unfortunately membrane proteins are difficult to crystalize [12,13]. The crystallization of membrane proteins was successful in only a few cases. Additionally, the acetylcholine receptor channel molecule is too large for its three-dimensional structure to be resolved by means of NMR techniques.

The most direct structural information about the acetylcholine receptor channel has been obtained by means of cryo-electron microscopy of crystalline layers of channel molecules [14–16] which provided a resolution above 10 Å. Recently, this resolution has been improved down to 9 Å [17].

More indirect methods have also been used to obtain structural information about ligand gated ion channels. There is a huge amount of amino acid sequence data about ligand gated ion channels [11,18]. Comparison of amino acid sequences of many similar channels indicates the importance of certain regions. The hydro-pho°bicity of segments in the amino acids chain provides hints whether a certain segment contributes to the surface or to the interior of the molecule [19]. Comparison with better known structural data from other less complicated channels may also help [20].

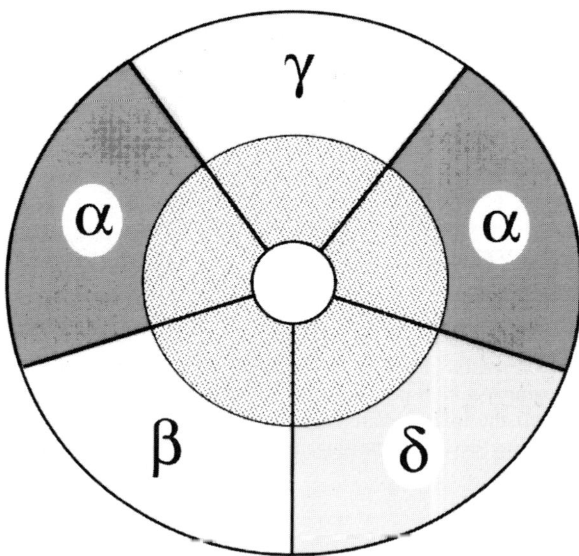

Fig. 2.   The acetylcholine receptor channel consists of five subunits, two of which are identical. The placing of the β and γ subunit is still under debate. The five subunits are placed in a pseudo-symmetric order; the channel follows the axis of pseudosymmetry.

Chemical modification of the channel protein, for instance via photolabeling [21] provides an other source of structural information. Only those amino acid side chains accessible to the reactant may be labeled by this method. In this way amino acids close to the protein surface or to a specific binding site are found.

Another powerful method is the use of single site mutations in a channel and to study the resulting change in its function [22]. By this means it is possible to change a cation-selective channel into a anion-selective one [23].

These methods provide a wealth of indirect structural data and were utilized in this model building study.

2.1. THE COMPOSITION OF THE ACETYLCHOLINE RECEPTOR CHANNEL

Acetylcholine receptor channels transfer excitation from synapses to muscle fibers. The excited presynaptic membrane secretes acetylcholine which activates the acetylcholine receptor channels in the post-synaptic membrane (see Figure 1) [7]. Additionally, there are also neuronal acetylcholine receptor channels [21]. The muscular acetylcholine receptor channels consists of five subunits of which four are different [24]; they have the composition $\alpha_2\beta\gamma\delta$. The order of the five subunits is still controversial. There is evidence for $\alpha\gamma\alpha\beta\delta$ [24] as well as for $\alpha\beta\alpha\gamma\delta$ [15]. The difference between the two models should have consequences only for finer details of the model because the five subunits are highly homologous. This view is supported by point mutations experiments [25] where amino acids at a homologeous position are exchanged between different subunits. In the model described below the $\alpha\gamma\alpha\beta\delta$ model is realized as shown in Figure 2.

```
         i        m     t      s1      l s2     v                e
α   DSG - EKMTLS I SVLLSLTVFLLVIVELIP
β   DAG - EKMGLS I FALLTLTVFLLLLADKVP
γ   KAGGQKCTVATNVLLAQTVFLFLVAKKVP
δ   DCG - EKTSVAISVLLAQSVFLLLISKRLP
```

Fig. 3. The amino acid sequence of the M2 segment of the acetylcholine receptor channel from rat muscles [33] using the single letter code. Highly conserved amino acids are italicized. The M2 segment most likely forms the pore of this ion channel. Because of the fivefold pseudosymmetry the amino acids at a given homologous position pointing into the channel interior are assumed to form rings: (i) the intra-cellular ring, (m) the intermediate ring, (t) the threonine ring, (s1) the first serine ring, (l) the leucine ring, (s2) the second serine ring, (v) the valine ring and (e) the extracellular ring.

2.2. THE PORE LINING PART OF THE ACETYLCHOLINE RECEPTOR CHANNEL

Each of the five subunit of the acetylcholine receptor channel consists of approximately 700 amino acids. Thus, the whole channel assembly represents a rather large protein. The wealth of data from single point mutations apply largely to those regions of the channel which determine the flow of ions through the channel.

Following the publication of the first amino acid sequences of subunits of acetylcholine receptor channels [26–28] four segments (M1, M2, M3 and M4) were hypothesized to fold into hydrophobic α-helices building the transmembrane part of the channel. Additionally, a putative amphipathic helix MA was proposed to line the pore with its hydrophilic residues [19,29].

By means of constructing chimeric ion channels consisting of the amino acid sequences from calf and torpedo channels, the M2 segment was shown to be a major determinant of ion flow through acetylcholine receptor channels [30]. These findings were supported by binding studies of a channel blocking drug QX-222 [31,32].

Several amino acids in the M2 segment have been shown to strongly influence the conductivity of acetylcholine receptor channels. The single-letter code of the amino acid sequence of the M2 segment from rat muscles [33] is shown in Figure 3. The fivefold pseudosymmetry means that the side chains of these amino acids are believed to form rings around the axis of symmetry. Three rings of charged amino acids – the intracellular, the intermediate and extracellular ring are found to be major determinants of the ionic conductivity [25,34,35]. Also, a ring of uncharged hydrophilic amino acids was found to influence the conductivity strongly: the threonine ring [36–38]. This threonine ring is assumed to be part of the pore constriction. Two serine rings are involved in the binding of the open channel blocker QX-222 [32,39]: chloropromazine photolabeled hydrophobic leucine and valine rings [21].

Due to the spacing of the influential amino acids (see Figure 3) the proposal

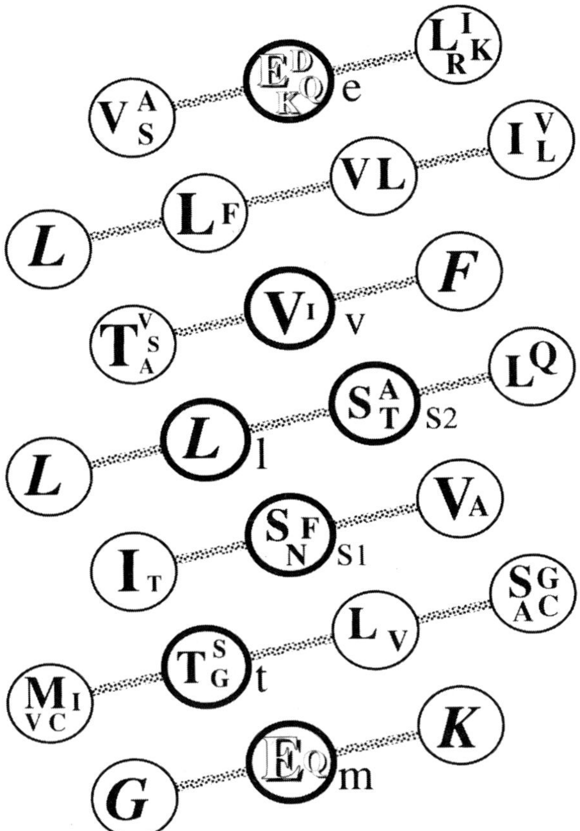

Fig. 4. The amino acid sequence of the M2 segment shown in form of an α-helix projected onto the surface of an unfolded cylinder given in the single letter code (see also Figure 3). The amino acids known for their reactivity with photolabels or open channel blockers, or for their strong influence on the single channel ionic conductivity are indicated by thick circles and grey background. The indices at these circles are explained in the legend of Figure 3. The size of the characters in the circles correlates with the occurrence of the respective amino acid. Amino acids strongly conserved on all acetylcholine receptor M2 segments are indicated by italicized characters. The amino acid of the intracellular ring is not included in this picture.

that the M2 segment builds an α-helix is supported. Figure 4 shows the amino acid sequence in the 'unfolded-cylinder' representation. The important amino acids are facing the same direction (emphasized circles in Figure 4). Highly conserved amino acids (italicized characters in Figure 4) are facing either to the side or to the back side. Single point mutations of the hydrophilic amino acids which presumably point in the direction of the lipid membrane are found to have no measurable effect on ion flow [Villarroel, personal communication]. Recent cryo-electron microscopic studies support the view that the walls of the acetylcholine receptor channels are formed by α-helices.

2.3. THE PICTURE OF THE ACETYLCHOLINE RECEPTOR CHANNEL BASED ON
     EXPERIMENTAL DATA

There are many lines of evidence that the pore of the acetylcholine receptor
channels is formed by five α-helices made by the M2 segments. This view is
supported by the influence of point mutations on the ionic conductivity, by binding
studies of blocking drugs and photolable experiments. Also 9 Å resolution cryo-
electron microscopic data are in agreement with this view. Photolabeling and point
mutation studies have been used to obtain structural information about the binding
site for the ligand in ligand gated channels [2,40].

Unfortunately, for other parts of the channel there is much less data, a situation
which at the moment makes structural hypotheses of the other regions in the
channel difficult and speculative. Therefore, in this article only the pore lining
part of the channel will be modeled.

3. Modeling the Acetylcholine Receptor Channel Pore

Since the publication of the first amino acid sequences of acetylcholine receptor
subunits [26–28] there have been many proposals for structural models of the
acetylcholine receptor channel created by means of computer simulation tools. In
early models the pore was assumed to be lined by the putative amphipathic helix
MA [19,29]. After the strong influence of the M2 region on the conductivity
became known [30], the M2 segment was utilized to form the pore [41–45].

Using hydrophobicity patterns of the amino acids sequence of the acetylcholine
receptor channel and related ion channels, the three segments M1, M2 and M3
of the acetylcholine receptor channel are assumed to form transmembrane α-
helices [11,26–28,40,41,46]. A comparison of several methods to predict protein
secondary structure has shown the assignment of hydrophobic helices in membrane
proteins to be relatively reliable as compared to crystallographic structures [47].
Together with other lines of evidence derived from experimental data it is reasonably safe to assume an α-helical structure for the M2 segment.

The M1 and M3 segments are assumed to form a wall between the M2 segments
and the lipid membrane [29,40,41]. However, the commonly proposed α-helical
structure for the M1 and M3 segments has recently been questioned [17]. A β-
sheet structure similar to that of verotoxin-1 from *E. coli* [48] was proposed
instead.

In addition to computer models of ion channels, synthetic peptides were designed with the purpose of forming ion channels after incorporation into lipid
membranes. These peptides consist of putative α-helices. The amino acid sequences are either taken from a minimalist approach [49–51] or from existing ion
channels [52–57]

Based on the experimental data described in the previous section an atomic
model for the acetylcholine receptor channel was designed using the amino acid
sequence of the acetylcholine receptor channel from rat muscles [33]. The aligned
parts of the amino acid sequences for the α to δ subunits of the M2 segments
including additional amino acids at the N-terminal side are given in Figure 3. It
is assumed that the M2 segments consists of α-helices (see below). α-Helices of

the appropriate amino acid sequence were built first, and then the close contacts between intrahelical side chains were removed using the QUANTA program package. To relax the internal strain and improve the side chain packing, the internal energy of the starting structures were minimized using the AMBER 3a program package [58]. In order to place the helices in the correct symmetry and orientation, a special program was used which allowed to introduce arbitrary rotations and translations of the individual helices.

3.1. CONSTRAINTS TO THE MODEL IMPOSED BY EXPERIMENTAL OBSERVATIONS

Most of the experimental data available for the acetylcholine receptor channel provides low resolution and often rather indirect structural information on the M2 segments. Therefore an atomic model of that segment was designed on the assumption that it forms the pore of the acetylcholine receptor channel, which is consistent with the experimental observations. According to the hypothesis of α-helix folding for the M2 segment, the negative charges that form anionic rings are located at both ends of the segment and in the center of the segment near the cytoplasmic side and should be accessible to the ions (Figures 3 and 4). The direction of the α-helix of the M2 segment is such that its N-terminal side is intracellular and its C-terminal side extracellular.

According to conductivity measurements [36–38] the threonine ring should be close to the narrowest part of the channel and the threonine should point into the channel lumen. QX-222 binding studies suggest that also the two serine rings should point into the channel. To account for the different influence of the three hydrophilic rings (threonine and two serine rings), the channel should be narrow at the threonine ring and widen towards the serine rings.

3.2. MODELLING THE M2 TRANSMEMBRANE SEGMENTS

An α-helix was first designed with the helix axis being the z-axis. The sequence starts one amino acid before the intracellular ring and ends at the terminal proline of the M2 segment. The first three amino acids preceding the M2 segment are given an extended conformation. The α-helix starts at the glycine before the intermediate ring and ends at the proline (see Figure 3). This helix was rotated around the z-axis in order to make the side chain of the threonine ring point to the negative x-direction (see Figure 3). This starting helix was used as a template to generate the remaining three amino acid sequences for the β, γ and δ subunits. Before the single helices where assembled together to build the pore of the channel, each helix was optimized separately using the AMBER program. The number of amino acids per turn are 3.72, a number slightly larger compared to the standard value of 3.61 amino acids per turn.

The pore of the channel was formed by rotating the helices by 10° counterclockwise around the y-axis, which results in an opening of the pore towards the extracellular side (see Figure 5). Next they were rotated by 20° clockwise (i.e. −20° according to mathematical sign conventions) around the x-axis, to bring all three hydrophilic rings facing into the pore. The helices were then shifted 7 Å in the positive x-direction. Finally the subunits α_1, γ, α_2, and β were rotated counter-

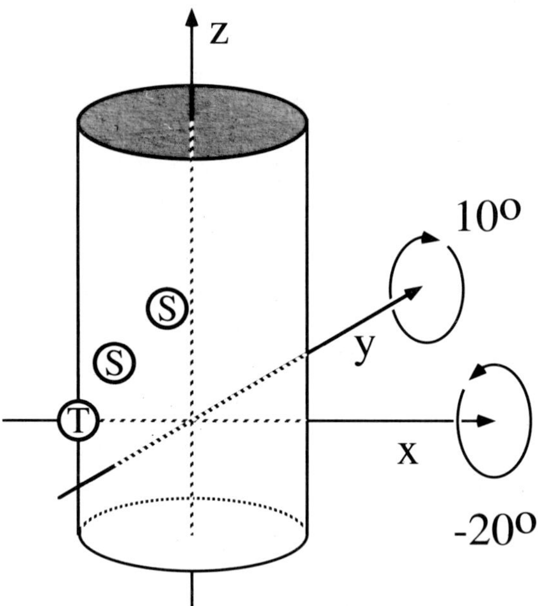

Fig. 5.   The orientation of the α-helices of the M2 segment in the model of the acetylcholine receptor channel. The pore is formed by five pseudosymmetrically placed helices. The axes of the helices form a right handed screw to allow the hydrophilic rings (see text) to point into the pore. The origin in orienting the helices is defined to be placed on the helix axis at the height of the C_α atom of the threonine in the threonine ring. The helices are rotated by $-20°$ around the x-axis and by $10°$ around the y-axis. The distance of the helices from the axis of pseudosymmetry is $7\,\text{Å}$.

clockwise around the z-axis by $0°$, $72°$, $144°$, $216°$ and $288°$, respectively. The five helices form a right-handed screw around the z-axis, the axis of pseudosymmetry.

The given values for the orientation of the α-helices and their distances from the central axis of fivefold pseudosymmetry were optimized to obtain dense packing of the helices at the level of the threonines, to avoid overlap of side chains, and to orient the threonine and the two serine rings residues so that their side chains face into the pore.

To further optimize side chain packing the whole structure was energetically optimized using the AMBER 3a software. Because the major part of the channel protein is missing, it cannot be expected that the M2 segments by themselves will produce a stable structure. Therefore the C_α atoms where constrained to their initial coordinates by a force constant of $1\ \text{kcal mol}^{-1}\ \text{Å}^{-2}$ and the remaining main chain atoms by $0.01\ \text{kcal mol}^{-1}\ \text{Å}^{-2}$. These constraints give sufficient flexibility to relax the structure internally and to improve side chain packing, keeping the overall shape of the structure. The relaxation was achieved by means of dynamic runs over a time of 50 ps. After the dynamic runs the energy of the structures was minimized. The structure with lowest energy was taken as the reference structure for the calculations described below.

A general view of the channel model is given in Figure 6. To make the α-helices more visible, the main chain atoms were enlarged to their van der Waals radii,

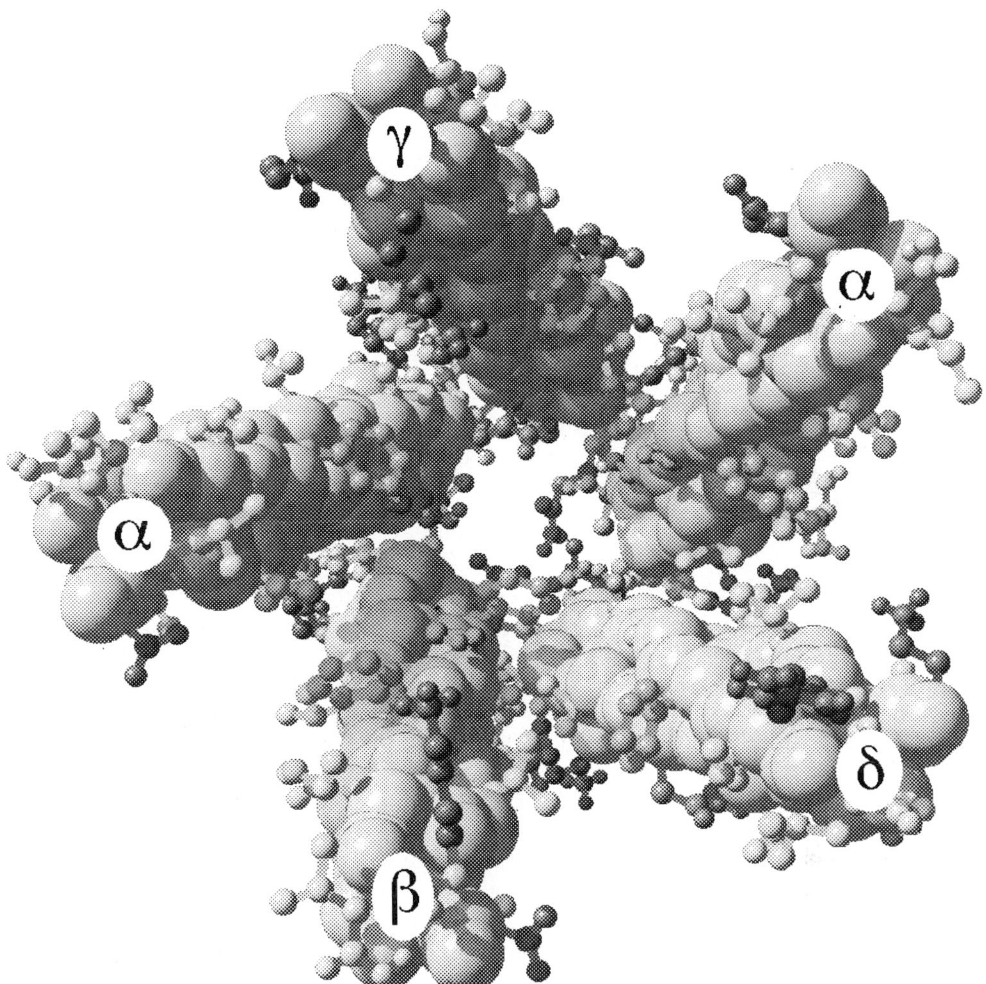

Fig. 6. A model of the pore of the acetylcholine receptor channel consisting in the M2-segments. The M2 segment is assumed to form an α-helix, the five subunits α, γ, α, β and δ are indicated in the figure. The radii of the atoms of the backbone are drawn using their van der Waals radii, whereas the side chains are drawn using a stick-and-ball modus of the SCHAKAL92 program.

whereas the side chain atoms are given in a stick and ball representation. The different segments are indicated in the figure.

4. Consequences for the Ionic Conduction Mechanism in the Acetylcholine Receptor Channel

The previous sections have presented a model of the pore of the acetylcholine receptor channel. What can we learn from such models? Do they tell us anything hitherto unknown? In modeling such molecules one can try to translate low resolution experimental data into molecular structures. How well is the structure determined by the data?

In the case of the acetylcholine receptor channel, experimental data which provide structural constraints for the molecule are available only for the pore and to a lesser extent for the acetylcholine binding site. Therefore, an overall shape of the pore can be derived. A detailed three-dimensional structure of the pore requires more specific and more direct information as the influence of single point mutations on the channel conductivity and on the affinity of various drugs can reveal.

In the case of ion channels there is a special problem. Much of the data consists of conductivity measurements of various point mutations of the channel. Unfortunately, there is no precise theory to translate molecular structure into ionic conductivities or *vice versa*. Although much work has been done in this field [5,44, 59–66], our knowledge about the conduction mechanism of ions through ion channels is still limited. The model of the acetylcholine receptor channel was used for further simulations which shed some light on the mechanism of ion conduction.

4.1. ENERGY PROFILES AS A FUNCTION OF IONIC RADII

An important structural parameter of an ionic channel is its radius. Experimentally, the radii of channel pores are estimated by using organic ions with a hydrophobic shell [67]. It is assumed that for these ions the hydration shell does not dominate the conductivity as might be expected from alkali or earth alkali metal ions.

To 'measure' the effective radius of the channel model, the energy profile of ions in the channel with different van der Waals radii were calculated. Initially the ion was placed 30 Å away from the confinement of the channel, and constrained to this position. The main chain atoms of the α-helices of the channel model were also constrained to their initial position as described above. The structure was relaxed using a dynamics run followed by a minimization. The structure of the resulting model was taken as the starting structure for the next ion position more close to the narrowest part of the channel. At every further step the structure of the previous calculation was used as the starting structure for dynamics and minimization for the next step. The step length through the channel depended on the position of the ion. Outside the channel the step length was large, inside the channel the step length was reduced to 1 Å for each optimization.

This 'tour through the channel' was done with van der Waals radii (or, more precisely, the distance of minimal van der Waals energy for the ion) of 2.3 Å, 3.7 Å and 4.9 Å. The resulting energy profiles are shown in Figure 7. That the ion with a radius of 3.7 Å still can pass the channel agrees nicely with the measurements of Dwyer *et al.* [7,67] using organic cations.

The negatively charged amino acids in the intermediate ring (see Figure 2, column 'm') follow the ion on its path through the pore of the channel. In Figure 8 the position of the ion on its way through the channel is plotted against the position of the C_δ atoms of the intermediate ring side chains. In the empty channel the γ and δ side chains point towards the extracellular side of the channel, whereas

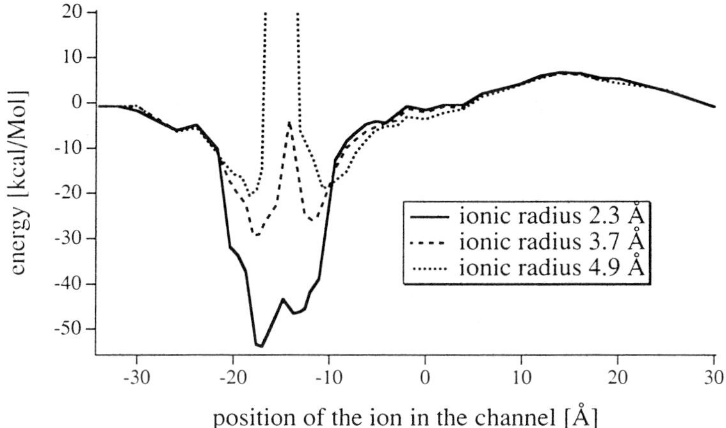

Fig. 7.   Energy profiles of the acetylcholine receptor channel model for ions with different radii as indicated in the figure legend. The radii indicate the position of lowest energy for the van der Waals interaction as used in AMBER3a.

that of α_1 points towards the intracellular side. α_2 and β occupy an intermediate position. When the small ion (2.3 Å radius) enters the confined part of the channel (Figure 8a), it first becomes solvated by the side chain of the amino acid at the intermediate ring of the δ subunit. This side chain follows the ion half the way through the confined part of the pore. After the first half the side chain of the intermediate ring of the α_1 subunit solvates the ion and follows it into the intracellular direction. In the case that water would be included in the calculation one might expect that the charged side chains may replace water molecule in the hydration shell of the ion.

The situation gets more crowded if the ionic radius increases (Figure 8b). In this case there is no sufficient space for a simple reorientation of the side chains. They have to repack to create a sufficiently large hole for the ion. The degree of repacking increases for the ion with a radius of 4.9 Å (data not shown).

4.2. THE NUMBER OF IONS IN THE CHANNEL

The pore of the acetylcholine receptor channel contains several negatively charged amino acid side chains. The charge of these amino acids is experienced by the ions as shown by point mutation experiments [25,34]. Changing glutamic acids into glutamine or aspartic acid into asparagine has a strong influence on the ionic conductivity. In the rat acetylcholine receptor channel three net charges are found in the intracellular ring, four in the intermediate and a single one in the extracellular ring (see Figure 3: please remember that the charges of the α-subunit have to be counted twice). Mutations of the strongly conserved lysine (single letter K) in the ring directly beneath the intermediate ring in the β, γ and δ subunits into glutamic acid does not influence the ionic conductivity [34]. These lysines appear not to compensate for the high negative net charge close to the channel confine-

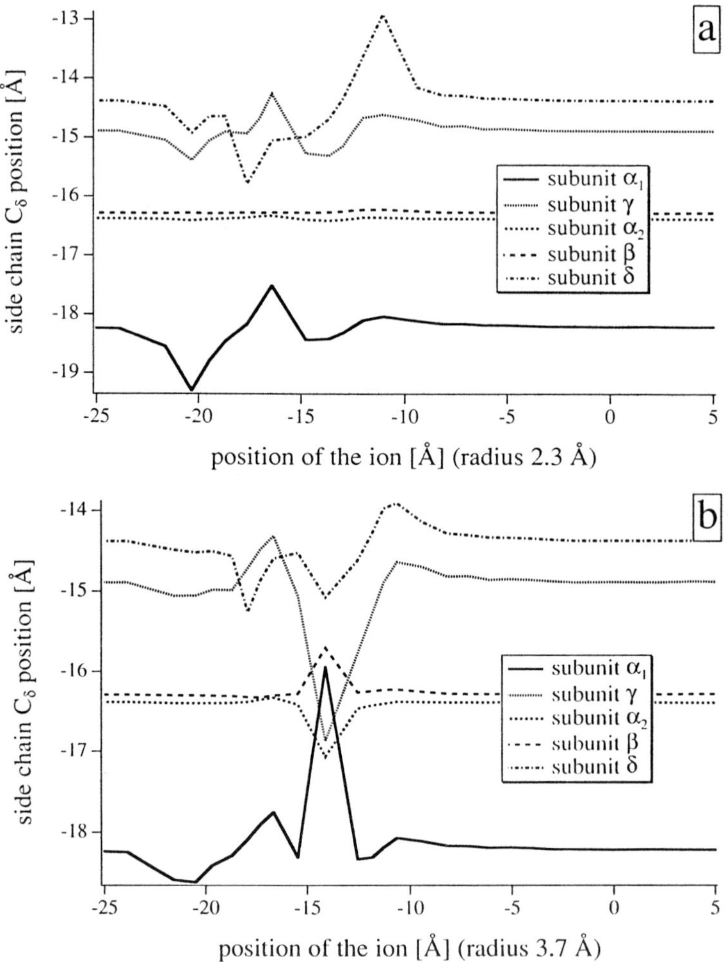

Fig. 8. The position of a single ion on its path through the acetylcholine receptor channel model plotted against the position of the C_δ atoms of the side chains of the amino acids in the intermediate ring (see Figure 3). In the case of the small ion (a) with a radius of 2.3 Å the negatively charged side chains solvate the ion and follow it through the channel. The larger ion (b) with a radius of 3.7 Å enforces reorganization of the side chains due to their larger radius.

ment. The same mutation applied to the α-subunit caused a failure to observe channel openings.

The acetylcholine receptor channel is generally considered to be a 'single ion channel' [7], i.e. that only a single ion may enter the pore of the channel. This hypothesis is derived from modeling ion transport properties by means of Eyring rate approaches [7,68]. In such an approach, the channel is believed to have a few binding sites which are separated by energy barriers. Thermally actived ions cross the barriers. This model requires that the binding energies of the ions are not too strong compared to the thermal energy, i.e. no more than about 5 kcal mol^{-1}.

The binding energy of a single ion in the model described in this article is found to be in the order of 50 kcal mol^{-1} (Figure 7). Similar binding energies are found in other modeling studies [44,45,69–71]. It is known from electrolyte theory that, at equilibrium, charges tend to become neutralized within a Debye length [72]. Therefore, the negative net charge in the channel should be compensated either by protons or, if abundantly available, by cations.

To shed some light on the question of ion occupancy, a multi-ion series of dynamics and optimization runs was performed with the 2.3 Å ion. The channel structure containing a single ion at the position of lowest energy was taken as the starting structure for the two ion simulation. The starting structure was first optimized with no constraints on the ion. Then the same procedure was applied for the calculation of the two ion energy profile as for the calculation of the energy profile of the single ion. The second ion was stepped through the channel with a positional constraint on it with 1 kcal mol^{-1} Å^{-2}. The first ion at its low energy position was left unconstrained. As in the single ion case an energy profile for the channel model as a function of the position of the second ion was calculated.

The structure with lowest energy in the double ion calculation was then used (after unconstrained dynamics and minimization) as the starting structure for the triple ion energy profile. The position of the 'moving' ion was constrained as in the previous cases, while that of the other ions remained unconstrained. This procedure was repeated up to the five-ion energy profile. The pore of the structure with lowest energy containing five ions is shown in Figure 9. Only four ions can be seen in the figure, the fifth is on the intracellular side. The n ion energy profiles with $n \in \{1,2,3,4,5\}$ are given in Figure 10. To make the curves comparable with each other the internal energy of the respective starting structure was taken as the reference energy in this plot.

Figure 10 clearly indicates that the model discussed in this article can accommodate up to four ions of the radius of 2.3 Å. The fifth ion did not find a position with a considerable binding energy. This study was done only for a single radius. If the radius of the ions were to be increased, fewer ions could enter the channel. In the case of smaller radii, the fifth ion could have found a binding site.

The position of the free ions is relatively stable for the two, three and four ion channel. In Figure 11 the position of the moving ion is plotted against the position of the other ions for the three and four ion channel. In the five-ion channel the movement of the ions in the channel becomes somewhat chaotic (data not shown). This indicates the instability of the situation. To accommodate the fifth ion into the channel larger rearrangements of the ion positions are required.

5. Discussion

This article has presented a model for the acetylcholine receptor channel pore based on experimental data. The implications of this model for the structure of the acetylcholine receptor channel in special and on the mechanism of ion conduction in general are discussed.

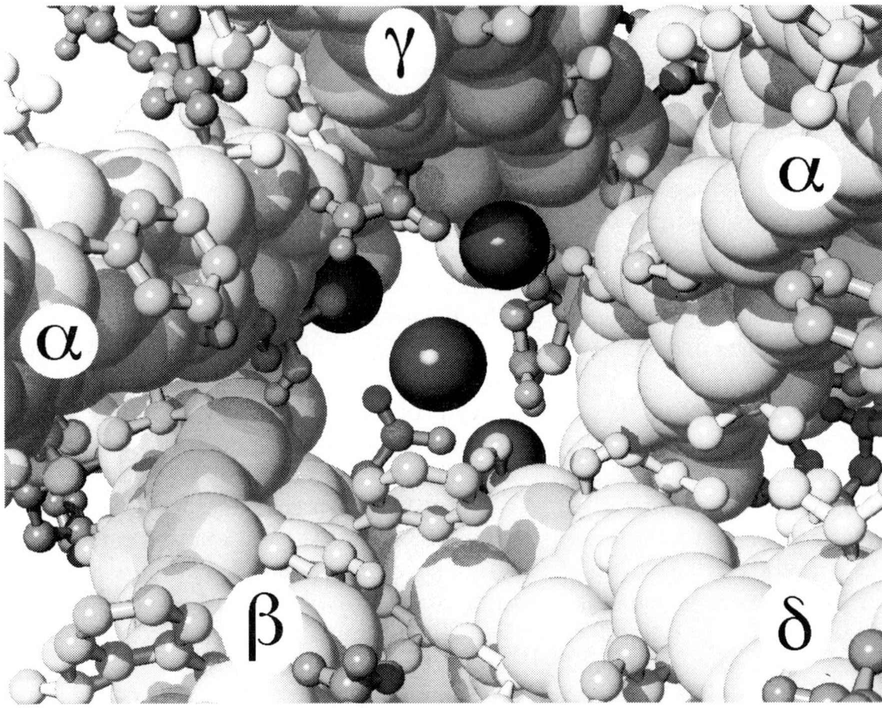

Fig. 9.   The pore model of the acetylcholine receptor channel is shown with five ions inside. To maintain mesoscopic electroneutrality the negative net charge of the charged rings in the M2 segment must be compensated for either by protons or by cations.

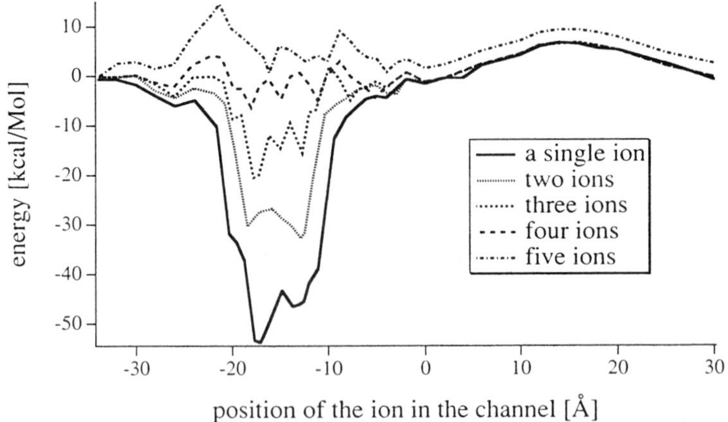

Fig. 10.   Energy profiles of ions in the acetylcholine receptor channel model for different occupation numbers. The energy profiles for up to five ions are shown. Only four ions find significant binding energies in the channel. For details of the calculation see text.

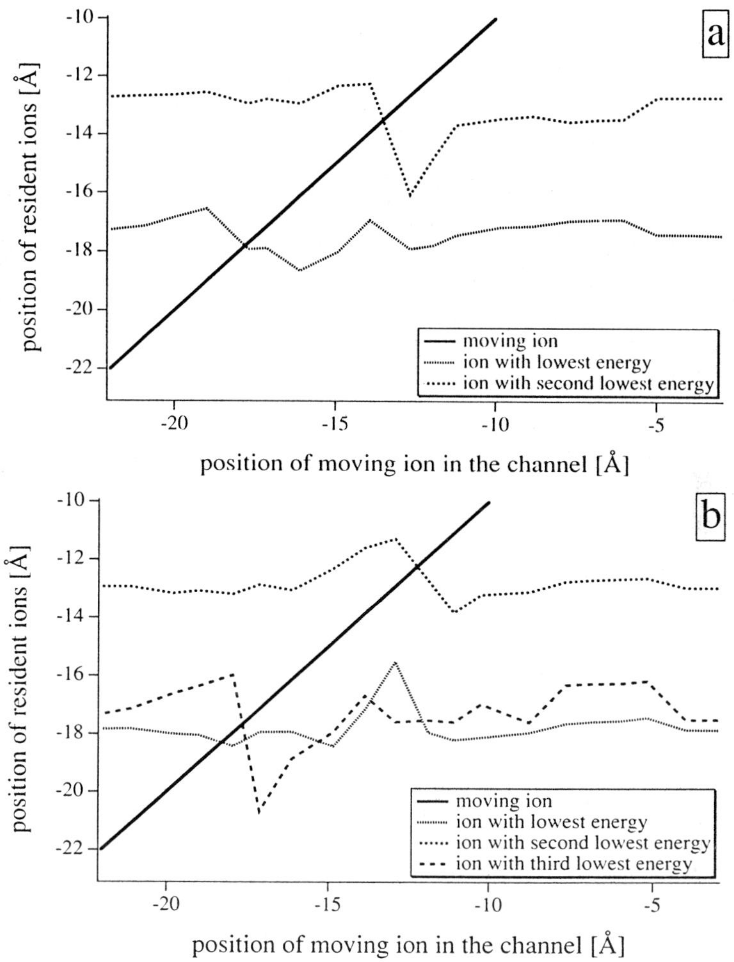

Fig. 11.   The position of the moving ion is drawn against the position of the ions residing in the channel for the three (a) and four (b) ion channel.

5.1. STRUCTURAL PROPERTIES OF THE MODEL

There is much evidence that the pore of the acetylcholine receptor channel is formed by the M2 segments forming α-helices. It is shown in this article that a close packing of the helices near the threonine ring gives rise to a sufficiently large pore. The experimentally estimated radius is 3.2 Å. In this model the ion with a radius of 2.3 Å fits well into this channel, but the ion with a radius of 3.7 Å feels considerable resistence.

The orientation of the helices can be estimated from the sensitivity of conductivity measurements and binding studies with respect to point mutations. Precise values for the orientational parameters of the M2 α-helices can only be obtained by studying the three-dimensional structure of the open acetylcholine receptor

channel by means of more direct experimental techniques like cryo electron microscopy.

This model is quite similar to an other computer model of the pore of the
acetylcholine receptor channel [70]. The major difference appears to be an additional tilt of the helices to form a right handed screw. This tilt moves the two
serine rings into the channel lumen.

A structure of the closed acetylcholine receptor channel at 9 Å resolution has
recently been published [17]. It probably differs from the model proposed above
in the position of the narrowest part of the channel. The narrowest part in the
cryo-electro microscopic structure consists in five kinked helices. If the helices are
assumed to be formed by the M2 segment, the narrow part would be close to
the leucine ring. On the other hand, this proposed topology cannot explain the
conductivity data, binding studies and photolabeling experiments for the open
channel. Based on this experimental study one can assume that the closing of the
acetylcholine receptor channel is achieved by kinking the α-helical M2 segments.
This kink may bring the leucine ring into the channel which closes the path for
the ions.

This view is supported by an additional experimental result. Mutating the leucine
in the leucine ring (see Figure 3) into a serine ring leads to an additional conducting
state. This conducting state seems to resemble a closed conformation of the
channel [40]. Replacement of the bulky leucine residues in the leucine ring appears
to be sufficient to create a new path for the ions.

5.2. CONSEQUENCES FOR THE MECHANISM OF ION CONDUCTION THROUGH THE ACETYLCHOLINE RECEPTOR CHANNEL AND RELATED CHANNELS

For the simulation of the energy profiles of the ions along the channel axis, a
very approximate approach was used. Water was not included in the simulation.
Furthermore, the long range dielectric influence of the membrane and the water
around the protein was also neglected. Therefore, these profiles cannot be interpreted quantitatively.

On the other hand, there are several aspects of the simulation which would
remain valid even if all these details would have been included in the simulation:

- The motion of the ions through the channel will strongly couple to the motion
  of charged and highly polar side chains.
- The negative charges in the channel require neutralization from counter ions.
  These counter ions may either be protons or if abundantly present cations.

The coupling of the motion of the ion to that of the environment (i.e. that of
water and the protein walls) was discovered in previous simulation studies [62,73].
This coupling increases the effective mass of the ion considerably. It also contrasts
to the very popular Eyring rate theories [7,68] commonly used to describe the
transport of ions through ion channels based on the assumption of ballistic trajectories of ions. Whereas for the electrically neutral gramicidine channel the motion
of the ions appears to be in the intermediate friction regime [65,73], due to the
strong coupling between channel protein and ion motion in channels like the

acetylcholine receptor channel, here the motion of ions may fall into the high friction regime.

The strong binding energies of single ions in the acetylcholine receptor channel places restrictions on the mechanism of ion transport through this channel. Strong binding energies due to long range electrostatic interaction imply that more than a single ion may occupy the channel. The strong local electric fields leads to a situation, where the commonly used assumption of constant electric field through the channel is not valid. A proper description of the distribution of ions may be obtained by means of mean field theories [60,61,74,75] or by Brownian dynamics simulations [59,65,66,76].

Acknowledgements

Thanks to Dr. A. Villarroel for the many inspiring discussions and Dr. L. Wollmuth for critically reading the manuscript. The figures displaying molecular structures where drawn using the SCHAKAL92 program written by Prof. E. Keller, University of Freiburg, Germany.

References

1. F. Franciolini and A. Petris: *Mol. Biol. Evol.* **6**, 503–513 (1989).
2. V. B. Cockcroft, D. J. Osguthorpe, E. A. Barnard, and G. G. Lunt: *Proteins* **8**, 129–169 (1990).
3. V. A. Parsegian: *Nature* **221**, 844–846 (1969).
4. V. A. Parsegian: *Ann. New York Acad. Sci.* **264**, 161–174 (1975).
5. R. S. Eisenberg: *J. Membrane Biol.* **115**, 1–12 (1990).
6. O. S. Andersen and R. E. Koeppe II: *Physiol. Rev.* **72**, S89–S158 (1992).
7. B. Hille: *Ionic Channels of Exitable Membranes*, Sinauer Associates Inc., Sunderland, Massachusetts (1992).
8. E. A. Barnard: *Trends Biochem. Sci.* **17**, 368–374 (1992).
9. W. A. Catterall: *Science* **242**, 50–60 (1988).
10. A. V. Maricq, A. S. Peterson, A. J. Brake, R. M. Myers, and D. Julius: *Science* **254**, 432 (1991).
11. V. B. Cockcroft, D. J. Osguthorpe, E. A. Barnard, A. E. Friday, and G. G. Lunt: *Mol. Neurobiol.* **4**, 129–169 (1990).
12. J. Deisenhofer, O. Epp, K. Miki, R. Huber, and H. Michel: *J. Mol. Biol.* **180**, 385–398 (1984).
13. H. Michel and J. Deisenhofer: *Chem. Scripta* **27B**, 173–180 (1987).
14. C. Toyoshima and N. Unwin: *Nature* **336**, 247–250 (1988).
15. N. Unwin: *Neuron* **3**, 665–676 (1989).
16. A. K. Mitra, M. P. McCarthy and R. M. Stroud: *J. Cell Biol.* **109**, 755–774 (1990).
17. N. Unwin: *J. Mol. Biol.* **229**, 1101–1124 (1993).
18. H. Betz: *Biochem.* **29**, 3591–3599 (1990).
19. J. Finer-Moore and R. M. Stroud: *Proc. Natl. Acad. Sci. USA* **81**, 155–159 (1984).
20. G. Spach, H. Duclohier, G. Molle and J.-M. Valleton: *Biochemie* **71**, 11–21 (1989).
21. J.-P. Changeux, J.-L. Galzi, A. Devillers-Thiry, and D. Bertrand: *Quart. Rev. Biophys.* **24**, 395–432 (1992)
22. C. Miller: *Neuron* **2**, 1195–1205 (1989).
23. J.-L. Galzi, A. Devillers-Thiry, N. Hussy, S. Bertrand, J.-P. Changeux, and D. Bertrand: *Nature* **359**, 500–505 (1992).
24. A. Karlin: *The Harvey Lecture Series* **85**, 71–107 (1991).
25. T. Konno, C. Busch, E. von Kitzing, K. Imoto, F. Wang, J. Nakai, M. Mishina, S. Numa, and B. Sakmann: *Proc. R. Soc. Lond. B* **244**, 69–79 (1991).
26. M. Noda, H. Takahashi, T. Tanabe, M. Toyosato, Y. Furutani, T. Hirose, M. Asai, S. Jinahama, T. Miyata, and S. Numa: *Nature* **299**, 793–797 (1982).

27. T. Claudio, M. Ballivet, J. Patrick, and S. Heineman: *Proc. Natl. Acad. Sci. USA* **80**, 1111–1115 (1983).
28. A. Devillers-Thiery, J. Giraudat, M. Bentaboulet, and J.-P. Changeux: *Proc. Natl. Acad. Sci. USA* **80**, 2067–2071 (1983).
29. H. R. Guy and F. Hucho: *Trends Neurosci.* **10**, 318–321 (1987).
30. K. Imoto, C. Methfessel, B. Sackmann, M. Mishina, Y. Mori, T. Konno, K. Fukuda, M. Kurasaki, II. Bujo, Y. Fujita, and S. Numa: *Nature* **324**, 670–674 (1986).
31. R.J. Leonard, C. G. Labarca, P. Chanet, N. Davidson, and H. A. Lester: *Science* **242**, 1578–1581 (1988).
32. P. Charnet, C. Labarca, R. J. Leonard, N. Vogelaar, L. Czyzyk, A. Gouin, N. Davidson, and H. Lester: *Neuron* **2**, 87–95 (1990).
33. V. Witzemann, E. Stein, B. Barg, T. Konno, M. Koenen, W. Kues, M. Criado, M. Hofmann, and E. Sakmann: *Eur. J. Biochem.* **194**, 437–448 (1990).
34. K. Imoto, C. Busch, B. Sackmann, M. Mishina, T. Konno, J. Nakai, H. Bujo, Y. Mori, K. Fukuda, and S. Numa: *Nature* **335**, 645–648 (1988).
35. F. Wang and K. Imoto: *Proc. R. Soc. Lond. B* **250**, 11–17 (1992).
36. A. Villarroel, S. Herlitze, M. Koenen, and B. Sakmann: *Proc. R. Soc. Lond. B* 243, 69–74 (1991).
37. K. Imoto, T. Konno, J. Nakai, F. Wang, M. Mishina, and S. Numa: *FEBS Letters* **289**, 193–200 (1991).
38. A. Villarroel and B. Sakmann: *Biophys. J.* **62**, 196–208 (1992).
39. H. A. Lester: *Ann. Rev. Biophys. Biomol. Struct.* **21**, 267–292 (1992).
40. A. Devillers-Thiery, J.-L. Galzi, J. L. Eisele, S. Bertrand, D. Bertrand, and J.-P. Changeux: *J. Membrane Biol.* **136**, 97–112 (1993).
41. F. Hucho and R. Hilgenfeld: *FEBS Letters* **257**, 17–23 (1989).
42. S. Furois-Corbin and A. Pullman: *Biochim. Biophys. Acta* **984**, 339–350 (1989).
43. R. M. Stroud, M. McCarthy, and M. Shuster: *Biochem.* **29**, 11009–11023 (1990).
44. G. Eisenman, A. Villarroel, M. Montal, and O. Alvarez: 'Energy profiles for ion permeation in pentameric protein channels: from viruses to receptor channels', in J. M. Ritchie, P. J. Magistretti and L. Bolis (eds.), *Progress in Cell Research*, Springer Verlag, New York, pp. 195–211 (1990).
45. A. Pullman: *Chem. Rev.* **91**, 793–812 (1991).
46. E. A. Barnard, M. G. Darlison, and P. Seeburg: *Trends Neurosci.* **10**, 502–509 (1987).
47. G. Fasman and W. A. Gilbert: *Trends Biochem. Sci.* **15**, 89–95 (1990).
48. P. E. Stein, A. Boodhoo, G. J. Tyrrell, J. L. Brunton, and R. J. Read: *Nature* **355**, 748–750 (1992).
49. J. D. Lear, Z. R. Wasserman, and W. F. DeGrado: *Science* **240**, 1177–1181 (1988).
50. W. F. DeGrado and J. D. Lear: *Biopol.* **29**, 205–213 (1990).
51. K. S. Akerfeldt, J. D. Lear, Z. R. Wasserman, L. A. Chung, and W. F. DeGrado: *Acc. Chem. Res.* **26**, 191–197 (1993).
52. S. Oiki, W. Danho, V. Madison, and M. Montal: *Proc. Natl. Acad. Sci. USA* **85**, 8703–8707 (1988).
53. S. Oiki, V. Madison, and M. Montal: *Proteins* **8**, 226–236 (1990).
54. A. Grove, J. M. Tomich, T. Iwamoto, and M. Montal: *Prot. Sci.* **2**, 1918–1930 (1993).
55. M. O. Montal, T. Iwamoto, J. M. Tomich, and M. Montal: *FEBS Letters* **320**, 261–266 (1993).
56. M. Oblattmontal, L. K. Buhler, T. Iwamoto, J. M. Tomich, and M. Montal: *Journal of Biological Chemistry* **268**, 14601–14607 (1993).
57. G. L. Reddy, T. Iwamoto, J. M. Tomich, and M. Montal: *J. Biol. Chem.* **268**, 14608–14615 (1993).
58. S. J. Weiner, P. A. Kollman, D. A. Case, U. C. Singh, C. Ghio, G. Alagona, S. J. Profeta, and P. Weiner: *J. Am. Chem. Soc.* **106**, 765–784 (1984).
59. K. Cooper, E. Jakobsson, and P. Wolynes: *Prog. Biophys. molec. Biol.* **46**, 51–96 (1985).
60. D. G. Levitt: *Biophys. J.* **59**, 271–277 (1991).
61. D. G. Levitt: *Biophys. J.* **59**, 278–288 (1991).
62. B. Roux and M. Karplus: *Biophys. J.* **59**, 961–981 (1991).
63. V. Barcilon, D. P. Chen, and R. S. Eisenberg: *SIAM J. Appl. Math.* **52**, 1405–1425 (1992).
64. V. Barcilon: *SIAM J. Appl. Math.* **52**, 1391–1404 (1992).
65. E. Jakobsson: *Int. J. Quant. Chem.* 25–36 (1993).

66. S. Bek and E. Jakobsson: *Biophys. J.* **66**, 1028–1038 (1994).
67. D.J. Adams, T. M. Dwyer, and B. Hille: *J. Gen. Physiol.* **75**, 493–510 (1980).
68. H. Eyring, R. Lumry, and J. W. Woodbury: *Rec. Chem. Prog.* **10**, 100–114 (1949).
69. S. Furois-Corbin and A. Pullman: *FEBS Letters* **252**, 63–68 (1989).
70. S. Furois-Corbin and A. Pullman: *Biophys. Chem.* **39**, 153–159 (1991).
71. G. Eisenman and O. Alvarez: *J. Membrane Biol.* **119**, 109–132 (1991).
72. C. Gruber, J. L. Lebowitz, and P. A. Martin: *J. Chem. Phys.* **75**, 944–954 (1981).
73. S.-W. Chiu, J. A. Novotny, and E. Jakobsson: *Biophys. J.* **64**, 98–109 (1993).
74. D. P. Chen, V. Barcilon, and R. S. Eisenberg: *Biophys. J.* **61**, 1372–1393 (1992).
75. F. F. Offner: *Biophys. J.* **62**, 281–283 (1992).
76. E. Jakobsson: *Biophys. J.* **64**, A201 (1993).

Simulation of a Fluid Phase Lipid Bilayer Membrane: Incorporation of the Surface Tension into System Boundary Conditions

S.-W. CHIU,[1] M. CLARK,[2] V. BALAJI,[1] S. SUBRAMANIAM,[1,3,4,5] H. L. SCOTT,[2] and E. JAKOBSSON[1,3,4,5]

[1]*National Center for Supercomputing Applications, University of Illinois, Urbana, IL 61801, USA*
[2]*Department of Physics, Oklahoma State University, Stillwater, OK 74078–0444, U.S.A.*
[3]*Department of Physiology*
[4]*Center for Biophysics and Computational Biology.*
[5]*Beckman Institute for Advanced Science and Technology, University of Illinois, Urbana, IL 61801, U.S.A.*

Abstract. Modeling membranes is not just modeling another kind of macromolecule, but modeling an entire environment for a large class of biomolecular processes. Membrane modeling poses quite a different set of technical problems and scientific isues from modeling proteins. This paper reviews some of these issues and suggests approaches that seem promising for resolving them based on work in our laboratories and that of others.

Key words. Surface tension, molecular dynamics, lipid membranes, membrane models, interfacial phenomena.

Introduction

DIFFERENCES BETWEEN MODELING MEMBRANES AND MODELING PROTEINS

A notable difference between modeling membranes and modeling proteins is that proteins are relatively rigid, whereas membranes in their biological state are liquid crystals. A lipid bilayer in the liquid crystal phase poses a number of different problems for the modeler than a soluble protein. One problem is that, in the plane of the membrane, the molecular structure to be modeled is essentially infinite in extent. *Any* truncation of the system cuts off the possibility of observing, for example, bending and splay motions of the system having characteristic wavelengths that are comparable to or greater than the size of the simulated system. By contrast, the accuracy of simulations of a soluble protein is not seriously compromised by being placed in a computationally tractable box of water. The membrane situation is anisotropic. The compressibility of the membrane is different in the dimension normal to the membrane plane as compared to the dimensions in the membrane plane. The effective pressure near the membrane is anisotropic. Whereas the pressure normal to the membrane plane is that pertaining to the laboratory, for example 1 atmosphere, the effective pressure parallel to the membrane plane at the water-lipid interface must be negative, based on measurements of surface tension [20]. To make matters even more complicated, the lateral pressure will be different right at the interface from what it is deep within the hydrocarbon region, and different yet in the electroyte some significant distance

A. Pullman et al. (eds.), Modelling of Biomolecular Structures and Mechanisms, 59–67.
© 1995 *Kluwer Academic Publishers. Printed in the Netherlands.*

from the membrane surface.* By contrast, the pressure around the soluble protein is isotropically the bulk pressure of the electrolyte in which it is residing. The precise structure of the membrane is labile and constantly changing. NMR studies have given us data on the mean orientations of hydrocarbon tails [26]. NMR and neutron and X-ray diffraction have given us data on the mean orientations of the head groups [23,30]. But one expects the fluctuations about the mean to be enormous, because liquid crystal theory tells us that large deformations of the membrane can be accomplished with thermal energy at physiological temperatures [18]. Indeed direct evidence from neutron and X-ray diffraction studies tells us that the distribution of positions along the normal to the membrane plane of particular phospholipid atoms is as much as 15 Å wide [30]. Thus the surface of the membrane is very rough. NMR studies suggest that 11–16 water molecules per phospholipid molecule are associated with the membrane, which probably means that they are interdigitating with the polar head groups in the interfacial region [6]. There is also an enormous electric field in the interfacial region. Measurements from several labs show that there is an electric potential difference of several hundred millivolts between the hydrocarbon interior of the membrane and the electrolyte [12]. This is equivalent to a mean electric field across the interfacial region well above the dielectric breakdown strength of a variety of materials. If we imagine a molecular landscape, the boundary region where the water and the membrane headgroups interact is the site of a violent electrical micro-storm! The surfaces of soluble proteins also exhibit significant motions and critical irregularities of shape, but they are not nearly as mobile and disordered as the membrane surface. Figure 1 shows in pictorial form some of the concerns mentioned above.

TECHNICAL ISSUES IN MODELING MEMBRANES

Because the membrane in its biological state is a liquid crystal, meaningful and useful models must be dynamic rather than static. It is possible to put phospholipids into crystalline solids [16], but the biological state of membranes is far from this solid crystalline state. Thus molecular dynamics, Monte Carlo, or stochastic dynamics methods are mandatory to describe biological membranes. This paper will focus on several important issues that arise in molecular dynamics simulations of membranes. The choice of the model system, i.e. number of lipid and solvent molecules, choice of boundary conditions, nature of interaction potentials and the method for equilibration bear significantly on the validity of computation and we will address these issues with a view to providing a well-defined set of validation criteria for membrane simulations.

Computations of membrane patches have been reported with stochastic boundary conditions [19], with periodic boundary conditions at constant volume [1,9,28], and with periodic boundary conditions at constant pressure [4,7]. Each choice is fraught with trade-offs. The appropriate choice may depend on the experimental conditions under consideration.

The stochastic boundary conditions suffer from being essentially artificial, especially in that they may inhibit chain tilting of the boundary lipid, which will in turn affect tilting in the rest of the simulation cell. In addition they fail to simulate

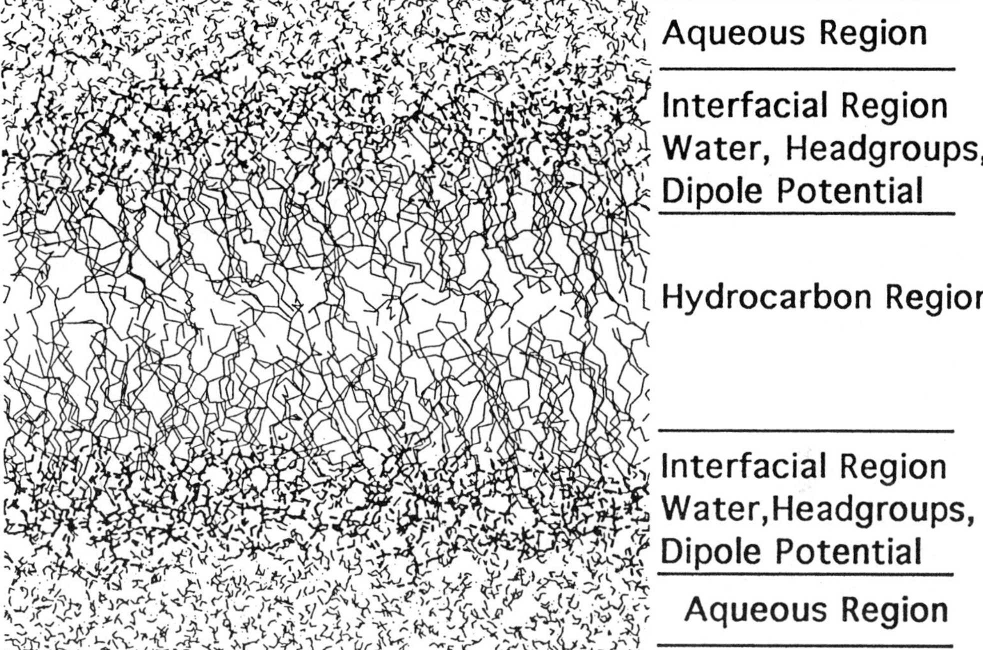

Fig. 1. Schematic diagram of the regions in a fluid phase membrane. The lipid visualization is a snapshot from a molecular dynamics simulation of DMPC at 325 K, done in our lab. The heavier lines in the molecular visualization are the more polar regions of the phospholipid, from the choline group at the outer end to the carbonyl groups at the heads of the hydrocarbon chains. The spatial distribution in the dimension normal to the membrane plane of atom types in the interfacial region is in close agreement with the neutron and X-ray diffraction studies of Wiener and White [30]. The order parameters of the hydrcarbon tails are in close agreement with experimental values for fluid membranes [26].

the essentially infinite bilayer system. However they have the advantage that, with a finite delimited computational space, it is feasible to eliminate cut-offs of electrostatic forces within the computational space by use of a multipole method [5]. But it is not clear that electrostatic forces longer than a reasonable cut-off distance, 15 Å or so, are a significant modulator of membrane dynamics and structure. However it probably is important to use neutral group-based rather than atom-based cutoffs. Atom-based cutoffs are likely to introduce spurious polarizations, based on single atoms of a group being on different sides of a cut-off distance. In using group based cut-offs, each water and phospholipid molecule may be treated as a group [9]. However it is more economical in non-bonded interactions to divide the phospholipid molecule into several neutral subgroups. This can generally be done with only minor adjustments to the partial charges, well within their inherent range of uncertainty. If it is deemed desirable to eliminate cut-offs altogether, a possible reasonably efficient technique to eliminate cut-offs in simulations with periodic boundary conditions is the particle mesh method for Ewald sums [10].

Periodic boundary conditions are natural for the membrane system. A concern however is that periodic boundary conditions may introduce spurious collective motions, for example an excessive collective tilt of the hydrocarbon chains. This effect might be mitigated by making the box size large enough that the boundaries would have little effect on the interior of the box, but of course this is at the cost of less economical computation. If the periodic boundary conditions are to be used at constant volume, it is necessary to be confident of the density of material in the lipid and at the lipid-electrolyte interface. This is possible in a multi-lammelar preparation of known water content, such as the DOPC membrane that is the subject of a series of papers from the laboratory of White [30]. In this case constant volume calculations would seem justified. However in simulating a situation where there is excess water it may be advisable to do the computations at constant pressure. In this method pressure components in each spatial dimension are calculated according to a virial expression, and the dimensions of the box are (slowly) adjusted during the simulation until the mean internal virial matches the external applied pressure, which is set as a boundary condition [4]. This method, with the external pressure at one atmosphere, has been used by the laboratory of Berendsen [3] and by us [7]. Recently in our laboratory we have been using a variant of this method in which the lateral pressure (in the plane of the membrane), rather than being set to the external laboratory pressure of one atmosphere, is set according to the measured surface tension at the membrane-water interface. This surface tension can be derived from the pressure-density curve of a monolayer in a Langmuir trough and the surface tension for a pure air-water interface. (For a clear discussion of surface tension at the membrane-water interface, see pp. 75–78 of Gennis [14]). For monolayers at the area/lipid characteristic of the fluid phase, this surface tension is of the order of a few tens dyne/cm. Thus the lateral pressure for the simulation is negative and, depending on the thickness of the simulated interface, of the order of a hundred atmospheres. These boundary conditions are shown in schematic form in Figure 2.

The force fields for the lipid and water molecules are also a matter of some concern. In a previous constant pressure simulation at one atmosphere, it was found that the partial charges in the lipid molecules had to be reduced in order to produce a fluid phase in a simulated DPPC membrane [4]. In the later part of this paper it will be seen that a constant pressure simulation as represented in Figure 2 can produce an appropriate fluid phase in a simulated PC membrane with full charges on the lipid molecules.

CRITERIA FOR A VALID MEMBRANE SIMULATION

There is by now a fairly wide range of experimental criteria for assessing whether a membrane simulation of the fluid state is a good representation of reality, based on experimental studies cited above. A partial list would include the following: The molecular order parameters of the hydrocarbon tails should show a plateau near the top part of the tails and decline sharply near the middle of the bilayer. (For PC membranes the plateau is at a molecular order parameter of about 0.4). There should be approximately 11 to 16 water molecules associated on the average with each lipid molecule, in a condition of excess hydration. For PC membranes

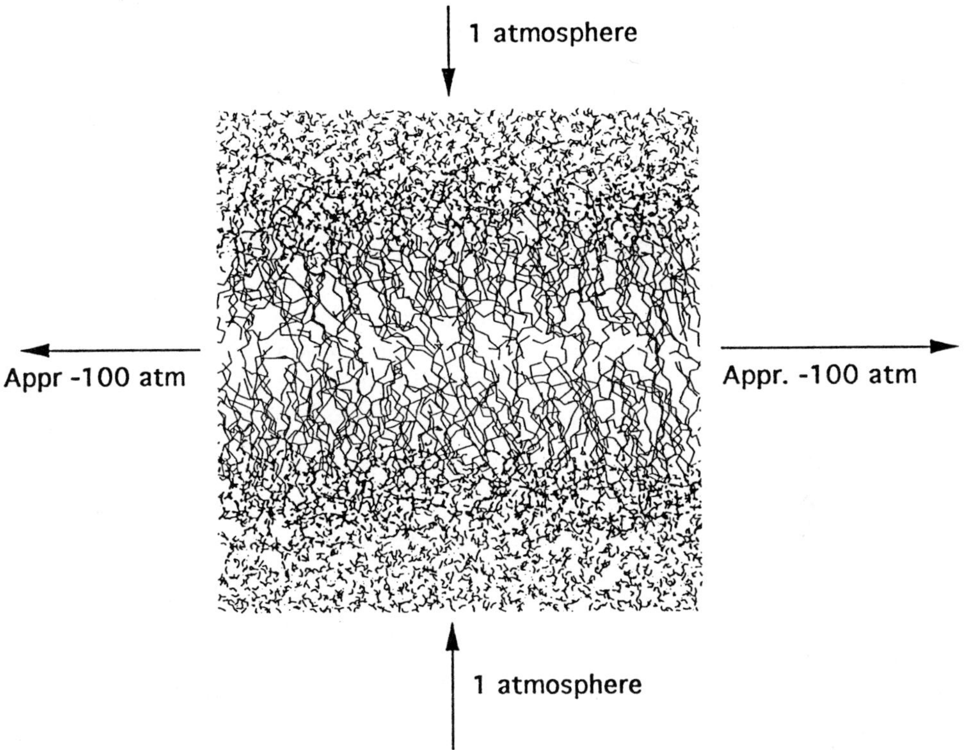

Fig. 2. Schematic representation of the constant pressure boundary conditions for computing a fully hydrated fluid phase membrane. The one atmosphere posititive pressure normal to the membrane plane is derived from the laboratory conditions. The negative pressure in the membrane plane is derived from the measured surface tension of a phospholipid monolayer.

in the fluid phase, the distribution of headgroup atoms in the direction normal to the membrane plane should be in agreement to that given in neutron and X-ray diffraction measurements. (With the possible caveat that the experimental results were obtained at a particular level of lipid hydration, and may not precisely pertain to different levels of hydration.) In particular, computed electron density profiles should show peaks in the head group regions, a 'methylene plateau' at a lower level representing the constant electron density of the methyls, and a 'methyl trough' representing the electron poor terminal methyl region. The interfacial region of the simulated phospholipid should exhibit a dipole potential of several hundred millivolts, in which the interior of the phospholipid bilayer is positive relative to the electrolyte. For PC membranes, the major part of the dipole potential must come from orientation of water molecules in the interfacial region, with some contribution also from orientation of the carbonyl groups at the head of each hydrocarbon chain. The orientation of the PC headgroup itself must be counter to the dipole potential, since the choline group carries primarily a positive charge and the phosphate group a negative charge. This interpretation is supported by the experimental observation that the dipole potential is smaller in magnitude in PC membranes than in other phospholipids [13].

The above criteria are not completely exhaustive, but a simulation that satisfied all of them would capture much of the essence of a fluid phase membrane of homogenous composition. To date, no simulations have been completely successful in reproducing all of the criteria.

Results of a Simulation of a DMCP Bilayer Using Anisotropic Constant Pressure Boundary Conditions

We have recently done in our laboratory a simulation of a DMPC bilayer in a fluid state using the constant-pressure boundary conditions shown in Figure 2. A full description of this work will be presented in a journal more suitable for specialized membrane biophysics. A summary of the computation and results is as follows:

The partial charges on the phospholipid molecules were calculated by ab initio electronic structure computations using the GAUSSIAN 92 program at Hanree-Fock SCF level using the 6–31G* basis set. The geometry of the lipid was taken to be that of the X-ray crystal structure. The partial atomic charges were extracted from the SCF total electron density by Mulliken population analysis. Other aspects of the phospholipid force fields were as in a previous simulation of a DPPC bilayer [11], including the use of the Ryckaert–Bellemans potential for the dihedral angles of the hydrocarbon chains. Unified atom representations were used for the carbons and hydrogens in the hydrocarbon chains. The SPC/E water model was used [3]. Neutral group-based cutoffs of 20 angstroms were used for the non-bonded interactions. The simulation contained 100 phospholipid molecules, 50 in each monolayer, and 2100 water molecules. The starting configuration of the lipid was the crystal structure of Hauser et al. [16] which has a membrane area per lipid of about 38 square angstroms. The pressure tensors were calculated by an internal virial [2]. A coupling time constant of 0.4 psec was used. The system was gradually heated to 325 K. After warming, the system had spread to a cross-sectional area of about 56 square angstroms per lipid molecule, and appeared to stabilize in its dimensions at that value. One change was made in the computational method after heating. When the system reached 325 K, it was seen that about 90% of the hydrocarbon chain dihedrals remained in the trans configuration. At this point the van der Waals interactions were removed for the 1–4 atoms in the hydrocarbon chains, and the simulation was restarted. After this change, the system relaxed to a trans/gauche ratio of about 3/1, appropriate for a fluid phase.

The hydrocarbon order parameters were in good agreement with those expected for a fluid phase PC membrane. Although the actual area of the system was 56 square angstroms/lipid, applying the formula of Nagle [22] to our order parameters indicated an area of 62 square angstroms. The distribution of particular groups in the direction normal to the membrane is quite similar to that seen experimentally by neutron and X-ray diffraction [30]. However it must be added that the neutron and X-ray diffraction experiments were done at a lower hydration, so the measurements may not be strictly comparable to our simulations.

The major discrepancy between our simulated results was the size of the computed dipole potential. This is an electrical potential of the hydrocarbon interior of the membrane positive relative to the external electrolyte [12]. It is largely due to orientations of the water molecules in the interfacial region, although the lipid charges make a contribution as well. In the case of the PC membrane, the contribution of the lipid is negative, because of the charge distribution in the headgroup in which the choline group carries a net postive charge and the phosphate a net negative charge. Experimentally the net dipole potential should be of the order of a few hundred millivolts. In our simulations we found a computed dipole potential of a few volts, about an order of magnitude too large. It appears our water orientations were too extreme. At this writing we believe this may be due to our periodic boundary conditions combined with not quite enough water in the system to meet the condition of excess hydration, but the precise explanation awaits further computation and analysis.

In summary, our results show for the first time that a simulated fluid phase phase membrane can be produced with parameters and boundary conditions that are derived from independent experiments and theory, rather than from parameters and boundary conditions that are explicitly designed to give the fluid phase. This has not previously been done. In previous fluid phase constant pressure simulations, the partial charges on the lipid atoms were arbitrarily reduced to produce the fluid phase [4]. In constant volume simulations, the lipid density has been set at a low enough value to ensure the fluid phase, and initial configurations have been randomized [1,9,28]. In our simulations the fluid phase emerged from lipids with partial charges that were computed by *ab initio* calculations, with the whole system under constant pressure boundary conditions derived from measured surface tension. Further our fluid phase was produced from the crystal structure as the initial conformation, to ensure that the system had no previous bias to be fluid.

It now seems reasonable to consider some future applications of simulated fluid membranes in attacking molecular engineering problems of biological significance.

Using Simulated Fluid Membranes to Understand the Interactions of Organic Molecules in Membranes

There are many interactions of organic molecules with membranes that are of critical biological importance. In each case, a simulated fluid membrane in which the underlying physics is reasonably correct can be used to explore the degree to which the organic molecule partitions into the membrane, the depth of such partitioning, and the orientation of the molecule in the membrane. Four categories of such molecules include peptides, experimentally used probes of membrane structure, pollutants (Proceedings of American Chemical Society Workshop: The Environmental Fate of Complex Organic Molecules, Airlie House, Virginia, May 9–12, 1993), and drugs [17,21]. Approximate methods have been developed to model solute partitioning into, and orientation in, the membrane-water interface [24]. These methods need to be verified and refined by explicit molecular dynamics simulations of these partitioning processes.

Interaction of peptides with membranes is of large interest for understanding protein transport as well as membrane protein structure. Most transported proteins have been found to contain an extension at the N-terminus, called the leader peptide, which is cleaved by specific peptidases after the protein is successfully translocated through the membrane. Recent experiments suggest a more direct functional role for the signal peptides, which insert into membranes to initiate the protein translocation. For a review of function and mechanism of signal peptides, especially as elucidated by experiments on model systems, see Tamm [27].

Another scientific issue that might be helped towards elucidation by membrane simulations is membrane protein structure prediction. That might be by looking at the penetration and orientation of relatively simple peptides in the membrane; it should be possible to assess the relative tendency of different amino acids to penetrate into different levels of the membrane. For example, an amino acid that will clearly be significant to study is tryptophan. In the membrane proteins for which reliable structures are known, there is a strong statistical tendency for tryptophans to reside within the interfacial region of the membrane [8,15,25]. This tendency is also seen in experiments on a synthetic peptide probe [29]. By membrane simulation the physics of this tendency should by capable of understanding, as well as comparable tendencies for other amino acids. Combining knowledge of these tendencies with sequence knowledge of membrane proteins should lead to constraints on amino acid positions that will help in the prediction of membrane protein structure. This is especially important because the traditional methods of X-ray crystallography and NMR spectroscopy have so far proved much less generally feasible for structure determination of membrane proteins than for soluble proteins.

It is of great interest to know with confidence how fluorescence and spin label probes orient themselves in the membrane. Some calculations with respect to this have been done with a Monte Carlo method in which the lipid atoms move on a lattice (Y. K. Levine, Proceedings of the 1994 Jerusalem Symposium). Detailed molecular dynamics simulations should be done to validate and refine this work.

Summary

We have succeeded in producing realistic fluid phase behavior and molecular orientations in a simulated membrane without a prior bias to fluidity, either by adjusting the lipid charges, or by adjusting the surface area per lipid of the simulated membrane, or by the initial conditions of the simulation. Because of this, we believe that we have produced reasonably well the physical interactions governing membrane organization. Therefore it is timely to begin using the simulation model to deal with issues of interaction of organic molecules with membranes, such as peptides, drugs, and organic pollutants.

Acknowledgements

We gratefully acknowledge grant support from the National Science Foundation and the use of the facilities of the National Center for Supercomputing Applications. Major computing time was granted from the National Science Foundation Supercomputing Metacenter, for time on the C90 at the Pittsburgh Supercomputing Center and the Convex C3 at the National Center for Supercomputing Applications.

Note

*Of course there is not actually any such physical quantity as "negative pressure". What is meant in this context is that the attractive hydrophobic surface tension at the membrane-water interface is translated formally into an equivalent negative pressure as constant pressure boundary conditions are defined in conventional molecular dynamics programs.

References

1. H. E. Alper, D. Bassolino-Klimas, and T. R. Stouch: *J. Chem. Phys.* **99**, 5547–5559 (1993).
2. H. J. C. Berendsen, J. P. M. Postma, W. F. van Gunsteren, A. DiNola, and J. R. Haak: *J. Chem. Phys.* **81**, 3684–3689 (1984).
3. H. J. C. Berendsen, J. R. Grigera, and T. P. Straatsma: *J. Chem. Phys.* **91**, 6289–6291 (1987).
4. H. J. C. Berendsen, B. Egberts, S.-J. Marrink, and P. Ahlstrom: in *Membrane Proteins: Structures, Interactions and Models*, A. Pulman, J. Jortner, and B. Pullman (Eds.), Kluwer Academic Publishers, the Netherlands, pp. 457–470 (1992).
5. J. A. Board, Jr., J. W. Causey, J. F. Leathrum, Jr., A. Windemuth, and K. Schulten: *Chem. Phys. Lett.* **198**, 89 (1992).
6. F. Borle and J. Seelig: *Biochim. Biophys. Acta.* **735**, 131–136 (1983).
7. S. W. Chiu, K. Gulukota, and E. Jakobsson: in *Membrane Proteins: Structures, Interactions, and Models*, A. Pullman, J. Jortner, and B. Pullman (Eds.), Kluwer Academic Publishers, the Netherlands, pp. 315–338 (1992).
8. S. W. Cowan and J. P. Rosenbusch: *Science* **264**, 914–916 (1994).
9. K. V. Damodaran and K. M. Merz: *Biophys. J.* **66**, 1076–1087 (1994).
10. T. Darden, D. York, and L. Pedersen: *J. Chem. Phys.* **98**, 10089–10092 (1993).
11. E. Egberts: Ph.D. Thesis, University of Groningen (1988).
12. R. F. Flewelling and W. L. Hubbell: *Biophys. J.* **49**, 541–552 (1986).
13. K. Gawrisch, D. Ruston, J. Zimmerberg, V. A. Parsegian, R. P. Rand, and N. Fuller: *Biophys. J.* **61**, 1213–1223 (1992).
14. R. B. Gennis: *Biomembranes. Molecular Structure and Function.* Springer-Verlag (1989).
15. D. V. Greathouse, J. F. Hinton, K. S. Kim, and R. E. Koeppe II: *Biochemistry* **33**, 4291–4299 (1994).
16. H. Hauser, I. Pascher, R. H. Pearson, and S. Sundell: *Biochim. Biophys. Acta* **650**, 21–51 (1981).
17. L. Herbette: *Pesticide Sci.* **35**, 63–68 (1992).
18. P. Helfrich and E. Jakobsson: *Biophys. J.* **57**, 1075–1084 (1990).
19. H. Heller, M. Schaefer, and K. Schulten: *J. Phys. Chem.* **97**, 8343–8360 (1993).
20. J. N. Israelachvili, S. Marcelja, and R. G. Horn: *Quart. Rev. Biophys.* **13**, 121–200 (1980).
21. K. Jorgensen, J. H. Ipsen, O. G. Mouritsen, and M. J. Zuckermann: *Chemistry and Physics of Lipids* **65**, 205–216 (1993).
22. J. F. Nagle: *Biophys. J.* **64**, 1476–1481 (1993).
23. C. R. Sanders: *Biophys. J.* **64**, 171–181 (1993).
24. C. R. Sanders and J. P. Schwonek: *Biophys. J.* **65**, 1207–1218 (1993).
25. M. Schiffer, C.-H. Chang and F. J. Stevens: *Protein Engineering* **5**, 213–214 (1992).
26. J. Seelig and J. L. Browning: *FEBS Lett.* **92**, 41–44 (1978).
27. L. K. Tamm: *Biochim. Biophys. Acta.* **1071**, 123–148 (1991).
28. R. M. Venable, Y. Zhang, B. J. Hardy, and R. W. Pastor: *Science* **262**, 223–226 (1993).
29. S. H. White and W. C. Wimley: *Curr. Opinion Struct. Biol.* **4**, 79–86.
30. M. C. Wiener and S. H. White: *Biophys. J.* **61**, 434–447 (1992).

Protein Dynamics: From the Native to the Unfolded State and Back Again

MARTIN KARPLUS, AMEDEO CAFLISCH, ANDREJ ŠALI, and
EUGENE SHAKHNOVICH
*Department of Chemistry, 12 Oxford Street, Harvard University, Cambridge, Massachusetts 02138,
U.S.A.*

Abstract. Simulations to study protein unfolding and folding were performed. The unfolding simulations make use of molecular dynamics and treat an atomic model of barnase in aqueous solvent. The cooperative nature of the unfolding transition and the important role of water are described. The folding simulations are based on a bead model of the protein on a cubic lattice. It is shown for the 27-mer model that a large energy gap between the lowest energy (native) state and the excited states is a necessary and sufficient condition for fast folding.

Key words. Molecular dynamics, barnase denaturation.

I. Introduction

The dynamics of proteins includes a wide range of length and time scales [1]. Figure 1 shows a schematic diagram of the energy of a protein as a function of a configurational coordinate, such as the radius of gyration. There is the native state, which includes fluctuations of up to 2 Å in length on a time scale in the picosecond to nanosecond range. Many molecular dynamics simulations have explored this region of conformational space [2], which is of importance for protein function. A second region involves the transition from the native state to a compact globule, which may be an intermediate of the 'molten globule' type that is of great current interest [3]. Here the length scale is in the range 2–5 Å and the time scale of the motions involved is in the nanosecond to microsecond range. Finally, there are the vast number of configurations of the denatured (coil) state and the folding process from the coil to the compact globule with near native conformation. In this region the length scales are on the order of 10–20 Å and the time scales are in the microsecond to millisecond range. This paper briefly reviews some recent studies that explore the native to globule transition in barnase [4, 5] and makes use of a lattice model to study the folding process from the denatured to the organized globule state [6, 7].

II. Barnase Denaturation

Molecular dynamics simulations of the initial stages of unfolding of barnase (a 110 amino acid ribonuclease from *Bacillus amyloliquefaciens*) have been made at high temperature in the presence of water [4, 5]. This protein is a particularly good system for such studies because transition states and pathways of barnase folding and unfolding have been investigated by protein engineering and NMR hydrogen-

A. Pullman et al. (eds.), Modelling of Biomolecular Structures and Mechanisms, 69–84.
© 1995 *Kluwer Academic Publishers. Printed in the Netherlands.*

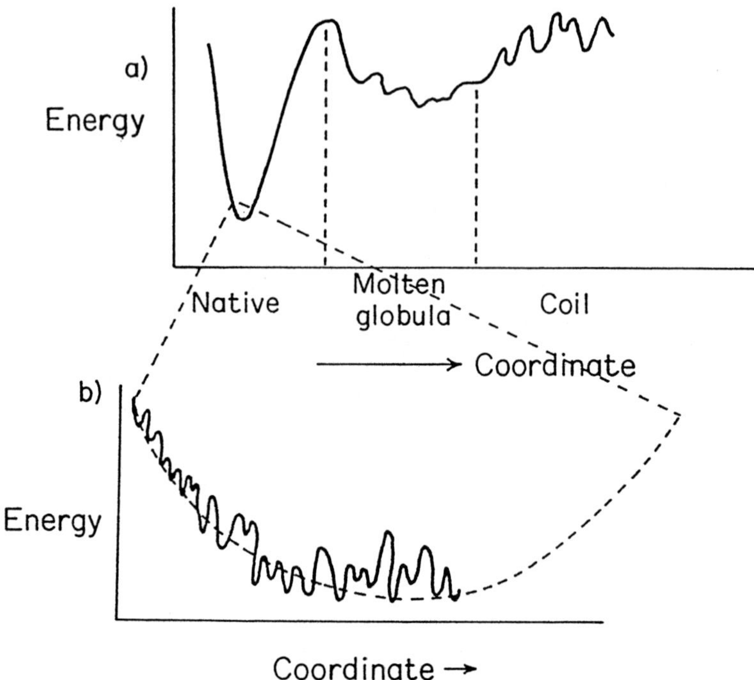

Fig. 1. Schematic representation of the configuration space of a protein giving the energy as a function of a configurational coordinate: (a) complete space; (b) enlarged view in the vicinity of the native state.

exchange trapping experiments [8]. The rate determining step for both folding and unfolding involves the crossing of a free energy barrier near the native state [9,10]. Figure 2 shows a schematic drawing of the barnase structural elements. The present simulations provide a detailed mechanism for solvent denaturation of the secondary structure and the hydrophobic cores.

The barnase denaturation simulations used a deformable boundary potential [10,11] and standard molecular dynamics methodology [2]. The system consisted of 1091 protein atoms and 3003 water molecules in a sphere of 30 Å radius. Two denaturation simulations were performed at 600 K (A600, 120 ps; R600, 230 ps) and a 300 K control trajectory was run for 250 ps. R600 was recently continued to 250 ps; in addition, a third simulation at 600 K (200 ps) was performed and analyzed (Caflisch & Karplus, in press). The denaturation process is similar in the three 600 K simulations. The 600 K temperature was used to speed up the unfolding transition. An increase in the unfolding rate by many orders of magnitude should result since the activation energy is expected to be large; the activation free energy for unfolding is 20 kcal/mol [12].

The radius of gyration (R_g) and heavy-atom root-mean-square deviation (RMSD) from the X-ray structure as a function of time are given in Figure 3. In A600 and R600, R_g starts to increase after 30 ps, while the RMSD increases immediately. Both R_g and RMSD then increase over most of the simulation.

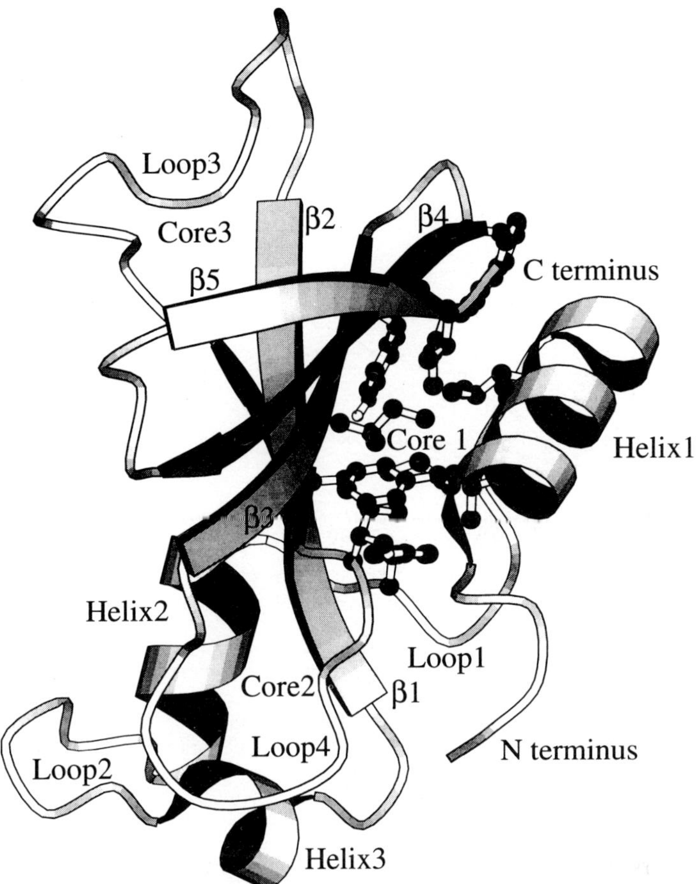

Fig. 2.   Schematic picture of the backbone of barnase, emphasizing the secondary structural elements; sidechains of hydrophobic core$_1$ are plotted in a ball–and–stick representation. The structural elements include the following residues: N-terminus (1–5), helix$_1$ (6–18), loop$_1$ (19–25), helix$_2$ (26–34), loop$_2$ (35–40), helix$_3$ (41–45), type II β-turn (46–49), strand$_1$ (50–55), loop$_3$ (56–69), strand$_2$ (70–75), loop$_4$ (76–84), strand$_3$ (85–90), type I β-turn (91–94), strand$_4$ (95–100), type III$'$ β-turn (101–104), strand$_5$ (105–108), C-terminus (109–110). Drawn with the program.

However, the increase is not uniform; e.g. in A600, R_g is nearly constant for 20 ps between 45 and 65 ps; this may be indicative of an intermediate [13]. In the control simulation at 300 K, R_g shows a very small increase (the average R_g is 13.7 Å, relative to the X-ray value of 13.6 Å); the RMSD from the X-ray structure is 1.9 Å (the mainchain atom RMSD is 1.5 Å) during the last 50 ps.

In the A600 simulation (120 ps) and the first half (115 ps) of the R600 simulation, there are similar structural changes. The N-terminus, loop$_1$ and loop$_2$ begin to unfold during the first 30 ps. This is followed by partial denaturation of the hydrophobic cores; core$_2$ denatures relatively rapidly, followed by core$_1$, core$_3$ and loop$_3$ in both 600 K simulations. The solvation of hydrophobic core$_1$ is coupled with a large distortion of helix$_1$ and of the edge strands of the β-sheet. Both helix$_1$

                                              MARTIN KARPLUS ET AL.

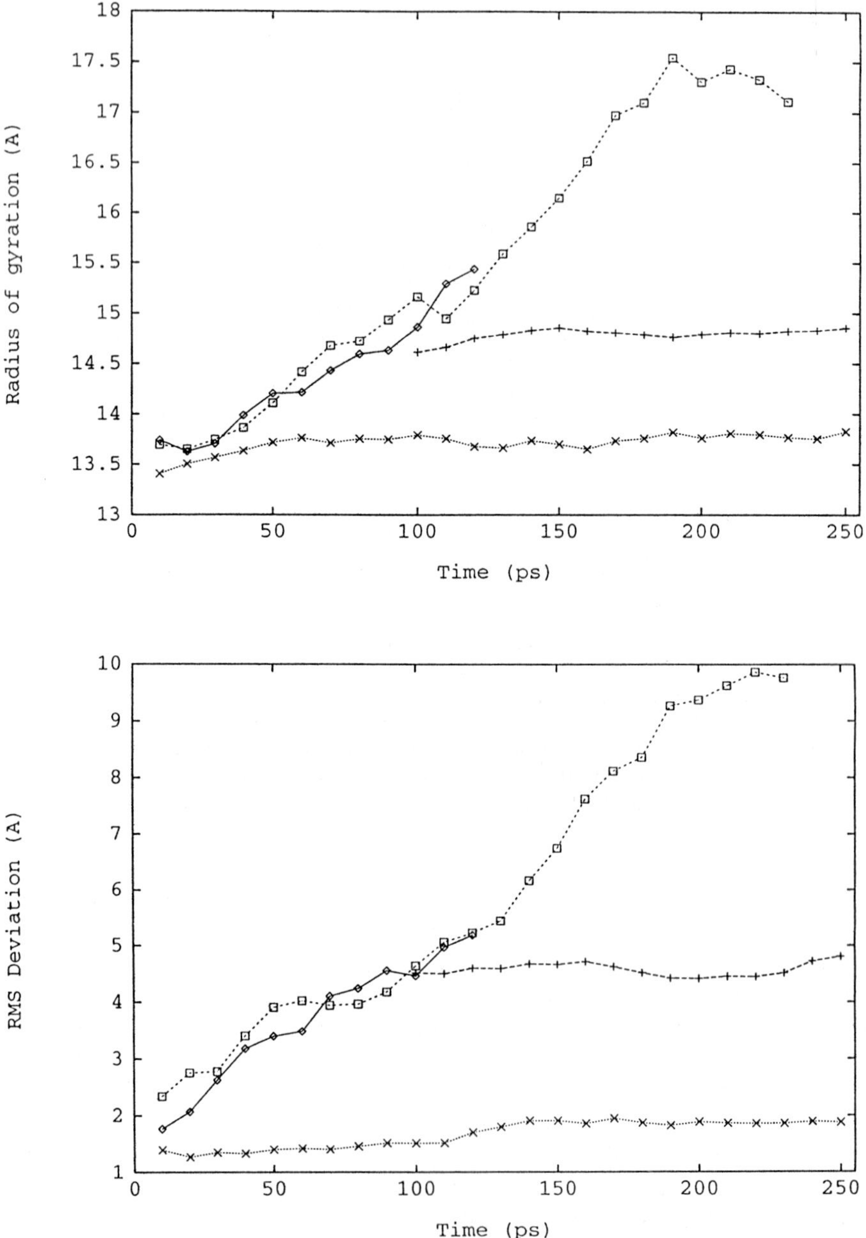

Fig. 3. (a) R_g as a function of simulation time averaged over 10-ps intervals. (b) RMSD from the X-ray structure as a function of simulation time averaged over 10-ps intervals: (——) A600, (--+--) B300, (--□--) R600, (··×··) control run at 300 K.

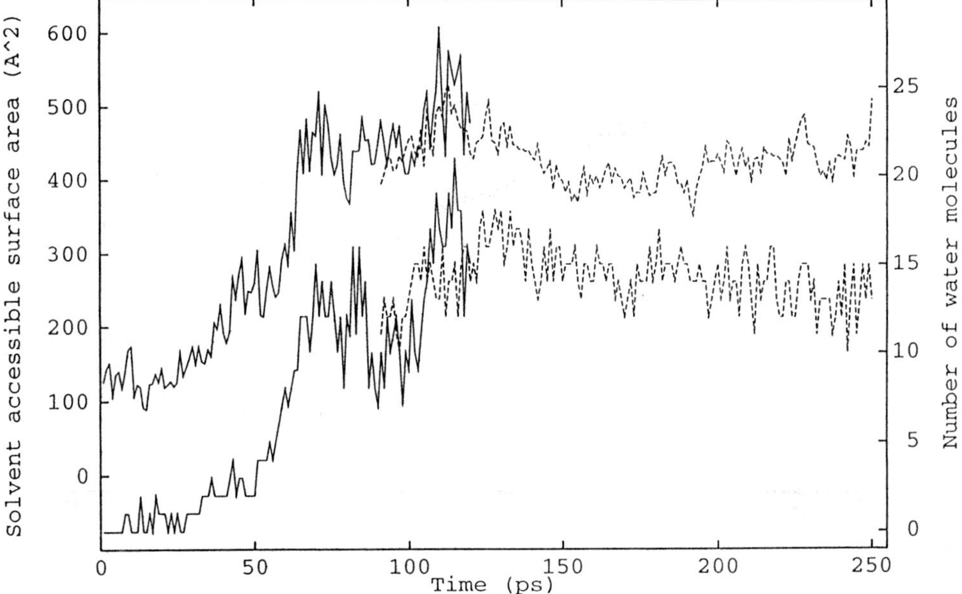

Fig. 4.   Solvent accessible surface area and number of water molecules for core$_1$ as a function of time.
(——) A600; (------) B300. For the exposed surface area, shown in the upper curve with the scale in
Å^2 on the left, the Lee and Richards algorithm (CHARMM implementation) and a probe sphere of
1.4 Å radius were utilized. For the number of water molecules, shown in the lower curve with the
scale on the right, those within 7 Å of the center of the core (the instantaneous center of geometry of
the carbon atoms of the sidechains of residues Phe 7, Val 10, Ala 11, Leu 14, Leu 20, Tyr 24, Ala 74,
Ile 76, Ile 88, Tyr 90, Trp 94, Ile 96, Ile 109) were included.

and helix$_2$ lose about half of the native α-helical hydrogen bonds; helix$_3$ unfolds
after about 20 ps. In the β-sheet, about half of the native interstrand hydrogen
bonds have disappeared after 100 ps; in R600 the β-sheet is fully solvated after
150 ps. During the last 50 ps of R600, the mainchain still shows essentially the
same overall fold as in the native structure, although the polypeptide chain is
almost fully solvated and all of the secondary structure is lost except for the last
two turns of helix$_1$.

Core$_1$, which is an important stabilizing element of barnase [8,10], is formed by
the packing of helix$_1$ against the β-sheet and is centered around the sidechain of
Ile 88 (Figure 2). Figure 4 shows the time dependence in A600 of the solvent
accessible surface area of the sidechains of core$_1$ and the number of water molecules in the core; similar behavior is seen during the first half of the R600
trajectory. Increase in accessible surface area and water penetration are nearly
simultaneous and begin at about 35 ps. Sixteen water molecules have penetrated
at 82 ps; this falls to 10 between 89–98 ps and increases to 17 in the period from
111 to 120 ps. Many of the solvating waters make hydrogen bonds to waters outside
the core. The accessibility of core$_1$ to water is coupled with the relative motion
of helix$_1$ and the β-sheet (see Figure 5); i.e., they begin to move apart at about
30 ps and their separation is continuous during the 30–80 ps period; between 80

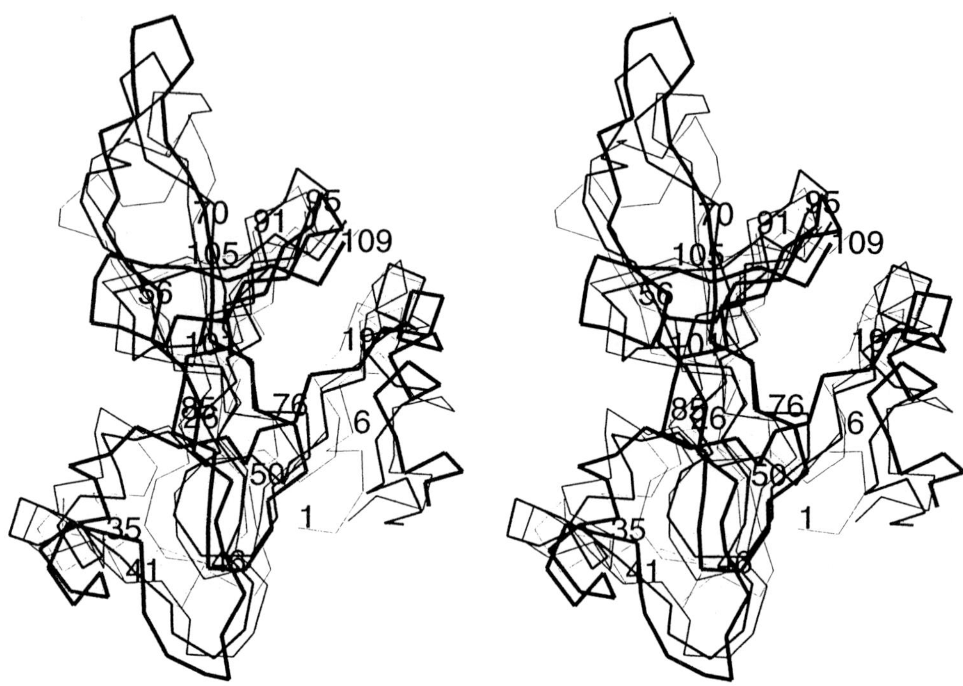

Fig. 5.   Stereo view of the barnase C_α atoms to illustrate the relative helix$_1$/β-sheet motion during A600. 1 ps (thin line and labels), 70 ps (medium line), 90 ps (thicker line), 120 ps (thick lines).

and 100 ps there is a small closing movement in accord with the decrease of water in the core, followed by expansion for the remainder of the simulation. During the B300 simulation the number of water molecules in core$_1$ and the solvent accessible surface area of its sidechains are nearly constant (see Figure 4); the average number of water molecules is 14. This steady state of solvation of the core is correlated with the nearly constant number of hydrogen bonds in helix$_1$ and the central part of the β-sheet (strands 2–3 and 3–4).

A number of detailed results of the present analysis are of interest because they may play a role in protein denaturation, in general. The polar OH and NH groups of tyrosines and tryptophan sidechains, respectively, play an important role in the penetration of water (from the top part of core$_1$), while the motion of a lysine sidechain helps the water molecules reaching the center of core$_1$ from the bottom part (Figure 6a,b). Clusters of hydrogen bonded water molecules surrounding hydrophobic sidechains participate in hydrogen bonds with polar groups of the backbone and/or of the sidechains (Figure 6c,d). The β-sheet disruption starts near the irregular element (β-bulge at residues 53–54) and at the edges (strands 1 and 5); it is promoted by an increase in the twist and an influx of water molecules, some of which insert between adjacent strands and participate in hydrogen bonds as both donors and acceptors with the mainchain polar groups (Figure 7). Water molecules act mainly as hydrogen bonding donors in the initial phase of solvation

(a)

(b)

(c)

(d)

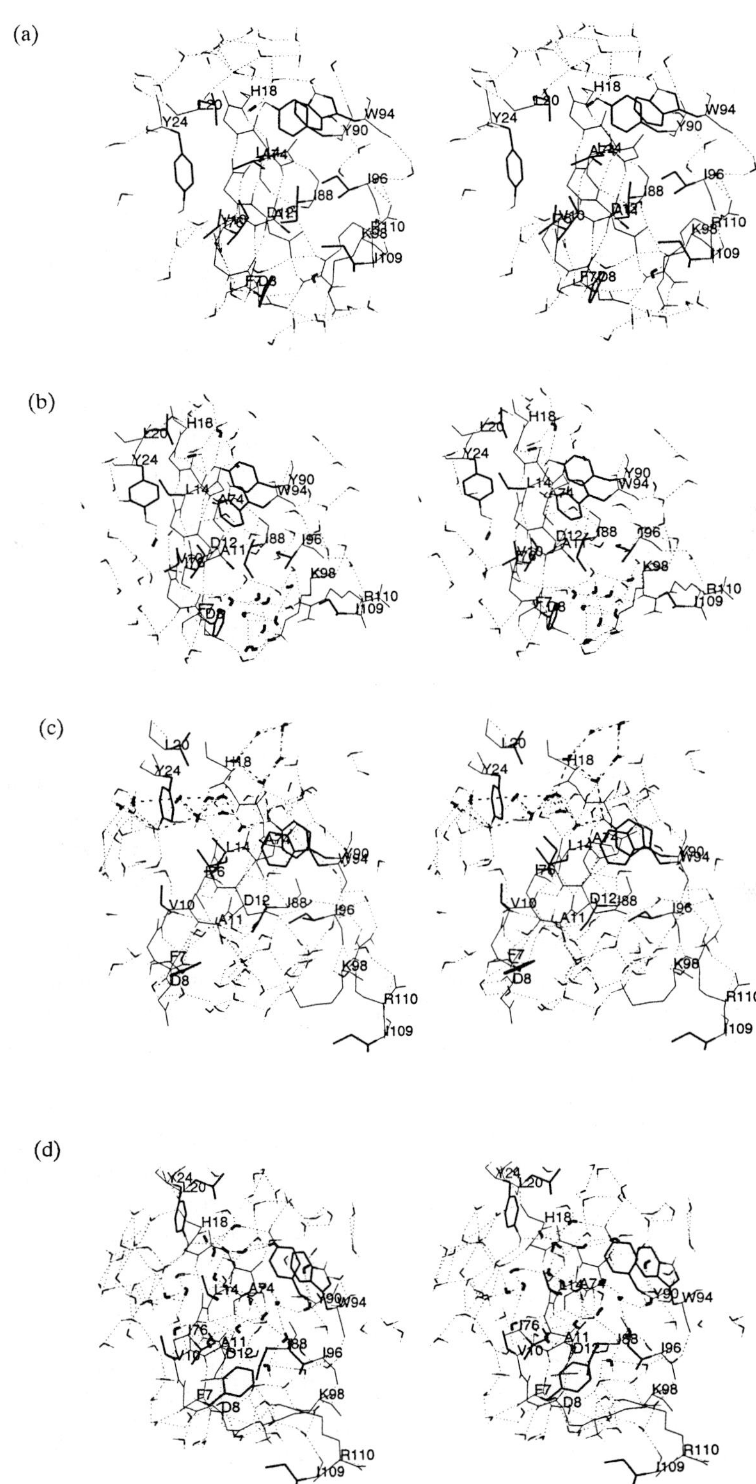

of the main helix; sometimes they insert and replace the helical hydrogen bond, as previously found in the fragment studies [14,15].

The present results suggest a possible mechanism for the solvation of hydrophobic cores and for the dissolution of secondary structural elements in protein denaturation. Very similar results have been obtained in a simulation of acid denaturation at 365 K (Caflisch & Karplus, in press). Testing of this mechanism is a challenge for experimentalists. Techniques that make use of photo-CIDNP [16], NOEs of water interactions with specific residues [17] and chemical markers [18] are possible approaches.

III. Lattice Model for Folding and Its Implications

The essential question is how a polypeptide chain is able to fold rapidly, in ms to s, to the native state despite the very large number of conformations that exist for the denatured chain (Levinthal paradox) [19]. To approach this problem, we use a simplified model that consists of a 27-bead self-avoiding chain on a cubic lattice (Figure 8). The native (lowest-energy) state can be determined exactly [20] and a survey of the folding behavior of many sequences is possible [6,7]. In addition, the full phase space density of the system can be obtained and the thermodynamic properties can be calculated as a function of the folding reaction coordinate, which is defined as the fraction of native contacts (out of a total of 28). The model is sufficiently complex that the resolution of the Levinthal paradox is required for folding; i.e., some sequences find the native state in only $\sim 10^7$ Monte Carlo (MC) steps even though there are $\sim 10^{16}$ conformations. Since the lattice simulation does not include the amino acid sidechains, the process considered here may correspond to the folding of real proteins to the molten globule stage; i.e., the molten globule, if it has a defined fold [3], is the native state in the present model.

In the first part of the analysis, 200 sequences with random interactions were generated and subjected to MC folding simulations [6]. Of these, 30 chains found the known native state in a short time. These chains correspond to actual protein sequences in the present model; the remaining sequences, which do not fold, do not correspond to protein sequences and serve as controls. The 30 folding sequences were analyzed and compared with the non-folding sequences. Several suggested mechanisms for resolving the Levinthal paradox do not apply to the present model; i.e., the features assumed to be responsible for rapid folding are found to be the same for the folding and non-folding sequences. These include a high number of short versus long range contacts in the native state [21], a high content of secondary structure in the native state [22], a strong correlation between the native contact map and the interaction parameters [23], and the existence of

---

Fig. 6.   Stereo views of A600 dynamics: (a) 9 ps (b) 39 ps, (c) 69 ps, (d) 115 ps. Hydrophobic sidechains of $core_1$ are shown as thick lines; α-helix backbone (residues 7–18), Asp 8, Asp 12, Lys 98, and Arg 110 sidechains are shown as thin lines. Water molecules within 12 Å of the center of geometry of $core_1$ are included. Hydrogen bonds are dotted (acceptor-hydrogen distance smaller than 2.5 Å; no angular criterion). Waters discussed in text are shown with thick lines.

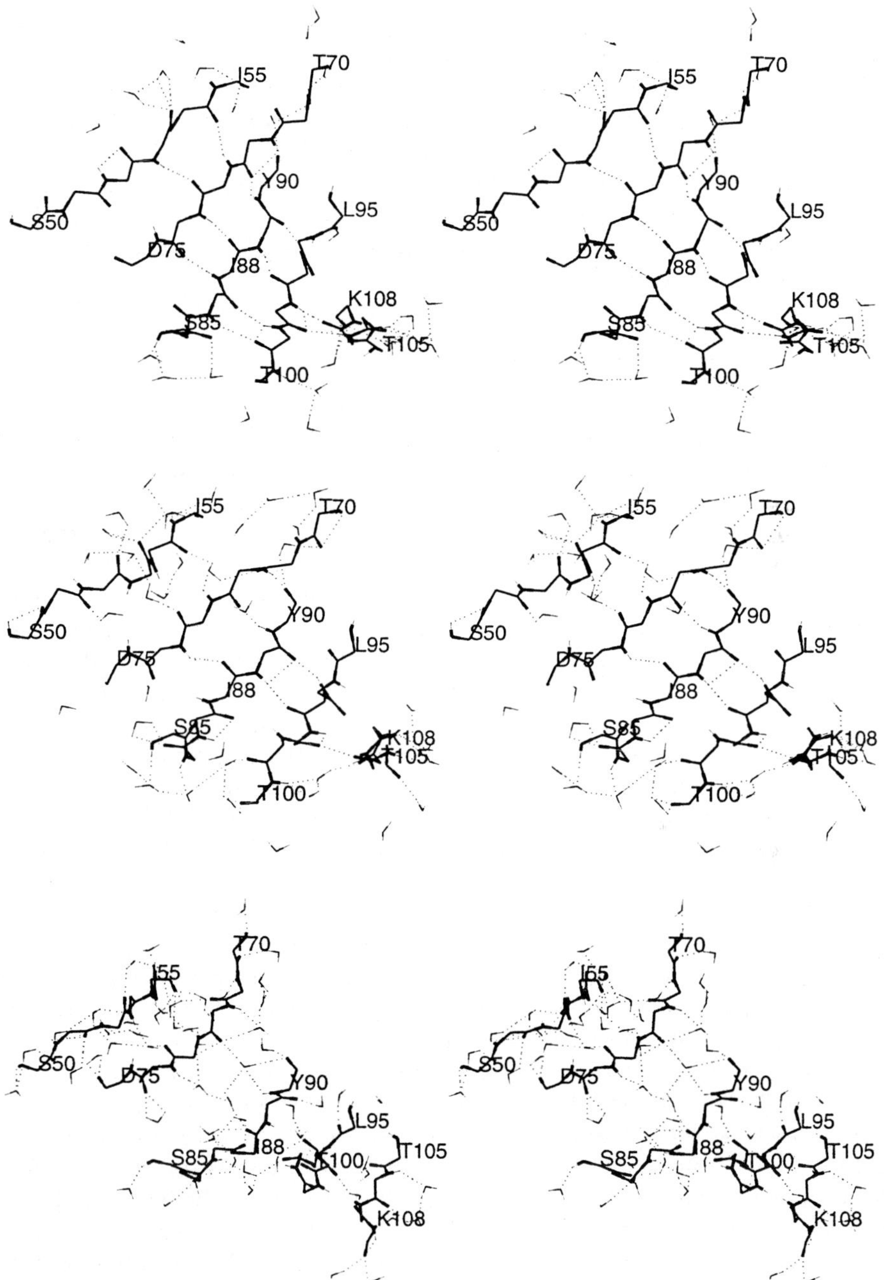

Fig. 7. Stereo views of water penetration into the β-sheet during R600. Mainchain N and O atoms are thick, hydrogens are thin and hydrogen bonds are dotted; water molecules within 3 Å of any mainchain atom of the β-sheet are shown. Top, 1 ps; middle, 60 ps; bottom 150 ps.

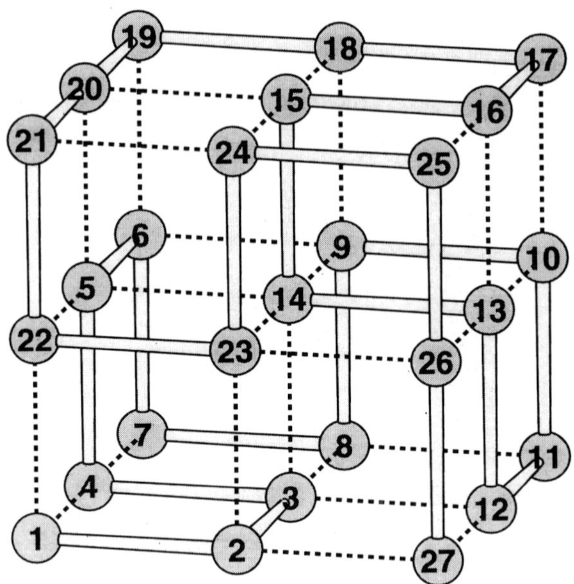

Fig. 8.   Lattice model of protein folding. An example of a compact self-avoiding structure of a chain of 27 monomers (filled numbered circles) with 28 contacts (dashed lines). The total energy of a conformation is the sum of contact energies: $E = \Sigma_{i<j} \Delta(r_i, r_j) B_{ij}$, where r_i are the positions of monomers i, B_{ij} are the contact energies for pairs of monomers i, j, and $\Delta(r_i, r_j)$ is 1 if monomers i and j are in contact and is 0 otherwise; two monomers are in contact if they are not successive in sequence and at unit distance from each other. The values of the B_{ij} are obtained from a Gaussian distribution with a mean B_0 and standard deviation σ_B. The parameter B_0 is an overall attractive term that emulates the hydrophobic effect observed in globular proteins. The *native* conformation is the compact self-avoiding chain with the lowest energy.

a high number of low energy states with near-native conformations [20]. Moreover, there is no repetitive trapping of the non-folding sequences in the same local minimum, so that the native state cannot be a metastable state [24]. The only significant difference between folding and nonfolding sequences is that the native state is at a pronounced energy minimum in the former. As can be seen in Figure 9, there is a large energy gap between the native and the 'excited' states in the sequences that fold; no such gap is present in the nonfolding sequences. This energy gap is the necessary and sufficient condition for a sequence to fold rapidly in the present model.

The time history of the folding process (Figure 10) shows that there is a rapid collapse in $\sim 10^4$ MC steps to a semi-compact random globule; i.e., the number of contacts increases, the energy decreases, while the fraction of native contacts remains below 0.3. In this way, the total of $\sim 10^{16}$ random-coil conformations is reduced to $\sim 10^{10}$ random semi-compact globule states. The fast collapse results from a large energy gradient and the presence of many empty lattice points. In the second, rate limiting, stage the chain searches for one of the $\sim 10^3$ transition states. The transition region consists of all states from which the chain folds rapidly

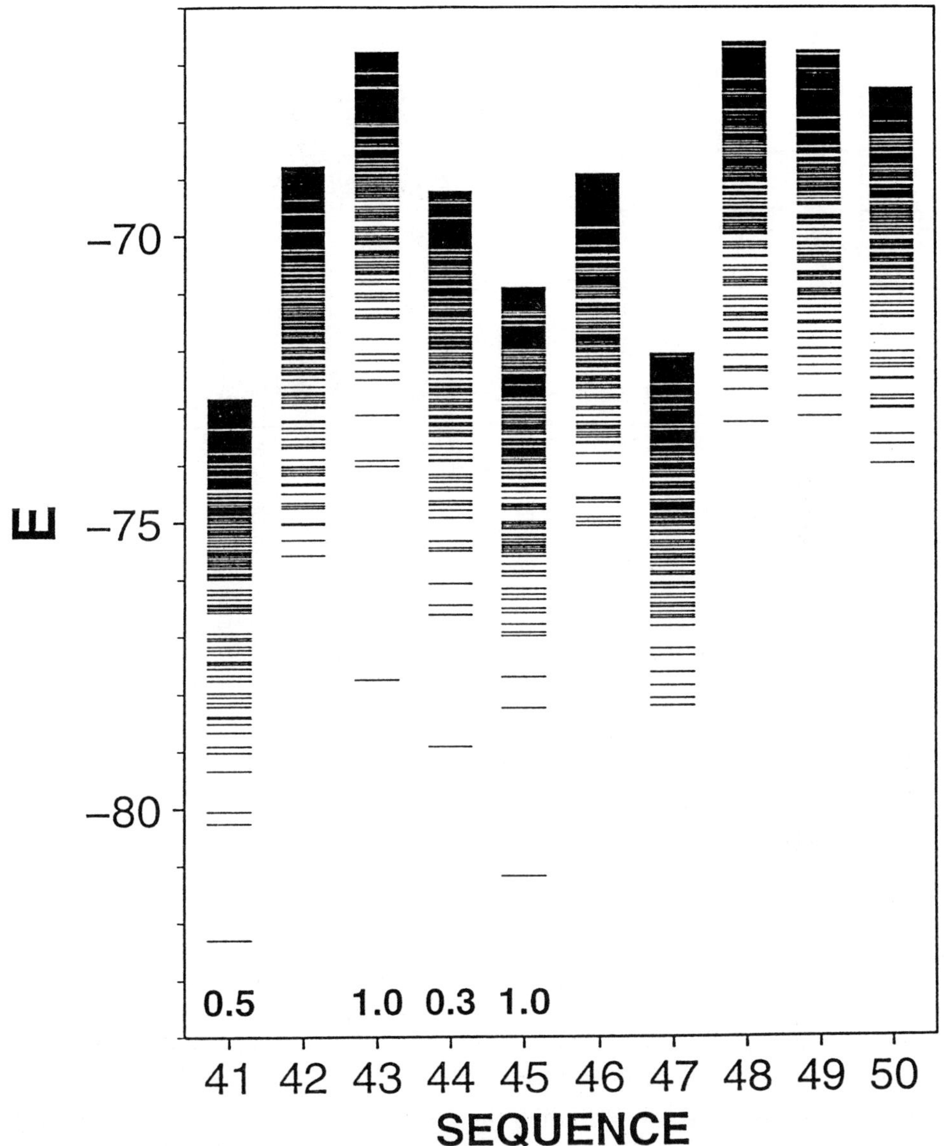

Fig. 9. Energy spectra for 10 folding and nonfolding random sequences. The energies of the 400 lowest compact self-avoiding conformations are shown. The native state corresponds to the bottom bar. The numbers below the spectra show the probabilities that the corresponding sequence will fold under the conditions of the simulation (6); if no number is given, the probability is 0.

to the native state. The transition states are structurally similar to the native state, with 23 to 26 of the native contacts. The mean first passage time, τ, for finding any of the n states among the total of N states by a random search which explores r states per unit of time is $N/(n \times r)$. For the present model, $\tau \sim 10^{10}/(10^3 \times 1) =$

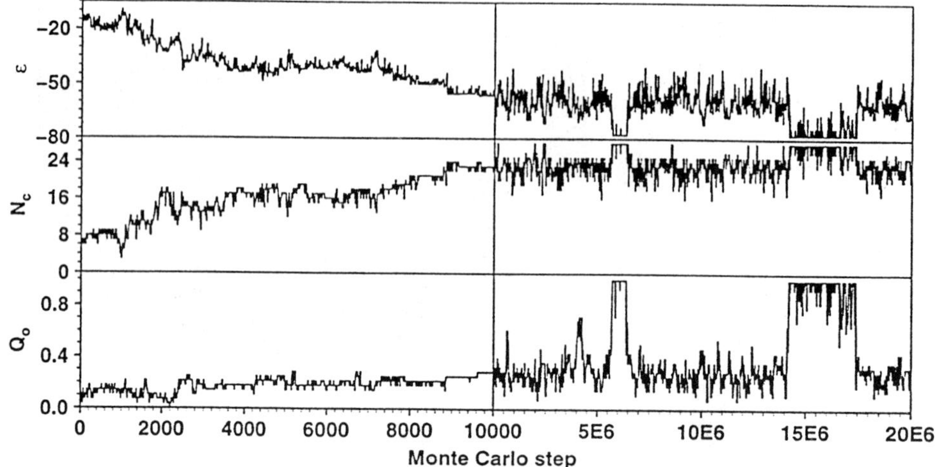

Fig. 10. Typical trajectory for a folding sequence ($T = 1.3$). Energy, ϵ (in units of $k_{\mathrm{B}}T$); the number of contacts, N_{c}; fraction of the number of contacts in common with the native state, Q_0. The instantaneous values of these quantities are plotted every 10 MC steps in the first part of the trajectory (≤ 10000 MC steps) and every 20000 steps in the subsequent part. The folding trajectory starts with a random-coil conformation and consists of local MC moves of one or two successive monomers that preserve bond lengths and avoid multiple occupancy of the lattice sites [6].

10^7, of the same order as the observed time scale (Figure 10). This indicates that the rate limiting stage in folding consists of a random search for a transition state in the semi-compact part of the phase space; i.e., a folding 'pathway' is not involved in finding the native state. In the third stage, the chain rapidly (within $\sim 10^5$ MC steps) attains the native conformation from any one of the transition states.

To examine the kinetics of the folding process, we write the unimolecular rate expression; i.e.,

$$\frac{\mathrm{d}C_{\mathrm{N}}}{\mathrm{d}t} = kC_{\mathrm{D}}$$

where C_{N} is the probability of finding the unique native state at time t, C_{D} is the probability of the denatured states (all other states; i.e., $C_{\mathrm{D}} = 1 - C_{\mathrm{N}}$) and k is the rate coefficient. Since the folding process takes place on a complex multiminimum surface, it would be possible for k not to be the usual rate constant, but a function of time [25]. To determine the kinetic behavior, we show in Figure 11 a plot of $\ln(1 - C_{\mathrm{N}}) = -kt$ versus the number of MC steps, which correspond to the time in the lattice model. The probability of finding the native state at time t is estimated from 100 independent folding simulations at several temperatures. It is calculated from the distribution of first passage times (i.e., when the native state is reached for the first time). It is evident from the figure that the coefficient k is indeed a rate constant; i.e., it is independent of time. To examine the dependence of the folding rate on temperature, $\log k$ is plotted versus $1/T$ in Figure 12. This plot

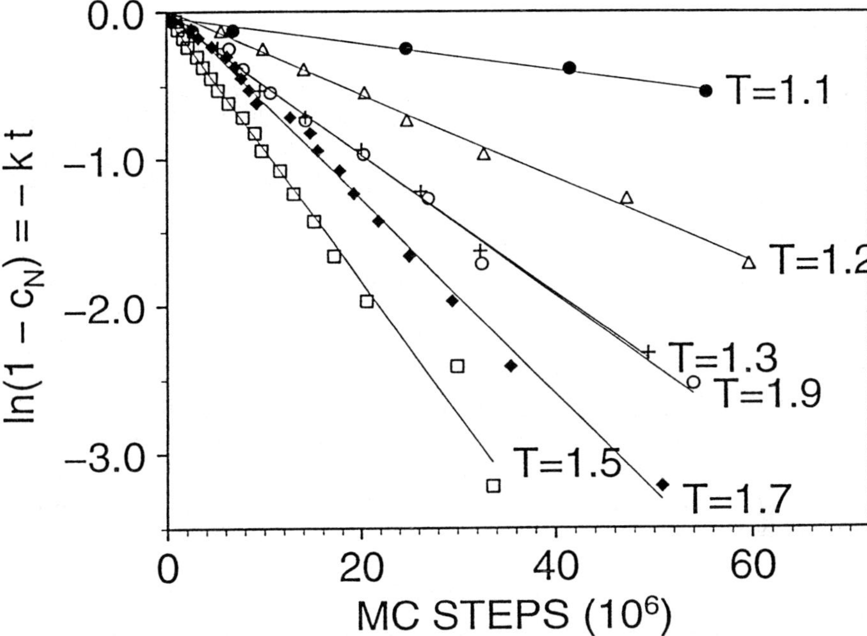

Fig. 11. Distribution of the mean first passage times for a sample folding sequence. The distribution of the mean first passage times is shown at the temperatures indicated on the plot. The lines are the linear least-squares fits to the points. 100 independent folding trials were done to obtain the points shown.

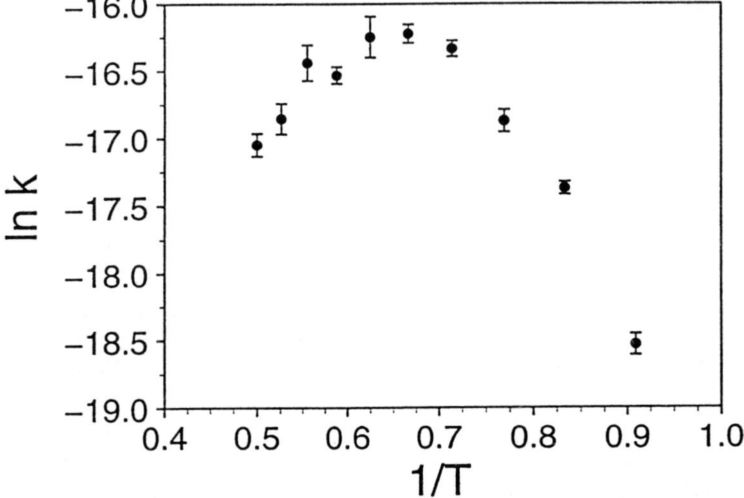

Fig. 12. Temperature dependence of the folding rate constant for a sample folding sequence. This Arrhenius plot shows the rate constants at different temperatures that were obtained from Figure 12. The standard error of this estimate was found by the jacknife test.

shows a strong deviation from standard Arrhenius type behavior. For low T $(0.90 < 1/T < 0.65)$ there is an increase in the folding rate, as expected from an activation energy limited process. However, at higher T $(0.65 < 1/T < 0.50)$ the rate decreases with increasing temperature as expected from an activation entropy dominated process. This is in accord with analysis of the configuration space of the system; i.e., at low temperatures, an increase in temperature makes it easier to get over the energy barriers which are rate limiting. However, at higher temperatures, a larger portion of the configuration space becomes accessible, which results in a slowing down of the folding process.

The pronounced energy minimum is the necessary condition for the folding of a 27-mer on a lattice because it guarantees that the native state is stable above the critical temperature, where the rearrangements of the structure with local unfolding required in the rate limiting stage are energetically possible. It is also a sufficient condition because a random search of compact globules with random structures can rapidly find a transition state that folds to the stable native state in a short time. While a non-folding sequence may also fold slowly to its native state above the critical temperature, it would not be stable and, therefore, could not correspond to a real protein.

The size of the search space for the native state is greatly reduced when the chain is semi-compact as it is in real proteins [26]. Nevertheless, in the 27-mer model as in real proteins, a random search of a collapsed globule cannot find the native state in the observed time. It is the existence of a transition region, consisting of a large number of states, that reduces the search time to realistic values when combined with the search of the random compact globule. Extrapolation of the folding time to real proteins suggests that the three-stage random search mechanism could be effective for small proteins (Figure 13). However, the mechanism breaks down for long chains because the folding time increases exponentially with chain length; i.e., the number of semi-compact states increases faster than the number of the transition states. Thus, a modification of the present mechanism is required for larger proteins.

It is likely that proteins existing early in evolution were small enough to fold according to the three-stage random search mechanism [27]. Since the pre-biotic and early biotic environment was hot, unusually thermostable proteins were required, such as those found in the most primitive bacteria that live at temperatures as high as 105° [28]. If so, the stability condition required a native state that corresponded to a low energy minimum for which the folding problem was solved simultaneously. As the evolution progressed, longer proteins evolved. These proteins had to fold on the same time scale. One way of achieving this is by evolving proteins with sequences that have a larger difference between the native and non-native contact energies than the random folding sequences of the present model. Such sequences would have an even more pronounced global minimum in their potential surfaces, in line with the 'consistency principle' [29] and the 'principle of minimal frustration' [30]. A factor may be the existence of a nucleus for folding [21], or the early appearance of secondary structural elements [31], neither of which are found in the present model. As a result, the collapse would not result in random semi-compact globules and the very favorable native contacts would

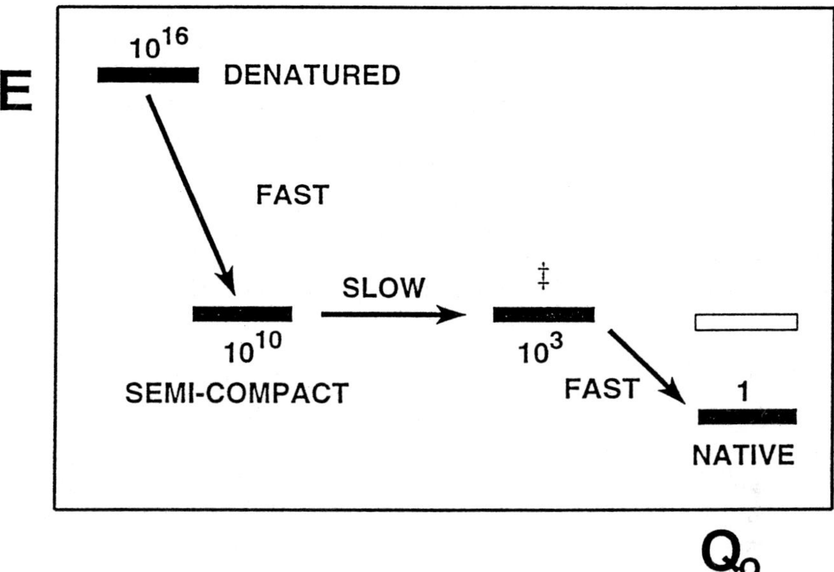

Fig. 13. Three-stage random search model of folding. The numbers indicate the geometrically possible conformations obtained from the density of states that was determined as a function of energy and reaction coordinate. The solid rectangles correspond to folding sequences. The empty rectangle indicates the energy of the native state of a nonfolding random sequence. The other states of a nonfolding sequence appear at the same positions as the corresponding states of a folding random sequence.

lead more directly to a native-like molten globule state. Actual proteins could use an intermediate mechanism that might vary with the external conditions.

The results of this study may have implications for the prediction of the structure of a protein from its amino acid sequence. The success of the three-stage random search in finding the pronounced global minimum on a potential surface suggests, at least for small proteins, that the bottleneck in structure prediction may be the derivation of a suitable potential function rather than the design of folding algorithms.

Acknowledgements

Most of the material in this paper is based on a lecture at the 27th Jerusalem Symposium (May 1994) and it was taken mainly from references 4, 6 and 7. A. Caflisch was supported by a grant from the Schweizerischer Nationalfonds. A. Sali is a Fellow of the Jane Coffin Childs Memorial Fund for Medical Research. Support was also provided by a David and Lucille Packard Fellowship (E.S.), and by a grant from the National Science Foundation and a gift from Molecular Simulations Inc. (M.K.).

References

1. M. Karplus and E. Shakhnovich: in *Protein Folding*, T. Creighton (Ed.), W.H. Freeman & Sons, p. 127 (1992).
2. C. L. Brooks III, M. Karplus, and B. M. Pettitt: *Proteins: A Theoretical Perspective of Dynamics, Structure, & Thermodynamics*, Adv. Chem. Phys. Vol. LXXI, John Wiley & Sons (1988).
3. O. B. Ptitsyn: in *Protein Folding*, T. E. Creighton (Ed.), W.H. Freeman, New York, p. 243 (1992).
4. A. Caflisch and M. Karplus: in *The Protein Folding Problem and Tertiary Structure Prediction*, K. Merz, Jr. and S. Le Grand (Eds.), Birkhäuser, Boston, MA, p. 193 (1994).
5. A. Caflisch and M. Karplus: *Proc. Natl. Acad. Sci. USA* **91**, 1746 (1994).
6. A. Šali, E. Shakhnovich, and M. Karplus: *J. Mol. Biol.* **235**, 1614 (1994).
7. A. Šali, E. Shakhnovich, and M. Karplus: *Nature* **369**, 248 (1994).
8. A. R. Fersht: *FEBS Letters* **325**, 5 (1993).
9. L. Serrano, A. Matouschek, and A.R. Fersht: *J. Mol. Biol.* **224**, 805 (1992).
10. L. Serrano, A. Matouschek, and A.R. Fersht: *J. Mol. Biol.* **224**, 847 (1992).
11. C.L. Brooks III and M. Karplus: *J. Chem. Phys.* **79**, 6312 (1983).
12. A. Matouschek, J. T. Kellis, Jr., L. Serrano, M. Bycroft and A. R. Fersht: *Nature* **346**, 440 (1990).
13. A. E. Mark and W. F. van Gunsteren: *Biochemistry* **31**, 7745 (1992).
14. J. Tirado-Rives and W. L. Jorgensen: *Biochemistry* **30**, 3864 (1991).
15. D. J. Tobias and C. L. Brooks III: *Biochemistry* **30**, 6059 (1991).
16. R. Kaptein, K. Dijkstra, and K. Nicolay: *Nature* **274**, 293 (1978).
17. G. Otting and K. Wüthrich: *J. Am. Chem. Soc.* **111**, 1871 (1989).
18. C. Ghelis: *Biophysical J.* **32**, 503 (1980).
19. C. Levinthal: in *Mossbauer Spectroscopy in Biological Systems, Proceedings of a Meeting held at Allerton House, Monticello, Illinois*, P. Debrunner, J. C. M. Tsibris, and E. Münck (Eds.), University of Illinois Press, Urbana, p. 22 (1969).
20. E. Shakhnovich, G. Farztdinov, A. M. Gutin and M. Karplus: *Phys. Rev. Lett.* **67**, 1665 (1991).
21. D. B. Wetlaufer: *Proc. Natl. Acad. Sci. USA* **70**, 697 (1973).
22. P. Kim and R. Baldwin: *Ann. Rev. Biochem.* **59**, 631 (1990).
23. N. Go and H. Abe: *Biopolymers* **20**, 1013 (1981).
24. J. D. Honeycutt and D. Thirumalai: *Biopolymers* **32**, 695 (1992).
25. Y. S. Bai and M. D. Fayer: *Phys. Rev.* **B39**, 11066 (1989).
26. K. A. Dill: *Biochemistry* **24**, 1501 (1985).
27. E. E. Di Iorio, W. Yu, C. Calonder, K. H. Winterhalter, G. De Sanctis, G. Falcioni, F. Ascoli, B. Giardina, and M. Brunori: *Proc. Natl. Acad. Sci. USA* **90**, 2025 (1993).
28. K. O. Stetter: in *Frontiers of Life*, J. K. Trân Thanh Vân, J. C. Mounolou, J. Schneider, and C. McKay (Eds.), Editions Frontières, Gif-sur-Yvette, France, p. 195 (1992).
29. N. Go and H. Abe: *Adv. Biophys.* **18**, 149 (1984).
30. J. D. Bryngelson and P. G. Wolynes: *Proc. Natl. Acad. Sci. USA* **84**, 7524 (1987).
31. H. Taketomi and N. Go: *Intl. J. Peptide Prot. Res.* **7**, 445 (1975).
32. M. Karplus and D. L. Weaver: *Protein Sci.* **3**, 650 (1994).

Essential Degrees of Freedom of Proteins

ANDREA AMADEI, ANTONIUS B. M. LINSSEN, BERT L. DE GROOT and
HERMAN J. C. BERENDSEN
*Department of Biophysical Chemistry and BIOSON Research Institute, the University of Groningen,
Nijenborgh 4, 9747 AG Groningen, The Netherlands*

Abstract. Analysis of extended molecular dynamics (MD) simulations of several proteins in aquous
solutions reveals that it is possible to separate the configurational space into two subspaces: (1) an
'essential' subspace containing only a few degrees of freedom in which anharmonic motion occurs that
comprises most of the positional fluctuations; and (2) the remaining space in which the motion has a
narrow Gaussian distribution and which can be considered as 'physically constrained'.

Key words. Molecular dynamics, essential degrees of freedom of proteins, lysozyme, haloalkane
dehalogenase, HPr.

1. Introduction

Functional proteins are generally stable mechanical constructs that allow certain
types of internal motions to enable their biological function. Functional internal
motions may be subtle and involve complex correlations between atomic motions,
but their nature is inherent in the structure and interactions within the molecule.
It is a challenge to derive such motions from the molecular structure and interactions, to identify their functional role, and to reduce the complex protein dynamics
to its essential degrees of freedom.

In 1993 we developed a new approach and methodology to identify an 'essential'
subspace in a protein configurational space [1]. In this method we analyse the
correlations between atomic positional fluctuations and diagonalize the covariance
matrix of atomic displacements. We find that most of the positional fluctuations
are concentrated in correlated motions in a subspace of only a few degrees of
freedom (not more than 1%), while all other degrees of freedom represent much
less important, basically independent, Gaussian fluctuations orthogonal to the
essential subspace. In our treatment we do not include atomic masses and we do
not require any harmonicity for the potential energy. We showed [1] that, if in
the configurational space ideal or approximate linear constraints for the motions
are present, we are always able, with this method, to identify them and define the
complementary subspace (essential subspace) where all the relevant motions
should be concentrated. We also do not require a complete equilibration of the
system. It is sufficient to have a trajectory that is extended enough to equilibrate
the near-constraint subspace and to produce motions in the essential subspace that
are large compared to the near constraints fluctuations. Since the near-constraints
coordinates have rather short relaxation times and fluctuation ranges, this is usually
accomplished within a reasonable simulation time (trajectories within 1 ns).

In this paper we show, in a more general way, the meaning of diagonalizing the
covariance matrix. We show the basic results obtained up to now and we describe
the new developments we are working on.

A. Pullman et al. (eds.), *Modelling of Biomolecular Structures and Mechanisms*, 85–93.
© 1995 *Kluwer Academic Publishers. Printed in the Netherlands.*

2. Theory

In an earlier publication [1] we described the theory which forms the basis of the essential degrees of freedom method. The standard procedure for analysis is as follows: First a simulation of several hundreds of picoseconds is performed. From the trajectory obtained, the overall translational and rotational motion is eliminated. Next a covariance matrix of the atomic displacements with respect to the average positions is built:

$$\mathbf{C} = \mathrm{cov}(\Delta\mathbf{x}) = \langle \Delta\mathbf{x} \otimes \Delta\mathbf{x} \rangle \tag{1}$$

Where $\otimes$ denotes a tensor product and $\Delta\mathbf{x}$ is given as:

$$\Delta\mathbf{x} = \mathbf{x} - \langle \mathbf{x} \rangle \tag{2}$$

By diagonalizing $\mathbf{C}$ we obtain a set of eigenvectors (we choose to sort the eigenvectors in order of decreasing eigenvalue) of which the first few define a subspace in which all the essential motions of the protein appear to occur. The eigenvalues are mean square displacements along the corresponding eigenvector. We now will describe a more general definition of the meaning of the eigenvectors of the covariance matrix $\mathbf{C}$. We will show that the first n eigenvectors define the best fitted hyperplane through the density of the probability distribution in configurational space. This implies that through the density in this space no set of orthonormal vectors can be defined in such a way that one of them has a positional fluctuation exeeding that of the first eigenvector. This means that the first eigenvalue is the maximum possible positional fluctuation along a single direction which was actually sampled. Alternatively, the total positional fluctuation in the $(N-1)$-dimensional plane (where N is the total number of dimensions of the system) orthogonal to this eigenvector is the minimum possible one. Figure 1 illustrates this for a two-dimensional case.

In general, in an N-dimensional space, the subspace orthogonal to the first n eigenvectors has the minimum possible positional fluctuation in respect of the positional fluctuations of any other $N-n$-dimensional subspace. We define the positional fluctuation in an $N-n$-dimensional subspace as:

$$\sigma^2_{N-n} = \sum_{\zeta=n+1}^{N} \langle [(\mathbf{x} - \langle \mathbf{x} \rangle) \cdot \boldsymbol{v}_\zeta]^2 \rangle \tag{3}$$

where $\{\boldsymbol{v}_\zeta\}$ is any set of N orthornormal vectors.

If we can prove that the vector along which the smallest positional fluctuation is observed, is the last eigenvector, then this can recursively be applied to the complementary $N-1$ dimensional subspace, and so on. The positional fluctuation of $\boldsymbol{v}_N$ is:

$$\sigma^2_1 = \langle [(\mathbf{x} - \langle \mathbf{x} \rangle) \cdot \boldsymbol{v}_N]^2 \rangle \tag{4}$$

$\boldsymbol{v}_N$ can always be written as a linear combination of eigenvectors $\boldsymbol{\eta}_i$:

$$\boldsymbol{v}_N = \sum_{i=1}^{N} \alpha_{iN} \boldsymbol{\eta}_i \tag{5}$$

Now $\sigma^2_{(N-1)}$ can be written as:

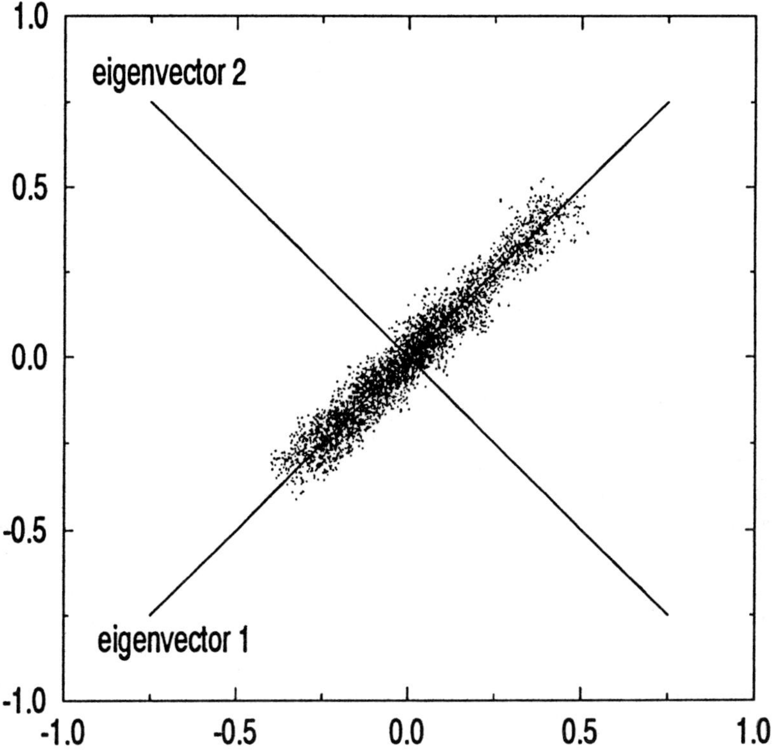

Fig. 1.   A two dimensional example of essential dynamics analysis.

$$\sigma_1^2 = \left\langle \left[\sum_{i=1}^{N} (\mathbf{x} - \langle \mathbf{x} \rangle) \cdot \alpha_{iN} \boldsymbol{\eta}_i \right]^2 \right\rangle \tag{6}$$

or

$$\sigma_1^2 = \left\langle \left[\sum_{l=1}^{N} \sum_{l'=1}^{N} [\alpha_{lN} \alpha_{l'N} \{ (\mathbf{x} - \langle \mathbf{x} \rangle) \cdot \boldsymbol{\eta}_l \} \{ (\mathbf{x} - \langle \mathbf{x} \rangle) \cdot \boldsymbol{\eta}_{l'} \} \right] \right\rangle \tag{7}$$

Since

$$\langle (\Delta \mathbf{x} \cdot \boldsymbol{\eta}_l)(\Delta \mathbf{x} \cdot \boldsymbol{\eta}_{l'}) \rangle = 0 \quad \forall l \neq l' \tag{8}$$

and

$$\langle (\Delta \mathbf{x} \cdot \boldsymbol{\eta}_i)^2 \rangle = \lambda_i \quad (\lambda_i \text{ is the eigenvalue corresponding to } \eta_i) \tag{9}$$

this can be written as:

$$\sigma_1^2 = \sum_{i=1}^{N} \alpha_{iN}^2 \lambda_i \tag{10}$$

We can now write the following inequality:

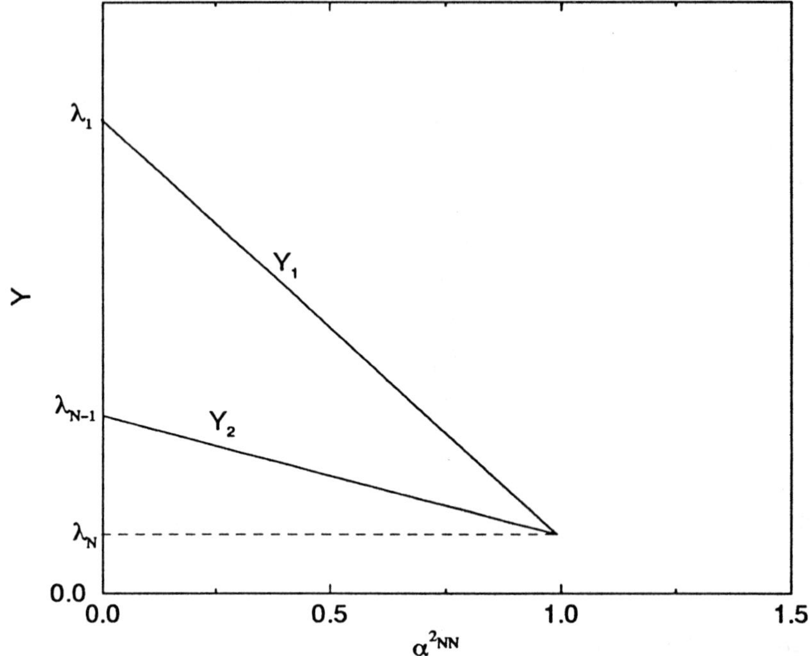

Fig. 2. $\sigma^2_{(N-1)}$ lies between y_1 and y_2.

$$\alpha^2_{NN}\lambda_N + \lambda_1 \sum_{i=1}^{(N-1)} \alpha^2_{iN} \geq \sigma^2_1 \geq \alpha^2_{NN}\lambda_N + \lambda_{(N-1)} \sum_{i=1}^{(N-1)} \alpha^2_{iN} \tag{11}$$

And since

$$\sum_{i=1}^{N} \alpha^2_{iN} = 1 \tag{12}$$

and

$$\sum_{i=1}^{(N-1)} \alpha^2_{iN} = 1 - \alpha^2_{NN} \tag{13}$$

this can be rewritten as

$$\sigma^2_{NN}\lambda_N + \lambda_1(1 - \alpha^2_{NN}) \geq \sigma^2_1 \geq \alpha^2_{NN}\lambda_N + \lambda_{(N-1)}(1 - \alpha^2_{NN}) \tag{14}$$

Since we want to find a minimum for σ^2_1 and its value lies between two boundaries, we must determine the minimal values of these boundaries. We can plot both boundaries as a function of α^2_{NN} (Figure 2) using the following properties:

$$\lambda_i \geq 0 \tag{15}$$

with $i = 1, 2, \ldots, N$

$$\lambda_{(N-1)} > \lambda_N \tag{16}$$

$$0 \leqslant \alpha_{NN}^2 \leqslant 1 \tag{17}$$

The upper boundary (as a function of α_{NN}^2) looks as follows:

$$y_1(\alpha_{NN}^2) = \alpha_{NN}^2(\lambda_N - \lambda_1) + \lambda_1 \tag{18}$$

and the lower boundary:

$$y_2(\alpha_{NN}^2) = \alpha_{NN}^2(\lambda_N - \lambda_{(N-1)}) + \lambda_{(N-1)} \tag{19}$$

As can be seen from Figure 2 σ_1^2 has a minimum for $\alpha_{NN}^2 = 1$, which means that

$$\boldsymbol{\nu}_N = \boldsymbol{\eta}_N \tag{20}$$

and

$$\sigma_1^2 = \lambda_N \tag{21}$$

This means that the vector for which a minimum mean square positional fluctuation is found, is the last eigenvector. This minimum value is exactly the eigenvalue corresponding to that eigenvector.

3. Results and Discussion

Up to now every protein to which our method was applied showed a very low dimensional essential subspace (which was always defined by less then ten eigenvectors). The kind of structural transitions involved in the essential motions are always connected to possible biological behaviour of the proteins. In lysozyme [1] we found a motion that opened the active site cavity of the enzyme described by the first eigenvector with all the other 'essential' ones involving smaller motions also in the active site region. In thermolysin (manuscript in preparation, in collaboration with D. van Aalten) we found two essential eigenvectors producing clear hinge bending motions between the two domains that enclosed the active site. This motion was already postulated on the basis of crystallographic data. Analysis of haloalkane dehalogenase (manuscript in preparation) revealed a possible entrance or exit of substrate or products. It is interesting to note that the motion along the first eigenvector also involved a tunnel that, on the basis of crystallographic data [2] was supposed to be the entrance as well as the exit. Both tunnels opened and closed in an anticorrelated way. Also the essential eigenvectors of HPr (manuscript in preparation, in collaboration with N. van Nuland, R. Scheek and G. Robillard) of which several structures based on NMR data were available [3] showed a hinge bending (opening and closing) motion between two α-helices enclosing the biological active region of the protein. In Figure 3 two superimposed structures from two (most distant) positions along the first eigenvector of HPr are shown. The hinge bending motion involving both helices can clearly be seen. Figure 4 shows the C_α-eigenvalue curves from three proteins, lysozyme, dehalogenase and HPr. It is evident that after the first few eigenvalues the remaining ones represent negligible motions for all three proteins. We found [1] that the essential eigenvectors derived from the all atoms covariance matrix produced backbone

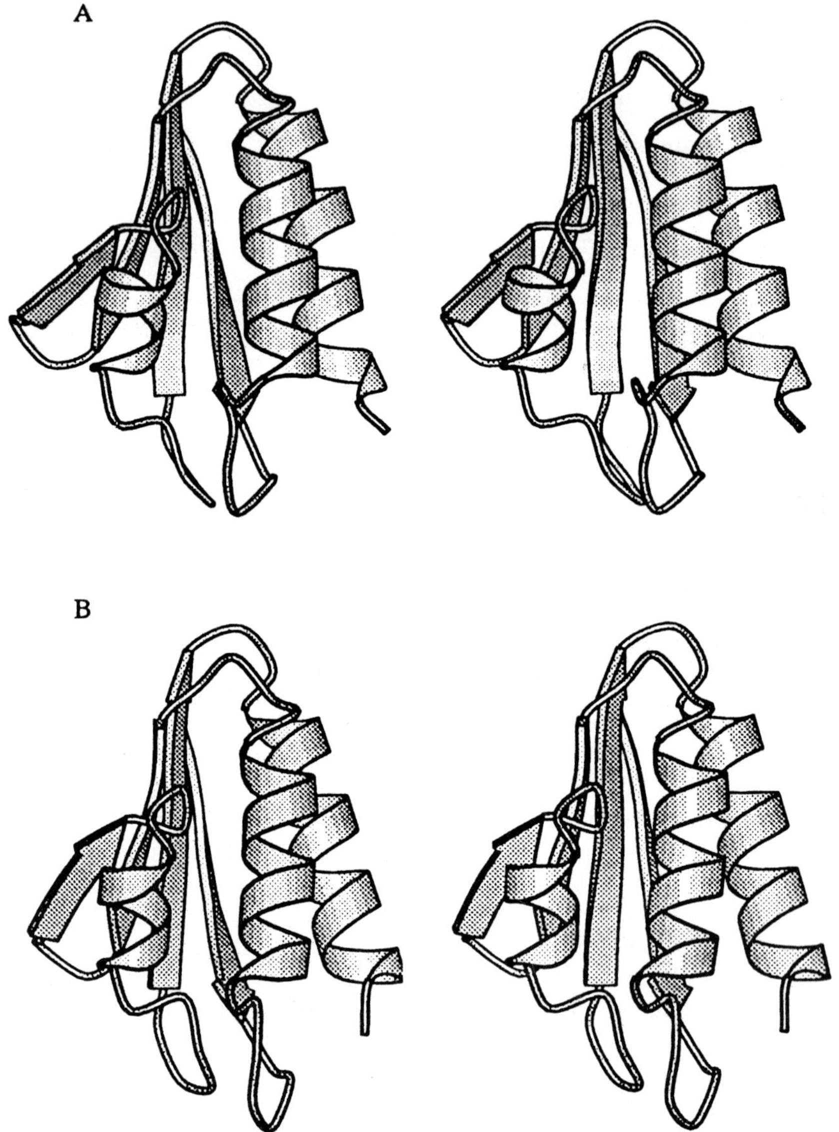

Fig. 3.   Stereoview of the closed structure (A) and the open structure (B) of HPr derived from the
displacement along the first eigenvector.

motions that were identical to the motions obtained from the essential C_α-eigenvectors.

In Figure 5 we show the projectons of three HPr trajectories (starting from three different NMR-structures) on the three planes defined by the first three C_α-eigenvectors, and one plane defined by two near-constraints eigenvectors (vectors 20 and 50). In the three 'essential planes' the density is far from being equilibrated but the possibility for large displacements is already evident. In fact the three

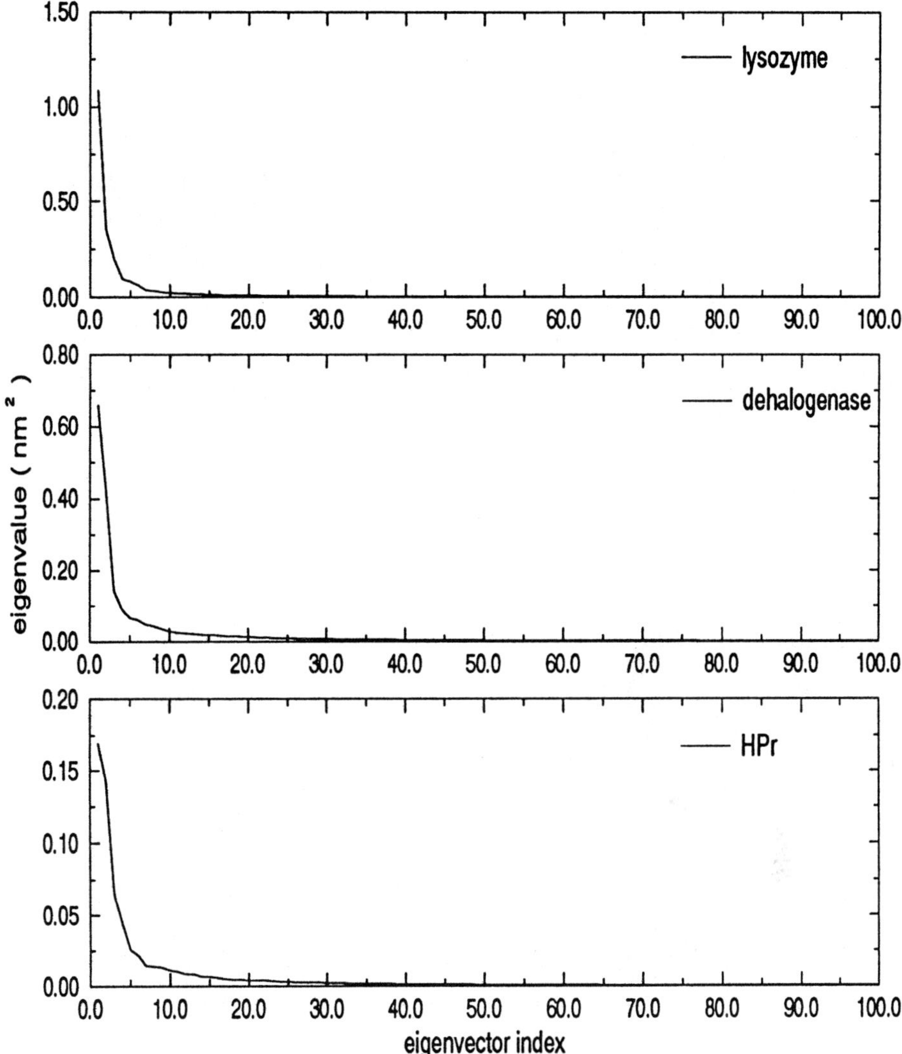

Fig. 4.   Eigenvalues of lysozyme, dehalogenase and HPr of C_α-analysis (see text).

trajectories on these essential planes span very different regions (although always in contact). On the contrary, in the near constraints plane the three trajectories fully overlap with the same very narrow region, indicating a full equilibration and a stable near-constraint behaviour.

4. Conclusions

The *essential degrees of feedom* method has, up till now, proved to be useful for detecting protein dynamical behaviour that is related to biological functions. This methodology can also be used as a powerful tool for comparing different dynamics

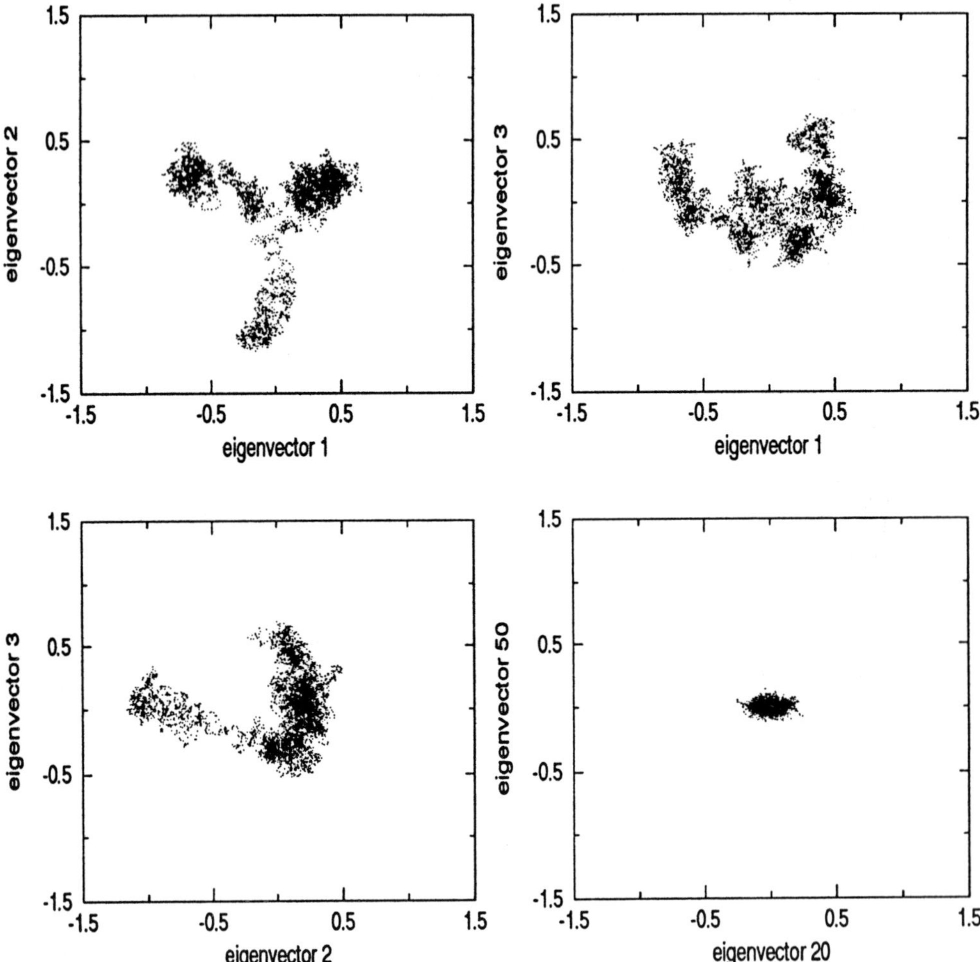

Fig. 5.   Three trajectories of HPr projected on four different planes (see text).

of different (but related) proteins. At this moment our main goal is the investigation of the essential subspace of HPr. The geometrical definition of such subspace gives us the capability to sample this subspace in a very efficient way. This can be accomplished by either moving the system along one of the essential eigenvectors or using more complex pathways that are not necessarily linear. If a sufficient sampling can be obtained this should produce all the configurations that are involved in the biological activity. This gives the possibility to predict several properties of the protein and also can give insight into the consequences of mutations.

Acknowledgements

We would like to thank Dr. A. Mark from ETH in Zürich who gave us a 1 ns trajectory of lysozyme, N. van Nuland who provided us with several Molecular Dynamics trajectories of HPr and Dr. K. Verschueren who was so kind to supply us with crystal structure coordinates of dehalogenase with and without substrate. The work concerning HPr was performed in close collaboration with Dr. R. Scheek and Prof. G. Robillard. Dr. A. Amadei was financially supported by 'Istituto Pasteur Fondazione Cenci Bolognetti'. Visualization of HPr was performed with MOLSCRIPT [4].

References

1. Andrea Amadei, Antonius B. M. Linesen, and Herman J. C. Berendsen: *PROTEINS: Structure, Function, and Genetics* **17**, 412–425 (1993).
2. Sybille M. Franken, Henriette J. Rozeboom, Kor H. Kalk, and Bauke W. Dijkstra: *EMBO J.* **10**, 1297–1302 (1991).
3. Nico A. J. van Nuland, Ilona W. Hangyi, René C. van Schaik, Herman J. C. Berendsen, Wilfred F. van Gunsteren, Ruud M. Scheek, and George T. Robillard: *J. Mol. Biol.* **237**, 544–559 (1994).
4. P. J. Kraulis: *J. Appl. Crystalogr.* **24**, 946–950 (1991).

De Novo Simulations of the Folding of GCN4 and Its Mutants

J. SKOLNICK,[a]* M. VIETH,[a] A. KOLINSKI,[a,b] and C. L. BROOKS, III[c]
[a]*Departments of Molecular Biology and Chemistry, The Scripps Research Institute, 10666 N Torrey Pines RD, La Jolla, CA 92037, U.S.A.*
[b]*Department of Chemistry, University of Warsaw, 02-093 Warsaw, Poland*
[c]*Departmenent of Chemistry, Carnegie Mellon University, Pittsburgh, PA 15213, U.S.A.*

Abstract. A hierarchical approach to protein folding is employed to examine the folding pathway and predict the quaternary structure of the GCN4 leucine zipper. Structures comparable in quality to experiment have been predicted. In addition, the equilibrium between dimers, trimers and tetramers of a number of GCN4 mutants has been examined. In most of the cases, the simulation results are in accord with the experimental studies of Harbury *et al.* [1].

Key words. Protein folding/GCN4, leucine zipper/computer, simulations/lattice, protein models.

Introduction

One of the most important unsolved problems in molecular biology is the prediction of protein structure from amino acid sequence [2–4]. As a step in this direction, we have examined the folding pathway and predicted quaternary structure of the GCN4 leucine zipper [5]. In addition, the equilibrium between various GCN4 mutants [1] has been explored. The method uses a hierarchical approach to protein folding [6,7]. The basic idea is to employ a reduced model to simulate the early stages of folding and to obtain the native state topology. Restrained molecular dynamics are then used to produce full atom models [5,7]. An overview of the entire procedure is presented in Figure 1.

Folding of the model chains commences on a very high coordination lattice, from a pair of random unfolded chains. Following dimer assembly to a parallel, left handed in register coiled coil, the resulting structures are refined on the lattice to produce a family of native structures whose rms from native [8] ranges from 2.3 to 3.7 Å [5]. Then, full atom models are built, the structures are solvated and refined using molecular dynamics with the CHARMM potential [9]. The resulting family of structures is indistinguishable from the native one when subjected to the molecular dynamics refinement protocol. The average structure from this family has a backbone heavy atom rms of 0.8 Å, all heavy atoms in the dimerization interface differ by 1.31 Å rms from native, and all heavy atoms differ from the crystal structure by 2.29 Å rms. These studies marked the first time that protein quaternary structures of this quality have been obtained from random unfolded conformations.

In a recent paper in Science by Harbury *et al.* [1], they examined the shift in equilibrium between parallel, coiled coil dimers, trimers and tetramers associated

---

★ Author for correspondence.

A. Pullman et al. (eds.), *Modelling of Biomolecular Structures and Mechanisms*, 95–98.
© 1995 *Kluwer Academic Publishers. Printed in the Netherlands.*

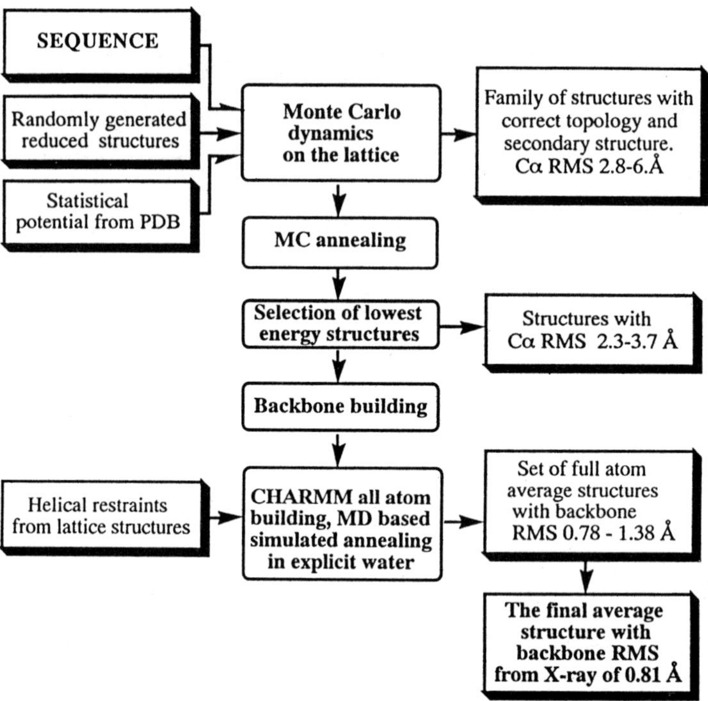

Fig. 1. An overview of the hierarchical approach. The left hand column displays the input information, the central column displays the protocol, and the right-hand column shows the resulting output.

with varying the identity of the residues of the two hydrophobic residues in the coiled coil heptad repeat. To further test the validity of the model, we prepared a series of dimers, trimers and tetramers for each of the eight species that have been experimentally studied. In order to compare with experiment, we have to calculate the equilibrium constants associated with the dimer, D trimer, T, and tetramer, R, species.

$$3D \leftrightarrow 2T \tag{1a}$$

$$2D \leftrightarrow R \tag{1b}$$

with $\{L\}$, the number of chains in species $L = D, T, R$. The equilibrium constants are

$$K_{DT} = \{T\}^2/\{D\}^3 = VZ_{T^2}/Z_{D^3} \tag{2a}$$

$$K_{DR} = \{R\}^3/\{D\}^2 = VZ_R/Z_{D^2} \tag{2b}$$

with V the total accessible volume of the system. Finally, the total concentration of chains, C, independent of their degree of association is

$$C = 2\{D\} + 3\{T\} + 4\{R\}. \tag{3}$$

Since Equations 2a, 2b, and 3 comprise three equations in three unknowns

Table I.
Cormparison or simulation with experiment on GCN4 mutants

Residues at positions		Experimental results	Simulation results		
a	d	Dominant species	Dominant species	Concentration dependence	
Wild	Type	2	2	99:1:0	*low
				94:6:0	high
I	L	2	3 !	13:87:0	low
				4:95:1	high
I	I	3	3	0:100:0	low
				0:100:0	high
L	I	4	4	0:4:96	low
				0:0:100	high
V	I	?	3	0:100:0	low
				0:100:0	high
L	V	3	3	1:99:0	low
				0.2:99.8:0	high
V	L	(2,3)	3 !	2:98:0	low
				0.1:99.9:0	high
L	L	3	3	0.1:99.9:0	low
				0:100:0	high

Low and high indicate $2\,\mu$M and $200\,\mu$M concentration of a peptide respectively. ! indicates those results which apparently disagree with experiment. In the VI case, the degree of association could not be assigned experimentally.

$(\{D\}, \{T\}, \{R\})$, these equations can be solved numerically. The key to the analysis is to employ a method based on Mayer and Mayer [10] which requires that in each species, γ the system be in a local energy minimum, with energy E_{avg}. Then, the internal partition function is given by

$$Z_{\mathrm{int},\gamma} = V_{\exp}(-E_{\mathrm{avg}}/kT)\left\{\frac{\Pi_N\,\mathrm{d}v_i}{P(R=0)}\right\}^{1-1/N} \tag{4}$$

with $\mathrm{d}v_i$ the infinitesimal volume element centered about the origin which corresponds to the minimum energy conformation, N is the number of groups in the species γ, and $P(R=0)$ is the probability that all the N groups in the molecule are in the minimum energy state. In our case, the groups correspond to the alpha carbons and side chains' center of mass positions.

The results of our simulations are shown in Table I, where rather good agreement with experiment is demonstrated. We found that antiparallel species do not contribute at all. Finally, to verify that our results are not limited to the GCN4 and its mutants, we examined the equilibrium degree of association of a sequence designed by DeGrado, Eisenberg and coworkers, coil-Ser [11]. In agreement with experiment, we predict that the dominant species is an antiparallel, three helix bundle.

Conclusions

These simulations on the folding of coiled coils demonstrate that it is possible to predict quaternary structure of simple folding motifs from amino acid sequence alone. Furthermore, the ability to predict the dominant species provides confidence that the phenomenological potentials have captured some aspects of the experimental situation. Future work will entail extension to larger coiled coils and extension of the model to predict the quaternary structure of some small globular proteins.

Acknowledgment

This research was supported in part by NIH Grant No. GM48835 and NIH Fogarty International Center Grant No. 1 R03 TW00418.

References

1. P. B. Harbury, T. Zhang, P. S. Kim, and T. Alber: *Science* **262**, 1401 (1993).
2. T. E. Creighton: *Biochem. J.* **270**, 131 (1990).
3. M. Levitt: *Current Opinion in Structural Biology* **1**, (1991).
4. O. B. Ptitsyn: *Journal of Protein Chemistry* **6**, 273 (1987).
5. M. Vieth, A. Kolinski, C. L. Brooks, III and J. Skolnick: *J. Mol. Biol.* **237**, 361 (1994).
6. A. Kolinski, A. Godzik and J. Skolnick: *J. Chem. Phys.* **98**, 7420 (1993).
7. J. Skolnick, A. Kolinski, C. L. Brooks and A. Godzik: *Current Biol.* **3**, 414 (1993).
8. E. K. O'Shea, J. D. Klemm, P. S. Kim, and T. Alber: *Science* **254**, 539 (1991).
9. B. R. Brooks, R. Bruccoleri, B. Olafson, D. States, S. Swaminathan, and M. Karplus: *J. Comp. Chem.* **4**, 187 (1983).
10. J. E. Mayer and M. G. Mayer: *Statistical Mechanics*, New York: John Wiley & Sons, Inc. (1963).
11. B. Lovejoy, C. Seunghyan, D. Cascio, D. K. McRorie, W. F. DeGrado, and D. Eisenberg: *Science* **259**, 1288 (1993).

A Model of HIV-I Reverse Transcriptase:
Possible Mechanisms for AZT Resistance

ROBERT F. SETLIK,[1] MASAYUKI SHIBATA,[1] RICK L. ORNSTEIN,[2] and
ROBERT REIN[1]★
[1]*Department of Biophysics, Roswell Park Cancer Institute, Elm and Carlton Streets, Buffalo, New York 14263, U.S.A.*
[2]*Molecular Science Reserch Center, Pacific Northwest Laboratory, Richland, Washington 99352, U.S.A.*

Abstract. In this report we present our preliminary results on the explanation of possible roles that mutations at positions 41, 67, 70, 215, and 219 of the p66 subunit of HIV-I reverse transcriptase play in AZT resistance. These explanations stem from molecular modeling studies of AZT in both the active site of our full-coordinate model, and in the channel to the active site. These results suggest the following: (a) that mutant residues Arg 70, Tyr 215, and possibly Leu 41, sterically block the opening to the active site and help reverse transcriptase selectively exclude AZT due to its larger 3′ group compared to normal nucleotides, (b) the loss of the positively charged lysine ammonium head group in the Lys 219 → Gln mutation removes a favorable monopole-dipole interaction with the 3′ azido group of AZT, and (c) the mutant residue Asn 67 can hydrogen bond to the single-stranded template of the template-primer complex and constrains the β3-β4 loop into a conformation which disallows AZT uptake.

Key words. HIV-I reverse transcriptase, AZT-resistance mutations, AZT resistance, AIDS.

Introduction

Reverse transcriptase (RT) is the retroviral enzyme responsible for the conversion of single-stranded RNA into double-stranded DNA [1,2]. This process is the first in the pathway of incorporation of the retroviral genome into the host cell's DNA genome. To date, a number of nucleoside analog drugs have been developed for the treatment of AIDS patients which inhibit the activity of RT, the most notable of these being AZT. These drugs inhibit the action of RT when incorporated onto the nascent DNA chain due to their lack of a 3′OH group necessary for the continuation of chain elongation. Unfortunately, over the course of drug therapy, mutations occur in the polymerase domain of RT which render these drugs ineffective as anti-AIDS agents.

In this study we utilized our full-coordinate model of the polymerase domain of RT [3] to discuss the possible molecular roles that mutations in the RT polymerase domain play in AZT resistance. These roles were determined, with respect to mutant and wild-type structures of RT, from observing changes in: (a) the electrostatic properties of the amino acid side-chains, (b) the size of the side-chains, and (c) the side-chain hydrogen bonding properties.

★Author for correspondence.

A. Pullman et al. (eds.), *Modelling of Biomolecular Structures and Mechanisms*, 99–106.
© 1995 *Kluwer Academic Publishers. Printed in the Netherlands.*

Materials and Methods

A detailed description of the model building of the RT/template-primer binary complex is presented elsewhere [3]. A brief overview of the molecular modeling protocol is as follows. Our full-coordinate model was constructed from the $C\alpha$ coordinates of Jacobo-Molina *et al.* [4] using the MAXSPROUT suite of programs [5]. Subsequent refinement and molecular modeling was accomplished using the SYBYL package, version 5.52 (obtained from Tripos Associates Inc.). A seven-base-pair dsA-DNA was positioned in the nucleic acid binding cleft of the polymerase domain (referred to as the polymerase cleft herein) to represent the template-primer complex. Positioning of this structure in the polymerase cleft was guided by the locations of phosphorus atoms of a 19-base/18-base dsDNA bound in the Jacobo-Molina structure. The missing thymidine base in the crystal structure was added in our model (referred to as T14). This base represents an incoming dTTP after addition to the 3′ end of the primer strand (to residue A13) and before translocation.

A single strand of six thymidine residues in the A conformation was added to the residue at the 5′ end of the template strand of the template–primer complex (residue A1) in order to observe the location of the single-stranded residues of the template in the RT structure. This conformation was used since Jacobo-Molina *et al.* reported that the nucleic acid complex bound to the polymerase cleft resembled the A conformation. The numbering of these residues (shown in Figure 3) is from −1 to −6 with T-1 connected to A1.

The AZT-resistance mutations were constructed from our original model by replacement of residues Met 41, Asp 67, Lys 70, Thr 215, and Lys 219 with amino acids Leu, Asn, Arg, Tyr, and Gln [6,7], respectively, using the BIOPOLYMER module of SYBYL. The orientations of the side chains of these residues were refined first by molecular modeling, and then by using ANNEAL with 2 Å and 6 Å cutoffs for the 'HOT' and 'INTERESTING' regions, respectively, with radii centered on the mutant residues. The AZT molecule was constructed using structural information from Parthasarathy and Kim [8]. AZT was positioned in the active site added to the 3′ end of the primer strand (to A13 of the template-primer complex model) and in various positions in the active site channel to determine its path into the active site relative to the AZT-resistance residues. The figures appearing in this paper were generated with the MOLSCRIPT program [9].

Results and Discussion

Principally, five mutations to RT are responsible for the onset of resistance to AZT (Met 41 → Leu, Asp 67 → Asn, Lys 70 → Arg, Thr 215 → Phe or Tyr, and Lys 219 → Gln; [6,7]). These residues are situated rather distant from the center of the active site (>9 Å from the O3′ of A13 in both wild-type and mutant; see Table I). When measured from N3 of the azido group of AZT (AZT substituted for T14), the side-chains of the above residues are 6.5 Å away or greater in the case of wild-type enzyme and greater than 9 Å away for the fully mutated form. The placement of these residues with respect to the active site suggest three

TABLE I

Distances for residues involved in AZT resistance from O3′ of A13 and from N3 of the 3′ azido group of AZT situated in the active site (AZT substituted for T14). These distances are measured for both wild-type and fully mutated RT (Å)

Wild-Type

A13 O3′ – Met 41 Cα	22.7	AZT N3 – Met 41 Cα	15.6
A13 O3′ – Met 41 Cε	18.6	AZT N3 – Met 41 Cε	11.4
A13 O3′ – Asp 67 Cα	20.1	AZT N3 – Asp 67 Cα	15.8
A13 O3′ – Asp 67 Cγ	18.9	AZT N3 – Asp 67 Cγ	15.5
A13 O3′ – Lys 70 Cα	21.5	AZT N3 – Lys 70 Cα	15.5
A13 O3′ – Lys 70 Nζ	26.3	AZT N3 – Lys 70 Nζ	20.2
A13 O3′ – Thr 215 Cα	14.1	AZT N3 – Thr 215 Cα	11.4
A13 O3′ – Thr 215 Oγ	15.5	AZT N3 – Thr 215 Oγ	11.7
A13 O3′ – Lys 219 Cα	12.4	AZT N3 – Lys 219 Cα	12.1
A13 O3′ – Lys 219 Nζ	9.7	AZT N3 – Lys 219 Nζ	6.5

Mutant

A13 O3′ – Leu 41 Cα	22.7	AZT N3 – Leu 41 Cα	15.0
A13 O3′ – Leu 41 Cγ	21.0	AZT N3 – Leu 41 Cγ	13.8
A13 O3′ – Leu 41 Cδ1	20.0	AZT N3 – Leu 41 Cδ1	13.0
A13 O3′ – Leu 41 Cδ2	20.3	AZT N3 – Leu 41 Cδ2	13.0
A13 O3′ – Asn 67 Cα	20.1	AZT N3 – Asn 67 Cα	15.8
A13 O3′ – Asn 67 Oδ	18.8	AZT N3 – Asn 67 Oδ	15.3
A13 O3′ – Asn 67 Nδ	19.7	AZT N3 – Asn 67 Nδ	16.7
A13 O3′ – Arg 70 Cα	21.4	AZT N3 – Arg 70 Cα	15.5
A13 O3′ – Arg 70 Cζ	25.8	AZT N3 – Arg 70 Cζ	20.0
A13 O3′ – Tyr 215 Cα	14.0	AZT N3 – Tyr 215 Cα	11.4
A13 O3′ – Tyr 215 OH	20.1	AZT N3 – Tyr 215 OH	15.9
A13 O3′ – Gln 219 Cα	12.4	AZT N3 – Gln 219 Cα	12.1
A13 O3′ – Gln 219 Nε	11.9	AZT N3 – Gln 219 Nε	9.3

possible effects these residues have on AZT resistance: (1) these residues may reduce the affinity of RT for AZT; (2) these residues may impede the path AZT takes into the active site; and (3) these residues most likely do not have a major direct effect on the mechanism of addition of AZT to the nascent DNA chain.

In our model, positions 41, 70, 215, and 219 are situated at the opening to the active site (Figure 1); that is, the space between the β3-β4 hairpin of the finger and residues 214–219 and 110–116 of the palm subdomains (see Reference 4 for a listing of residues to secondary structure elements). Assuming that the incoming nucleotide enters the active site in the proper orientation for chain elongation, molecular modeling with AZT indicates that residues 70 and 215 are situated adjacent to the path of the azido group as AZT enters the active site; with position 70 on the endo side of the dideoxyribose sugar, and 215 on the exo side (Figures 1 and 2). These two residues mutate to bulkier residues over the course of AZT treatment. Since the azido group is roughly double the size of the normal 3′ hydroxy group, it is possible that these mutations may act to impede the path AZT takes into the active site by sterically blocking the entrance. Both residues 70 and 215 exist on exposed loop structures which very likely exhibit large dynamic motions. The dynamic flexibility of these regions coupled with the incorporation

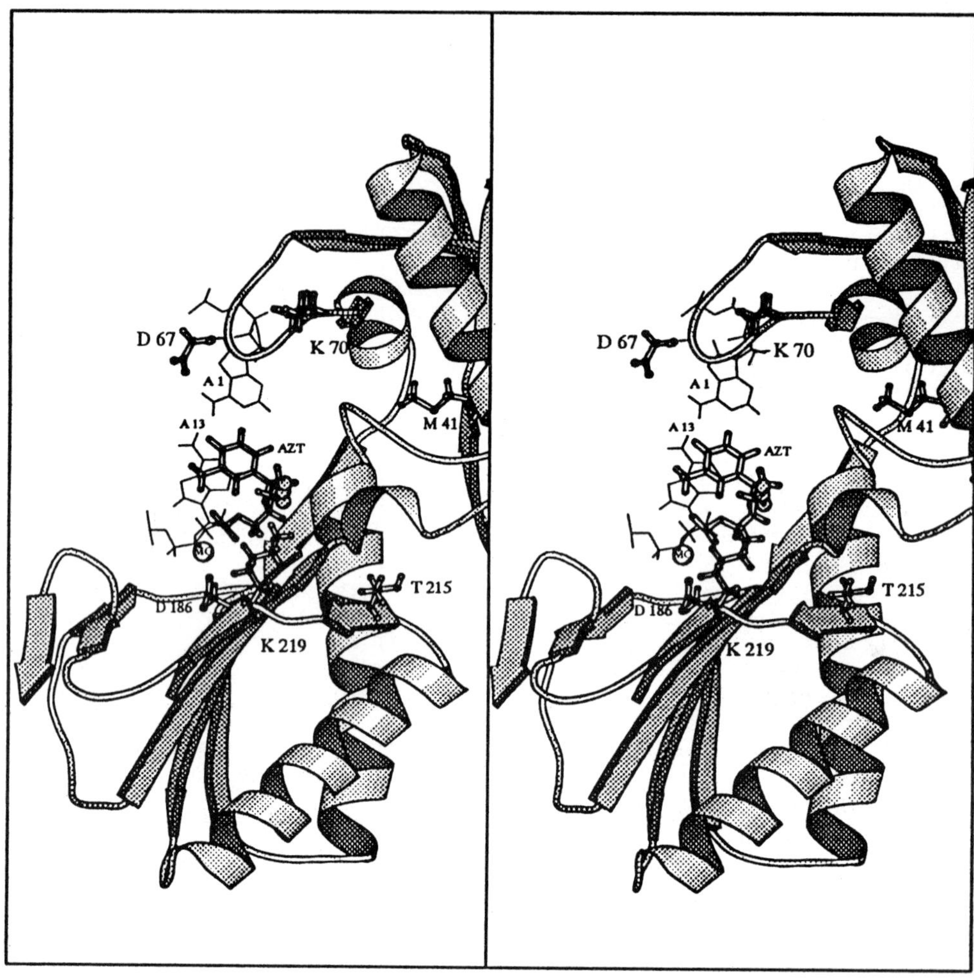

Fig. 1.   Positions of the five residues involved in AZT resistance. (a) Wild-type. (b) Full mutant. In this figure, AZT is substituted for T14 of the primer strand and the 3′ azido group is shown in CPK. Also shown are: A13 which is the residue at the 3′ end of the primer strand before the addition of AZT, Mg1 which is one of the metal ions involved in catalysis (see reference 3), one of the catalytic triad residues D 186, and the template base Al which pairs with T14 (or AZT as shown here). The location of the single-stranded template is shown in Figure 3.

of bulkier residues would have a pronounced effect on the larger substrate. This explanation of the role of residues 70 and 215 appears to be consistent with the observation that the drugs ddI and ddC (with two hydrogens bound to C3′) are effective inhibitors both of HIV isolates from patients who have developed AZT resistance and displayed a combination of the aforementioned mutations, and mutated RT's obtained through cloning studies [11]. No clear role for the mutation at position 41 is shown by our model. However, it is possible that this residue similarly adds steric bulk to the opening of the active site.

Fig. 1(b).

An alignment of twenty retroviral *pol* sequences and HIV-I RT in Setlik *et al.* [3] shows that lysine is well conserved at the positions homologous to residue 70. The substitution of an arginine for the lysine at this position is then reasonable in order to satisfy the apparent requirement for a positively charged residue at this site. In contrast, both alignments of Setlik *et al.* [3] and Poch *et al.* [12] show little, if any, conservation at the other mutation positions 41, 67, 215, and 219. This data suggests either the relative unimportance of these residues in the structural/functional properties of RT or structural differences amongst the various RT's.

The mutation at position 219 changes the residue from a positive and basic, to a neutral and polar residue. The Lys 219 side-chain, in our model, sits 6.5 Å from the azido group of AZT when incorporated onto the primer strand (Table I). As AZT enters the active site, assuming a path as described above (Figure 2), molecu-

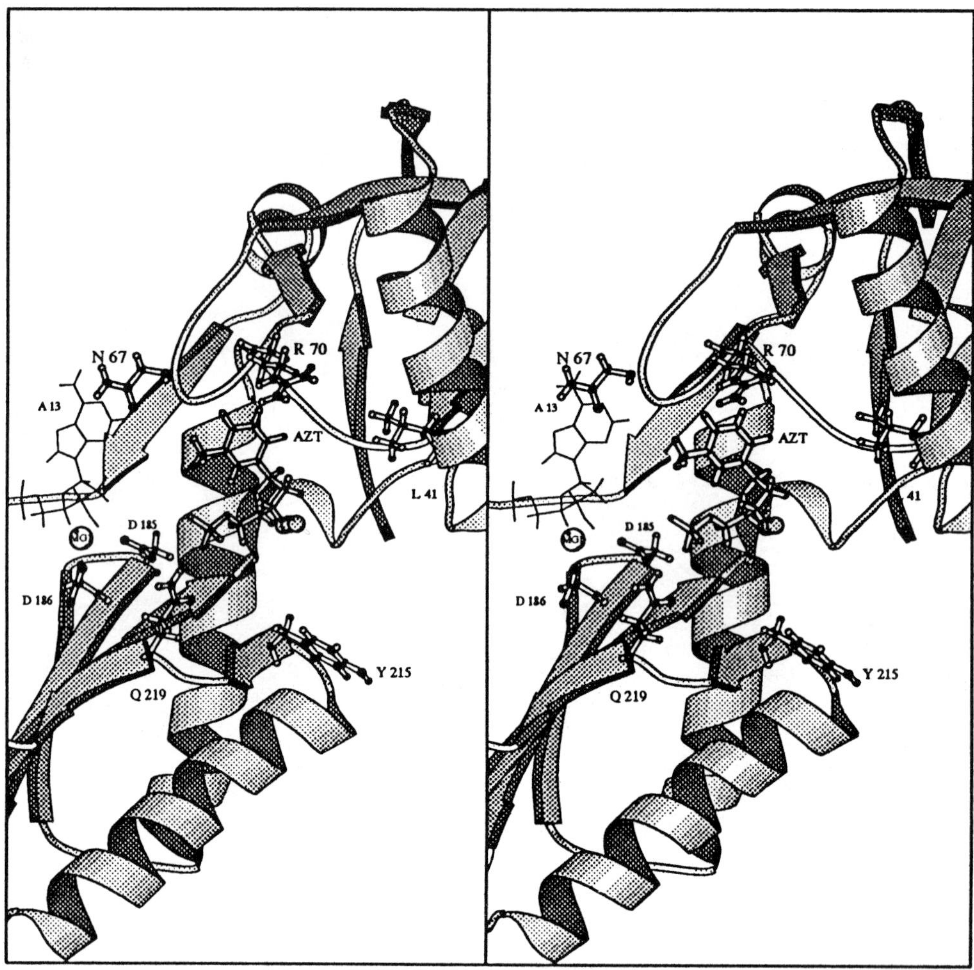

Fig. 2.   The orientation of AZT-TP entering the active site. (Only the α phosphate of the triphosphate group is shown.) Also shown is the azido group in CPK, Mg1, A13, and the side chains of D 185 and D 186 of the catalytic triad, and of the five AZT-resistance mutations.

lar modeling indicates that the azido group passes close to this residue. Since the azido group has a strong dipole moment due to the positive and negative formal charges on nitrogens 2 and 3, respectively ($-N_1{=}\overset{+}{N_2}{=}\overset{-}{N_3}$), the lysine ammonium group and the azido group would form a favorable monopole-dipole interaction. This interaction would be lost with the Lys $\rightarrow$ Gln mutation which, in turn, would reduce the affinity of RT for AZT.

A possible role for the Asp $\rightarrow$ Asn mutation at position 67 is not readily apparent, since it lies at the bend of the β3-β4 hairpin. The side chain of this residue is not oriented toward the active site channel, unlike the other AZT resistance residues, and does not seem able to interfere directly with the uptake

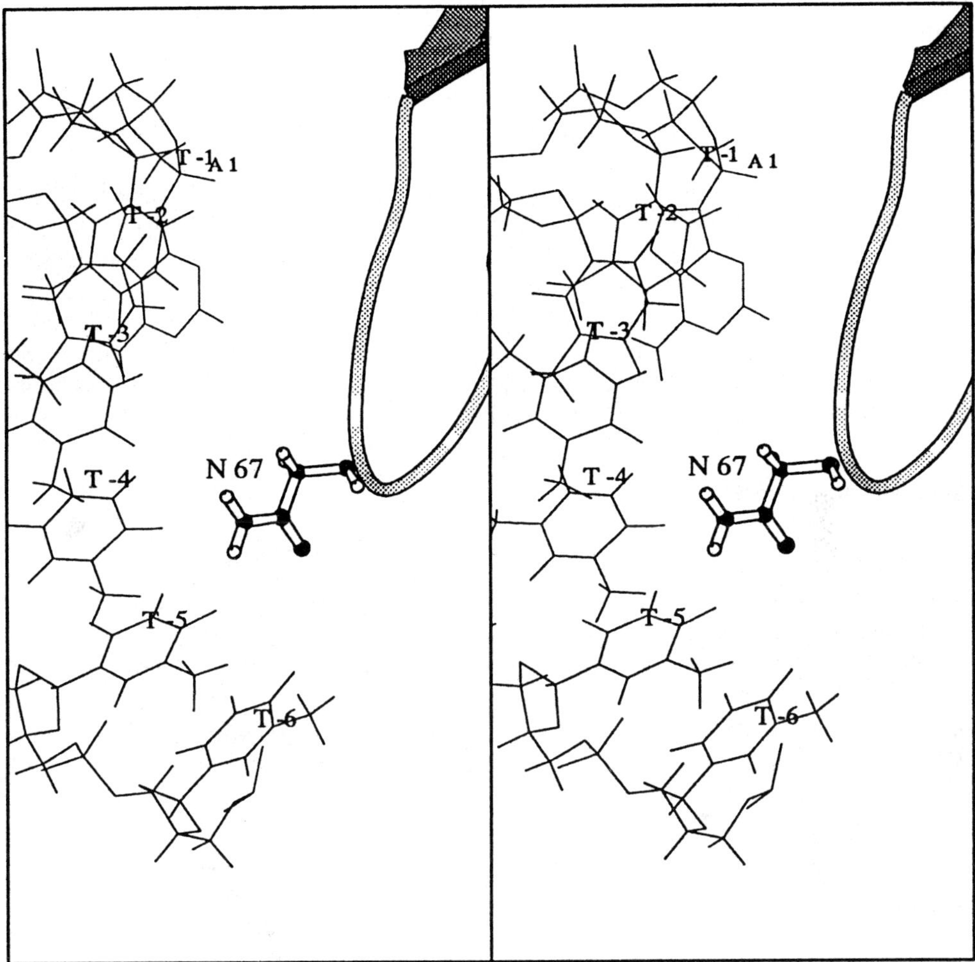

Fig. 3.   D 67 → N mutation and its proximity to the single-stranded template. Shown is the template base A1 and the single-stranded template residues T-1 to T-6. As shown in this figure and in Figure 1, residue 67 is oriented away from the opening to the active site, as opposed to residues 41, 70, 215, and 219. The proximity of this residue to the hydrogen bonding atoms of the single-stranded template bases T-5 and T-6 (see text) and the ability of this residue to hydrogen bond to any of the four bases suggests that this residue could attach the β3-β4 hairpin to the template strand. This could serve to constrain the β3-β4 hairpin into a conformation which narrows the opening to the active site, causing RT to selectively exclude AZT.

of substrate (Figures 1 and 3). However, the base pairing atoms of T-5 and T-6 of the single-stranded portion of the template are situated near to this residue. The distances between residues T-5, T-6 and Asn 67 are as follows: Oδ1–N3 (T-6) 6.7 Å, Oδ1–N3 (T-5) 6.1 Å, Nδ2–O2 (T-6) 8.6 Å, Nδ2–O2 (T-5) 6.5 Å, Nδ2–O4 (T-6) 7.1 Å, Nδ2–O4 (T-5) 6.2 Å. The distances of Asn 67 to an equivalently placed poly(G) template are roughly an angstrom less. It has been shown that an asparagine can form hydrogen bonds with any of the four bases [13]. Although these distances are out of the range for hydrogen bonding, the energetics of hydrogen bond formation and the proximity of this residue to the single-stranded template combined with the flexibility of both the single-stranded template and the β3-β4 hairpin could easily allow these two structures to reorient closer to each other, allowing hydrogen bonds to form between Asn 67 and either T-5 or T-6. This attachment could then constrain the β3–β4 hairpin into a conformation which narrows the opening to the active site and selectively excludes AZT from entering the polymerase active site. Figure 3 shows the Asp 67 → Asn mutation and its proximity to the single-stranded template.

Work is currently in progress to test these hypotheses through more detailed docking and modeling studies. We are also in the process of planning molecular dynamics simulations to observe the dynamic behavior of the loop regions mentioned above in both wild-type and mutant RT.

Acknowledgements

The authors would like to thank Drs. Holm and Sander for supplying the MAXSPROUT program, Dr. P. Kraulis for supplying the MOLSCRIPT program, and Dr. Edward Arnold for providing material regarding his work on reverse transcriptase. This work was supported by a grant from NASA (NAGW-1546), the Office of Health and Environmental Research of the U.S. Department of Energy grant KP0402 (RLO), the Laboratory Directed Research and Development Program of the Pacific Northwest Laboratory (RLO), and by the National Foundation of Cancer Research. The Pacific Northwest Laboratory is operated for the U.S. Department of Energy by the Battelle Memorial Institute under Contract DE-AC06-76RLO 1830.

References

1. H. Varmus: *Sci. Am.* **257**, 56 (1987).
2. E. Arnold and G. F. Arnold: *Adv. Virus. Res.* **39**, 1 (1991).
3. R. F. Setlik, D. J. Meyer, M. Shibata, R. Roskwitalski, R. L. Ornstein, and R. Rein: *J. Biomol. Str. Dyn.* **12**, 37–60 (1994).
4. A. Jacobo-Molina, J. Ding, R. G. Nanni, A. D. Clark, X. Lu, C. Tantillo, R. L. Williams, G. Kamer, A. L. Ferris, P. Clark, A. Hizi, S. H. Hughes, and E. Arnold: *P.N.A.S. USA* **90**, 6320 (1993).
5. L. Holm and C. Sander: *J. Mol. Biol.* **218**, 183 (1991).
6. B. A. Larder and S. D. Kemp: *Science* **246**, 1155 (1989).
7. P. Kellam, C. A. B. Boucher, and B. A. Larder: *P.N.A.S. USA* **89**, 1934 (1992).
8. R. Parthasarathy and H. Kim: *Biochem. Biophys. Res. Commun.* **152**, 351 (1988).
9. P. J. Kraulis: *J. Appl. Cryst.* **24**, 946 (1991).
10. L. A. Kohlstadt, J. Wang, J. M. Friedman, P. A. Rice, and T. A. Steitz: *Science* **256**, 1783 (1992).
11. M. H. St. Clari, J. L. Martin, G. Tudor-Williams, M. C. Bach, C. L. Vavro, D. M. King, P. Kellam, S. D. Kemp, and B. A. Larder: *Science* **253**, 1557 (1991).
12. O. Poch, I. Sauvaget, M. Delarue, and N. Tordo: *EMBO J.* **8**, 3867 (1989).
13. W. Saenger: *Principles of Nucleic Acid Structure*, C. R. Cantor (Ed.), Springer-Verlag, New York (1984).

Fold Recognition

MANFRED J. SIPPL, SABINE WEITCKUS, and HANNES FLÖCKNER
Centre for Applied Molecular Engineering, Institute for Chemistry and Biochemistry, University of Salzburg, Salzburger TechnoZ, Jakob Haringer Straße 1, A-5020 Salzburg, Austria

Abstract. Proteins frequently adopt similar three-dimensional folds even if there is no detectable homology on the sequence level. Therefore, it is likely that a substantial fraction of known sequences will have native folds which resemble some known structure. The task of fold recognition is to identify such coincidences, thereby predicting the unknown folds of known amino acid sequences. We present the principles of a fold recognition technique based on mean force potentials and apply the procedure in two case studies, demonstrating that the method is able to recognize native-like sequence structure pairs in cases where no significant sequence homology is detectable.

Key words. Protein folding, protein modeling, prediction of protein structures, sequence structure alignment, molecular force field.

1. Introduction

The recognition of errors in protein folds is a necessary step on the way to a solution of the protein folding problem. It is very unlikely that protein structures can be successfully predicted if the methods employed are unable to discover errors or if they can be fooled by faulty structures [1,2]. In recent years several groups have concentrated their efforts on the development of reasonable energy models for protein solvent systems (for reviews see for example [3–6]) resulting in computational tools which can be applied to real problems in protein structure theory. The new algorithms and programs can be used to detect incorrect protein folds and they are able to spot faulty parts in experimentally determined conformations or modeled structures [2,7].

The prediction of structures solely from the information contained in amino acid sequences is of course more demanding. Even if a reasonable model of protein solvent systems is available there is still the problem of locating low energy structures in conformation space. With the accelerated rate of experimental structure determinations it became clear that a new structure often resembles some known fold, even if there is no detectable homology on the sequence level. Therefore, many of the proteins whose sequences are stored in the rapidly growing data bases will have folds which are to some extent similar to a fold already known from experiment.

These observations immediately suggest a strategy which, at least in principle, has the potential to discover the unknown folds for a subset of proteins of known sequence. By combining sequences and structures it should be possible to recognize native-like sequence-structure pairs or, in other words, to determine whether or not a sequence is compatible with a given fold [8–10].

Over the last few years several groups have started to develop fold recognition and threading techniques. The first results obtained were already quite encouraging, but some of the initial excitement was damped when it became clear that

A. Pullman et al. (eds.), *Modelling of Biomolecular Structures and Mechanisms*, 107–118.
© 1995 *Kluwer Academic Publishers. Printed in the Netherlands.*

many of the results obtained by fold recognition can be obtained by conventional sequence alignment techniques and that none of the methods had successfully passed a blind test, where a structure was predicted ahead of experiment and later confirmed to be correct.

In fold recognition one has to deal with several theoretical and practical problems, which perhaps are not obvious at first sight. The aim of this paper is to summarize some of these problems, to describe an algorithm which circumvents at least one important obstacle and to describe the results obtained in two case studies.

2. A Recipe for Fold Recognition

In fold recognition we have to deal with two basic problems. First we have to align (or thread) a sequence to a given fold and second, we have to determine whether or not the aligned sequence-structure pair is native-like (i.e. whether or not the sequence adopts this fold).

To tackle the first problem–aligning a sequence to a structure–we have to specify what we mean by the fitness or quality of an alignment and we have to use algorithms which are able to find optimal alignments or at least alignments, which are close to the optimal solution. In addition the algorithms have to be computationally efficient since the goal is to align one sequence to a large database of structures or an even larger data base of sequences to one structure.

The current structure data bases hold in the order of 1000 structures. The sequence data bases storing 30000 sequences are considerably more complex. To align one sequence to all available structures, therefore, requires the calculation of 1000 alignments. The second variant, determination of all sequences which are compatible with a given fold, is even more demanding.

In summary, a recipe for the alignment of sequences to structures requires four basic ingredients: (1) specify what is meant by the fitness of a given sequence structure alignment; (2) find optimal alignments using this criterion; (3) determine whether or not the alignment obtained corresponds to a native-like sequence–structure pair; and (4) try to be as computationally efficient as possible.

3. Implementation of Fold Recognition Techniques Based on Mean Force Potentials

As already indicated fold recognition can be applied in two different ways. The first variant tries to identify a fold (or several closely related folds) which is compatible with a given sequence, thereby predicting the unknown structure of this protein. The second variant tries to identify all known sequences which adopt a given fold, predicting the unknown folds of one or several proteins, all having a similar architecture. In the present study we apply fold recognition in both ways and discuss the results obtained but first we give a brief summary of the techniques employed (e.g. [11]).

The most critical part of fold recognition techniques is the fitness function employed. To be successful in fold recognition this function has to capture the most important characteristics of protein solvent systems. Several parameter sets are currently in use, such as secondary structure preferences, surface accessibilities, knowledge based potentials, etc. The implementation investigated in the present study is based on mean force potentials extracted from a set of known protein structures [12,13].

The set of known protein structures contains an enormous amount of information on protein solvent systems and on the relationship between amino acid sequences and their native folds. The forces which stabilize these folds can be extracted from the data base using basic principles of statistical mechanics. The result is a set of potentials of mean force describing the interactions between protein atoms in an average protein solvent environment [5]. As has been demonstrated in several cases, force fields constructed in this way have considerable predictive power (e.g. [14,15]), and they can be applied to spot and correct errors in protein structures [2].

Force fields of this type should be applicable in fold recognition problems, and one of the goals during recent last years was to develop threading techniques based on these potentials, where the mean force energy acts as a criterion for optimal alignments.

When a sequence is aligned to a structure the fold acts as a framework and the sequence has to be arranged along this scaffold in some optimal or reasonable way. When using pair interactions one of the technical difficulties is that the interaction energy of a particular residue in a given fold cannot be evaluated as long as the locations of other residues are unknown (e.g. [16]). The trick we employ here is to use the field produced by the original sequence belonging to the fold. To evaluate the fitness of residue A_i of sequence A at position B_j of conformation B, the original amino acid at B_j is replaced by A_i and its interaction energy e_{ij} with all other residues in conformation B is calculated. The process is repeated for all sequence positions i and structure positions j, yielding a matrix whose elements describe the fitness of the amino acid residues of A with respect to all possible positions in B.

The resulting energy matrix is used as input to conventional dynamic programming algorithms [17]. The output is an optimal path through the matrix corresponding to an optimal alignment of sequence A and structure B. However, one has to be fully aware of the approximation used here. The alignment of sequence A to conformation B is obtained by assuming that the force field produced by the sequence of B is a good approximation to the field produced by sequence A. This assumption may be a good one if the proteins A and B have similar energetic architecture, but it may fail if the conformations of A and B, although similar in three dimensions, are stabilized by different mechanisms.

The main advantage is that we are able to construct efficient algorithms. Using this assumption the time complexity is of the order N^2 (N average sequence length of A and B), i.e. it is quadratic in sequence length. In contrast, the complexity of the original problem is at least N^4. The resulting algorithm is fast enough to align all sequences of current sequence data bases like swissprot to one structure ($\approx 30\,000$ alignments) within a few hours of CPU time on typical workstations (e.g. Silicon Graphics Indy).

The alignment obtained corresponds to a first model for the unknown fold of the respective amino acid sequence. In general the algorithm opens gaps in the sequence and/or structure and the preliminary model consists only of those positions in the structure which have been mapped to sequence positions in the alignment process. In most cases the model is incomplete since parts of the sequence have been deleted in generating the alignment.

In fold recognition we implicitly assume that the unknown fold of the sequence is exactly the same as the structure we use in the alignment. The structure acts as a rigid template and there is no room for structural variations. But even if the unknown fold is similar in architecture to the fold used in the alignment the two structures will differ to some extent. Hence the result produced by our algorithm is incomplete and unrefined.

Alignments can be produced for any sequence structure pair. Most of the alignments obtained, either by aligning a set of sequences to one structure or, as in the second variant, by combining one sequence with a data base of structures, will be non-native. The most important question is whether or not a given alignment corresponds to a native-like sequence structure pair, or in other words, whether or not a particular model obtained is similar to the native structure of the sequence used in the alignment.

To answer this question the total mean force energy is of little significance as pointed out previously [5]. It is necessary to compare the energy of a particular model to the energy distribution of the associated sequence in conformation space. The mean $\bar{E}$ and standard deviation σ obtained this way are then used to transform the energy E of the model to a z-score: $z = (E - \bar{E})/\sigma$. The range of native-like z-scores is known from previous studies (e.g. [2]) so that this parameter can be used to judge the quality of the method. This final step requires some computational effort since the energies for some 10^5 conformations have to be calculated in order to estimate $\bar{E}$ and σ. Depending on the size of a particular model the calculation of these parameters requires several seconds CPU time, or a total of 10 to 20 hours CPU time in case of a complete sequence data base search (i.e. $\approx 30\,000$ models).

In summary the alignment technique used in this study consists of the following parts: (1) generate a substitution matrix e_{ji} by combining all sequence positions with all structure positions; (2) find an optimal path through this matrix and hence an optimal alignment by dynamic programming; (3) generate a model by clipping all positions in the sequence and the structure which were not aligned; (4) calculate the z-score for the model.

4. Case Studies

We present two cases studies. In the first example we align a sequence to a data base of structures, in the second we align all sequences in the swissprot data base to a given structure. In both cases the main goal is to investigate the ability of our fold recognition technique to recognize distant folds and to get some idea about the limitations and difficulties encountered.

In our first example we investigate whether or not the procedure is able to identify distantly related folds in cases where there is no recognizable homology on the sequence level. We choose nitrogenase molybdenum-iron protein from *Azotobacter vinelandii*. The structure of this protein was determined by Kim and

Rees [18] to a resolution of 2.2 Å (Brookhaven code 1min). The molecule is a tetramer. Chains *a* and *c*, and chains *b* and *d* are identical. The sequence homology between *a* and *b* (and *c* and *d*) is only 17%, but they have similar architecture. 391 residues (of 522) of the *a* and *b* chains can be superimposed to an rms-error of 3 Å (Figure 1).

The question is whether or not our fold recognition technique is able to recognize that the sequence of subunit *a* (1min-a) adopts a fold which is similar to subunit *b* (1min-b). The 1min molecule was not included in the data base used to derive the potentials. This ensures that the results produced are not biased towards the prior knowledge of this structure. In other words the test corresponds to a realistic application.

In order to simulate a typical fold recognition study the sequence of 1min-a is aligned to 238 known structures. The set of structures contains 1min-a, 1min-b and 1mio-c, the *c*-chain of a nitrogenase molybdenum iron protein from *Clostridium pasteurianum*, having 39% sequence homology (and structural homology) to 1min-a. The result is a set of 238 structurual models for the 1min-a sequence obtained from aligning this sequence with all 238 structures. The z-score for each model is then calculated as reported previously [15]. Table I assembles the results. The models are sorted by their z-score. The most significant score (-13.42) is obtained for the 1min-a structure, i.e. the combination of the 1min-a sequence with its native fold yields the most significant score.

The combination of 1min-a (sequence) and 1mio-c (structure) yields the second best score of -10.59. The result is correct, but the relationship between 1min-a and 1mio-c can be inferred from conventional sequence alignment techniques due the significant homology of -40% on the sequence level.

The 1min-a-sequence when aligned to the 1min-c structure yields a score of -9.73 occuring on third position (Table I). We know from previous studies [2,14,15] that scores of this magnitude point to native-like sequence structure combinations. The subsequent entries in Table I have less significant scores. Scores in the range of -8 to -6 indicate that the corresponding models have some similarity to the native fold of 1min-a (e.g. secondary structure) but they also indicate that in general there are significant differences. We do not intend to analyse these models in detail here, since this task is beyond the scope of this presentation. The main point is that the fold recognition technique is indeed able to recognize 1min-b as an approximate model for the 1min-a sequence, although there is no significant homology on the sequence level.

In our second case study we use a member of the GTPase family, H-RAS p21 protein 5p21 [19], residues 1–166, as shown in Figure 2. The goal is to align all sequences in the current release of the swissprot data base to this structure and to investigate whether or not protein sequences known to belong to the GTPase superfamily are recognized using this fold.

The approximately 27800 sequences were aligned to 5p21 and the scores for the resulting models were calculated. Table II assembles the results for the top scoring sequences. The most significant scores are obtained for sequences strongly homologous or identical to 5p21, followed by sequences whose homology to 5p21 is in the range of 60–30%. The scores are smaller in absolute value, as compared to the nitrogenase results, since z-scores of native-like sequence structure pairs

1MIN-A

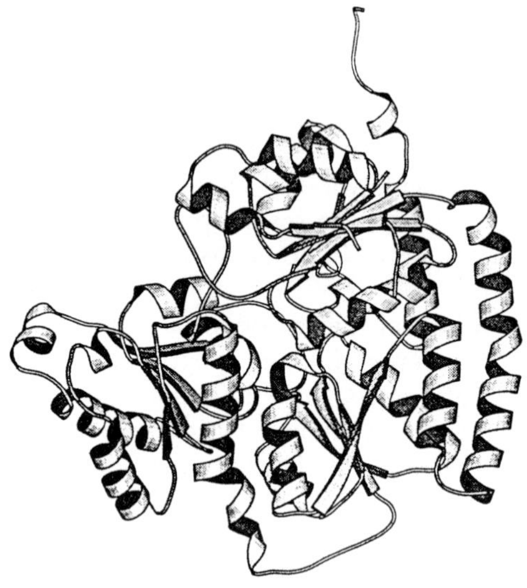

1MIN-B

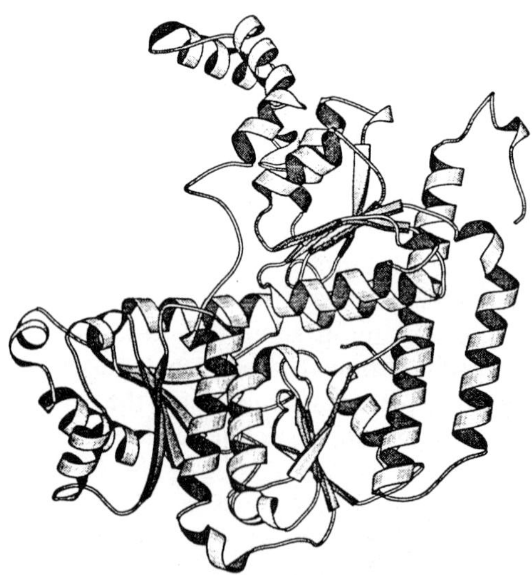

Fig. 1.   Chains *a* and *b* of nitrogenase molybdenum-iron protein from *Azotobacter vinelandii*. The three-dimensional folds of these subunits are similar. Roughly two thirds of the structures can be superimposed to an rms error of 3 Å. The homology on the sequence level is only 17%.

TABLE I
1min-a sequence aligned to 238 structures

structure	z-score	%ID[a]	%ID[b]	length[c]
1min-a	−13.42	100	100	468
1mio-c	−10.59	34	39	472
1min-b	−9.73	12	17	481
3tgl	−7.53	7	13	265
1pii	−6.89	9	13	452
1ldm	−6.86	9	12	329
1fha	−6.85	9	13	170
1mrr-b	−6.70	7	12	325
9rub-b	−6.69	8	15	431
2pol-a	−6.55	9	14	361
8cat-b	−6.40	8	13	449
1rve-a	−6.39	7	16	244
1lh1	−6.29	7	18	153
2had	−6.24	8	16	310
1sdh-a	−6.21	10	14	146
1sas	−6.20	6	16	185
1abm-a	−6.07	7	9	198
2pmg-a	−6.05	10	15	417
1nip-a	−6.02	9	17	283
1gd1-p	−5.94	8	16	302

[a]Number of identical residues obtained from sequence-sequence alignment.
[b]Number of identical residues obtained from sequence-sequence alignment.
[c]Number of aligned residues.

5P21

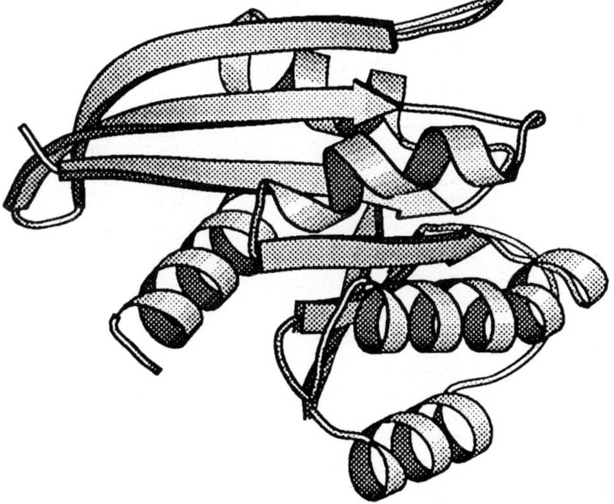

Fig. 2. Three-dimensional structure of H-RAS p21.

TABLE II

rank	swissprot code	z-score	%ID[a]	f[b]	h[c]		r[d]
1	RASL_HUMAN	−10.49	94.0	*	*		r
2	RASK_HUMAN	−10.44	94.0	*	*		r
3	RASK_MSVKI	−10.44	91.0	*	*		r
4	RASL_MOUSE	−10.42	93.4	*	*		r
5	RAS_CARAU	−10.41	93.4	*	*		r
6	RASN_CAVPO	−10.38	91.6	*	*		r
7	RASN_HUMAN	−10.38	91.6	*	*		r
8	RASN_MOUSE	−10.38	91.6	*	*		r
9	RASK_MOUSE	−10.37	93.4	*	*		r
10	RASH_CHICK	−10.13	98.8	*	*		r
11	RASH_MSVHA	−10.11	98.2	*	*		r
12	RASH_RRASV	−9.68	98.2	*	*		r
13	LT60_CAEEL	−9.66	81.9	*	*		r
14	RASH_MSV	−9.66	99.4	*	*		r
15	RASH_MSVNS	−9.66	99.4	*	*		r
16	RASH_HUMAN	−9.64	100.0	*	*		r
17	RASH_RAT	−9.64	100.0	*	*		r
18	RASI_DROME	−9.53	85.5	*	*		r
19	ARAS_ARATH	−8.90	29.5	*			r
20	RABI_DISOM	−8.86	30.1	*	*		r
21	RABBRAT	−8.83	28.9	*	*		r
22	YPTI_MAIZE	−8.71	30.1	*	*		r
23	RABI_HUMAN	−8.68	29.5	*	*		r
24	SEC4_YEAST	−8.66	31.3	*	*		r
25	RAS3_RHIRA	−8.62	61.4	*	*		r
26	RAS2_DICDI	−8.54	70.5	*	*		r
27	RAS_LENED	−8.51	62.7	*			r
28	RAB_1RAT	−8.47	29.5	*	*		r
29	RAP2_HUMAN	−8.41	44.6	*	*		r
30	RAS_SCHPO	−8.37	65.1	*	*		r
31	RAS1_RHIRA	−8.29	61.4	*	*		r
32	RAS2_DROME	−8.22	56.6	*	*		r
33	RASL_NEUCR	−8.20	66.3	*	*		r
34	YPTI_SCHPO	−8.20	29.5	*	*		r
35	RAPI_DICDI	−8.15	52.4	*	*		r
36	RRAS_HUMAN	−8.13	56.0	*	*		r
37	IF2_BACST	−8.11	13.3			INIT.FACT.	r
38	RAB7_CANFA	−8.08	31.3	*	*		r
39	IF2_ENTFC	−8.04	13.9			INIT.FACT	r
40	RAL_CALJA	−8.03	49.4	*	*		r
41	P23_RAT	−7.97	31.3	*	*		r
42	RALA_HUMAN	−7.97	49.4	*	*		r
43	RAP3_HUMAN	−7.97	43.4	*	*		r
44	YPTI_YEAST	−7.90	31.3	*	*		r
45	RASI_YEAST	−7.84	58.4	*	*		r
46	RRAS_MOUSE	−7.74	56.0	*	*		r
47	RAS2_YEAST	−7.73	59.6	*	*		r
48	ARA3_ARATH	−7.71	31.3	*			r
49	RTCI_HUMAN	−7.71	56.0	*	*		r
50	ARF2_BOVIN	−7.69	13.3			ADP−RIBOSYL FACT	r
51	RABX_CANFA	−7.68	31.3	*	*		r
52	RALB_HUMAN	−7.65	49.4	*[b]	*[c]		r[d]
53	RABB_HUMAN	−7.63	27.7	*	*		r

TABLE II. Continued

rank	swissprot code	z-score	%ID[a]	f[b]	h[c]		r[d]
54	RABA_DROME	−7.62	25.9	*	*		r
55	ARF2_YEAST	−7.59	13.9			ADP−RIBOSYL. FACT	r
56	P25B_BOVIN	−7.59	27.1	*	*		r
57	ARF3_HUMAN	−7.56	13.3			ADP−RIBOSYL. FACT	r
58	ARA4_ARATH	−7.53	28.3	*			r
59	YPT2_SCHPO	−7.53	28.9	*	*		r
60	P25A_BOVIN	−7.52	25.9	*	*		r
61	ARF5_HUMAN	−7.50	13.9			ADP−RIBOSYL.FACT	r
62	RASI_DICDI	−7.49	68.1	*	*		r
63	RAS3_DROME	−7.48	56.0	*	*		r
64	RAS_ARTSA	−7.48	77.1	*	*		r
65	ARFI_YEAST	−7.47	13.9			ADP−RIBOSYL.FACT	r
66	RAPA_DISOM	−7.47	53.6	*	*		r
67	SAS2_DICDI	−7.40	29.5	*	*		r
68	ARF_HUMAN	−7.38	13.9			ADP−RIBOSYL.FACT	r
69	RAN_CANFA	−7.38	19.9	*			r
70	RAB4_CANFA	−7.37	27.7	*			r
71	IF2_ECOLI	−7.35	12.7			INIT.FACTOR	r
72	RAB3_RAT	−7.35	25.9	*	*		r
73	RABA_HUMAN	−7.35	25.9	*	*		r
74	RAL_DISOM	−7.34	48.8	*	*		r
75	RAPB_HUMAN	−7.32	54.2	*	*		r
76	RAN_HUMAN	−7.31	19.3	*			r
77	SAS1_DICDI	−7.31	29.5	*	*		r
78	RAB9_CANFA	−7.29	24.7	*			r
79	ARF_CANAL	−7.28	13.3			ADP-RIBOSYL. FACT	r
80	RAS_GEOCY	−7.28	15.1	*	*		r
81	ARA1_ARATH	−7.25	27.1	*			r
82	ERA_ECOLI	−7.23	15.7			GTP−BIND.ERA.PROT	r
83	ARF_GIALA	−7.20	15.7			ADP−RIBOSYL.FACT	r
84	EXOY_RHISN	−7.19	7.2			POLYSACCH. PRODUCT.PR	r
85	RABY_HUMAN	−7.15	31.3	*	*		r
86	RAPA_HUMAN	−7.15	53.6	*	*		r
87	SPI_SCHPO	−7.14	20.5	*			r
88	IF2_BACSU	−7.13	11.4			INIT.FACT	r
89	RAB6_HUMAN	−7.08	24.1	*	*		r
90	RA03DISOM	−7.01	29.5	*	*		r
91	P25C_BOVIN	−6.96	26.5	*	*		r
92	RHAB_SALTY	−6.92	6.6			RHAMNULO- KINASE	r
93	SOPA_ECOLI	−6.92	5.4			SOPA.PROTEIN	r
94	DYN_DROME	−6.91	12.7			DYNAMIN	r
95	SFUB_SERMA	−6.86	4.8			IRON−TRANSP.PROT	r
96	ARA2_ARATH	−6.83	27.1	*			r
97	RAB8_CANFA	−6.83	31.9	*	*		r
98	3BHS_MACMU	−6.82	8.4			STER.DE- HYDROGENASE	r
99	RAB2_HUMAN	−6.82	30.1	*	*		r
100	OBG_BACSU	−6.78	14.5			SPOOB−ASSOC.PROT	r
101	3BHI_HUMAN	−6.77	8.4			STER.DE- HYDROGENASE	r
102	ARF6_CHICK	−6.77	14.5			ADP−RIBOSYL.FACT	r

TABLE II. Continued

rank	swissprot code	z-score	%ID[a]	f[b]	h[c]		r[d]
103	RA02_DISOM	−6.77	33.7	*	*		r
104	RYHL_SCHPO	−6.77	23.5	*	*		r
105	ARF6_HUMAN	−6.76	13.9			ADP−RIBOSYL.FACT	r
109	RAB2_RAT	−6.74	30.1	*	*		r
110	RAB4_RAT	−6.70	27.7	*			r
115	RAB4_HUMAN	−6.64	27.7	*			r
120	RAO1_DISOM	−6.61	31.3	*	*		r
130	RGPL_ORYSA	−6.51	27.7	*	*		r
133	RABS_HUMAN	−6.50	25.9	*	*		r
134	RH04_YEAST	−6.50	21.1	*			r
139	RSRL_YEAST	−6.47	50.0	*	*		r
155	RABS_CANFA	−6.36	25.9	*	*		r
161	YPT3_SCHPO	−6.33	25.9	*	*		r
169	RAS2_RHIRA	−6.25	48.8	*			r
273	YCR7_YEAST	−5.90	26.5	*	*		r
495	RAM_RAT	−5.58	24.7	*	*		r
622	RTCO_HUMAN	−5.48	25.9	*	*		r
716	RHOB_HUMAN	−5.40	24.1	*	*		r
732	RHO1_CANFA	−5.39	24.1	*	*		r
752	RHO_DISOM	−5.37	24.1	*	*		r
978	G25B_HUMAN	−5.20	28.3	*			r
1002	G25P_HUMAN	−5.19	27.7	*			r
1010	RAC2_HUMAN	−5.19	28.9	*	*		r
1028	RAC1_HUMAN	−5.18	27.1	*	*		r
1731	RHOL_YEAST	−4.86	25.9	*	*		r
1732	RH02_YEAST	−4.86	24.1	*	*		r
1756	RHOA_HUMAN	−4.85	27.1	*	*		r
1971	RHOC_HUMAN	−4.77	25.9	*	*		r
1993	RHO_APLCA	−4.76	26.5	*	*		r
2258	CC42_YEAST	−4.66	27.7	*	*		r
2392	RASK_RAT	−4.62	56.0	*	*		r
3866	RH03_YEAST	−4.04	22.3	*			r

[a]Number of identical residues derived from sequence structure alignment.
[b]Relationship Detected by fasta (else tblank).
[c]Relationship Detected by hssp (else blank).
[d]Sequence has been Reported to belong to the GTPase family (else blank).

increase (in absolute value) with the size of the protein. On position 37 (Table II)
we find the first sequence, IF2_BACST, an initiation factor for protein synthesis,
having a significant score of −8,11 but no significant sequence homology to 5p21
(13%). However, it is known that this type of initiation factor is related to the
GTPase family [20]. On position 39 (Table II) we find a similar initiation factor
IF2_ENTFC.

The following entries are either RAS-related proteins or proteins which are
known to be members of the GTPase family, like ADP-ribosylation factors [20]
and GTP-binding ERA protein from *E. coli* [21]. There are no false positives.
Exopolysaccharide production protein EXOY, EXOY_RHISN, position 84 in
Table II, (score −7.19) is the first protein sequence we encounter, which is not
related (or at least not known to be related) to the GTPase family. The score of

−7.19 is still in a range of native-like sequence structure combinations. There are several other protein sequences having similar scores which have not been reported to be GTPase related, like rhamnulokinase, sopa protein, or dynamin (Table II).

5. Discussion

We have applied our current fold recognition technique in two realistic test cases. In the first example the technique is able to recognize the relationship between the nitrogenase molybdenum protein *a* and *b* chains. These two proteins have no detectable sequence homology. In the second test we identified sequences related to the GTPase family in a large sequence data base. Again the method was able to detect relationships in the absence of significant sequence homology, as revealed by the significant scores obtained for initiation factors and ADP-ribosylation factors.

Perhaps the most interesting question is whether or not the score of −7.19 obtained for the exopolysaccharide production protein sequence when combined with 5p21 can be interpreted as a prediction for the unknown structure of this protein. We know that scores in this range point to a native like sequence structure pair, but there is an additional complication. The results obtained from the fold recognition technique used in this study are preliminary and correspond to incomplete models.

In general, parts of the sequence are deleted in order to optimally align a sequence to a structure. But if we change sequences we are unable to predict the consequences. It could be that the changed sequence actually adopts this (or a closely related) fold, but the complete sequence may fold to a different structure [11]. We have to be aware of such effects when interpreting the results of fold recognition studies.

To derive a complete model additional steps are required. Deleted parts of the sequence have to be incorporated and the structural framework has to be varied in order to find the most relaxed conformation, since the structure used as a scaffold and the sequence's genuine native fold are generally not identical. If it is possible to derive a complete and refined structure having a native-like score from the preliminary model then we have a more reliable basis for correct predictions.

In view of these restrictions it is remarkable that in both test cases we do not observe false positives having native-like scores. In fact the results obtained for the nitrogenase sequence is quite clear. The same is true for the sequences yielding significant scores when combined with 5p21. There is of course a twilight zone where no clear distinction is possible between native-like and non-native like scores (−7 to −5 for the models derived from 5p21). Models in this range can be very similar to the genuine native fold but they also can be significantly distinct. Subsequent model refinement could resolve many of the ambiguities in this range.

There are several RAS-related sequences which give rather poor scores, when combined with 5p21. In such cases the alignment is (at least in part) incorrect. In addition some sequences which are known to be GTPase related, like elongation factors, are not recognized. Part of the problem is, that the alignment algorithm has a few parameters like maximum gap lengths. Such parameters help to suppress artefacts. In the present study the maximum gap length was set to 10 residues.

However, proper alignments of elongation factor sequences to 5p21 require longer gaps.

The performance of the alignment technique strongly depends on the predictive power of the mean force potentials employed. The quality of these potentials improves steadily due to the growing number of available protein structures and by advances and improvememts in the techniques used to extract mean force potentials from structure data bases. In addition there is room for improvements in the alignment techniques employed. In summary we conclude that even at the present stage of development the fold recognition technique discussed in this study can be used to recognize and predict native structures. In fact in a recent blind prediction the method was able to successfully identify the unknown structure of the catalytic core of xylanase as a TIM-barrel fold.

Acknowledgement

This work was supported by the Fonds zur Förderung der wissenschaftlichen Forschung, grant P09661-MOB. Figures 1 and 2 were generated using Molscript.

References

1. J. Novotny, R. Bruccoleri and M. Karplus: *J. Mol. Biol.* **177**, 787 (1984)
2. M. J. Sippl: *Proteins* **7**, 355 (1993a).
3. S. J. Wodak and M. J. Rooman: *Curr. Opp. Struct. Biol.* **3**, 247 (1993)
4. J. U. Bowie and D. Eisenberg: *Curr. Opp. Struct. Biol.* **3**, 437 (1993)
5. M. J. Sippl: *J. Comput. Aided. Mol. Design* **473**, (1993b)
6. J. S. Fetrow and S. H. Bryant: *Biotechnology* **11**, 479 (1993)
7. R. Luthy, J. U. Bowie and D. Eisenberg: *Nature* **356**, 83 (1992).
8. K. E. Drexier: *Proc. Natl. Acad. Sci. USA* **78**, 5275 (1981).
9. J. W.Ponder and F. M. Richards: *J. Mol. Biol.* **193**, 775 (1987).
10. J. U. Bowie, R. Liithy and D. Eisenberg: *Science* **253**, 164 (1991).
11. M. J. Sippl, H. Flockner and S. Weitckus: 'In search of protein folds', in K. Merz and S. LeGrand, Eds.*The Protein Folding Problem and Tertiary Structure Prediction*, Birkhäuser, Boston, in press (1994a).
12. M. J. Sippl: *J. Mol. Biol.* **213**, 859 (1990).
13. M. Hendlich, P. Lackner, S. Weitckus, H. Flöckner, R. Froschauer, K. Gottsbacher, G. Casari, and M. J. Sippl: *J. Mol. Biol.* **216**, 167 (1990).
14. M. J. Sippl, M. Jaritz, M. Hendlich, M. Ortner and P. Lackner: 'Applications of knowledge based mean fields in the determination of protein structures', in S. Doniach, Eds., *Statistical Mechanics, Protein Structure and Protein-substrate Interactions*, Plenum Publishers, in press (1994b).
15. M. J. Sippl and M. Jaritz: 'Predictive power of mean force pair potentials', in H. Bohr and S. Brunak, Eds., *Protein Structure by Distance Analysis*, IOS Press, Amsterdam, pp. 113–134 (1994).
16. D. T. Jones, W. R. Taylor, and J. M. Thornton: *Nature* **358**, 86 (1992).
17. S. B. Needleman and C. D. Wunsch: *J. Mol. Biol.* **48**, 443 (1970).
18. J. Kim and D. C. Rees: *Science* **257**, 1677 (1992a).
19. I. Schlichting, S. C. Almo, G. Rapp, K. Wilson, K. Petratos, A. Lentfer, A. Wittinghofer, W. Kabsch, E .F. Pai, G. A. Petsko, and R. S. Goody: *Nature* **345,** 309 (1990).
20. H. R. Bourne, D. A. Sanders, and F. McCormick: *Nature* **349**, 117 (1991).
21. J. Ahnn, P. E. March, H. E. Takiff, and M. Inouye: *Proc. Natl. Acad. Sci. USA* **83**, 8849 ((1986).

Modelling the Interactions of Protein Side-Chains

JOHN B. O. MITCHELL and JANET M. THORNTON
Biomolecular Structure and Modelling Unit, Department of Biochemistry and Molecular Biology, University College London, Gower St., London WC1E 6BT, U.K.

SARAH L. PRICE
Department of Chemistry, University College London, 20 Gordon St., London WC1H 0AJ, U.K.

Abstract. The study of small model molecules containing the relevant functional groups can help us to understand the interactions between side-chains in proteins. *Ab initio* quantum chemical techniques allow the interactions between the model molecules to be studied with much greater accuracy than is possible for an entire protein, where the use of simple empirical potentials is the norm. In particular, the use of *ab initio* methods on model molecules permits us to incorporate the atom-atom anisotropic directionality of these interactions. We survey various methods of obtaining the components of the *ab initio* interaction energy. These are then applied to three systems of biological interest. The first of these is the arginine/aspartate pair found in salt bridges, which involves hydrogen bonding between two charged species. Secondly, we look at the arginine/phosphotyrosine interaction found in complexes between SH2 domains and peptide ligands: here we find that the arginine/phosphate part of the interaction is energetically far more important than the arginine/aromatic part. Finally, we describe a detailed study of amino/aromatic interactions in proteins: 'unconventional hydrogen bonds' are found to be remarkably uncommon relative to stacked geometries, and the reasons for this are examined.

Key words: Molecular modelling, protein side-chains, hydrogen bonding, electrostatic energy, distributed multipole analysis, amino/aromatic interactions, intermolecular perturbation theory.

Introduction

As the number of known three-dimensional protein structures continues to increase exponentially, the opportunities to understand the general principles of protein structure become ever greater. One can now take a non-homologous [1] dataset, containing in the region of 100 high resolution structures, from the Brookhaven Protein Databank [2], and obtain sufficient data to analyse statistically the geometrical distributions of contacts between each possible pair of side-chains.

This immediately lays down a challenge to the theoretician; to predict how each of these distributions should look, and to identify the energetic effects responsible for the observed trends in the experimental data. Experiment and theory can work together, comparison with experimental results providing an invaluable aid in developing new and appropriate theoretical methods, while the theoretical calculations help us to understand the experimental data better.

Here we focus on the use of small model molecules to represent the functional groups involved in interactions between protein side-chains. This work draws together analysis of experimental data with calculations of interaction energies, and particularly their variation with intermolecular geometry. Such an approach also allows us to study the competition between different possible interactions, an understanding of which is essential if we are to make sense of both the theoretical and the experimental data. Firstly, we will look at various methods of calculating

A. Pullman et al. (eds.), Modelling of Biomolecular Structures and Mechanisms, 119–135.
© 1995 *Kluwer Academic Publishers. Printed in the Netherlands.*

 JOHN B. O. MITCHELL ET AL.

interaction energies; then we will study three particular examples of the application of these methods to biological molecules.

Methods of Calculating Interaction Energies

We can calculate interaction energies using *ab initio* quantum mechanical wavefunctions. If we simply wish to know the overall interaction energy between two model molecules, we can carry out a supermolecule calculation, the result being the difference between the total energy of the dimer and the sum of the energies of the two monomers. This method has the advantage that it is conceptually simple. It is, however, expensive, since a dimer calculation is needed at each geometry for which the interaction energy is required. Even at the dispersion-free SCF level, the expense of the calculation is approximately proportional to the fourth power of the total number of basis functions. It also suffers from basis set superposition error, the correction of which by the Counterpoise method [3] requires expensive 'ghost orbital' calculations. Furthermore, the result of a supermolecule calculation is a single interaction energy, lacking the physical insight of a partitioning into its different components, although this can be achieved, albeit at greater cost, through 'Morokuma Analysis' [4,5].

An alternative is to use a perturbation theory approach, such as Hayes–Stone Intermolecular Perturbation Theory (IMPT) [6]. This partitions the total interaction energy into its components. These are the electrostatic energy, the short range exchange-repulsion incorporating the effects of the Pauli exclusion principle, the polarization energy due to the distortion of one molecule's charge distribution by the electric field arising from the other, and the charge transfer term resulting from the transfer of electrons between the moieties. It is, however, expensive to go beyond the usual 'single excitation' level and calculate electron correlation dependent terms such as the dispersion energy. IMPT involves a dimer calculation, but incorporates approximate corrections for the basis set superposition error, which is in any case confined to the charge transfer term. In fact, Stone has recently published a method of eliminating this error entirely [7]. Nonetheless, there are difficulties with this approach. In particular, the partition between the charge transfer and polarization energies is physically ill-defined, since they become equivalent in the large basis set limit [7]. We have previously used the IMPT method in a detailed study of N—H----O=C hydrogen bonding in the formamide/formaldehyde model system [8], and also in an investigation of the relative strengths of amide/amide and amide/water hydrogen bonding [9].

In many systems, especially charged ones or those exhibiting hydrogen bonding, the electrostatic energy is the dominant component of the interaction, controlling the directionality [5,8,10]. Thus, there are many situations where knowing the electrostatic energy is the key to understanding the interaction. Often it will not be possible to carry out full dimer calculations, particularly for finding minima, characterising regions of potential energy surfaces (which requires a large number of point calculations), or where the molecules are too large. In these cases, a monomer based model concentrating on the electrostatic term is often appropriate. While many packages represent the electrostatic energy by atomic point ('partial') charges, we will see that this is not a satisfactory description.

When two atoms come together to form a chemical bond, the charge distribution around each nucleus inevitably distorts away from spherical symmetry. Thus, the charge distribution around each nucleus in a molecule is anisotropic. The *ab initio* wavefunction naturally reflects this in its description of the charge density. Thus, any attempt to express the charge density in terms of atomic point charges will lead to a significant loss of the very information which one has just obtained from the *ab initio* calculation, probably at considerable expense. The use of potential derived charges [11–13], which is conceptually reasonable because the partial charge of an atom is not a uniquely defined property, is one way of reducing this loss; but there is a better approach, which effectively eliminates it altogether.

Rein's suggestion of representing the charge distributions around nuclei in molecules not simply by charges, but by atom-centred multipole series [14], represented the key breakthrough. It was now possible to produce a representation of the molecular charge distribution, the accuracy of which is essentially that of the *ab initio* (or semi-empirical) wavefunction. It is interesting that, even in 1973, Rein recognised the enormous potential for application to biomolecular modelling [14]. While there have been several procedures published for deriving these multipoles, the two most widely disseminated ones are probably the 'Cumulative Atomic Multipole Moments' (CAMM) of Sokalski *et al.* [15,16], and Stone's 'Distributed Multipole Analysis' (DMA) [17,18]. These two methods apparently produce results which are numerically equivalent [16,19].

Buckingham and Fowler [20,21] used Stone's DMA, together with a simple hard sphere repulsion, to calculate gas phase structures for a number of small molecule complexes. The success of the Buckingham–Fowler model showed that it is an excellent tool for modelling systems where the electrostatic energy is the dominant component of the interaction. In particular, where the electrostatics dominate the directionality of the interaction, as in hydrogen bonding [4,8,10], then this is a simple and inexpensive way of predicting the equilibrium geometry of a complex. We used the same model to study the biologically important N—H⋯O=C hydrogen bond in a number of small model systems [22].

In the general case, however, we obviously cannot assume that the electrostatics dominate every interaction [10,23]. Furthermore, even where the electrostatic term is the most important, we may wish to study the behaviour of the other contributions. Thus we desire a model incorporating an accurate description of each term in the potential. For most purposes, the expense of obtaining this from IMPT, or from Morokuma Analysis, is excessive, due to the *ab initio* dimer nature of these methods and the large number of points for which the energy is likely to be required. What we need, therefore, is an *ab initio* monomer based description of each of the physically distinct components of the interaction energy. When taken together, these constitute a 'synthetic' [24] or 'systematic' [25] potential.

Unfortunately, such models are not yet available for all the terms. The multipolar electrostatic component presents no problem, since it can be described with arbitrary accuracy by DMA. The penetration part of the electrostatic energy, which is not described by conventional DMA, can be included if 'Gaussian Multipoles' are used in place of the more usual point multipoles [26,27]. This procedure takes account of the overlap between charge distributions in a natural manner; however, it is still under development. The exchange-repulsion can also be modelled in an atom-atom anisotropic manner. One approach is to obtain the exchange-repulsion energy *via* its relationship with the penetration [27,28], using the 'Gaussian Multipole' approach outlined above. An alternative, which has been developed further, is to obtain the exchange-repulsion from approximate models of its dependence on the overlap of the monomer wavefunctions [29]. The dispersion can be modelled by a variety of approaches, varying from simple atom-atom isotropic 'C6' models [30], through those including anisotropy or 'C8' terms [25], to integrals of polarizabilities at imaginary frequencies [24]. The polarization energy can be described by 'Distributed Polarizabilities' [31–34], although this is expensive and further development work remains to be done. The representation of the polarization energy is complicated by the importance of many-body terms; it is highly non-additive. Little work seems to have been done on modelling the charge transfer energy [6–8], but its variation is probably similar to that of the exchange-repulsion.

We now move on to looking at three specific examples of applications of theoretical methods to biological molecules. These mainly involve the use of Buckingham–Fowler type DMA based models, with IMPT calculations at a few significant geometries. We hope in future to be able to apply some of the more sophisticated methods discussed in the previous two paragraphs; however, the models we have used, despite their limitations, do give many valuable insights which could be obtained neither from experimental data alone, nor from empirical force fields.

Arginine/Aspartate

The interaction between arginine and carboxylate ions [35] is particularly important because of its role in forming salt bridges in protein structures. Because it is an ion pair interaction, its gas phase strength is an order of magnitude greater than that of a hydrogen bond between neutral groups. In addition, because these groups can form two separate formal hydrogen bonds, the interaction is much more specific in its directionality than would otherwise be the case. This strong and directional interaction is often found to stabilise protein/substrate interactions at the active site [36,37].

Since this interaction is primarily electrostatic, a Buckingham–Fowler type model, with a DMA representation of the electrostatic energy, together with hard sphere repulsion, would be expected to give an accurate description of the directionality of the interaction energy. Thus we choose to use such a model. This, of course, describes the gas phase electrostatic energy; the effective electrostatic energy in solution is much smaller. There are a number of methods which try to model this overall effective interaction energy [38]. However, we choose to consider only the gas phase electrostatic energy, because it is well-defined, easily accessible, and should be responsible for the overall directionality of the interaction. It is the relative energies of different geometries, not the absolute values, which are of interest to us.

In order to represent arginine, we choose the methylguanidinium ion $[(NH_2)_2CNHCH_3]^+$. We represent aspartate (carboxylate) by acetate $[CH_3COO]^-$. We use 6-31G* basis sets to calculate the distributed multipoles,

Minimum A – Side On Symmetric

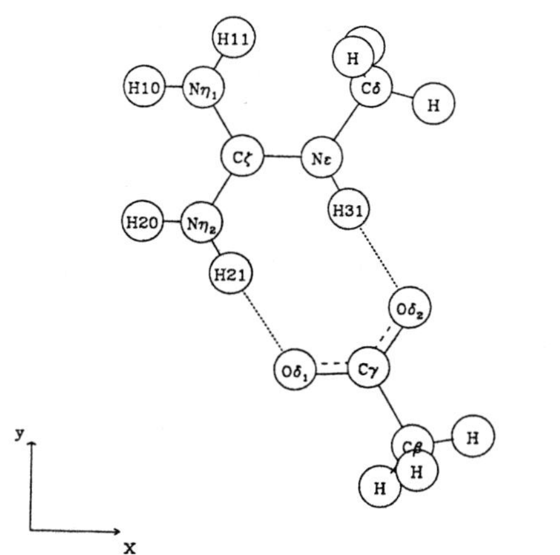

Minimum B – End On Symmetric

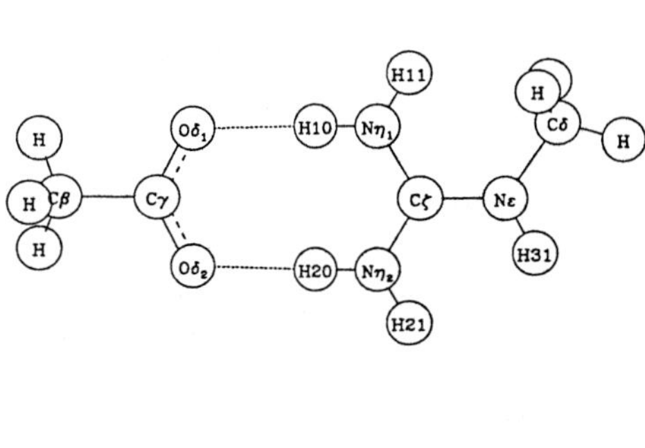

Minimum C – Side On Staggered

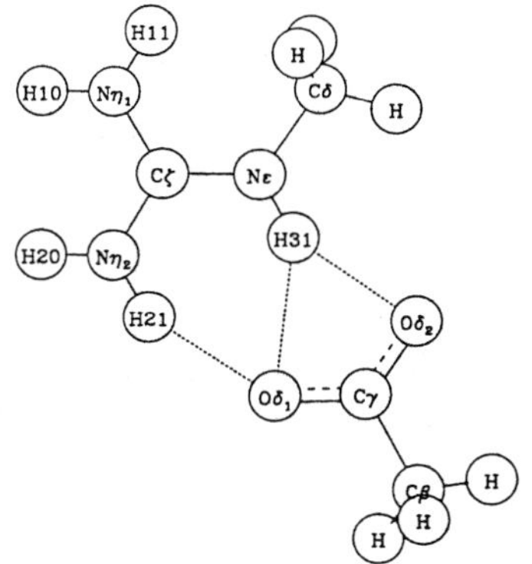

Minimum D – End On Staggered

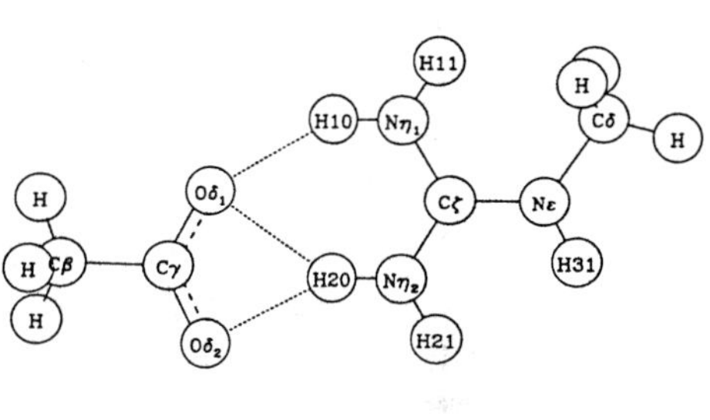

Fig. 1(a–d).

and standard Pauling [39] van der Waals radii for the hard spheres. Using around 40 different starting positions, we minimise the interaction energy in order to find the significant global and local minima.

The seven significant minima [35] are shown in Figure 1(a)–(g); we call these minima A to G, respectively. Minima A to E are significantly lower in energy than the others, as shown in Table I. These minima can be subdivided into those which are 'side on' contacts (A & C), and the 'end on' contacts (B, D & E). Although the breaking of threefold symmetry by the methyl substituent on one nitrogen (corresponding to N_ϵ in arginine) means that we can distinguish between the different structures in our model, we do not expect the small energy differences between model 'side on' and 'end on' structures to have any significance in the context of real structures. There is also a further distinction between 'symmetric' structures (A & C), and 'staggered ones' (B, D & E). While the 'symmetric'

Minimum E – End On Staggered   Minimum F –   Side On Twisted

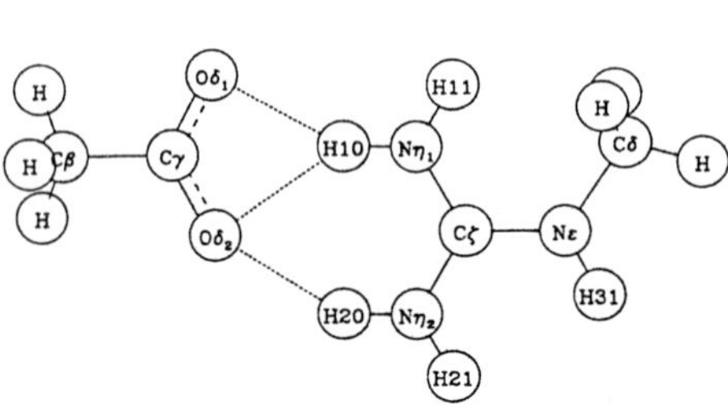

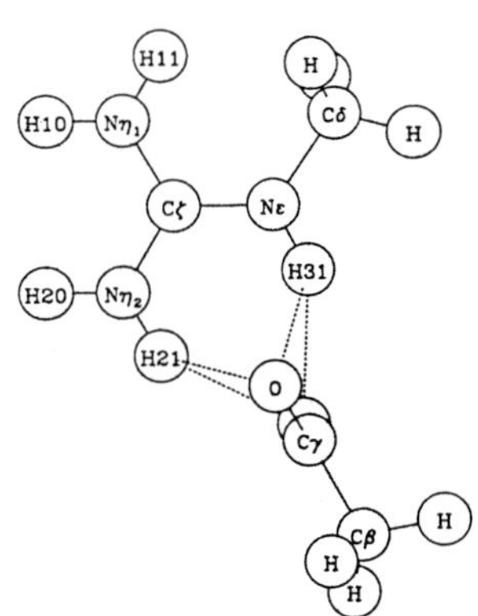

Minimum G – End On Twisted

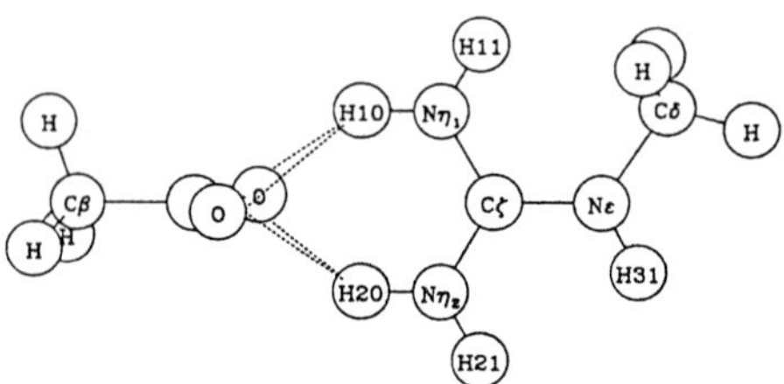

Fig. 1.   The structures of the seven lowest energy minima found for methylguanidinium/acetate using the 6-31G* DMA plus van der Waals hard spheres model. (a) Minimum A: side on symmetric; (b) Minimum B: end on symmetric; (c) Minimum C: side on staggered; (d) Minimum D: end on staggered; (e) Minimum E: also end on staggered; (f) Minimum F: side on twisted; (g) Minimum G: end on twisted.

TABLE I

The HS models consist of the given representation of the electrostatic energy plus slightly softened hard spheres, with radii adapted from Pauling [39]. The values used are: C 1.6 Å (when not bonded to H) or 2.0 Å (when bonded to H)); O 1.4 Å; N 1.5 Å; H 0.0 Å (contained in heavy atom sphere). In the Reiher models, the hard spheres are replaced by the Reiher [46] 12-6 potential. The CHARMM calculations were performed with the CHARMM modelling package [47] as described in the text. The electrostatic contributions, tabulated here, control the relative total energies of the structures. The use of parentheses denotes that the given energy does not correspond to a local minimum

Model (Electrostatic + Repulsion)		Min A Side Sym	Min B End Sym	Min C Side Stag	Min D End Stag	Min E End Stag
6-31G* DMA	+ HS	−439.36	−438.41	−436.13	−432.04	−432.90
3-21G DMA	+ HS	−435.21	−434.50	−434.22	−430.43	−431.31
6-31G* Mulliken	+ HS	−414.63	−421.23	−415.71	−420.25	−422.47
6-31G* DMA	+ Reiher	−454.22	−451.46	(−445.64)	(−439.30)	(−440.18)
6-31G* Mulliken	+ Reiher	(−422.53)	(−427.42)	−430.34	−434.07	−435.95
CHARMM		−211.51	−234.37	−209.49	−234.53	−234.06

structures have been investigated before [40–43], the 'staggered' ones found by our DMA calculations seem not to have been considered previously. It is interesting that the five lowest minima all have the two molecules coplanar. This is consistent with the favourability of hydrogen bonding in the oxygen's 'lone pair plane' found for numerous small molecule systems, both experimentally [44] and theoretically [8,22].

Table I also shows comparisons with a number of other models. In these, the 6-31G* DMA electrostatic model is replaced either by a 3-21G DMA or by 6-31G* Mulliken [45] charges. The alternative to the hard sphere repulsion is the 12-6 potential of Reiher [46]. We also compare the results produced by the CHARMM molecular modelling package [47]. We find that the relative DMA energies of the different geometries are very similar for the two basis sets, and the geometries of the minima are virtually identical. Substituting Mulliken point charges for the DMAs, however, has the effect of producing preferences for 'end on' and for 'staggered' structures. Replacing the hard sphere repulsion by Reiher's 12-6 potential reduces the number of minima found, although there is little effect on the relative total energies of the configurations. Although not shown in Table I, the total energies follow the same trend as the electrostatic energies. This is also true for CHARMM, which gives essentially the same structures; however, the CHARMM model assigns a different charge (-0.4) to N_ϵ than to either of the N_η atoms (-0.45) and the 'end on' structures therefore come out with rather lower energies than the 'side on' ones.

From these electrostatically based calculations, we have identified the structures which are of interest. We investigate these more fully using IMPT, which allows us also to calculate the non-electrostatic contributions to the interaction energy. These calculations are carried out for structures A to E within the CADPAC suite of *ab initio* programs [48]. As a further comparison, we also perform Direct SCF [49,50] supermolecule calculations at each of these geometries, using the Cambridge Direct SCF program [50]. Both these methods give the total intermolecular interaction energy at the SCF level, with no dispersion term.

The results of these *ab initio* calculations are shown in Table II. The relative total energies of the different structures follow their electrostatic components very closely [5,8,10,35], an observation which justifies our previous use of the DMA plus hard spheres (Buckingham–Fowler) model. The *ab initio* results show that the DMA model is far more accurate than Mulliken charges (Table I), which emphasizes the importance of including atomic anisotropy, in the form of multipoles, in our model of the electrostatic interaction. The closeness of the energies of the 'side on symmetric' and 'end on symmetric' structures is in agreement with work on similar systems by other authors [40–43], although they did not consider the 'staggered' structures suggested by our DMA model. The absolute energies, however, are very basis set dependent, and it is only the relative values to which one should attach any great significance. We conclude that the most favourable structures are the two 'symmetric' ones; the gap between them is so small as to be negligible, and furthermore the chemical difference between the sites in the model probably fails accurately to reflect that in arginine. The 'staggered' structures too are low in energy, so a range of in-plane orientations is likely to be accessible in proteins.

TABLE II

The IMPT results refer to Hayes–Stone Intermolecular Perturbation Theory calculations using 6-31G* basis sets and Möller–Plesset denominators [6]. These are uncorrelated (free of dispersion) and should thus be directly comparable with the supermolecule results. The DMA and IMPT electrostatic energies differ by a relatively small quantity known as the penetration energy [27]. The charge transfer (CT) term is the only contribution affected by basis set superposition error (BSSE). In order to estimate this uncertainty, CT is calculated firstly with no correction (CT(0)), and then also with each of the Fock (CT(F)) and Eigenvalue (CT(E)) BSSE corrections. Thus, three values are quoted for the CT and total energies at each of the five geometries. The supermolecule calculations use identical 6-31G* basis sets to IMPT, and were performed by the Direct SCF method [49,50] with no BSSE correction

Minimum	A	B	C	D	E
Side/γnd On	Side	End	Side	End	End
Sym or Stag	Sym	Sym	Stag	Stag	Stag
DMA 6-31G*					
Electrostatic	−439.36	−438.41	−436.13	−432.04	−432.90
IMPT6-31G*					
Electrostatic	−447.86	−446.84	−445.19	−441.45	−442.47
Exch-repulsion	46.34	45.70	43.25	44.29	44.57
Polarization	−34.06	−32.60	−34.32	−32.57	−32.60
CT (0)	−62.92	−63.45	−51.40	−48.58	−48.17
CT (F)	−60.67	−61.23	−48.97	−46.50	−46.10
CT (E)	−54.14	−54.87	−42.76	−41.58	−41.20
Total (0)	−498.51	−497.19	−487.67	−478.30	−478.67
Total (F)	−496.26	−494.98	−485.24	−476.22	−476.60
Total (E)	−489.73	−488.61	−479.02	−471.30	−471.70
Supermolecule 6-31G*					
Supermol	−484.99	−483.87	−478.09	−470.97	−471.59

We next compare our theoretical predictions of the favoured interaction geometries for these groups with experimental protein structure data from the Brookhaven Protein Databank [2]. We take data from 150 arginine/aspartate interactions: 109 geometries from pairs of residues in the same chain ('intramolecular' interactions, taken from 52 high resolution structures) and 41 from pairs in different chains ('intermolecular' interactions, taken from 35 high resolution structures) [35], all pairs containing atom----atom contacts within (sum of van der Waals radii plus 1 Å). These experimental results are shown in Figure 2(a)–(b). They clearly follow the theoretical prediction, with the 'side on' and 'end on' regions being the most populated. In each region, the cluster of structures includes both 'symmetric' and 'staggered' geometries, as can be seen from Figure 2(a)–(b) and also from Table III, which categorises the different observed contacts. Twin contacts in the 'side on' and 'end on' regions account for some 35% of the 84 formally hydrogen bonded pairs (or 19% of all contacting pairs).

The aspect of the experimental results which is initially most puzzling is the very strong tendency for intramolecular interactions to be 'side on' [52] and also for the intermolecular ones to be 'end on'. This is beyond the scope of our model, which does not distinguish between the intra- and intermolecular geometries. Intermolecular interactions occur between the surfaces of the two separate molecules, which interact once each molecule has adopted its native conformation. In

109 Intramolecular ARG–ASP Contacts

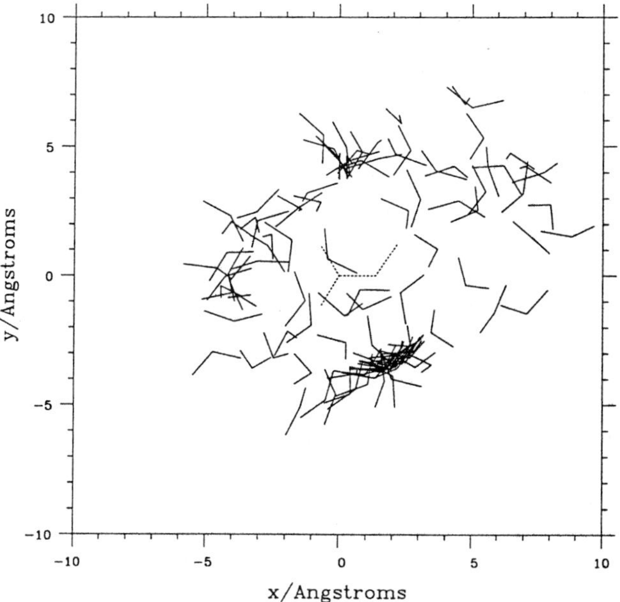

41 Intermolecular ARG–ASP Contacts

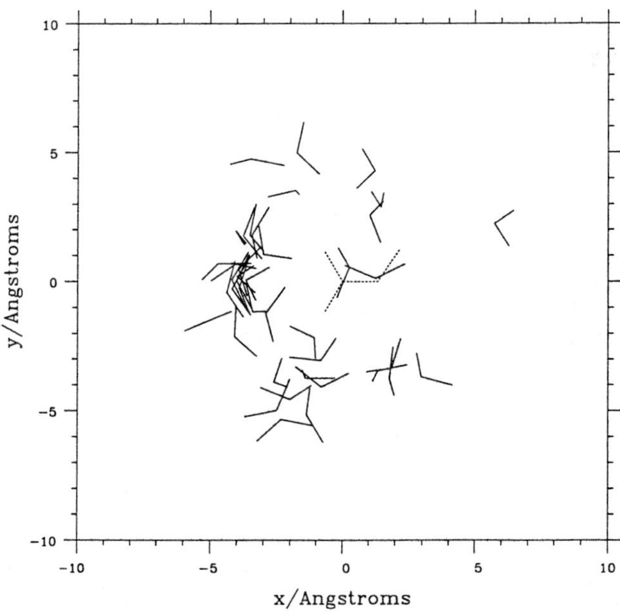

Fig. 2. The distributions of (a) 109 intramolecular (same chain) and (b) 41 intermolecular (different chains) aspartate COO⁻ groups around arginine. The arginine position is indicated by the broken lines.

TABLE III

The 109 intramolecular and 41 intermolecular experimental arginine/aspartate interactions are classified according to their hydrogen bonding patterns. Any O----H contact less than 2.5Å is considered to constitute a hydrogen bond. The structure C′ is related to minimum C in the same way as minima D and E are related, but it does not correspond to a DMA local minimum

Hydrogen H-Bonded to				Occurrences	
Oδ1	Oδ2	Min	Type	Intra-	Inter-
H21	H31	A	Side, Sym.	11	1
H21 & H31	H31	C	Side, Stag.	4	0
H21 & 1131	H21	C′	Side, Stag.	3	0
H10	H20	B	End, Sym.	2	6
H20 & H10	H20	D	End, Stag.	0	1
H20 & H10	H10	E	End, Stag.	0	1
H20 & H21	H21	–	'1 to 2'	0	1
H10	H10	–	'1 to 2'	1	2
H11	H11	–	'1 to 2'	0	0
H10 & H11	H10	–	'1 to 2'	1	0
H10	H11	–	'1 to 2'	1	0
H20 & H10	–	–	'2 to 1'	1	3
H21 & H31	–	–	'2 to 1'	5	1
H20	–	–	'1 to 1'	6	0
H21	–	–	'1 to 1'	6	2
H10	–	–	'1 to 1'	5	2
H11	–	–	'1 to 1'	11	2
H31	–	–	'1 to 1'	2	1
H20 & H21	–	–	'1 to 1'	0	1
H10 & H11	–	–	'1 to 1'	1	0
H21 & H31	H21 & H31	F	Side, Twist.	0	0
H20 & H10	H20 & H10	G	End, Twist.	0	0
–	–	–	none	49	17
Total				109	41

TABLE IV

The data are taken from the IDITIS^{TM} database (Oxford Molecular Ltd.). In four cases, $N_{\eta1}$ and $N_{\eta2}$ are missing from the database, which explains the small discrepancy in the numbers of observations

	N_ϵ	$N_{\eta1}$	$N_{\eta2}$
Mean Accessibility (/Å^2)	6.19	22.65	29.12
% of Zero Accessibilities	20.8%	9.5%	6.1%
Number of Observations	5834	5830	5830

contrast, the intramolecular ones are involved in forming the native structures. The difference between the two sets of experimental data can be explained by observing that the distal nitrogen atoms ($N_{\eta1}$ and $N_{\eta2}$) are usually on the exposed surface, while N_ϵ is likely to be more buried; this is shown for a large sample of over 5800 arginines in Table IV. Thus the 'end on' position is more exposed, and thus more suited to intermolecular interactions. Furthermore, the 'side on' position is naturally favourable for intramolecular contacts where the residues are very close in the sequence; indeed 7 of the 18 'side on' intramolecular contacts are between residues with an $i \leftrightarrow i + 2$ relationship. In these cases, the aspartate can hydrogen bond in the 'side on' geometry, but the 'end on' position is sterically hindered because the aspartate side-chain is not long enough to reach it in any reasonable conformation.

Thus we have seen that the DMA model can successfully predict the most favourable interaction geometries for arginine/aspartate pairs, an interaction which is dominated by electrostatics. Our DMA model also identified 'staggered' minima, and indeed the experimental clusters incorporate both 'symmetric' and 'staggered' geometries. The CHARMM calculations performed reasonably well, but their results were very sensitive to the details of the electrostatic model. Nonetheless, it is clear that environmental effects and competing interactions are also important since, despite the clusters in the minimum energy positions, the majority of pairs interact in other geometries. The contrast between the 'side on' preference of the intramolecular contacts and the 'end on' preference of the intermolecular contacts can be explained on the basis of the differences between the average accessibilities of the nitrogens involved.

Arginine/Phosphotyrosine

In 1992, Waksman et $al.$ published the crystal structure of a complex between v-src SH2 domain and a (non-cognate) peptide ligand [52]. There was particular interest in the interaction of an arginine from the SH2 domain (residue 155 in their numbering scheme) and the phosphotyrosine of the ligand, where an amino/aromatic hydrogen bond [53,54] was observed between the arginine $N_{\eta 2}$ and the aromatic ring.

We investigate this arginine/phosphotyrosine interaction. While Waksman et $al.$ [52] reported the amino/aromatic interaction, and also a conventional hydrogen bond between N_ϵ of arginine and the phosphate group, we have used the program HBPLUS [55] to identify a further hydrogen bond [56]. This involves the same nitrogen ($N_{\eta 2}$) and proton as the amino/aromatic interaction, with the acceptor being a phosphate oxygen (the N----O distance is 3.05 Å). All three interactions are shown in Figure 3.

In order to analyse the contributions of these different interactions to the overall electrostatic energy, we have prepared three different models, using 6-31G* DMAs throughout. Because of the dominant role of the electrostatic energy in hydrogen bonding, especially between charged species, we can safely use such electrostatically based calculations. In the first instance, we model arginine by methylguanidinium and the aromatic ring by toluene, thus representing only the arginine/aromatic interaction. In the second case, we replace toluene by p-cresol, so that we now include the O_η substituent atom. Finally, we use p-phosphotoluene as our model, so that we now represent the whole interaction including the two arginine/phosphate hydrogen bonds. The electrostatic energies of these model interactions are -25.0, -23.2 and -695.2 kJ/mol, respectively. Thus the major contribution to the total energy comes from the arginine/phosphate hydrogen bonds, with the arginine/aromatic interaction playing a relatively small energetic role.

Although these are gas phase energies, the conclusion is clear. However interesting the arginine/aromatic interaction may be, it is energetically secondary to the

v-src SH2 Domain / Peptide A Complex
ARG 155 - PTY Interaction

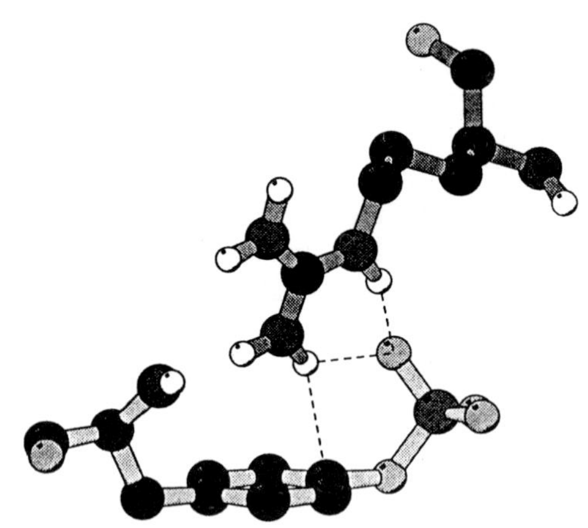

6-31G* DMA Electrostatic Model Energies:

Methylguanidinium / Toluene -25.0 kJ/mol

Methylguanidinium / Phosphotoluene -695.2 kJ/mol

Fig 3. The arginine/phosphotyrosine interaction, incorporating an amino/aromatic hydrogen bond, reported by Waksman *et al.* [52] in their v-src SH2 domain/non-cognate peptide complex structure. The dashed lines represent both conventional and amino/aromatic hydrogen bonds between the two residues. This plot was generated using P. J. Kraulis' MOLSCRIPT program.

arginine's interaction with the phosphate group, and must be understood in this context. We consider amino/aromatic interactions in more detail below.

Amino/Aromatic Interactions

The arginine/phosphotyrosine interaction referred to above is just one example of an amino/aromatic interaction. There has been considerable interest in these interactions since 1988, when Levitt and Perutz [53] suggested that a hydrogen bond like geometry should be favoured. Although some examples of amino/aromatic hydrogen bonds have been reported, subsequent studies have suggested that an alternative stacked geometry, where the planes of the sp^2 nitrogen and the aromatic ring are approximately parallel, is in fact much more common [57,58]. However, these studies have either been limited to a small subset of the possible donor nitrogens [57], or carried out on data containing homologous protein structures [58].

We analyse [56,59] the interactions between sp^2 nitrogens, whether from main-chains or side-chains, and phenylalanine or tyrosine residues. We use a non-

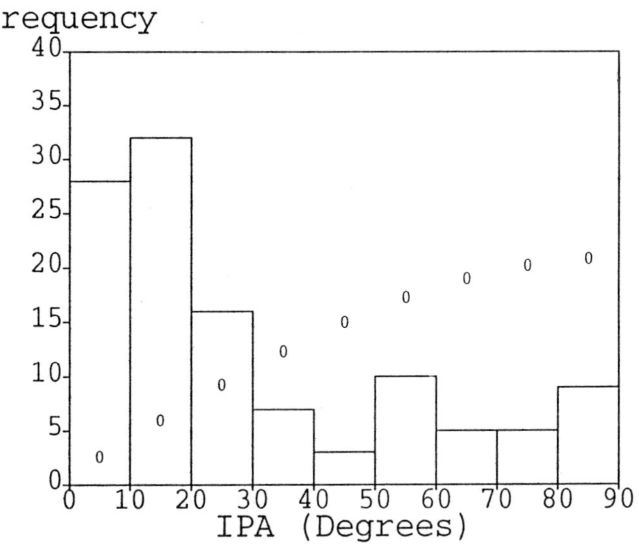

Fig. 4. The distribution of the interplanar angle (IPA) between sp^2 nitrogen and an aromatic ring for 115 interactions in which the nitrogen lies above the ring face. This is defined by requiring the nitrogen to be within 20° of being directly above either a ring carbon or the ring centre, and no more than 3.8 Å away from the same site. The small oval symbols represent the expected sinusoidal distribution of the angle between two planes (see text).

homologous [1] sample of 55 chains from high resolution protein crystal structures, taken from the Brookhaven Protein Databank [2]. There is a total of 1108 residue/residue contacts where the nitrogen is within 3.8 Å of an aromatic carbon (excluding immediately adjacent residues). Since we are looking at the relative favourabilities of amino/aromatic hydrogen bonded and stacked structures, we are mainly interested in those structures where the nitrogen lies above the aromatic ring. There are 115 such 'Above Ring' structures (the criteria for defining these are set out more precisely elsewhere [56,59]).

For each of these 115 interactions, we calculate the interplanar angle between the planes of the sp^2 nitrogen and the aromatic ring. All interactions with interplanar angles less than 30° are categorised as stacked. When the interplanar angle is 30° or more, we test for the presence of an amino/aromatic hydrogen bond by calculating the angle ∠N—H----C for all possible aromatic carbon acceptors within 3.8 Å of the nitrogen (the hydrogens are added to the structures by the program HBPLUS [55]). An angle of 120° or more is required to indicate an amino/aromatic hydrogen bond. Those remaining residue/residue interactions not qualifying as stacked or hydrogen bonded are classified simply as 'others.'

The distribution of interplanar angles is shown in Figure 4, and contrasted with

TABLE V

Categorization of 'Above Ring' nitrogen/aromatic interactions. Amino/aromatic hydrogen bonded interactions are those with an N----C distance of 3.8 Å or less, an interplanar angle not less than 30°, and also an $\angle$N—H----C angle of 120° or more. Any interaction with an interplanar angle less than 30° is classified as stacked. The others are those 'Above Ring' interactions falling into neither of the previous categories

	H-bonded	Stacked	Others
Side-chain N/Phe	8	14	2
Main-chain N/Phe	5	19	1
Side-chain N/Tyr	6	15	3
Main-chain N/Tyr	11	28	3
Total	30	76	9

the sinusoidal one which would be expected simply by chance. This figure shows a striking preponderance of small interplanar angles (stacked structures), and a corresponding paucity of larger angles (which mostly correspond to amino/aromatic hydrogen bonds) in the experimental data. χ^2-analysis shows that the difference between the expected and observed distributions is massively statistically significant. The breakdown into different categories (Table V) demonstrates the same trend, showing 76 stacked geometries and only 30 amino/aromatic hydrogen bonds. Thus our analysis agrees with other work [57,58] in finding that stacking, and not amino/aromatic hydrogen bonding, is the predominant geometry for the 'Above Ring' contacts.

The 30 amino/aromatic hydrogen bonds represent 26% of the 'Above Ring' structures, and only 2.7% of all amino/aromatic residue/residue contacts. The 70 stacked structures, by way of comparison, constitute 61% of all 'Above Ring' contacts. Nonetheless, 26 out of the 55 protein chains, or 47%, contain at least one amino/aromatic hydrogen bond if we also include interactions involving lysine (if only sp^2 nitrogen donors are allowed, this figure falls to 22/55, or 40%). Thus, although they are disfavoured relative to stacked geometries, a good proportion of protein structures will nevertheless contain at least one amino/aromatic hydrogen bond.

The explanation for the relative favourabilities of the different geometries lies in the balance between two effects. One is the advantage of the stacked structure for forming conventional hydrogen bonds. In a stacked structure, all the protons on the nitrogen-bearing residue are free to form conventional hydrogen bonds. In contrast, one of the protons of an amino/aromatic hydrogen bonded donor residue is tied up in this unconventional hydrogen bond, and so is not available for conventional hydrogen bonding. Our work [56] has shown that, while the situation is not in practice quite as straightforward as suggested above, protons involved in amino/aromatic hydrogen bonds do form significantly fewer conventional hydrogen bonds than protons in stacked structures (the average deficit is around two thirds of a conventional hydrogen bond).

The second effect is the advantage of the amino/aromatic hydrogen bond in terms of gas phase interaction energy. For the typical case of arginine/phenylalanine, we use 3-21G Hayes–Stone IMPT [6], together with an empirical dispersion

TABLE VI

Gas phase interaction energies of the guanidinium/benzene model system in geometries corresponding to experimental arginine/phenylalanine contacts. The energies are calculated using Hayes–Stone IMPT [6] with 3-21G basis sets

Code[a]	Arg/Phe	Type	Elect[b]	Disp[c]	Rest[d]	Total
3FGF	107/95	Stacked	−7.90	−11.70	−3.92	−23.51
2FB4	H19/H80	Stacked	−13.26	−11.01	−3.12	−27.40
5TMN	E260/E178	Stacked	−11.79	−13.03	−3.09	−27.91
1SNC	81/76	H-bond	−10.98	−7.99	−10.46	−29.43
2CYP	127/276	Stacked	−16.15	−14.23	−2.01	−32.39
6XIA	176/41	Stacked	−15.99	−17.55	−1.98	−35.53
5TMN	E101/E40	H-bond	−18.12	−13.19	−4.41	−35.73
4PTP	65A/82	Stacked	−15.84	−19.19	−1.23	−36.25
2HMZ	A110/A107	Stacked	−15.76	−19.27	−1.47	−36.51
2FB4	H72/H29	H-bond	−19.95	−18.00	−3.02	−40.97
2FB4	H98/H27	H-bond	−22.87	−10.75	−8.50	−42.12
Average of seven Stacked			*−13.81*	*−15.14*	*−2.41*	*−31.36*
Average of four H-bonds			*−17.98*	*−12.48*	*−6.60*	*−37.06*

[a]The identifier code from the Brookhaven Protein Databank [2]. The crystal structure references are: 3FGF, recently superseded by 4FGF, [61]; 2FB4 [62]; 5TMN [63]; 1SNC [64]; 2CYP [65]; 6XIA [66]; 4PTP [67]; 2HMZ [68].
[b]The IMPT 3-21G electrostatic energy [6]. This is not the same as the Distributed Multipole Analysis electrostatic energy [7], since the IMPT energy also includes a penetration term. However, it is equivalent to the Gaussian Multipole electrostatic energy of Wheatley [26]. All energies are quoted in kJ/mol.
[c]Dispersion energy from the C_6 parameters in 'Set II' given by Huiszoon and Mulder [30].
[d]The sum of the exchange-repulsion, polarization and charge transfer energies, including the 'Fock' basis set superposition error correction [6,8].

term [30,56] to calculate partitioned *ab initio* energies for guanidinium $[(NH_3)_3C]^+$/benzene in geometries corresponding to the 11 experimental structures. Of these geometries, 7 are stacked and 4 are amino/aromatic hydrogen bonded. The results (Table VI) show a small gas phase energetic favourability for amino/aromatic hydrogen bonds over stacked structures, amounting to a 5.6 kJ/mol difference between the best energies for each type, and 5.7 kJ/mol between the average energies. The distributions of gas phase interaction energy for each kind of geometry overlap significantly, so that four stacked structures are lower in energy than the worst hydrogen bonded one.

The gas phase energetic favourability of amino/aromatic hydrogen bonding is, overall, insufficient to overcome the conventional hydrogen bonding advantage of stacking. So although the gas phase minimum is, as we would expect on the basis of the Legon–Millen rules [60], an amino/aromatic hydrogen bonded structure, our theoretical results are consistent with the observed strong preference for stacked geometries.

Conclusions

The examples given above show the power of theoretical and experimental methods in tandem. In each case, we are able to compare our energy calculations with experimental data and, where there are apparent discrepancies, this allows us to understand the limitations of our models. On the other hand, given experimental data, we can use theory to determine the most powerful energetic influences which lie behind it.

We find that our theoretical models give valuable descriptions of the preferred directionalities of intermolecular interactions, and also of the relative energies of possible competing interactions. It is essential to consider such competing interactions, and also the effects of the environment, before using these small molecule theoretical calculations to interpret experimental data.

This kind of experience is crucial as we learn the strengths and weaknesses of our models, as well as understanding more about the interactions of biomolecules. Such experience will be essential in the development of methods which will allow biomolecules to be modelled with an accuracy reflecting that which is now routinely achieved for smaller systems, such as the representative fragments considered here. This will involve the incorporation of the other important components of the intermolecular potential, with an accuracy consistent with that which DMA has achieved for the long range electrostatic energy. Such methods will open up exciting new possibilities, particularly in the area of drug design.

References

1. C. A. Orengo, T. P. Flores, W. R. Taylor, and J. M. Thornton: *Prot. Eng.* **6**, 485 (1993).
2. F. C. Bernstein, T. F. Koetzle, G. J. B. Williams, G. F. Mayer, M. D. Brice, J. R. Rodgers, O. Kennard, T. Shimanouchi, and M. Tasami: *J. Mol. Biol.* **112**, 535 (1977).
3. S. F. Boys and F. Bernardi: *Mol. Phys.* **19**, 553 (1970).
4. K. Kitaura and K. Morokuma: *Int. J. Quant. Chem.* **10**, 325 (1976).
5. H. Umeyama and K. Morokuma: *J. Amer. Chem. Soc.* **99**, 1316 (1977).
6. I. C. Hayes and A. J. Stone: *Mol. Phys.* **54**, 83 (1984).
7. A. J. Stone: *Chem. Phys. Lett.* **211**, 101 (1993).
8. J. B. O. Mitchell and S. L. Price: *J. Comp. Chem.* **11**, 1217 (1990).
9. J. B. O. Mitchell and S. L. Price: *Chem. Phys. Lett.* **180**, 517 (1991).
10. G. J. B. Hurst, P. W. Fowler, A. J. Stone, and A. D. Buckingham: *Int. J. Quant. Chem.* **29**, 1223 (1986).
11. D. E. Williams and D. J. Craycroft: *J. Phys. Chem.* **89**, 1461 (1985).
12. T. R. Stouch and D. E. Williams: *J. Comp. Chem.* **14**, 858 (1993).
13. K. M. Merz: *J. Comp. Chem.* **13**, 749 (1992).
14. R. Rein: *Advan. Quant. Chem.* **7**, 335 (1973).
15. W. A. Sokalski and R. A. Poirier: *Chem. Phys. Lett.* **98**, 86 (1983).
16. W. A. Sokalski and A. Sawaryn: *J. Mol. Struct.* **256**, 91 (1992).
17. A. J. Stone: *Chem. Phys. Lett.* **83**, 233 (1981).
18. A. J. Stone and M. A. Alderton: *Mol. Phys.* **56**, 1047 (1985).
19. M. A. Spackman: *J. Chem. Phys.* **85**, 6587 (1986).
20. A. D. Buckingham and P. W. Fowler: *J. Chem. Phys.* **79**, 6426 (1983).
21. A. D. Buckingham and P. W. Fowler: *Canad. J. Chem.* **63**, 2018 (1985).
22. J. B. O. Mitchell and S. L. Price: *Chem. Phys. Lett.* **154**, 267 (1989).
23. E. M. Duffy, P. J. Kowalczyk, and W. L. Jorgensen: *J. Amer. Chem. Soc.* **115**, 9271 (1993).
24. A. J. Stone: in *Hydrogen Bonded Liquids*, J. C. Dore and J. Texeira, Eds., Kluwer Academic Publishers, Dordrecht, Netherlands, 25–47 (1991).
25. R. J. Wheatley and S. L. Price: *Molec. Phys.* **71**, 1381 (1990).
26. R. J. Wheatley: *Molec. Phys.* **79**, 597 (1993).
27. R. J. Wheatley and J. B. O. Mitchell: *J. Comp. Chem.* **15**, 1187 (1994).
28. J. N. Murrell and J. J. C. Teixeira-Dias: *Mol. Phys.* **19**, 521 (1970).

29. R. J. Wheatley and S. L. Price: *Molec. Phys.* **69**, 507 (1990).
30. C. Huiszoon and F. Mulder: *Mol. Phys.* **38**, 1497 (1979).
31. A. J. Stone: *Mol. Phys.* **56**, 1065 (1985).
32. A. J. Stone: *Chem. Phys. Lett.* **155**, 102 (1989).
33. A. J. Stone: *Chem. Phys. Lett.* **155**, 111 (1989).
34. C. R. Le Sueur, A. J. Stone and P. W. Fowler: *J. Phys. Chem.* **35**, 3519 (1991).
35. J. B. O. Mitchell, J. M. Thornton, J. Singh, and S. L. Price: *J. Mol. Biol.* **226**, 251 (1992).
36. J. F. Riordan, K. D. McElvany, and C. L. Borders: *Science* **195**, 884 (1977).
37. J. F. Kirsch, G. Eichelle, G. C. Ford, M. G. Vincent, J. N. Jasonius, H. Gehring, and P. Christen: *J. Mol. Biol.* **174**, 497 (1984).
38. M. E. Davis and J. A. McCammon: *Chem. Rev.* **90**, 509 (1990).
39. L. Pauling: *The Nature of the Chemical Bond*, 3rd. edition, Cornell Univ., p. 260 (1960).
40. S. Nakagawa and H. Umeyama: *J. Amer. Chem. Soc.* **100**, 7716 (1978).
41. A. M. Sapse and C. S. Russell: *Int. J. Quant. Chem.* **26**, 91 (1984).
42. A. M. Sapse and C. S. Russell: *J. Mol. Struct.* **137**, 43 (1986).
43. D. W. Deerfield, H. B. Nicholas, R. G. Hiskey, and L. G. Pedersen: *Proteins* **6**, 168 (1989).
44. R. Taylor, O. Kennard, and W. Versichel: *J. Amer. Chem. Soc.* **105**, 5761 (1983).
45. R. S. Mulliken: *J. Chem. Phys.* **23**, 1833 (1955).
46. W. E. Reiher: *Theoretical Studies of Hydrogen Bonding*, Ph.D. Thesis, Harvard Univ., U.S.A. (1985).
47. B. R. Brooks, R. E. Bruccoleri, B. D. Olafson, D. J. States, S. Swaminathan, and M. Karplus: *J. Comp. Chem.* **4**, 187 (1983).
48. R. D. Amos and J. E. Rice: *CADPAC, the Cambridge Analytical Derivatives Package*, issue 4.0, Dept. of Chemistry, University of Cambridge, U.K. (1987).
49. J. Almlöf, K. Faegri, and K. Korsell: *J. Comp. Chem.* **3**, 385 (1982).
50. C. W. Murray, J. S. Andrews, and R. D. Amos: *Cambridge Direct SCF Program*, University of Cambridge, Dept. of Chemistry, U.K. (1991).
51. J. Singh, J. M. Thornton, M. Snarey, and S. F. Campbell: *FEBS Lett.* **224**, 161 (1987).
52. G. Waksman, D. Kominos, S. C. Robertson, N. Pant, D. Baltimore, R. B. Birge, D. Cowburn, H. Hanafusa, B. J. Mayer, M. Overduin, M. D. Resh, C. B. Rios, L. Silverman, and J. Kuriyan: *Nature* **358**, 646 (1992).
53. M. Levitt and M. F. Perutz: *J. Mol. Biol.* **201**, 751 (1988).
54. M. F. Perutz: *Phil. Trans. Roy. Soc. A* **345**, 105 (1993).
55. I. K. McDonald, D. T. Jones, D. Naylor, and J. M. Thornton: *HBPLUS*, a computer program for calculating potential hydrogen bonds in protein structures, Dept. of Biochemistry, University College London, U.K. (1993).
56. J. B. O. Mitchell, C. L. Nandi, I. K. McDonald, J. M. Thornton, and S. L. Price: *J. Mol. Biol.* **239**, 315 (1994).
57. M. M. Flocco and S. L. Mowbray: *J. Mol. Biol.* **235**, 709 (1994).
58. J. Singh and J. M. Thornton: *J. Mol. Biol.* **211**, 595 (1990).
59. J. B. O. Mitchell, C. L. Nandi, Shamimara Ali, I. K. McDonald, J. M. Thornton, S. L. Price, and J. Singh: *Nature* **366**, 413 (1993).
60. A. C. Legon and D. J. Millen: *Acc. Chem. Res.* **20**, 39 (1987).
61. A. E. Eriksson, L. S. Cousens, L. H. Weaver, and B. W. Matthews: *Proc. Natl. Acad. Sci. U.S.A.* **88**, 3441 (1991).
62. M. Marquart, J. Deisenhofer, R. Huber, and W. Palm: *J. Mol. Biol.* **141**, 369 (1980).
63. H. M. Holden, D. E. Tronrud, A. F. Monzingo, L. H. Weaver, and B. W. Matthews: *Biochem.* **26**, 8542 (1987).
64. P. J. Loll and E. E. Lattman: *Proteins* **5**, 183 (1989).
65. B. C. Finzel, T. L. Poulos, and J. Kraut: *J. Biol. Chem.* **259**, 13027 (1984).
66. Z. Dauter, H. Terry, H. Witzel, and K. S. Wilson: *Acta Cryst. B* **46**, 833 (1990).
67. J. L. Chambers and R. M. Stroud: *Acta Cryst. B* **35**, 1861 (1979).
68. M. A. Holmes and R. E. Stenkamp: *J. Mol. Biol.* **220**, 723 (1991).

Dynamic Domains: A Simple Method of Analysing Structural Movements in Proteins

KRYSTYNA ZAKRZEWSKA
Laboratoire de Biochimie Théorique (URA77 CNRS), Institut de Biologie Physico-Chimique, 13, rue Pierre et Marie Curie, 75005 Paris, France

Abstract. A method is presented to analyse conformational changes in enzymes upon coenzyme and/or substrate binding. The method is based on a rigid body approximation for secondary structures and defines a dynamic domain as a collection of secondary structures which move together. The technique has been tested on a number of enzymes for which the conformation transition is well characterised. It was then applied to the analysis of two theoretically derived transition pathways for citrate synthase.

Key words. Enzymes, transition pathways, conformational analysis, protein domains.

Introduction

Understanding the conformational changes that an enzymatic protein undergoes along its reaction pathway is helpful in understanding enzyme functioning. Unfortunately this information is very difficult to obtain. In a rather small number of cases two forms of the same enzyme have been crystallised, both the free holoenzyme and its complex with its natural substrate or with an inhibitor. The analysis of such pairs of structures has shown that enzymes are often constructed from structural domains (for reviews see [1, 2]) and that the conformation change resulting from coenzyme and/or substrate binding can be described by the movement of these domains, treated as effectively rigid bodies, with respect to one another. In addition, the nature of the movement occurring can, in many cases, be reduced to a simple rotation around a localised hinge point. Amongst the classical examples of proteins analysed in this way are liver alcohol dehydrogenase, hexokinase and citrate synthase, which were all studied by Lesk [3, 4]. In all these cases the conclusions reached were based only upon the analysis of changes in the positions of secondary structure motifs between the 'open' and 'closed' forms of the enzymes in question.

The existence of detailed transition pathways for citrate synthase obtained by the theoretical simulations of Durup and El Kettani [5,6], provided us with much richer data for analysing domain movements by filling in the otherwise unknown conformational states (at a total of 15 intermediate points) between the two extreme forms of the enzyme in question. This analysis however also required a faster and more objective way of locating structural domains. We thus set about developing a simple method based on secondary structure movements which would enable us to follow the dynamic evolution of domains along the reaction pathway.

In a previous publication [7] we analysed the theoretically generated citrate synthase pathways concentrating mainly on the evolution of local data such as the main chain angles ϕ and ψ of individual residues, but also looking more globally at the deformation of constituent α-helices and angles and distances between these

A. Pullman et al. (eds.), Modelling of Biomolecular Structures and Mechanisms, 137–149.

secondary structural elements. These techniques were nevertheless inappropriate for summarising the global movements of 20 α-helices composing citrate synthase. In the present paper we are able to present a complementary study, based on treating citrate synthase as an ensemble of rigid secondary structures, defined with the help of the P-curve algorithm [8,9]. We develop the concept of dynamic domains which are defined as groups of secondary structures which move together. It is shown that this method is equally applicable to the two-state data available for several enzymes and to the virtually continuous movements occurring along the citrate synthase transition pathways.

Method

In the following analysis, all secondary structures have been defined using the P-curve algorithm [8,9], which has been described elsewhere and has already served in our earlier studies of citrate synthase. P-curves determines an optimal global axis (a segmented curve) for a protein in such a way as to minimise the departure of the system from helical symmetry. The axis is designed to optimally distribute structural irregularity between differences in the location and orientation of successive peptide planes with respect to their local helical axis segment and misalignments of the axis segments. This procedure involves a least squares optimisation. Once the helical axis has been found, each pair of successive peptide planes is characterised by a so-called DIF value (dimeric irregularity function), which summarises the local departure from helical symmetry. Since these values are small within regular secondary structures (α-helices and β-sheets) they can be used as a means to automatically locate such structural motifs. The maximum value of DIF allowed within a secondary structure is a function of the length of the segment in question as this value has been obtained by a statistical analysis of available protein conformations [9]. Once the presence of a secondary structure zone has been detected in this way, the values of helicoidal parameters Xdisp and Ydisp, which define the translation of the peptide plane with respect to the local helical axis system, allow easy identification of the type of secondary structure.

In the present model each secondary structure is treated as a rigid body. It is thus represented simply as a set of four points defining its centre of gravity, its N-terminal end, its rotational position via a point located on the dyad axis passing through the centre of mass and a point located on the axis perpendicular to the principal and the dyad axes. For each pair of secondary structures, the set of distances shown in Figure 1 was calculated. We define two secondary structures as moving together during the transition between two conformations if the average distance between them, defined using the simplified point representation, does not change. For N secondary structures, in two protein conformations, we thus obtain an $N-1$ dimensional triangular matrix of the differences in the inter-structure distances. To analyse this matrix, we begin with the smallest element of the distance matrix and work towards a structural arborescence by replacing the rows and columns of the matrix with their corresponding mean values. At the end of this procedure, we obtain $N-1$ arborescence levels which characterise the relative movements of the secondary structures treated as rigid bodies. In order to characterise the relative movements of groups of secondary structures (sub-domains) at

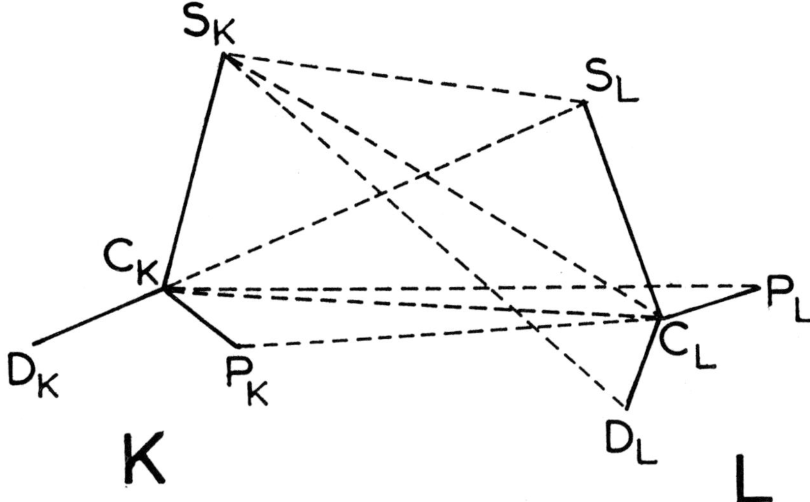

Fig. 1. Point representation of secondary structures and calculated distances. S: starting point, C: centre of inertia, D: point on the dyad axis of the central residue at 1.4 Å from C, P: point perpendicular to the principal inertial axis and the dyad axis.

each level of the tree, we calculate the best rotation matrix (using the superposition method of McLachlan [10]) and the corresponding rotation angle and axis as well as the displacement vector of the centre of mass of the sub-domains.

Results and Discussion

We first describe tests of the new domain analysis method on proteins conformations which have already been analysed by manual techniques.

LIVER ALCOHOL DEHYDROGENASE

The structure of liver alcohol dehydrogenase (LADH) has been solved for the apo enzyme and for a number of coenzyme and/or substrate complexes. A schematic drawing of the apo form (file 8ADH in the Brookhaven Data Bank [11]), obtained using the Molscript program [12] is given in Figure 2. Each monomer has 22 secondary structures, 9α-helices and 13β-strands. The structure has been described as consisting of two domains: a catalytic domain formed by residues 1–175 and 319–374 and the coenzyme binding domain formed by residues 176–318. The active site of the enzyme is at the bottom of a cleft between these two domains. This enzyme exists as a dimer, bound together by the hydrogen bonds between the β-sheets of the coenzyme binding domains. The analysis of the apo and holo crystal structures of the enzyme has led Colonna-Cesari *et al.* [13] to describe the reaction pathway as consisting of a rotation of the catalytic domain by 10° without translation. Our analysis technique (using the structures 6ADH and 8ADH) gives the results shown in Figure 2. The division of the enzyme into two domains is clearly visible. In agreement with the earlier analysis the rotation

Liver Alcohol Dehydrogenase

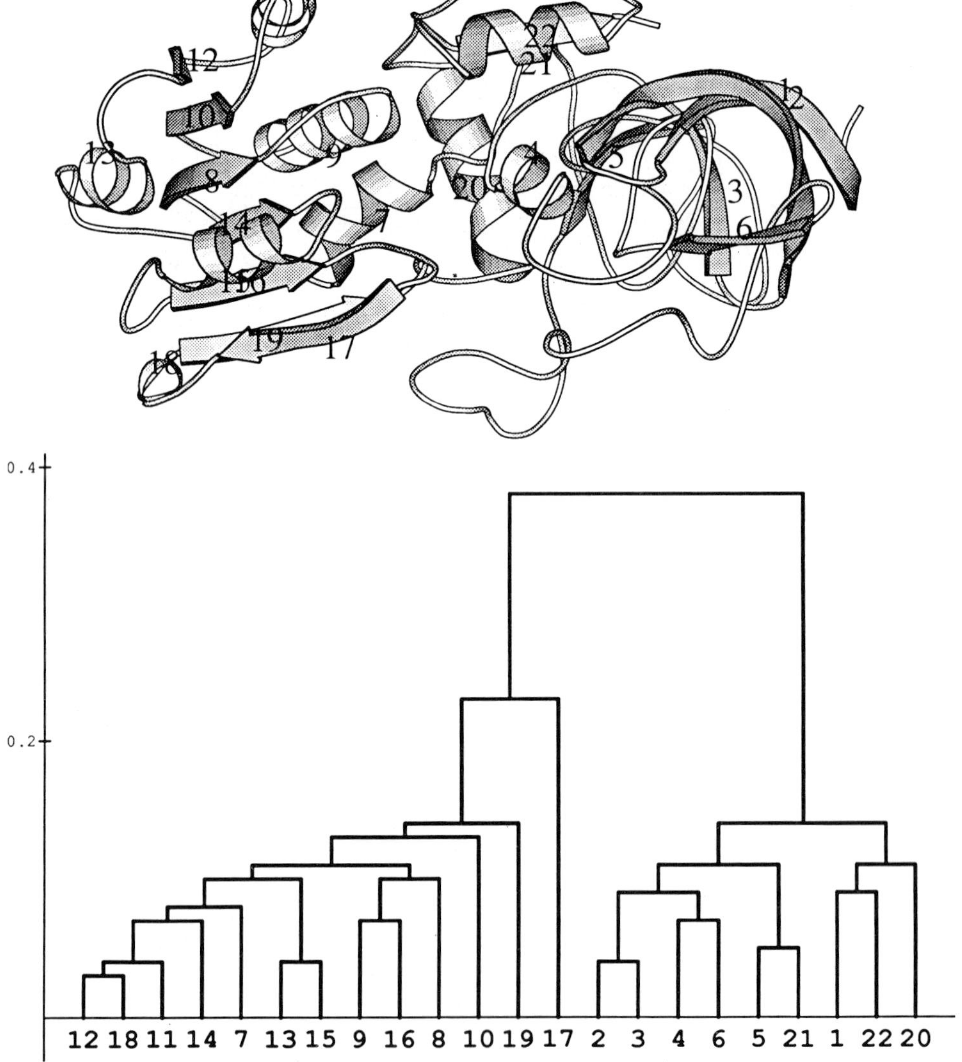

Fig. 2.   Molscript representation and dynamic domains of liver alcohol dehydrogenase.

angle of the catalytic domain is calculated as 10°, but this is found to be associated
with a translation of the centre of mass of 2.4 Å.

D-GLYCERALDEHYDE-3-PHOSPHATE DEHYDROGENASE

The two structures used for this analysis are the holo (1GD1) and apo (2GD1)
forms. The Molscript diagram of the holo form is given in Figure 3 together with

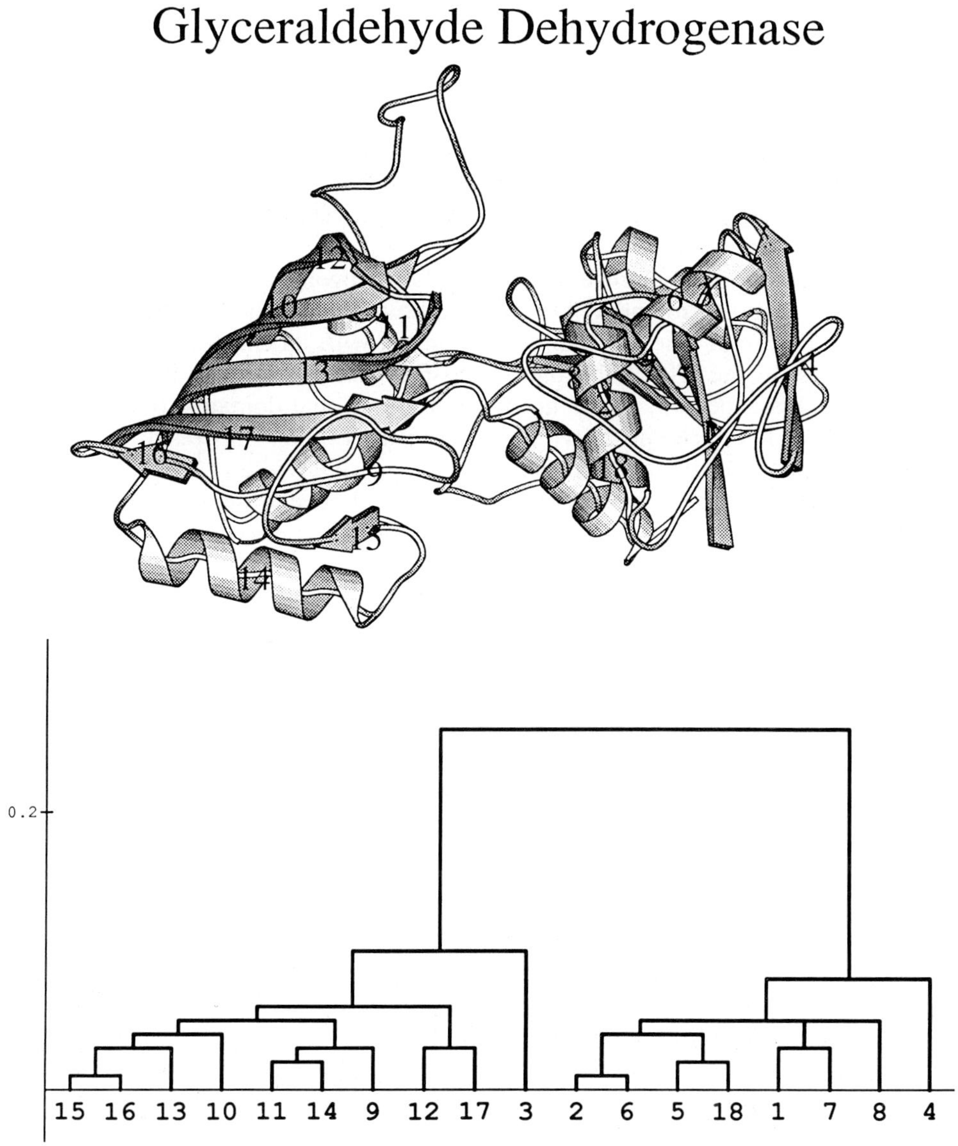

Glyceraldehyde Dehydrogenase

Fig. 3.   Molscript representation and dynamic domains of glyceraldehyde-3'-phosphate dehydrogenase.

the analysis of its domains. Two domains are again clearly visible: a catalytic domain consisting of secondary structures 1–8 and 18 and a larger NAD (nicotinamide adenine dinucleotide) binding domain. The movement observed between these domains is a 6.4° rotation and a 1 Å translation. It is interesting to note that, in our analysis, helix 3, which structurally belongs to the catalytic domain, is actually assigned to the NAD binding domain. Support for this unusual finding comes from a recent analysis of the same structural transition [14, 15] which also concluded that the catalytic domain rotates by 5.5° with the exception of the helix 3 which rotates by only 1°.

L-LACTATE DEHYDROGENASE

Lactate dehydrogenase has been included in our test set of proteins because the conformation change between its apo and holo forms does not in fact involve domain movement as such.

Two conformations of dogfish M_4 lactate dehydrogenase were used for the test (6LDH for the apo form and 1LDM for the holo form). The latter conformation is shown in Figure 4. The NAD binding domain with its characteristic system of beta sheets is on the left of the figure and the catalytic domain is on the right. Our analysis, as expected, shows no division into global domains, but does show a movement of helices 8 and 19 with respect to the others. This result agrees with the analysis of the enzyme conformational change by White *et al.* [16], in which it was found that upon coenzyme binding the loop between the strand 7 and helix 8 changes its conformation in order to seal the bound coenzyme.

HEXOKINASE

Another example of an enzyme in which the conformational change is very impor-tant for the catalytic reaction is hexokinase. This protein contains 457 peptides and 20 secondary structures of which 12 are α-helices and 8β-sheets. The Molscript schematic diagram in Figure 5 shows the complex of form B hexokinase with *o*-toluoylglucosamine (2YHX), corresponding to the apo form of the enzyme. Our analysis of the dynamic domains was made with respect to the glucose complex (1HKG), which is also shown in Figure 5. It can be seen that the principal division separates a large domain (containing most of the protein) from a small domain containing structures 3–8. This division agrees with earlier descriptions [17]. The small domain movement can be described by a 12° rotation and a 2 Å translation. The rotation angle again agrees with previous estimations. It can be seen in Figure 5 that the large domain of hexokinase is actually further subdivided into two parts corresponding to the left lobe and the central part of the U-shaped enzyme. These sub-domains undergo a relative rotation of 3° and the shift of 1.5 Å between the two conformations studied.

CITRATE SYNTHASE

We now turn to the case of citrate synthase where we possess a more detailed model of the enzymatic pathway. However, we will begin our analysis with a simple comparison of the crystallographic open and closed forms [18–21]. The conformational change between these two forms has been studied by Lesk [3,4] and the results frequently cited. For this reason, we maintain an identical α-helix lettering scheme, to that originally given by Remington *et al.* [18]. Note, however, that our P-curves analysis does not in fact classify the B and C helices as secondary structures [7]. Our helices are also systematically shorter than those of Remington *et al.* since we adopted as secondary structures only those segments which retained

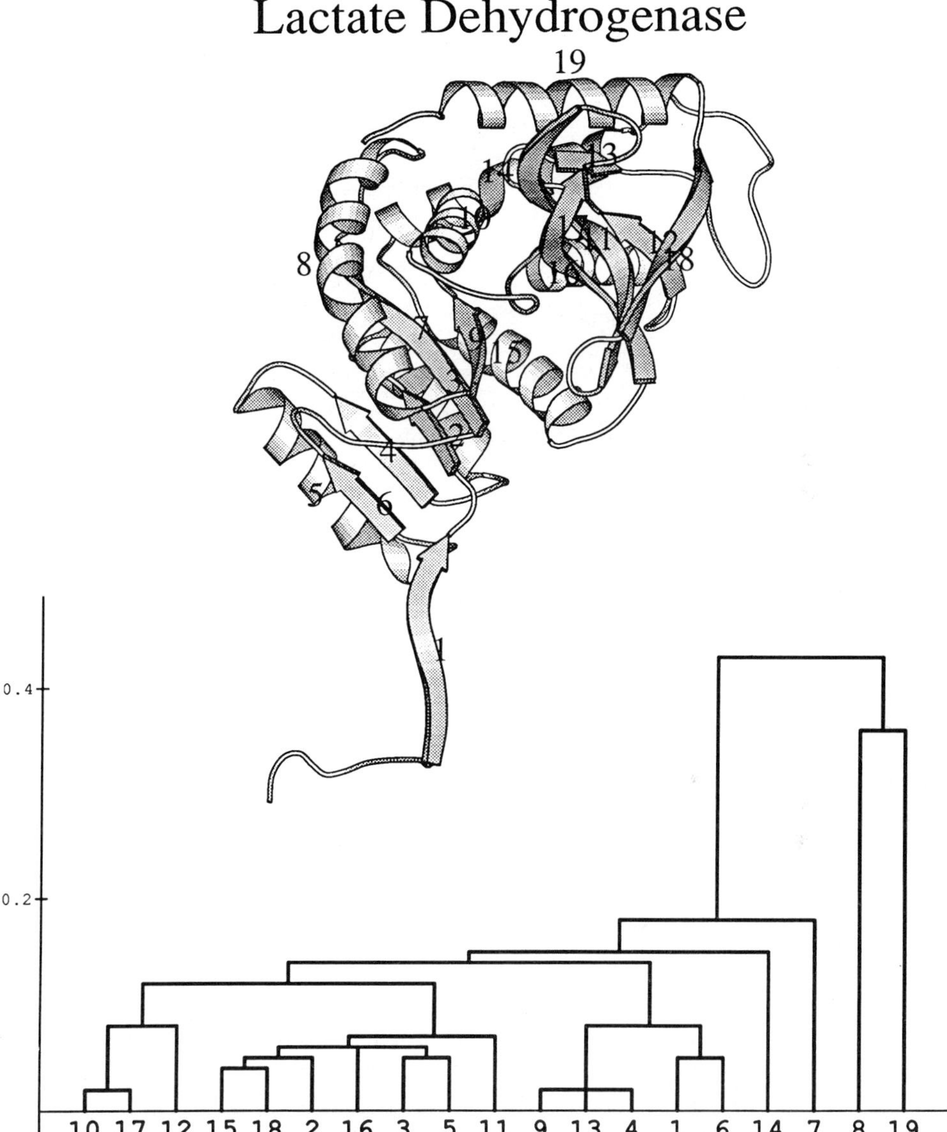

Fig. 4. Molscript representation and dynamic domains of lactate dehydrogenase.

this classification throughout the transition pathway. In this way we were able to limit the effects of the distortion of the ends of helices which can be quite important along the transition path, particularly close to the active site [7].

In Figure 6 we give the Molscript diagram of the open form of citrate synthase, together with the dynamic domain structure. As expected, there is a clear separation into two domains, a small domain composed of four helices N, O, P and Q and a large domain composed of the rest of the enzyme. The large domain can

Hexokinase

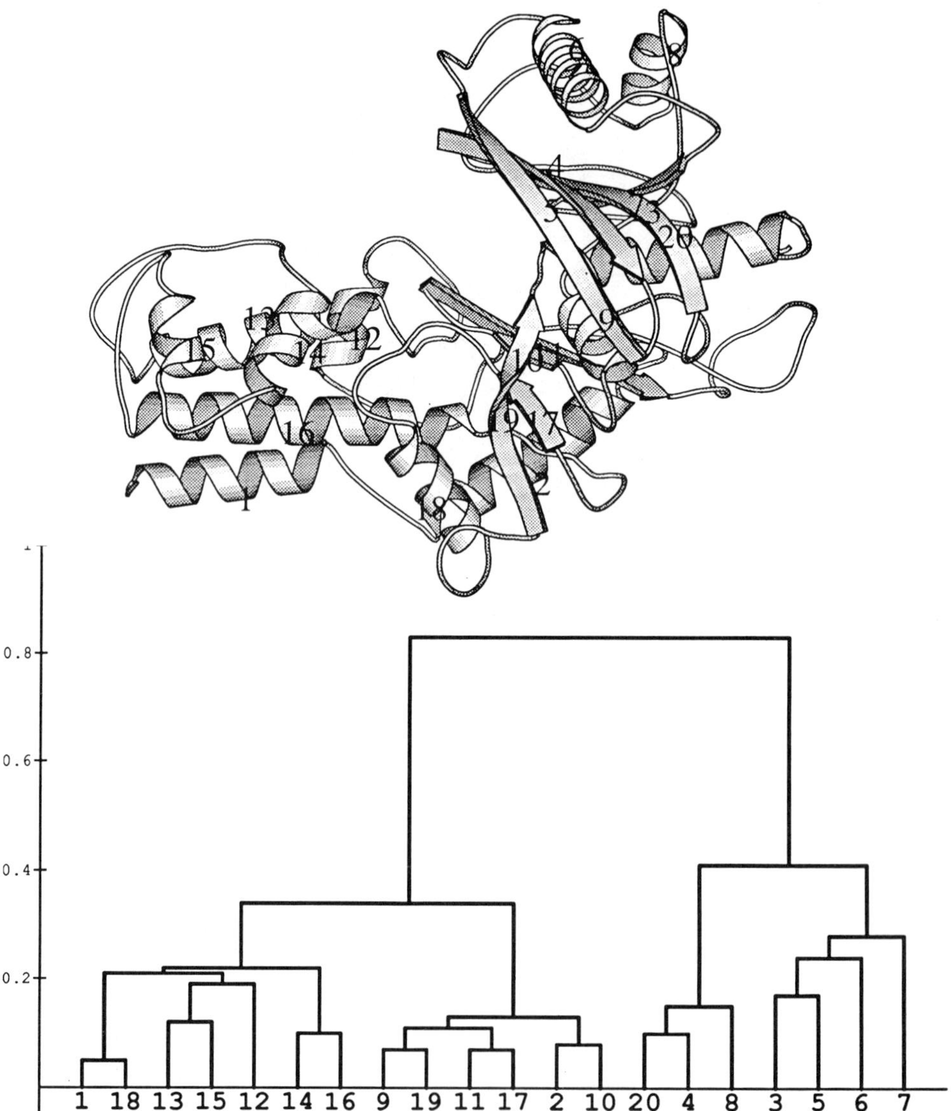

Fig. 5.   Molscript representation and dynamic domains of hexokinase.

be further divided into at least three sub-domains, which can be roughly described
as 'slices' through the enzyme. Table I lists the composition and relative rotations
and translations occurring between the sub-domains. It can be seen that the
internal rearrangements within the primary domains are quite important as the
enzyme passes from its open to its closed form. The largest rotation of the small
domain with respect to the large domain is 21°. Globally, our description of the
citrate synthase transition agrees well with that of Lesk, with however one impor-

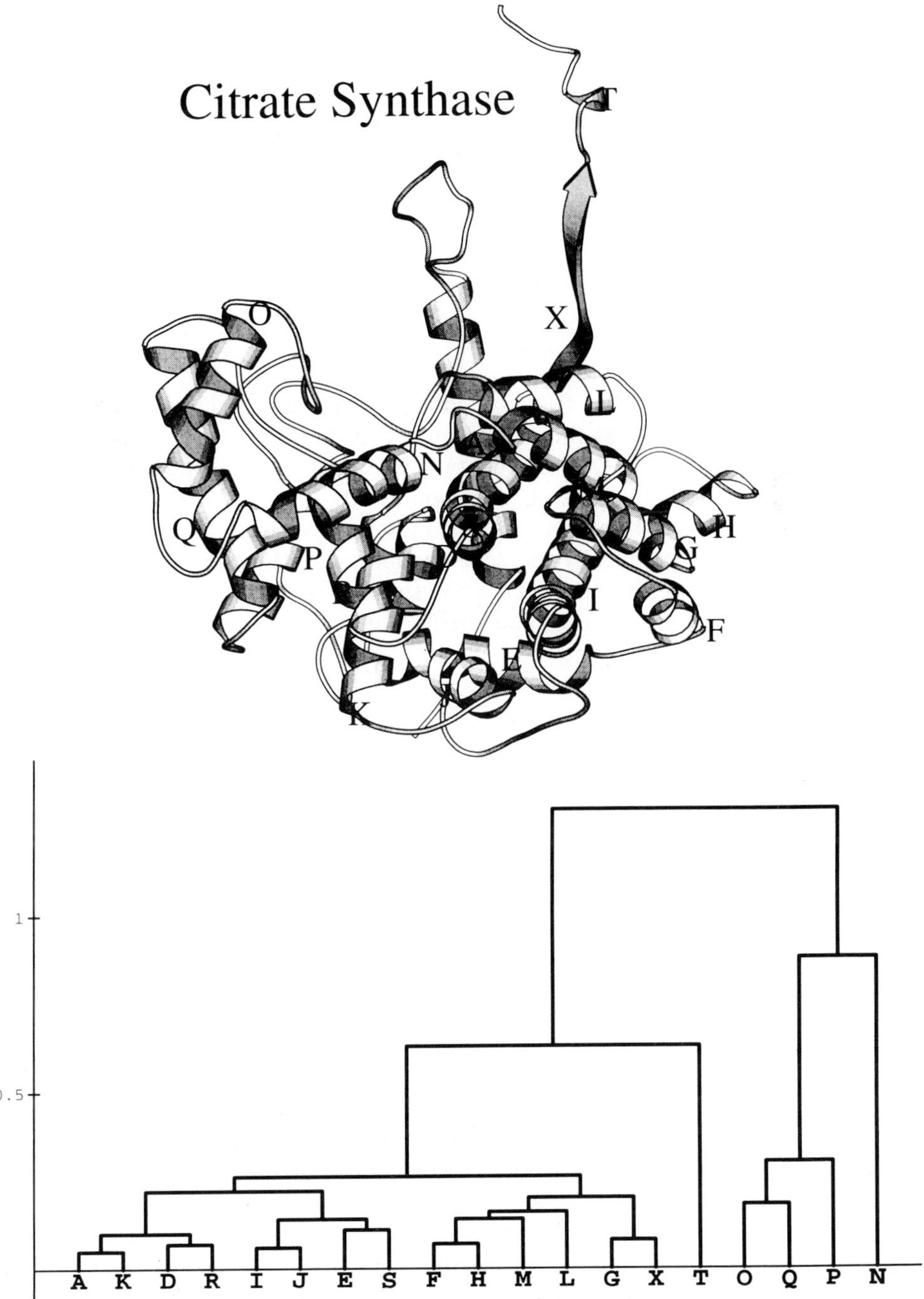

Fig. 6.   Molscript representation and dynamic domains of citrate synthase.

TABLE I

Rotation angles and centre of mass translations within the sub-domain structure of citrate synthase

Helices	Helices	Rotation angle (°)	Translation (Å)
G,X	→ F,H,M,L	3.3	0.6
A,K,D,R	→ I,J,E,S	1.	0.5
A,K,D,R,I,J,E,S	→ F,H,M,L,G,X	6.0	1.
T	→ Rest of big domain	27.4	2.7
O	→ Q	18.4	1.3
P	→ O,Q	18.6	1.8
N	→ O,P,Q	23.1	2.6
O,Q,P,N	→ Big domain	20.8	5.8

tant distinction involving the R helix which belongs to the large domain in our analysis, but was assigned to the small domain by Lesk.

We now pass to an analysis of the two model transition pathways of citrate synthase. To simplify the graphic representations of these pathways we group together those helices whose average distance change does not exceed 0.2 Å. In Figure 7 we present graphically the case of the so-called interpolated pathway [5] and in Table II we give the numerical analysis of both pathways. In each case, we indicate the helices which moved and their rotations and translations with respect to the static part of the enzyme.

Two important conclusions can be drawn from this analysis. Firstly, at each step along the transition pathway only a limited number of helices move. Secondly, the moving helices do not exclusively belong to the small domain. This latter observation corresponds to our earlier analysis at the level of individual ϕ and ψ angles. Using this data, we located a number of residues which had similar conformations in both the open and close forms, but nevertheless changed along the pathway, often passing temporarily into the forbidden zones of the Ramachandran map. We termed such effects transitory movements. We believe that these movements, whose existence can now be confirmed at the level of secondary structure displacements, are probably necessary to overcome steric hindrances which would otherwise build up along the pathway.

Conclusions

A simple and objective method has been developed to analyse conformational changes within proteins at the level of secondary structure movements. This method uses P-curves analysis to detect secondary structures which are subsequently treated as rigid bodies. A hierarchical analysis of the relative movements of these rigid bodies then allows structurally stable domains to be defined. The method has been successfully tested on a number of enzymatic proteins whose conformation changes have previously been characterised by manual techniques.

The application of our new method to two model transition pathways for citrate synthase shows that the passage between open and closed forms of this enzyme is more complex than the classical view involving only the rotation of a small domain with respect to a large domain. It is seen that several sub-domains can be detected

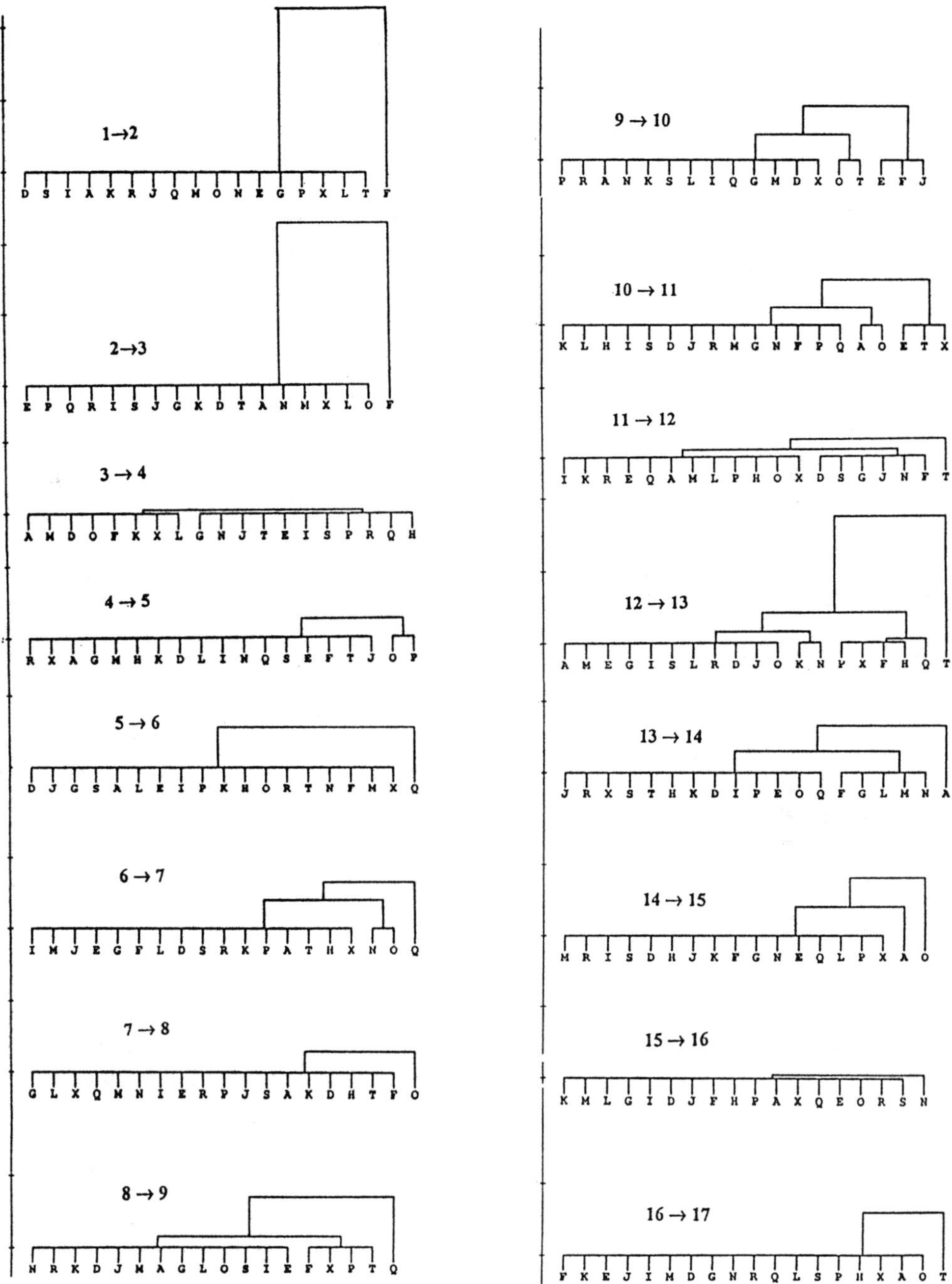

Fig. 7. Simplified dynamic domains of citrate synthase within the transition pathway between the closed and open forms.

TABLE II
Sub-domain movement along the citrate synthase transition pathway

Interpolated pathway				Directed pathway		
Step	Helices	Rotation (°)	Translation (Å)	Helices	Rotation (°)	Translation (Å)
$1 \rightarrow 2$	F	−21.6	1.8			
$2 \rightarrow 3$	F	27.2	1.8			
$3 \rightarrow 4$				X	4.7	0.5
$4 \rightarrow 5$	O,P	4.5	0.5	G,L,T	2.2	1.0
				F,N	4.0	0.4
$5 \rightarrow 6$	Q	5.8	0.8	O	8.3	1.0
				E,H,K,L,X	1.0	0.4
$6 \rightarrow 7$	N,O	2.4	1.0	F,G	4.6	0.5
				H	15.4	3.2
				A,T,L	2.6	1.1
$7 \rightarrow 8$	O	4.6	1.1	X	4.0	0.8
				A	6.1	0.4
				F,Q	2.9	0.8
$8 \rightarrow 9$	F,X,P,T	0.9	0.5	I	1.0	0.3
	Q	6.2	1.0	O,Q	9.1	0.8
$9 \rightarrow 10$	O,T	3.6	0.9	J,P,F,T	1.5	0.7
	E,F,J	4.5	0.5	A	4.4	1.2
$10 \rightarrow 11$	A,O	2.7	0.3	A,T	1.1	0.9
	E,T,X	2.0	0.5			
$11 \rightarrow 12$				O	11.3	0.7
$12 \rightarrow 13$	T	22.4	2.2	L,M,X,O	2.0	0.4
$13 \rightarrow 14$	A	10.0	0.7	D,E,J,K O,P,Q,R	2.8	0.7
$14 \rightarrow 15$	A	3.2	0.5	O	15.5	1.8
	O	28.6	1.4			
$15 \rightarrow 16$				R	6.1	0.5
				M,N	3.3	0.8
				O	15.7	0.7
$16 \rightarrow 17$	T	20.9	1.0	O	4.1	1.2

and that transitory movements of these sub-domains occur within the large domain. Since similar transitory movements were detected earlier at the level of individual amino acid residues, the present analysis appears to support the notion that such transitory effects may be a significant feature of enzymatic reaction pathways.

References

1. J. Janin and S.J. Wodak: *Prog. Biophys. Molec. Biol.* **42**, 21 (1983).
2. W. S. Bennett, Jr. and R. Huber: *CRC Crit. Rev. Biochem.* **15**, 291 (1984).
3. A. M. Lesk and C. Chothia: *J. Mol. Biol.* **174**, 175 (1984).
4. C. Chothia and A. M. Lesk: *Trends Biol. Sci.* **10**, 116 (1985).
5. M. A. Ech-Cherif El-Kettani and J. Durup: *Biopolymers* **32**, 561 (1989).
6. J. Durup: *J. Phys. Chem.* **95**, 1817 (1991).
7. M. A. Ech-Cherif El-Kettani, K. Zakrzewska, J. Durup, and R. Lavery: *Proteins: Structure, Function, and Genetics* **16**, 393 (1993).
8. H. Sklenar, C. Etchebest, and R. Lavery: *Protein: Structure, Function and Genetics* **6**, 46 (1989).

9. K. Zakrzewska and R. Lavery: in *Modelling of Molecular Structures and Properties*, J.-L. Rivail (Ed.), Studies in Physical and Theoretical Chemistry, Elsevier, Amsterdam, Vol. 71, p. 81 (1990).

10. A.D. McLachlan: *J. Mol. Biol.* **128**, 49 (1978).

11. F. C. Bernstein, T. F. Koetzle, G. J. B. Williams, E. F. Meyer, M. D. Brice, J. R. Rodgers, O. Kennard, T. Shimanouchi, and M. Tasumi: *J. Mol. Biol.* **112**, 535 (1977).

12. P. J. Kraulis: *J. Appl. Crystallogr.* **24**, 946 (1991).

13. F. Colonna-Cesari, D. Perahia, M. Karplus, H. Eklund, C. I. Branden, and O. Tapia: *J. Biol. Chem.* **261**, 15273 (1986.)

14. A. G. W. Leslie and A. J. Wonacott: *J. Mol. Biol.* **178**, 743 (1984).

15. T. Skarzynski and A. Wonacott: *J. Mol. Biol.* **203**, 1097 (1988).

16. Ch. M. Anderson, R. E. Stenkamp, and T. A. Steitz: *J. Mol. Biol.* **123**, 15 (1978).

17. W. S. Bennett, Jr. and T. A. Steitz: *J. Mol. Biol.* **140**, 211 (1980).

18. G. Wiegand and S. Remington: *Ann. Rev. Biophys. Biophys. Chem.* **15**, 97 (1986).

19. S. Remington, G. Wiegand, and R. Huber: *J. Mol. Biol.* **158**, 111 (1982).

20. G. Wiegand, S. Remington, J. Deisenhoffer, and R. Huber: *J. Mol. Biol.* **174**, 205 (1984).

21. M. Karpusas, B. Branchaud, and S. J. Remington: *Biochemistry* **29**, 2213 (1990).

Applications of Empirical Amino Acid Potential Functions

R. L. JERNIGAN,[1] L. YOUNG,[1,2] D. G. COVELL,[2] and S. MIYAZAWA[1,3]
[1]*Section on Molecular Structure, Laboratory of Mathematical Biology, National Cancer Institute, NIH, Bethesda, MD 20892, U.S.A.*
[2]*PRI-Dyncorp, Biomedical Supercomputing Laboratory, Bldg. 430, Frederick, MD 21701, U.S.A.*
[3]*Gunma University, Faculty of Technology, 1-5-1- Tenjin, Kyuryu, Gunma 376, Japan*

Abstract. Many characteristics of protein folds can be described at less than atomic detail. We have been pursuing a reductionist approach to represent proteins in which residues, rather than atoms, are considered as single points. Direct averages of protein crystal structures provide empirical potential functions that describe residue–residue interactions. These functions have averaged away many details but include essential features with attractive and repulsive terms, that are overall akin to a pair-wise hydrophobicity. Most computational evaluations of protein structures are deficient in not considering a broad enough range of possible conformations; this less detailed approach makes possible a more thorough consideration of conformations. We are developing a new lattice approach to generalize and improve the efficient generation of folded protein conformations, and these empirical potentials are an important part of this approach. Applications of these potential functions presented here include the location of binding sites for peptides on proteins based on geometry and hydrophobicity, specification of similarities of sequence substitutions, and investigations on the variability of the hydrophobicity of a given type residue averaged over occurrences in individual proteins. Recent progress in these areas will be described. This new approach is leading us to the ability to consider more complete sets of protein conformations.

Key words. Protein binding, empirical potential energies, amino acid substitutions, protein fold assessment, hydrophobicity.

Introduction

One major goal of structural biology is to understand how molecules recognize and interact with one another. Comprehension of the important interactions could enhance the understanding of numerous biological processes, as well as provide a sound basis for drug design. For example, it would lead to more rational design of improved inhibitors or substrates for enzymes because it would permit better use of the remarkable tools of site-directed mutagenesis and other gene splicing techniques. Theory can play a role in providing a deeper understanding to aid in developing new approaches.

An eventual goal is the direct calculation of active, stable macromolecular conformations, from their sequences. Calculations precise enough to place many thousands of atoms at their best positions are not feasible at present. There have been numerous demonstrations that the full atomic details of even small proteins cannot be calculated directly. The development of successful higher order approaches to molecular structure is essential before we can achieve a complete understanding of all the complexities of biological macromolecules themselves, as well as their interactions with other molecules and their assembly into biological structures. Even today, it is difficult to foresee the ability to determine rapidly the

A. Pullman et al. (eds.), Modelling of Biomolecular Structures and Mechanisms, 151–166.
© 1995 *Kluwer Academic Publishers. Printed in the Netherlands.*

most stable conformation of the largest protein, by calculating interactions among all atoms. Specifying all details of protein self-assembly or their assembly into larger structures would be even more difficult.

The approach that we have been pursuing is to develop ways to treat a more complete set of conformations, but with less detail, with the intention of obtaining approximately correct overall folds. Following this it is possible, as has been demonstrated, to complete the atomic details by refinement of the overall trajectory of the protein backbone. This approach has been successfully demonstrated, and the tactic has also been taken up by other research groups. The main tool in this process is a set of empirical residue–residue potential functions derived directly from structural data. In our case [1] these potentials are derived simply by counting in crystal structures the number of close approaches of all pairs of residues of different types.

Here we are focusing on several applications of these residue–residue potential functions that had been derived previously. These concepts have been applied to locate peptide binding sites on the surfaces of proteins [2], to develop a matrix relating the ease of amino acid substitutions [3], to evaluate the quality of non-native folds [4], and to develop some other simple models of proteins to answer questions such as: Is the observed native state identical to the lowest energy conformation [5]? How easy is it to make large scale transitions in proteins [6]? Below, we will summarize and discuss in detail the peptide–protein binding site results. Successes with all of these applications are encouraging us to pursue some further improvements to these residue–residue potential functions. More detailed residue–residue mean field interaction potentials are going to be developed, to include details related to the dense packing of residues. Further important factors to include in such potential functions are near neighbor interactions within short chain segments and an overall effect caused by the constraint of compactness which has recently been investigated [5]. It was found that the longer the separation in sequence, the more favorable the interaction because of the compactness.

Previously, in treating globular protein conformations [4], the protein had been confined to a space exactly the size and shape of the known native conformation, and all conformations were generated in that space. It was remarkable that, by this enforced constraint it became possible to generate thousands of highly varied conformations, in the approximation of one point per residue on a lattice. These were evaluated with the same residue–residue interaction potentials mentioned above. By applying the interaction values to all of the compact conformations enumerated for several small proteins, the native conformation was always found among the best few per cent of all the forms, and by use of these potentials most incorrect folds could be discarded.

Interactions in known protein–ligand and protein structures have been analyzed extensively to develop a tool to aid inhibitor design. We have found that hydrophobicity is the substantial determinant of binding strength, and hence the location of the binding site, in a verification of the biochemist's intuitive view of 'sticky patches'.

I. Residue–Residue Potentials

The aim here is to discuss a range of applications that demonstrate the utility of residue–residue potentials and to outline new improvements to these potentials.

METHOD

Our previously derived residue–residue potential functions were obtained from the frequencies of non-bonded proximate residues in crystal structures, on the assumption that these placements, on average, must reflect their effective energies of interaction. (See Reference 1 for the details of the method.) The basic underlying assumption, in treating the interactions between protein residues in this way, is that the close residues are those with the strongest interactions, and that we can ignore, to good approximation, longer distance interactions. The values clearly indicate the importance of hydrophobic interactions, as well as specific preferences for charged pairs of opposite sign. They indicate that the strongest interactions take place among the hydrophobic types of residues, and that the most specific, but weaker, interactions are between polar residues. See Figure 1.

There is extensive experience in the field of polymers for treating interaction energies among groups of atoms as parameters to be evaluated from experimental data on small analogous molecules or even by fitting physical measurements of large macromolecules [7]. (As an example, see Reference 8.) Such approaches were utilized in developing the energy parameters for diverse polymer repeat units with rotational isomeric state models. Part of the rationale for treating these interaction terms as adjustable parameters is that the relative populations of conformers can change, depending on the experimental conditions. For example, studies by Mizushima [9] demonstrated changes in the relative likeliness of isomeric states for different conditions; these results suggest the approach of adjusting potential functions to mimic the specific conditions. In the case of proteins, it is possible to use crystal structures directly to develop relative interaction preferences based upon the observed frequencies of occurrences of residue–residue contact pairs. The direct approach taken in extracting such functions from crystal structures is described in detail in Reference 1. Previously Tanaka and Scheraga [10] had performed a similar averaging of structural data. Subsequently, others including Gregoret and Cohen [11], Sippl [12] and Hinds and Levitt [13] have made similar analyses. Our residue–residue interaction energies resemble a pair-wise hydrophobicity index with the reference state being the residues individually in contact with water. The most important features of the interactions can be obtained only if averaging of the data is performed, to remove errors in the structures as well as to remove individual aberrant cases. Use of all of the structural data is less likely to yield ultimate success than using results after judicious averaging. We have also looked at such interactions in more general ways [5] as a uniform homogeneous term, the same for all interactions, and have probed individual positions to see where changes in interaction strengths would most easily cause changes in conformation [6].

These potential functions have numerous uses for quickly assessing conformational quality. They are useful to select sets of good conformations from large

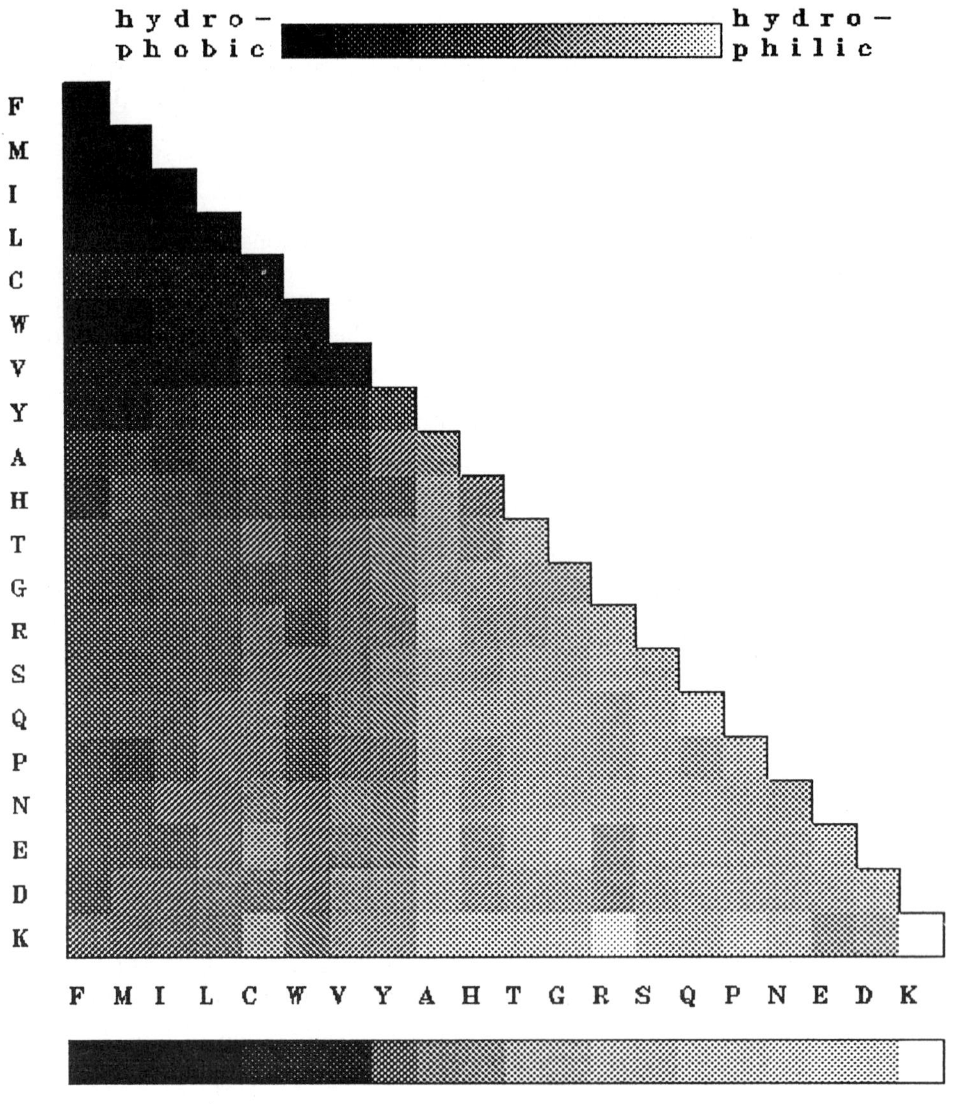

Fig. 1.   The residue–residue interaction energies from Reference 1; residues are designated by one letter codes. The darker squares are the lowest interaction energies and the lighter square the highest.

sets of conformations [4], and to derive amino acid substitution matrices for use in sequence comparisons [3]. Another use of the potentials is to evaluate the effects of amino acid substitutions on stability [14]. An important further area of application of potential functions is for the evaluation of different sequences threaded onto known structures [15,16]. In a similar way, hydrophobicities can be used to locate peptide binding sites on protein surfaces (see Section II below). The successes achieved with all of these applications encourage us to pursue

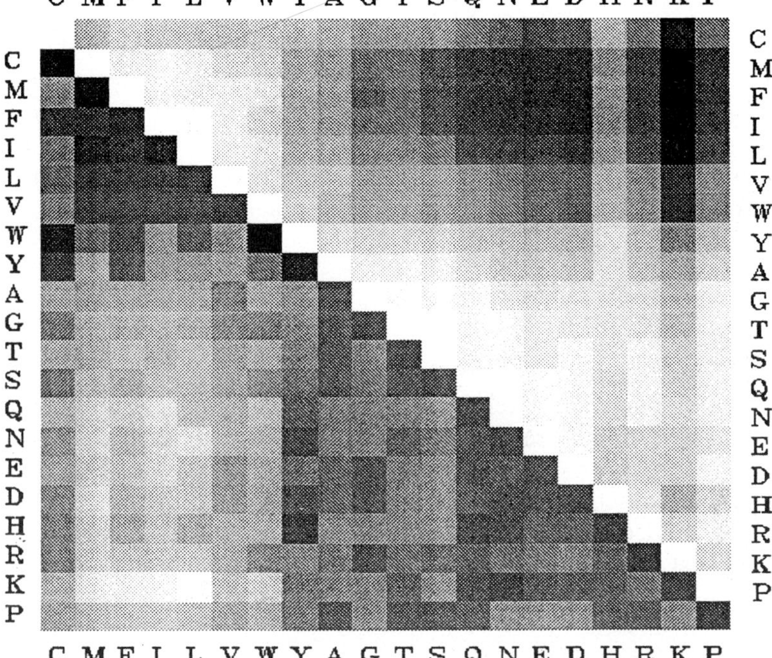

Fig. 2.   The change in energy upon exchange of two types of residues is shown in the upper diagonal half. White corresponds to the most favorable and black to the most disfavored exchanges. Generally, exchanges are more favored within the same class of hydrophobic or hydrophilic residues. The lower diagonal half displays the log relatedness odds matrix corresponding most closely to the PAM 250 matrix. Most similar substitutions are black, intermediate ones gray, and least similar ones white. The pair contact energies [1] and codon frequencies were utilized in its derivation [3].

several additional improvements. To this end, we are beginning the development of a newer generation of residue–residue potential functions.

Results

Amino Acid Substitutions

An amino acid substitution matrix to indicate the relative conservativeness of a substitution was derived. The bases for this table are the set of residue–residue potentials derived from known structures and the probabilities for amino acid codon substitutions. These structurally derived results exhibit a strong correlation with the Schwartz and Dayhoff numbers based on sequences of protein families. Their results have found extensive application for finding similarities among protein sequences. Our results from Reference 3 are shown in Figure 2.

Effects of Compactness

An additional aspect that has been pursued is the development of the long range components of residue–residue potential functions arising simply from the constraints of compactness [5]. The result of compactness is that the intrinsic interactions between residues become monotonically more favorable with an increase in their sequence separation. Hence, the largest effects of compactness are found to favor interactions between the two chain ends, in agreement with observations on known protein structures [4].

CONCLUSIONS

1. The most feasible amino acid substitutions are among hydrophobic residues and among hydrophilic residues. Changes between these two classes are less acceptable.
2. Compactness causes the frequencies of interactions to increase directly with increase in sequence separation. Consequently, interactions between the chain ends are most favored.

SECOND GENERATION OF RESIDUE–RESIDUE POTENTIALS

Since our original derivation of residue–residue potentials [1], the number of diverse protein crystal structures has increased rapidly. One problem with including all known structures is that many of them are closely related to one another. For this purpose, we have devised a weighting scheme that gives greater weights to structures with the most diverse sequences. In 1985 we had simply selected 42 diverse high quality structures for our data collection; now we have used, together with this weighting scheme, 1168 separate protein structures with a total of 1661 subunit structures, each of which is longer than 49 residues and has a stated resolution of 2.5 Å or better. This includes more than 54 000 effective residues, an increase of about 6-fold over the earlier data collection. A comparison is shown in Figure 3 between some of the old and new data that has been collected.

One modification to the potential functions will be the addition of a repulsive term to the previous attractive term. Again, the repulsive part of the residue–residue potential is being derived statistically from known structures in the X-ray crystal database. This additional term will complement the previously derived attractive component that was found to be effective in determining proper or improper chain folds. Results with this previous attractive potential were often too compact [17]. This new version should improve evaluations of residue packing in proposed structures. Other modifications will include some of the short range interactions which previously were ignored. These are collected from average side chain positions that are sequence neighbors, including their virtual bond lengths, virtual bond angles and virtual torsion angles. The sequence dependence of such terms should reflect the effects of solvent and the interior environment of the protein better than have previous potentials.

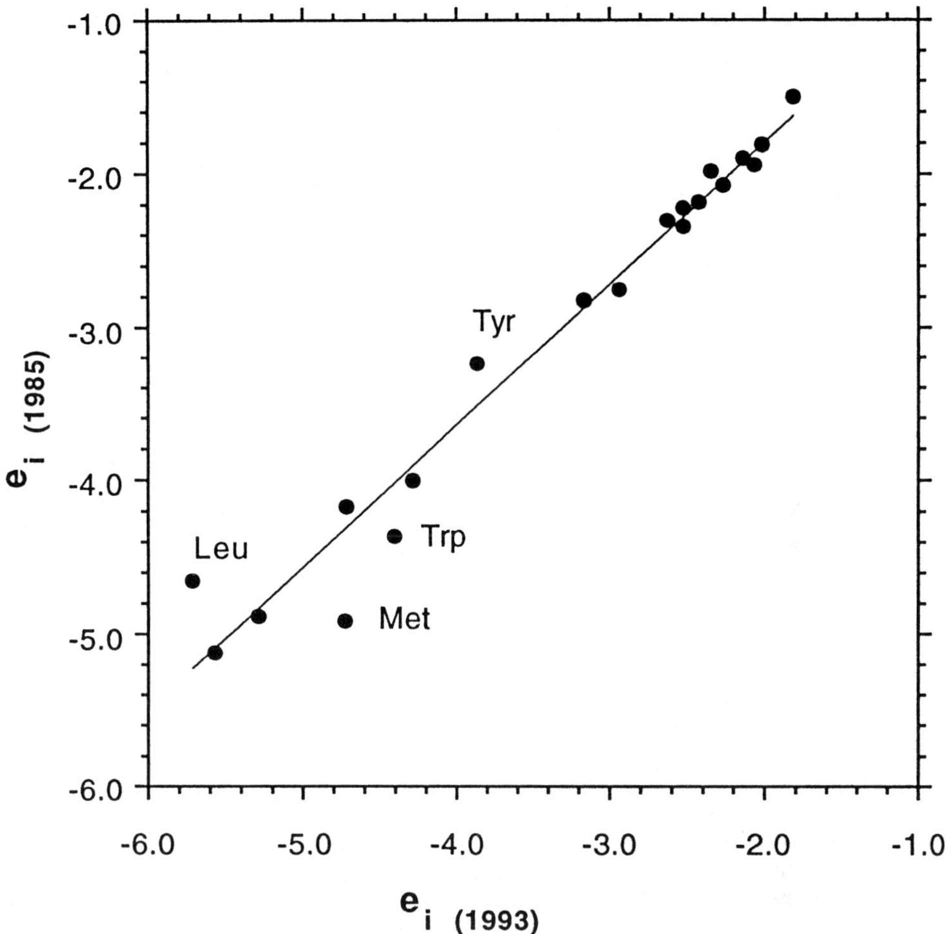

Fig. 3. Comparison between the hydrophobicity scale derived from 42 protein structures ('85) and the new ones derived from 1661 different protein subunits, by using sequence similarity weights ('93). This scale is calculated by Equation (1) from the pairwise contact energies. There are some expected changes; for instance, the most infrequent residues Trp and Met have substantial changes, presumably because of the increased amount of data. Substantial changes are also observed in the values for Tyr and Leu. The latter is especially surprising since all of the other high frequency residues exhibit only small changes.

DISCUSSION

One of the questions to investigate is the extent of burial or exposure of hydrophobic and polar residue on a local basis. This could be investigated initially, for instance, by determining whether or not polar residue pairs close in sequence are physically closer than polar-hydrophobic pairs. Biased conformations derived from such an investigation could be used statistically as conformational tiles in an overall conformation generation. These considerations of short range interactions differ

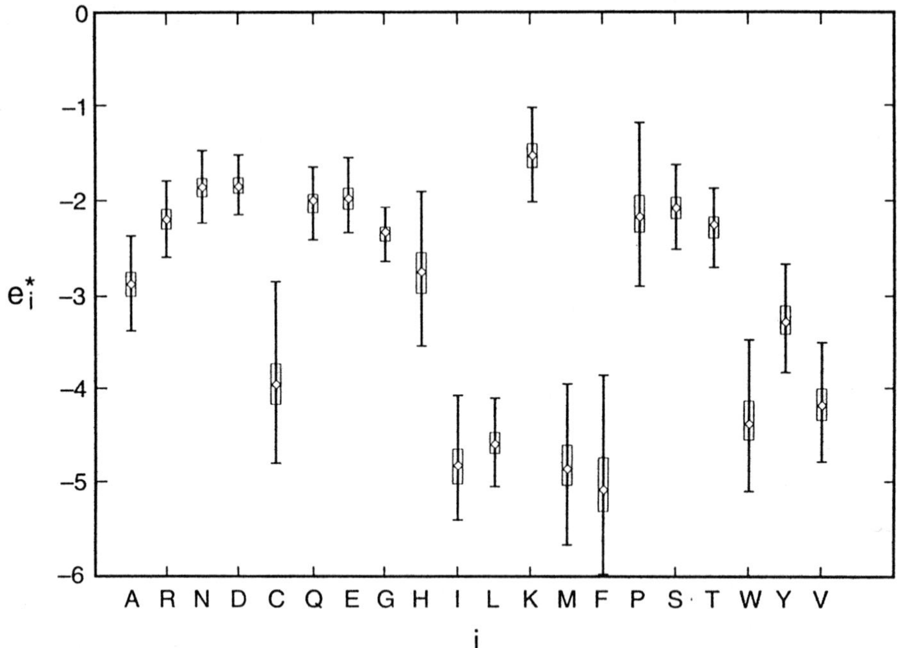

Fig. 4.   Variability of environmental interaction energies. These were calculated from the e_{ij} values in Reference 1 by summing over all neighboring residues within 6.5 Å in 709 protein subunit crystal structures. Means of the values for each protein are given as central points; half of the occurrences are within the boxes and 99.8% of all occurrences are within the range of the bar.

substantially from previous studies where short range interactions were usually calculated only in vacuo between nearest neighbor residues.

The regularity of the positions of residues around a central residue is striking [1]. The coordination number is not so variable, and also the hard core radii do not vary so widely. And, as we will see, the environment of a given residue type is remarkably constant.

HOW CONSTANT ARE ENVIRONMENTS OF A GIVEN RESIDUE TYPE?

In an attempt to investigate this question, we undertook the following data collection: For each of 709 crystal subunits, we have calculated the average interaction energy per interacting pair. In Figure 4 is shown the distribution of these values for the set of proteins for each type of residue.

$$e_i^* \text{ (each protein)} = \sum_j n_{ij} e_{ij} \bigg/ \sum_j n_{ij}$$

where n_{ij} is the number of i-type, j-type pairs having the mean of their side chain atoms within spheres of 6.5 Å radius for one protein. The nearest sequence neighbors have been excluded in the counts.

The values reflect the hydrophobicities; the polar residue types have weaker interactions and usually lie in a band between a value between -1.5 and -2.5 for e_i^*. The non-polar types have lower values, typically between -4 and -5. Several residue types fall between these two ranges, notably alanine, cysteine, histidine, proline and tyrosine.

The fascinating part of this result is that, even over the wide range of proteins that are included in the sample, there are clearly separable regions of total hydrophobicity of the environments for all occurrences of a residue type. Polar residue types have weaker total interactions but usually lie within a relatively narrow band of values. The non-polar types have lower values, but larger variabilities. This smaller variability of the hydrophobic environment for polar residues indicates that they would be particularly good residues for specifying their most probable location on the protein globule surface.

Another important class of interactions is between ions and proteins. Understanding their specific roles in stabilizing conformations and the sensitivities of conformation to the coordination with different ions is important. We have begun work on this problem by investigating the relative stabilities of K^+ and Na^+ ions in four stranded DNA structures. The stabilizing effects of Zn^{2+} in some protein structures and the regulating effect of Ca^{2+} on protein function are exciting problems to pursue [18,19]. Tools to place such ions, based on their coordination geometries, could be developed and included in lattice simulations. Treating the relative stabilities of protein or nucleic acid ion coordinated structures may, however, require better treatments of solvent and electric calculations. A better understanding of specific ion effects in protein structures ought to make possible the addition of ion coordination complexes as important construction elements in protein design.

II. Using Hydrophobicities to Locate Peptide Binding Sites on Proteins

The aim here is to develop a method to scan rapidly the surface of a protein of known structure to locate probable peptide or protein binding sites. The hypothesis is that it ought to be possible to select target binding sites on a macromolecule on the basis of its surface geometry and potential interactions, especially those of a hydrophobic nature.

METHOD

A new method to locate the most probable peptide binding sites on protein surfaces is presented. This simple method permits rapid screening of molecular surfaces for sites that would interact strongly with any hypothetical peptide. Favorable binding positions exterior to the surface of the target protein are derived on the basis of the total hydrophobicity of neighboring protein residues. When tested on 23 available peptide-protein X-ray structures, the observed sites always fall among the best 0.7% of all possible binding sites, with one exception. *These results strongly support the view that the hydrophobic interactions are responsible for the strength of protein–protein association and that the hydrophobic surface clusters can be used to choose small sets of surface loci for ligand attachment.*

Previous methods to locate important surface targets that include both active and non-active sites have typically focused on identifying regions with high densities of hydrogen bond sites [20,21], regions with desirable electrostatic properties [22], regions with appropriate features of surface curvature [23], and regions of high surface complementarity [24].

We fit the C^α positions of the target protein to a regular lattice and use a set of lattice points immediately exterior to the protein to define the binding space. Covell and Jernigan [4] found that the lattice providing the best fits of virtual bond and torsion angle geometry in the C^α backbone was the face-centered cubic lattice with unit edge 3.8 Å, corresponding to the fixed virtual bond distance between neighboring $C^{\alpha'}$s. With this lattice, an accurate model for the C^α positions of the target molecule is obtained, with overall fits of about 1 Å RMS deviation from the crystallographic protein structure. We simply extend the same lattice used to define the C^α backbone into the region surrounding the protein [25]. These lattice points define a region exterior to the protein surface (between 6.1 and 9.0 Å) that is appropriate for placement of an inhibitor peptide. The scheme in Figure 5 describes the geometry and shows how the clusters of points are developed. The size of the cluster of points reflects the concavity of the surface.

The hydrophobicity of a given protein cluster is taken to be simply the sum of the hydrophobicities of its constituent residues. The hydrophobicity scale used is based upon the contact energies calculated by Miyazawa and Jernigan [1] given at the bottom of Figure 1. This statistical study of non-bonded residue contact frequencies in protein structures in the Brookhaven Protein Data Bank (PDB) [26,27] investigated the local neighborhoods of the 20 residue types in 42 high resolution crystal structures and used a quasi-chemical model of the pairwise interactions between the different types of amino acids and solvent to calculate a set of residue–residue contact energies e_{ij}. These pairwise contact energies were derived for a model using a single point representation of the amino acids in which the point is placed at the center of side chain atom positions and all interactions within a distance of 6.5 Å, are counted [1].

Averages of these pairwise contact energies are used as the hydrophobicities for each residue type, i:

$$e_i = \frac{\Sigma_{j=1}^{20} e_{ij} n_{ij}}{\Sigma_{j=1}^{20} n_{ij}} \tag{1}$$

where the contact energy e_{ij} is the Miyazawa and Jernigan energy difference between an ij amino acid pair. i and j are each one of the standard 20 types and solvent. n_{ij} is the number of ij contacts in the set of structures used to calculate e_{ij}. These contact energies can be understood in terms of the hydrophobic-hydrophilic designations of amino acids and the pairings that contribute most to protein stability:

strongest	hydrophobic–hydrophobic
intermediate	hydrophobic–hydrophilic
weakest	hydrophilic–hydrophilic

This hydrophobicity scale in Equation (1) shows a strong correlation with the Nozaki–Tanford scale [28] and others, as shown by Cornette *et al.* [29]. These hydrophobicities are used to define the energy for a protein cluster which is then

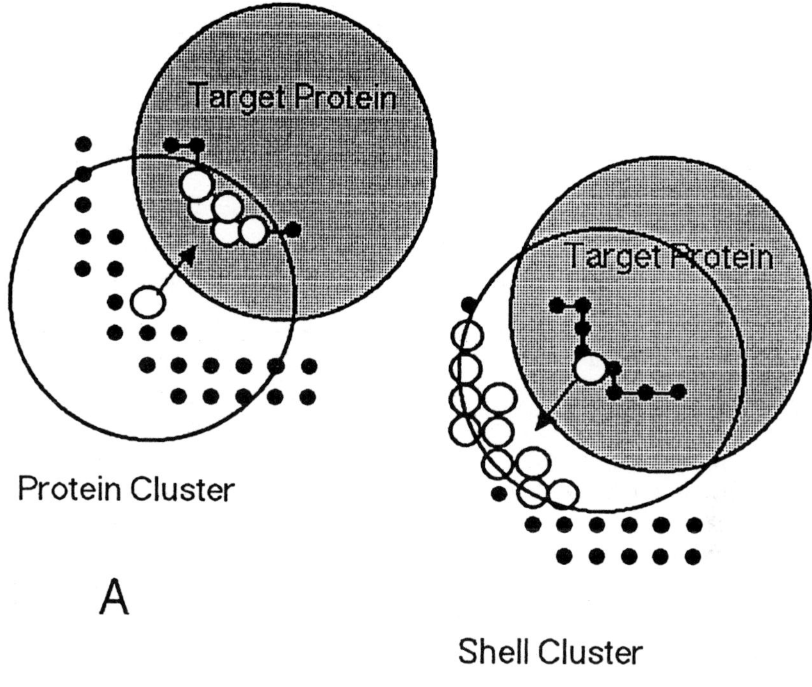

Fig. 5.   Two stage method for defining clusters of protein residues associated with a single exterior binding point (A) and the associated cluster of binding points centered about this point (B). The connected dots represent a portion of the protein under investigation and the unconnected dots the external potential binding points. The open circle with attached arrow in A is an exterior point which defines a cluster of protein residues, the small open circles located within the large circle centered upon the exterior point. In B each of the designated protein points is used together to define the cluster of exterior points shown here for one protein point.

$$E_{\text{cluster}} = \sum_{k=1}^{m} e_k$$

where m is the number of residues in the cluster.

PREDICTION OF BINDING SITE

The lattice points exterior to the protein are limited to a shell of points. Protein clusters are defined by centering a sphere of 9 Å radius on each of these shell points (Figure 5A); the number of lattice points composing the shell is usually a few thousand. The protein clusters are then ranked according to their cluster hydrophobicity as explained above. Clusters of binding points exterior to each protein cluster are then identified (Figure 5B).

Accessible surface areas are used to compare the results of the calculation with

TABLE I
Examples of best calculated surface protein clusters

		Overlaps	
	% rank	M_1	M_2
		with respect to	
		X-ray	Calculated
Protein/ligand			
Trypsin serine protease			
Anhydro-trypsin/pancreatic trypsin inhibitor	0.40	74.7	65.5
β-trypsin/pancreatic trypsin inhibitor	0.40	70.8	68.4
Trypsinogen/porcine pancreatic trypsin inhibitor	0.12	49.8	64.0
Trypsinogen/pancreatic trypsin inhibitor	0.55	72.2	68.1
Serine proteinase/potato inhibitor	0.13	83.7	55.9
Kallikrein A/bovine pancreatic trypsin inhibitor	0.17	60.6	44.6
Chymotrypsin serine protease			
α-chymotrypsin/turkey ovomucoid third domain	0.06	53.7	33.3

experiment. Actual and predicted binding sites are defined as the accessible surface area of the protein buried by the ligand (A_actual) and binding cluster ($A_\text{predicted}$), respectively. A comparison of the two should account not only of the extent of coincidence, but also for their total size, to account for any overprediction.

The best overlap between these two surface areas is used to identify which of the binding clusters corresponds most closely to the actual binding site. Ideally, the calculation would yield a predicted binding site exactly the same size as the actual binding site, with the two in perfect register. Two measures M_1 and M_2 of percentage overlaps have been used:

$$M_1 = \frac{A_\text{predict} \cap A_\text{actual} \times 100}{A_\text{actual}}$$

$$M_2 = \frac{A_\text{predict} \cap A_\text{actual} \times 100}{A_\text{predict}}$$

The set theoretic operator $\cap$ indicates the overlap between the two areas on each side of it. The larger the measure M_1, the better is the coverage of the actual binding site by the calculated cluster. These measures of percentage overlap for the two surfaces – actual and predicted – together indicate the relative sizes of the predicted and actual binding sites. The ideal is 100% for both M_1 and M_2, corresponding to the situation where the exterior cluster buries precisely the same protein surface as the ligand in the known structure.

RESULTS

Table I shows for some examples the percentile ranking of the surface cluster most similar to the actual structure as well as the measures M_1 and M_2. For 23 protein–peptide complexes in the protein data bank, the binding sites of 22 com-

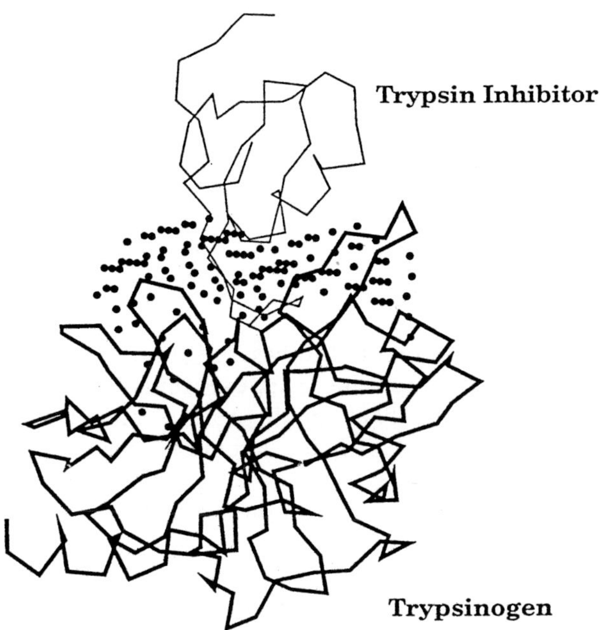

Fig. 6.   Example of a strongly hydrophobic protein binding site for the trypsinogen/trypsin inhibitor complex. The trypsinogen is shown in thick lines and the inhibitor in thin lines. The dots indicate the second-most hydrophobic cluster. Note its good coincidence with the actual binding surface.

plexes that are most similar to the actual sites are found within the top 0.7% of the predicted sites. For example, examination of the protein clusters for 1TPA, the first on the list in Table I, indicates that the seventh ranked site corresponds to the actual site of the ligand. For α-chymotrypsin/turkey ovomucoid third domain 1CHO, the actual site is the top ranking predicted site. These results represent a 2 to 3 order of magnitude reduction in the numbers of surface locations necessary to be examined before finding the native binding target. Usually the known binding site is found among the best ten or so calculated sites. Figure 6 shows one example of the correspondence between most favorable binding sites on the receptor and ligand and the actual position of the inhibitor. The number of lattice points contained in the binding sites defined by the clusters ranked for the 22 structures, ranges from 40 to 179, with an average of 96.

The two rightmost columns of Table I present the overlaps of the actual and predicted binding sites. In all cases, the predicted binding sites substantially overlap the actual binding sites. More complete comparisons show that the portion of the actual binding site included in the overlap ranges from 45.6 to 99.7%, with an average of 67.9%. This indicates that around two-thirds of the actual ligand binding site are found in the predicted site. The portions of the predicted binding sites included in the actual site range from 3.7 to 72.3% with an average of 44.5%. These results indicate that the predicted sites tend to cover a larger surface than the actual site. The extent of this overprediction does represent, however, only a small fraction of the protein's total accessible surface. Calculating the size of the binding sites as a percentage of the total protein surface area yields an average of only 9.2% for the predicted binding sites, the largest being 19.0% and the smallest, 4.6%. Similar results are obtained for the actual binding sites, the average size being 5.8% of the receptor surface area with values ranging from 0.92 to 14.0%. There is only one anomalous case: the result for one of the lysozyme/antibody structures HyHEL-5 (2HFL) is quite poor apparently because of the strong electrostatic character of its interactions [30].

OTHER STRUCTURES

It is also possible to apply the method to cases of a known protein structure but unknown ligand–protein structure if there are experiments that can indicate which are the binding residues. Useful data could consist of binding studies of proteins with modified sequences studied with the intention of mapping out the ligand binding site.

We have considered two cases where proteins are known to form complexes with peptides or other proteins, the human histocompatibility antigen HLA-2 (3HLA) (class I encoded in the MHC complex) and the fragment CD4 (2CD4) receptor. Both of these molecules are of considerable therapeutic interest as potential immunological targets. The analysis of HLA considered the α_1 and α_2 domains (residues 1 to 182), which are the two alpha helices above a beta sheet that form a cleft for the expected binding site [31]. The analysis of CD4 considered the N-terminal fragment comprising two domains (V1, V2) of the four predicted immunoglobulin-like extracellular domains [32,33]. This portion of CD4 is reported to be a receptor for the HIV gp120 fragment of he HIV coat protein.

For HLA, 1773 shell points surround the lattice model of the antigen binding domains referred to as $\alpha_1\alpha_2$. Based on the results for crystal complexes, the best 0.7% of the ranked $\alpha_1\alpha_2$ clusters are examined. These twelve clusters as ranked by their hydrophobicities bury 38.6% of the available surface area of $\alpha_1\alpha_2$. The strongest clusters are located in the cleft formed by $\alpha_1\alpha_2$ and bear a resemblance to the extra electron density found in the crystal structure [31]. The cleft itself is convexly curved, and fewer template points are found near its center. Hence there appear to be anchoring residues at the ends of peptides of different lengths that bind to the cleft with a varying bulge in the center of the peptide [34,35]. The three strongest $\alpha_1\alpha_2$ clusters are located in the binding site where the structures for two different viral peptides [36] have been solved. Considerable agreement is found between the list of residues defining the three strongest clusters and residues with atoms observed in van der Waals contact with these peptides. Another somewhat less strongly interacting region is at the interface of the $\alpha_1\alpha_2$ and β_2m domains. Choosing the best 0.7% of the 1773 points requires examining twelve clusters, and six of these are found to have substantial overlap with other clusters. It is useful to eliminate the weaker cluster if it has more than 50% of its points in common with another.

A total number of 1728 shell points surrounds the lattice model of CD4, and examining the best 0.7% of the associated clusters corresponds to examining twelve clusters. Three of these can be eliminated because of overlap. The remaining ones bury 64.4% of the total accessible surface of the CD4 two-domain fragment, nearly twice the total found for the HLA domains. This unusually large fraction of its surface area predicted to be involved in binding is consistent with studies of interactions between CD4 and gp120, monoclonal antibodies and class II MHC molecules [32,33,37–39]. These findings indicate an unusually large fraction of the surface of CD4 to be potentially interactive with other molecules.

DISCUSSION

Locating the most likely binding sites on a selected protein is an important initial step in designing a new ligand. For 22 of 23 enzyme-inhibitor complexes found in the Brookhaven Protein Data Bank, simple measures of hydrophobicity applied to an alpha-carbon model can be used to locate the correct receptor binding site within the top 0.7% of all possible sites. The procedure requires no prior knowledge of the ligands; only the structure of the target enzyme is required. The success of this procedure to locate surface targets lends additional support to the notion that surface hydrophobic interactions participate strongly in molecular recognition, confirming the biochemist's often stated intuition about 'sticky patches'. Furthermore, the present success provides evidence that peptides interact with similar strengths, whether intramolecularly in the interior of a folded protein or intermolecularly on its surface. And in both cases the largest stabilization comes from burial of hydrophobic residues.

Are these results simply a function of geometry so that one need only find the lattice point which defines the cluster with the largest number of residues, that is, a concave site on the protein? Even though the most favorable clusters often have a larger number of protein residues than most other clusters, selecting targets on the basis of the number of residues alone usually yields several clusters with similar numbers of residues. Geometric considerations alone do not provide a basis for choosing among these and would require that a larger number of clusters be examined as possible binding sites. The present approach of combining a lattice model with calculation of hydrophobicities to predict the binding site, includes such geometric effects implicitly, and is more restrictive [2].

This method of exhaustive lattice searching to explore the enzyme surface is novel in its thoroughness and in its use of surface hydrophobicity to identify possible interaction sites. Success in the use of surface hydrophobicity implies that interaction energies in the interior of a protein behave the same as those between peptide and protein surface residues.

The present analysis has focused on locating probable binding sites on a target protein and represents an initial step in a new approach for the design of peptide inhibitors. After a binding site has been chosen, peptide size and the composition of the prospective inhibitor can be addressed. A new approach attempts to combine the results described here, with previous efforts to identify the proper folds of simplified models of globular proteins. A preliminary analysis indicates that, for lattice models of peptides up to seven residues in length, the complete set of conformations can be enumerated [25,40] and examined for interactions with a target site. Once candidate peptides are identified, more detailed methods can be used to extend the single point representation to include all peptide atoms.

References

1. S. Miyazawa and R. L. Jernigan: *Macromolecules* **18**, 534 (1985).
2. L. Young, R. L. Jernigan, and D. G. Covell: *Prot. Sci.* **3**, 717 (1994).
3. S. Miyazawa and R. L. Jernigan: *Prot. Eng.* **6**, 267 (1993).
4. D. G. Covell and R. L. Jernigan: *Biochemistry* **29**, 3287 (1990).
5. I. Bahar and R. L. Jernigan: *Biophys. J.* **66**, 454 (1994).
6. I. Bahar and R. L. Jernigan: *Biophys. J.* **66**, 467 (1994).
7. P. J. Flory: *Statistical Mechanics of Chain Molecules*, Hanser Publ., Munich (1969).
8. R. S. Bohacek, U. P. Strauss, and R. L. Jernigan: *Macromolecules* **24**, 731 (1991).
9. S.-I. Mizushima: *Structure of Molecules and Internal Rotation*, Academic Press, N.Y. (1954).
10. S. Tanaka and H. A. Scheraga: *Macromolecules* **9**, 945 (1976).
11. L. M. Gregoret and F. E. Cohen: *J. Mol. Biol.* **211**, 959 (1990).
12. M. J. Sippl: *J. Mol. Biol.* **213**, 859 (1990).
13. D. A. Hinds and M. Levitt: *Proc. Natl. Acad. Sci. USA* **89**, 2536 (1992).
14. S. Miyazawa and R. L. Jernigan: *Prot. Eng.*, in press (1994).
15. J. U. Bowie, R. Lüthy, and D. Eisenberg: *Science* **253**, 164 (1991).
16. R. L. Jernigan: *Curr. Opin. Str. Biol.* **2**, 248 (1992).
17. D. G. Covell: *Proteins* **14**, 192 (1992).
18. R. L. Jernigan, G. Raghunathan, and I. Bahar: *Curr. Opin. Str. Biol.* **4**, 256 (1994).
19. L. Regan: *Annu. Rev. Biophys. Biomol. Struct.* **22**, 257 (1993).
20. D. J. Danziger and P. M. Dean: *Proc. Roy. Soc. Lond.* **B236**, 101 (1989).
21. D. J. Danziger and P. M. Dean: *Proc. Roy. Soc. Lond.* **B236**, 115 (1989).
22. P. Goodford: *J. Med. Chem.* **28**, 849 (1985).
23. A. Nicholls, K. A. Sharp, and B. Honig: *Proteins* **11**, 281 (1991).
24. B. K. Shoichet, D. L. Bodian, and I. D. Kuntz: *J. Comput. Chem.* **13**, 1 (1992).
25. R. L. Jernigan, H. Margalit, and D. G. Covell: 'Coarse graining conformations: A peptide binding example', in: D.L. Beveridge and R. Lavery (Eds.), *Theoretical Biochemistry and Molecular Biophysics*, Vol. 2, Adenine Press, Schenectady, New York, p. 69 (1991).
26. E. E. Abola, F. C. Bernstein, S. H. Bryant, T. F. Koetzle, and J. Weng: in: F. H. Allen, G. Bergerhoff, and R. Sievers (Eds.), *Crystallographic Databases–Information Content, Software Systems. Scientific Applications*, Data Commission of the International Union of Crystallography Bonn and Cambridge and Chester, p. 107 (1987).
27. F. C. Bernstein, T. F. Koetzle, G. J. B. Williams, E. F. Meyer, Jr., M. D. Brice, J. R. Rodgers, O. Kennard, T. Shimanouchi, and M. Tasumi: *J. Mol. Biol.* **112**, 535 (1977).
28. Y. Nozaki and C. Tanford: *J. Biol. Chem.* **246**, 2211 (1971).
29. J. L. Cornette, K. B. Cease, H. Margalit, J. L. Spouge, J. A. Berzofsky, and C. DeLisi: *J. Mol. Biol.* **195**, 659 (1987).
30. S. Sheriff, E. W. Silverton, E. A. Padlan, G. H. Cohen, S. J. Smith-Gill, B. C. Finzel, and D. R. Davies: *Proc. Natl. Acad. Sci. USA* **84**, 8075 (1987).
31. M. A. Saper, P. J. Bjorkman, and D. C. Wiley: *J. Mol. Biol.* **219**, 277 (1991).
32. J. Wang, Y. Yan, T. P. J. Garrett, J. Liu, D. W. Rodgers, R. L. Garlick, G. E. Tarr, Y. Husain, E. L. Reinherz, and S. C. Harrison: *Science* **348**, 411 (1990).
33. S. Ryu, P. D. Kwong, A. Truneh, T. G. Porter, J. Arthos, M. Rosenberg, X. Dai, N. Xuong, R. Axel, R. W. Sweet, and W. A. Hendrickson: *Science* **348**, 419 (1990).
34. P. Parham: *Nature* **360**, 300 (1992).
35. H. Guo, T. S. Jardetzky, T. P. J. Garrett, W. S. Lane, J. L. Strominger, and D. C. Wiley: *Nature* **360**, 364 (1992).
36. D. H. Fremont, M. Matsumura, E. A. Stura, P. A. Peterson, and I. A. Wilson: *Science* **257**, 919 (1992).
37. L. J. Clayton, N. Sieh, D. A. Pious, and R. L. Reinherz: *Nature* **339**, 548 (1989).
38. D. J. Capon and R. H. R. Ward: *Annu. Rev. Immunol.* **9**, 649 (1991).
39. G. J. Szabo, P. S. Pine, J. L. Weaver, P. E. Rao, and A. Aszalos: *J. Immunol.* **149**, 3596 (1992).
40. R. L. Jernigan: 'Generating general shapes and conformations with regular lattices, for compact proteins', in: R.H. Sarma and M.H. Sarma (Eds.), *Structure and Function: Proceedings of Seventh Conversation in Biomolecular Stereodynamics*, Vol. 2, Adenine Press, Schnectady, NY, p. 169 (1991).

Molecular Dynamics Study of the Dissociation of an Antigen–Antibody Complex in Solution

JEAN DURUP and FABIENNE ALARY
Laboratoire de Physique Quantique,[★] *IRSAMC, 118 route de Narbonne, 31062 Toulouse, France*

Abstract. Preliminary results are reported of a molecular dynamics calculation of free energy variations during the dissociation of an antigen–antibody complex, hen egg-white lysozyme – Fab D1.3, using atomic coordinates determined by the group of R. J. Poljak, and explicit handling of solvent molecules. After equilibration of the complex in solution at 300 K, a dissociation path was generated by a 'directed dynamics' protocol. Then the thermodynamic perturbation method was used for computing the derivative of the free energy of the system with respect to dissociation coordinate, both for the undissociated complex and in two points along the path. 200-ps molecular dynamics simulations were carried out at each of these points. The results obtained are discussed, with special emphasis on the role of interstitial water in the appearance of a hydrophobic activation free energy.

Key words: Free energy, antigen–antibody complex, hydration forces, dissociation, molecular dynamics.

Introduction

One of the most fundamental processes in biology is *molecular recognition*, whereby a complex is formed between two molecules, usually in an aqueous environment. In an increasing number of instances the structure of a complex, and often of its separate partners, is known from X-ray crystallography and sometimes from NMR spectroscopy, whereas the free energy (or more exactly Gibbs energy) variation in the process may be obtained from the measurement of association-dissociation equilibrium constants.

In contrast, the *path* followed during association or dissociation, as in most conformational transitions in biopolymers, escapes experimental observation: neither the evolution of the system in configuration space nor the associated energy or free energy profile can be obtained through experiments, except sometimes for rate constant measurements, whose interpretation in terms of activation free energy will be discussed later. Therefore, computer simulation using structural data from experiment is presently the most efficient way of studying the dynamical aspects of molecular recognition, and in particular to test the relevance of classical schemes [1] such as (i) the lock-and-key model, (ii) induced fit, (iii) two-steps schemes, like 'proofreading' or like Colman's 'handshake' [2].

Following an extensive computer simulation [3–4] of the fluctuation dynamics of an antigen–antibody complex, both in the crystal phase and in solution, with explicitly handled solvent molecules (water and counter-ions), we proceeded to the determination of a path for the separation of the complex in solution, and started a computation of the free energy variation along this path.

The model we used for the description of the complex in solution is the so-

★ Université Paul Sabatier, and URA 505 of C.N.R.S.

A. Pullman et al. (eds.), *Modelling of Biomolecular Structures and Mechanisms*, 167–180.
© 1995 *Kluwer Academic Publishers. Printed in the Netherlands.*

called 'egg' model [3,5], where the protein complex is immersed in the solvent within an ovoid box, designated as zone 1, with a 4 Å-wide outer shell of solvent molecules (zone 2), and two further shells of virtual molecules obtained from zone-2 ones by inversion-symmetry operations (zones 3 and 4). This model allows one to easily change the shape or size of the box when necessary, as well as to add water molecules if cavities appear, without introducing significant discontinuities in the dynamics simulation.

All programs we used for force fields, dynamics, sorting facilities, etc., were from the CHARMM-20 package developed by the group of M. Karplus in Harvard [6]. For nonbonding interactions we used a cutoff radius of 8 Å, with a switching function for van der Waals interactions and 'shifted electrostatics' [6] for Coulomb interactions, the dielectric constant being fixed to unity. The time step was 1 fs. No constraint was applied, except for the integration of nonpolar H atoms into CH, CH_2, CH_3 'extended atoms' [6].

The system under study is the complex between hen egg-white lysozyme (HEL) and the Fab fragment of antibody D1.3 (Fab D1.3). This complex was prepared, crystallized, and analysed by X-ray crystallography and thermochemical measurements, by T. Fischmann *et al.* in the group of R. Poljak at Institut Pasteur, Paris, and by R. Poljak *et al.* at CARB, Rockville, Maryland [7–10].

Determination of a Dissociation Path

We calculated a conformational path for the dissociation of the HEL-Fab D1.3 complex in solution by an implementation of the 'directed dynamics' method, which was first developed for the determination of a path for a conformational transition in an enzyme, citrate synthase [11].

In its present application, we started from a previously described [3] 93-ps simulation of the fluctuation dynamics of the complex in solution at 300 K. Then we gave a tiny additional impulse to all protein atoms in the direction of separation of the complex, by adding small velocity components in opposite directions along the line of centres of mass of antigen and antibody molecules. The new kinetic energy thus reached corresponded to a temperature of 300.5 K. Obviously, however, the additional 0.5 K are not thermal energy but *ca.* 50 kcal mol^{-1} concentrated on a single, collective degree of freedom.

We then allowed the system to evolve freely, while steadily bringing the 'temperature' back to 300 K by uniform rescalings of all atomic velocities. There occurred an extension of the complex by 0.21 Å, followed by a retraction almost back to the original centre-of-mass distance (47.70 Å), leaving, however, a gain of 0.03 Å. We reiterated the same protocol 13 times. However, the restoring force increased with separation of the complex, so that we had to raise the additional kinetic energy from 0.5 K up to 20 K; thereafter we could reduce it again, since the restoring force decreased when the complex was partly separated.

Various tests of this protocol, using varying amounts of additional kinetic energy at a given step, showed that a constant time of 400 fs was needed by the system to reach its maximum extension, whereas this maximum lengthening was proportional to the square root of the energy added. In other words in this protocol the complex behaves in a harmonic way, with a restoring force partly due to

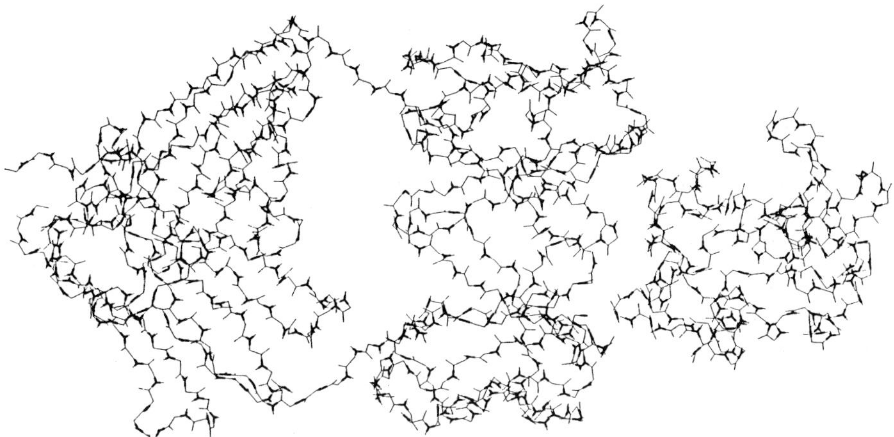

Fig. 1. Backbone of the HEL-Fab D1.3 complex after 93-ps equilibration in solution, while starting the 'directed dynamics' separation protocol. The projected centre-of-mass distance Z is 47.69 Å.

the binding forces between antigen and antibody and partly to the water mattress around the complex. However, the retraction of the complex requires more time than its extension, and clearly the overall relaxation which takes place during this retraction is what allows us to gain a few tenths of an Ångstrom at each iteration of the protocol.

All along the dissociation process there is free space appearing between antigen and antibody, and normally it will be filled by the solvent. This spontaneous filling could have taken place in our simulation if we had allowed for a relaxation time of at least a few tens of picoseconds at each step. Even so, some doubt would have remained since cavities might appear which communicate poorly with the bulk phase. Therefore we preferred to make use of the following protocol of 'rehydration', which we had developed in our previous work [3].

In the regular protocol of our molecular dynamics simulations using the CHARMM programs, there is a periodic updating of the list of nonbonding interactions between pairs of atoms separated by a distance lower than the cutoff radius. We proceed to this updating each 50 fs. In the simulations using our 'egg' model [3], we simultaneously update the lists of solvent molecules (H_2O and counter-ions) located in either zone 1 or zone 2, and we exchange those having diffused from zone 2 into zone 3 for their images, which then are located in zone 2. When we want to take care of the appearance of cavities in the solvent we complete the periodic tasks described above by moving over the system a cubic box filled with pre-equilibrated water molecules (at 300 K and unit density), and releasing into the system any H_2O molecule whose location is such that no atom is present at a distance less than the sum of its van der Waals radius and a conventional van der Waals radius of 1.6 Å for the water molecule. In practice, 1 or 2 water molecules are thus added each 50 fs to the *ca.* 10 000 already present.

Figures 1, 2a and 3a show the backbones of our complex before dissociation and at two points along the dissociation path obtained as described. Figures 2b, 2c and 3b, 3c show, as will be explained now, the result of further relaxation of

(a)

(b)

(c)

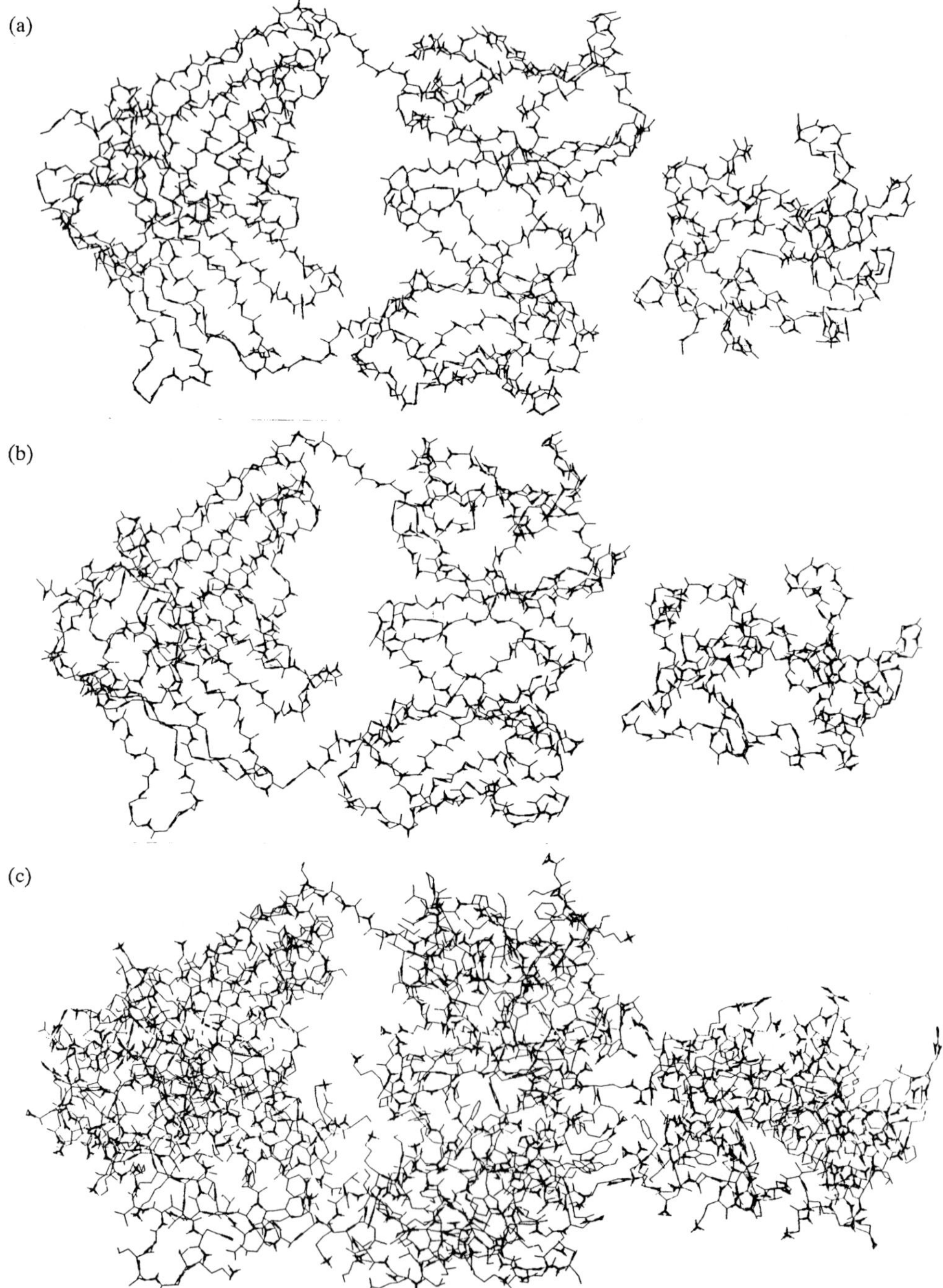

Fig. 2.   The HEL-Fab D1.3 complex in the middle of the separation protocol. The projected centre-of-mass distance Z is held at 48.41 Å. (a) Backbone of the complex before relaxation. (b) Backbone of the complex after 200-ps relaxation. (c) Full representation of the complex after 200-ps relaxation.

(a)

(b)

(c)

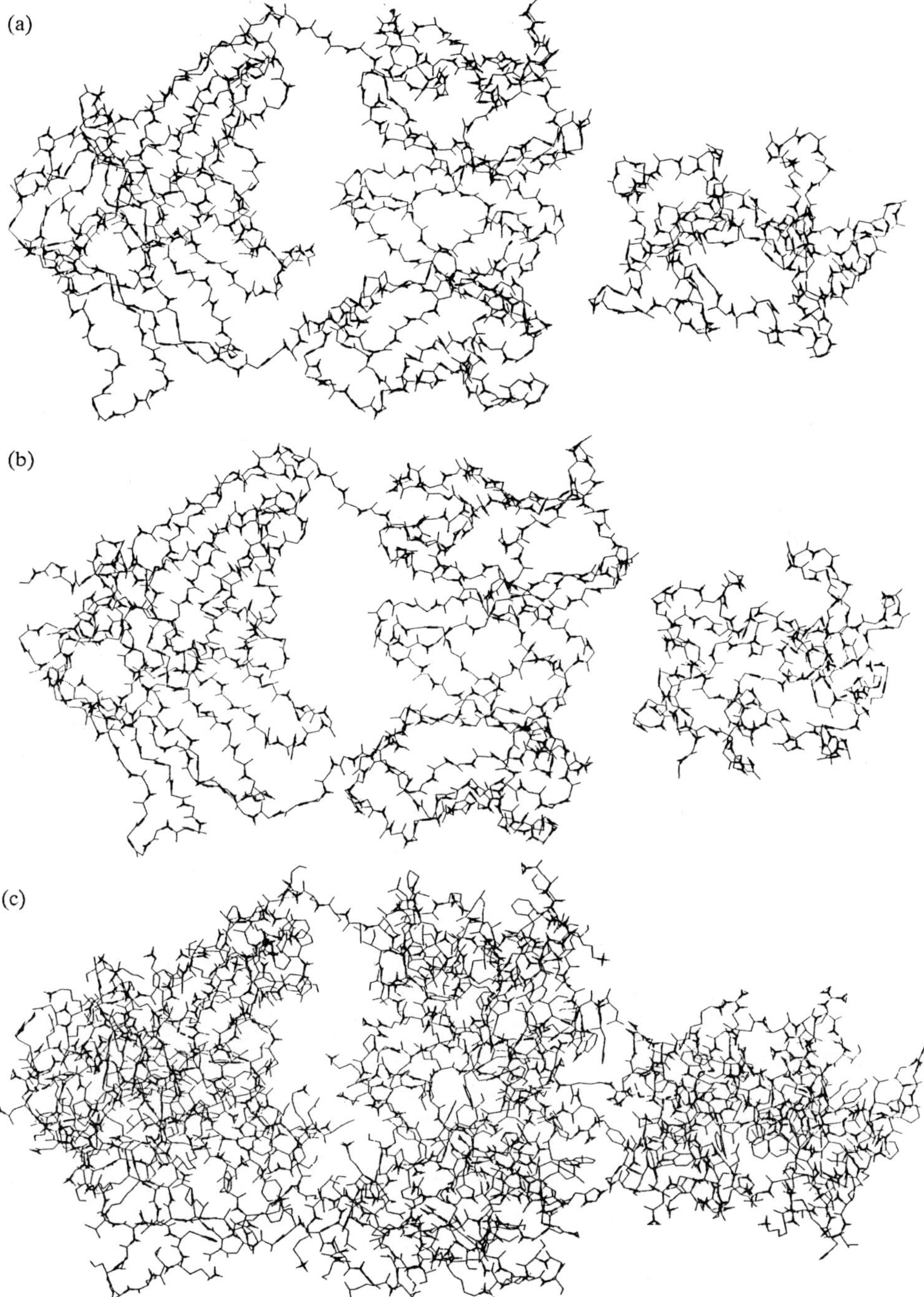

Fig. 3. The HEL-Fab D1.3 complex largely separated. The projected centre-of-mass distance Z is held at 52.90 Å. (a) Backbone of the complex before relaxation. (b) Backbone of the complex after 200-ps relaxation. (c) Full representation of the complex after 200-ps relaxation.

the complex at fixed distance between the centres of mass of antigen and antibody molecules.

Free Energy Computations along the Dissociation Path

The free energy A of a macromolecule in a given state at a given temperature T may in principle be obtained from the molecular partition function Q by using the familiar formula

$$A = -k_{B}T \ln Q, \tag{1}$$

with the semi-classical expression

$$Q = \frac{1}{h^{N}} \int d\mathbf{q} \, d\mathbf{p} \, \exp\left(-\frac{H(\mathbf{q},\mathbf{p})}{k_{B}T}\right), \tag{2}$$

where k_{B} and h are Boltzmann and Planck constants, N is the number of degrees of freedom, $\mathbf{q}$ and $\mathbf{p}$ are collective notations for coordinates and conjugated momenta, H is the classical Hamiltonian, and the integration is performed in the canonical ensemble. Using the ergodic theorem, we may replace the integral over the accessible phase space by an average over a trajectory run at constant temperature T during an infinite time. Furthermore, if only free energy *variations* are required, the kinetic energy part of H will cancel out if N is fixed and large, so that potential energy terms only may have to be considered.

This scheme raises a basic difficulty. If, for example, one is concerned with the free energy variation in *protein folding*, the initial unfolded state normally lives no more than perhaps a microsecond, and consequently the trajectory scanning its phase space should not be run during more than that time. Even the comparatively stable native folded state has a lifetime limited to hours or days and, obviously, within the infinite time necessary in principle for the ergodic theorem to be satisfied, any absurd conformation of the protein and even any of its isomers would be included in the accessible phase space since the thousands of kilocalories per mole of thermal energy available at 300 K are large enough to break any bond. Thus a meaningful use of Equations (1)–(2) requires that the relevant time scale be stated. Unfortunately indeed, the simulation times attainable with presently available computer power are even further below the biologically relevant time scales, so that the main question in free energy calculations on biopolymers by molecular dynamics is whether or not the interesting phase space has been conveniently sampled in the simulation.

Most recent calculations of *differences* in free energy variations in large molecular complex formation have made use of the *thermodynamic perturbation* method applied to 'alchemical' processes, i.e. slow mutations or transmutations, introduced by Wong and McCammon [12]. A review of the recent work using this method was given by Kollman [13]. This strategy permits the circumvention among other difficulties, of those due to the big variations in solvent arrangement when comparing a complex with its separated partners. In some calculations of the *potential of mean force* (p.m.f.) between pairs of small hydrocarbon molecules in solution, however, a direct use of Equations (1)–(2) has been made [14–17].

In our problem, which was to assess the free energy at given points along a dissociation path, we could not escape the above-mentioned difficulty. We hoped to be able to use the *thermodynamic perturbation* method for computing the *derivative* $\partial A/\partial Z$ of the free energy of our system with respect to the projection Z of the vector joining the antigen and antibody centres of mass on the initial direction $\mathbf{Z}_0$ of this vector, a direction which also was the direction of separation in our 'directed dynamics' protocol.

Starting from the following equation (see McQuarrie [18])

$$Q_{Z+dZ} = Q_Z \left\langle \exp\left(- \frac{U_{Z+dZ} - U_Z}{k_B T} \right) \right\rangle_Z,$$ (3)

where U is the potential energy of the system, and the average is taken over the canonical ensemble restricted to a fixed value Z, and making use of Equation (1), we readily obtain:

$$\frac{\partial A}{\partial Z} = \left\langle \frac{\partial U}{\partial Z} \right\rangle_Z.$$ (4)

In this mean force (in the sense of the p.m.f.), the energy part of A is in U and its entropic part is in the averaging which weights the potential hypersurface according to phase space volume.

We calculate $\partial U/\partial Z$ at given points of our dissociation path in the following way: if Z_G is the coordinate of the whole protein complex, Z_i the coordinate of atom i, and ζ_i its coordinate with respect to the centre of mass of the complex partner it belongs to, all these coordinates being taken along the $\mathbf{Z}_0$ vector, then

$$\frac{\partial U}{\partial Z} = \frac{M_L}{M_L + M_F} \sum_F \frac{\partial U}{\partial z_i} - \frac{M_F}{M_L + M_F} \sum_L \frac{\partial U}{\partial z_i},$$ (5)

where F and L stand for Fab and lysozyme, and the Ms are masses. The partial derivative in the l.h.s. is taken at Z_G and all ζ_i fixed, whereas in the r.h.s. the partial derivatives are at z_j ($j \neq i$) fixed.

The simplicity of this equation comes from the fact that both the Jacobian of the transformation $\{Z_i\} \rightleftarrows \{Z_G, Z, \zeta_i\}$ and the nonlinearity bias correcting term discussed by Depaepe *et al.* [19] here are constant.

It must be pointed out that the solvent here intervenes through the forces it exerts upon the protein atoms, whereas the dynamics of the solvent itself does not appear explicitly in Equation (5). It appears implicitly in Equation (4) through its effect on the phase space sampling.

In the molecular dynamics runs carried out for the implementation of Equation (4), no water was added systematically during the simulation.

The first test we performed of this method was its application to the complex before the dissociation protocol was started. Figure 4 displays the evolution of the computed $\partial A/\partial Z$ during the course of the molecular dynamics simulation. One sees the steady improvement of the statistics. Since the complex had been previously equilibrated, no separation force was expected, and the final value should have been zero. The plateau obtained after 100 ps is at 0.55 ± 0.1 kcal mol^{-1} Å^{-1}. If

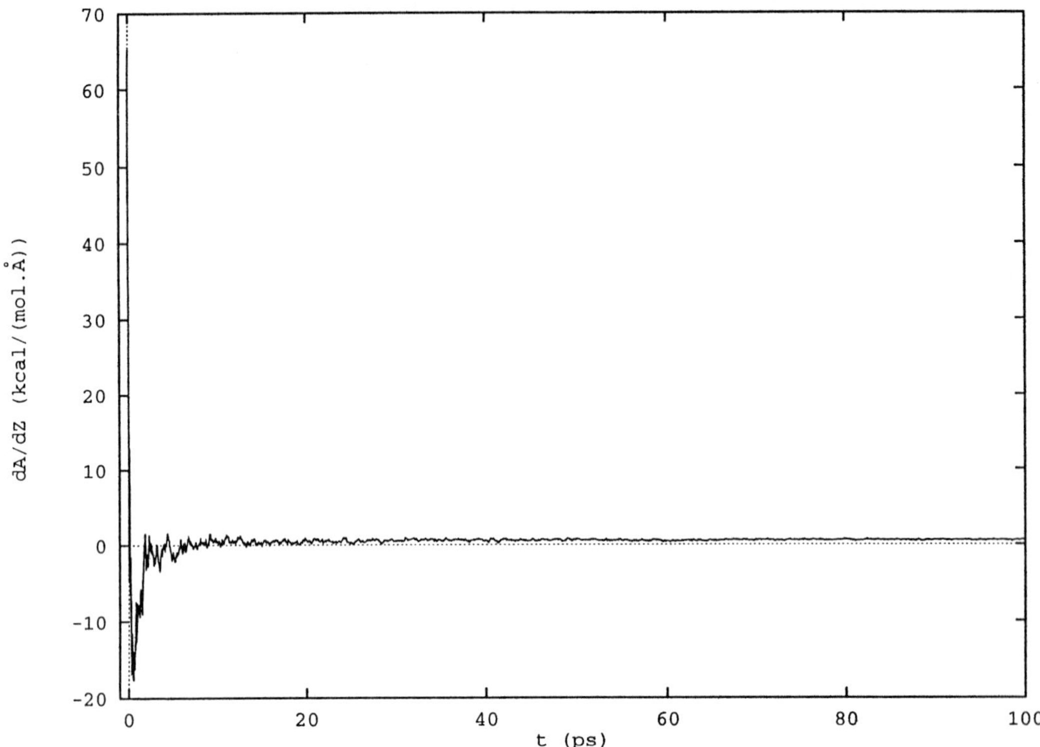

Fig. 4. Build-up of the computed $\partial A/\partial Z$ for the undissociated complex in a 100-ps molecular dynamics simulation.

the statistics is limited to the last 50 ps, the value obtained is 0.56, which shows that the departure from 0 is not due to an insufficient equilibration at the beginning of our reference period, but probably reflects the overall accuracy of the method. Keeping in mind that the total cohesion energy (van der Waals plus electrostatic potential energy) is around $-170\,000$ kcal mol^{-1} and the total kinetic energy $32\,500$ kcal mol^{-1}, the error observed appears acceptable.

We thus were encouraged to test the method at two points along the dissociation path (see Figures 2 and 3). The point at $Z = 48.41$ Å lies close to the midpoint of the separation path, where the restoring force is expected to be maximum. The point at 52.90 Å is on the side of decreasing restoring force. Because of this force opposing to separation it was necessary, for sampling the phase space restricted to a fixed Z, to make use of an artificial constraint. This can be done in essentially two ways, which were shown by Depaepe *et al.* [19] to be equally valid: either to run the dynamics with a harmonic constraint potential added to the potential energy function (the *umbrella sampling* method), or to readjust Z to its initial value at each step (the so-called *blue-moon* ensemble sampling). We chose the latter method.

Figures 2b, 2c and 3b, 3c show the complex at the end of the 200-ps molecular dynamics simulations carried out for obtaining $\partial A/\partial Z$ at these two points. Figures

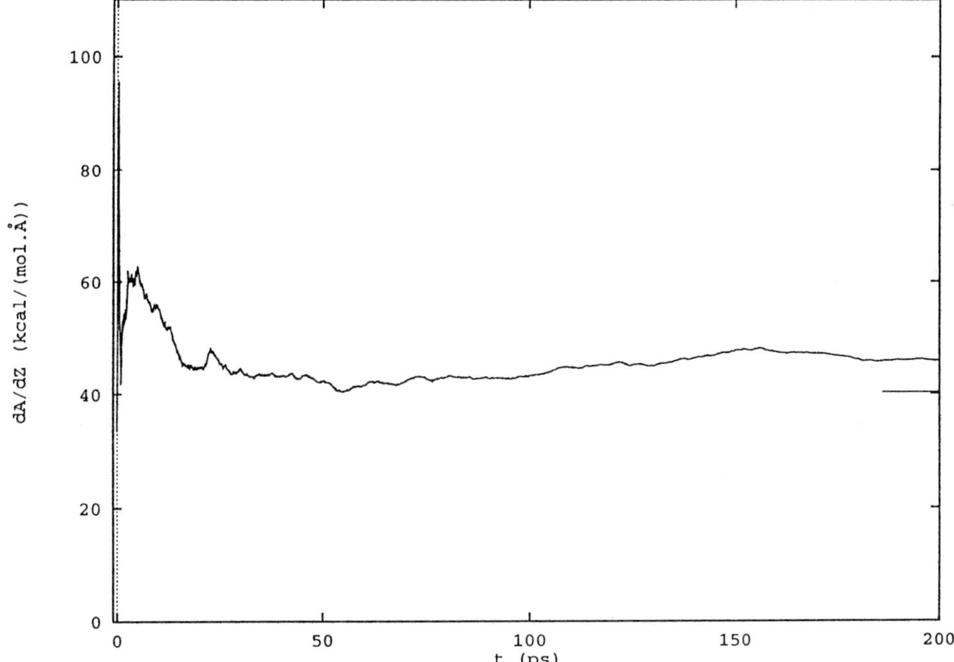

Fig. 5. Build-up of the computed $\partial A/\partial Z$ for the partly dissociated complex at a projected centre-of-mass distance Z held at 48.41 Å. The horizontal segment shows the value of $\partial A/\partial Z$ when the averaging is performed over the last 50 ps only.

5 and 6 display the evolution of the computed $\partial A/\partial Z$ along the simulation, as well as the value corresponding to a statistics limited to the last 50 ps.

In both cases there obviously occurs an initial relaxation due to the constraint imposed upon the system during its preparation by the 'directed dynamics' method. In contrast to what was observed for the undissociated complex, and although the total simulation time (taking into account the preequilibration of the complex) was the same, the curves in Figures 5 and 6 do not seem to reach a plateau value.

Discussion

The order of magnitude to expect for the free energy variations studied in our simulations may be derived from the experimental rate constants obtained by Ward et al. [20] for the complex between hen egg-white lysozyme and the VH fragment of antibody D1.3 (a fragment which includes only half of the interaction site of Fab D1.3), at 298 K:

$$\begin{cases} k_{\mathrm{on}} = 3.8 \times 10^{6}\,\mathrm{mol}^{-1}\,\mathrm{l}\,\mathrm{s}^{-1} \\ k_{\mathrm{off}} = 7.5 \times 10^{-2}\,\mathrm{s}^{-1}, \end{cases} \tag{6}$$

where 'on' and 'off' stand for the formation and dissociation of the complex.

From these values one gets the overall Gibbs free energy of dissociation:

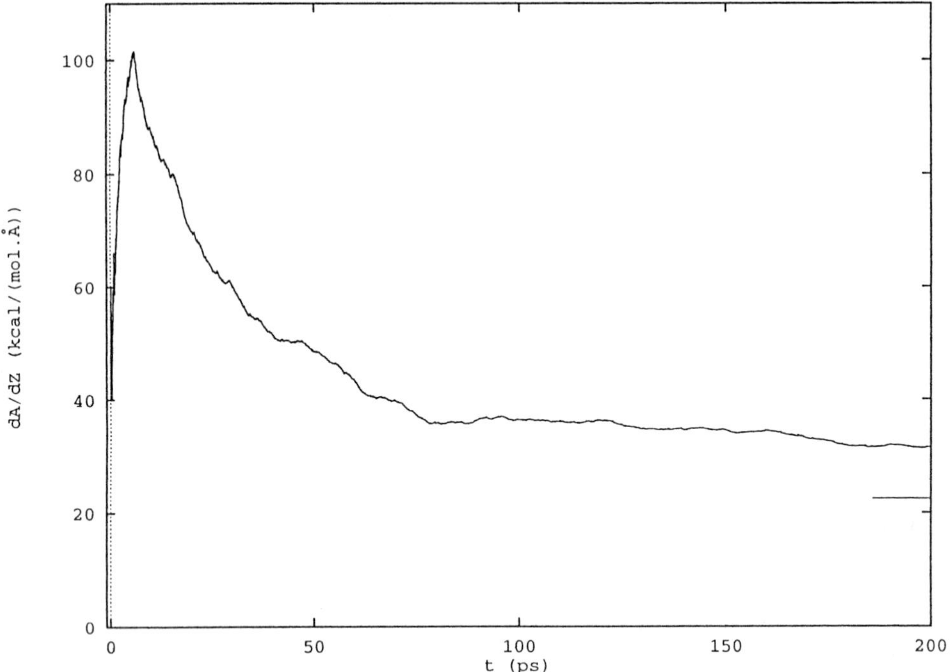

Fig. 6.   Build-up of the computed $\partial A/\partial Z$ for the largely dissociated complex at a projected centre-of-mass distance Z held at $52.90\,\text{Å}$. The horizontal segment shows the value of $\partial A/\partial Z$ when the averaging is performed over the last 50 ps only.

$$\Delta G^{\circ} = RT \ln (k_{\mathrm{on}}/k_{\mathrm{off}}) = 10.5 \text{ kcal mol}^{-1}. \tag{7}$$

Transition state theory allows one in principle to obtain a Gibbs *activation* energy for dissociation through the following equation [21]:

$$k_{\mathrm{off}} = \frac{k_{\mathrm{B}}T}{h} \exp\left(-\frac{\Delta G'^{\neq}}{RT}\right), \tag{8}$$

with

$$\Delta G'^{\neq} = \Delta G^{\neq} + TS_Z^{\neq}, \tag{9}$$

where $\Delta G'^{\neq}$ is the Gibbs energy difference between transition state and initial state, and $S_Z^{\neq}$ is the contribution, to the entropy of the transition state, of the degree of freedom associated with the reaction coordinate Z.

From Equations (6) and (8) we get an experimental

$$\Delta G'^{\neq} = 19.2 \text{ kcal mol}^{-1}. \tag{10}$$

Thus, since the separation of the complex takes place over a few Å, one expects from Equations (7) and (10) that $\partial A/\partial Z$ should not exceed a few $\text{kcal mol}^{-1}\,\text{Å}^{-1}$ along the path. The discrepancy by an order of magnitude between our results and this expectation will be fully discussed later. Moreover, it is quite possible that the activation barrier responsible for the experimental rate constants be associated with macroscopic phenomena like diffusion and docking (in the direction of complex formation), which have been successfully modelled by Northrup *et al.* [22]. If this is the case, the experimental $\Delta G'^{\neq}$ given by Equation (10) is an upper limit of the barrier for the microscopic process of complex dissociation we are considering here.

The origin of Equation (9) lies in the scheme

$$AB \underset{2}{\overset{1}{\rightleftharpoons}} AB^{\neq} \underset{4}{\overset{3}{\rightleftharpoons}} A + B, \tag{11}$$

where the dissociation rate constant is calculated as

$$k_1 = k_2\,K_{12} = k_2 \exp\left(-\Delta G^{\neq}\right), \tag{12}$$

and it turns out that k_2 includes in its denominator the phase space volume associated with the reaction coordinate, which therefore may be removed from $\exp\left(TS^{\neq}\right)$, thus yielding Equations (8)–(9). The physical significance of this cancellation of terms is that the equilibrium constant K_{12} is proportional to the density of states ρ of the motion along the reaction coordinate at the transition state, whereas the dissociation rate constant k_2 is inversely proportional to ρ, so that the association rate constant k_1 is independent of ρ. This density of states ρ, for an *isolated system*, essentially is proportional to the square root of the reduced mass associated with the reaction coordinate. Equation (9) further requires that the reaction coordinate Z be a normal mode, obtained by diagonalization of the Hessian at the geometry of the transition state, and thus being essentially uncoupled to the other degrees of freedom.

Now when we apply this scheme to our system, we have to take into account that the separation of the complex along Z is associated with a penetration of water into the interfacial region. This water becomes more or less confined, especially when there are links persisting between antigen and antibody, as apparent from Figures 2c and 3c. Thus the familiar idea that the formation of a protein complex in aqueous phase is partly driven by the *hydrophobic effect*, in the sense that some water has to move from the surface of the interaction sites towards the bulk, thus gaining entropy, should be extended to the intermediate situation where, in addition to the hydration shells of both protein molecules, a number of water molecules are confined between the interaction sites and have reduced translational and rotational mobilities with respect to bulk molecules. In other words, the activation free energy of complex formation or dissociation in solution includes a *hydrophobic negentropic* part, which may also be described in terms of 'osmotic stress' [23].

We may perform an order-of-magnitude calculation of this hydrophobic part of the activation free energy. For each confined water molecule a free energy cost between 0 and 2 (kcal mol^{-1} is expected at 300 K [24]. Let us try 0.6 kcal mol $^{-1}$, i.e. RT. The cross section of the antigen-antibody interaction region in our complex is around 1000 Å^2. Thus, since at unit density each water molecule occupies 30 Å^3, 33 additional molecules will be confined when the gap increases by 1 Å. Therefore in our system the hydrophobic part of the activation free energy should be in the order of

$$0.6 \times 33 = 20 \text{ kcal mol}^{-1} \text{ Å}^{-1}. \tag{13}$$

This value clearly is in the order of magnitude of the excessively high value which we obtain for $\partial A/\partial Z$ from Figures 5 and 6 in the last 50 ps of our simulations: 40.27 and 22.45 kcal mol^{-1} Å^{-1}, respectively.

In this discussion we do not consider the difference between (Helmholtz) free energy and Gibbs energy, which is very small in condensed phase.

The above analysis raises two questions.

(i) Is our protocol adapted to include this hydrophobic activation free energy ? If the answer is no, our results would reflect either the various imperfections of our model, or an insufficient sampling time. It must be recalled that in the reference simulation of the undissociated complex the starting point was the *experimental* structure of the complex in the crystal phase (which is likely to be close to its structure in solution), whereas our directed dynamics dissociation is artificial.

We think that the answer to question (i), however, is probably yes. In effect, confinement negentropy means that the statistical weight of the conformations with more water molecules in the interfacial region, and therefore a lower attraction between complex partners, is comparatively lower than that of the conformations with less interfacial water and therefore a higher binding force. Now from our simulations of the complex in solution we determined [25] self-diffusion coefficients of water molecules ranging from 0.27 Å^2 ps^{-1}, close to the protein surface, to 0.35 Å^2 ps^{-1}, far from the protein. Thus our water molecule are able to diffuse by 8 Å in average during 200 ps, which allows for a significant exchange between interfacial region and bulk.

(ii) Should the hydrophobic activation free energy appear in the actual rate constant?

If the answer is yes, the estimate we made in Equation (13) is far too high to account for the observed rate constant and the resulting $\Delta G'^{\neq}$ [see Equation (10)]. Then the assumption we used of a confinement free energy of the order of RT per water molecule is wrong and the actual value is much lower.

If the answer is no, our calculations are consistent with experiment, using the assumption leading to Equation (13). Then the simulation should be completed by a detailed analysis of the confinement negentropy, permitting to compute its contribution to $\partial(TS_Z^{\neq})/\partial Z$ and to subtract it from our $\partial A/\partial Z$ in order to get a meaningful comparison with experiment.

We presently have no clear answer to issue (ii).

Conclusions

The issue of this preliminary study was the possibility, and the cost in terms of computer time, of a realistic direct determination of the dissociation rate constant of a protein complex, starting from a three-dimensional conformation provided by the experimentalists, then applying a 'directed dynamics' protocol for getting a dissociation path after molecular dynamics equilibration of the complex in solution, and finally using the thermodynamic perturbation method for obtaining a profile for the derivative of the free energy with respect to dissociation coordinate.

In the present work a 40 000-atom system (protein complex + solvent) was studied. Each 200-ps simulation, using a time step of 1 fs and a force field with a cutoff radius at 8 Å for nonbonding interactions, required 125 CPU hours on a Cray C98.

These 200 ps appeared to be enough for convergence of $\partial A/\partial Z$ in the simulation of the undissociated complex, but probably insufficient for points taken along the path. However, we think that the correct method would have been to proceed to the 200-ps relaxation and to the concomitant free energy computation all along the generation of the path by the 'directed dynamics' separation protocol. In this way it might be expected that e.g. 10 points along the path, with a cost of 1250 CPU hours of the same machine, would provide the desired information.

More questionable are the improvements needed in the dynamics simulation parameters. We did not use any constraint, such as the SHAKE protocol which would have reduced the computer time by a factor of 2; however we made use of the inclusion of nonpolar H atoms into 'extended atoms' CH, CH_2, CH_3. The main weakness of our protocol is the comparatively low cutoff radius (8 Å), which biases the long-range electrostatic interactions within each protein molecule. We think, however, that this bias on the conformational dynamics of our proteins is far less severe than what several authors [26–27) have shown to occur with small peptides, whose conformations obviously are far more flexible than those of medium- or large-size proteins. Nevertheless, the effect of increasing the cutoff radius in free-energy calculations on large systems should be inspected. Also, in a study of myoglobin, Loncharich and Brooks [28] drew attention to the advantage of using a distance-dependent dielectric constant. This certainly is valuable for the description of the interactions within each protein molecule but raises other difficulties when applied to the solvent, whose force field parameters have been optimized for the use of a unit dielectric constant.

We keep being strongly in favor of the explicit treatment of the solvent molecules, because of the paramount importance of the first hydration layers. As regards the interaction between protein and solvent, the problem of the cutoff radius appears as minor with respect to the details of the force field used. A considerable literature has been devoted to molecular dynamics simulations testing force fields against experimental quantities. However, as earlier discussed by us [3], there is very little experimental information on the variation of the microscopic density of water molecules and of their transport properties near the protein surface. This in our opinion is the bottleneck for the improvement of semi-empirical potentials.

Our results let appear the difficult problem of the rôle of interfacial water in the free energy profile along the dissociation path of a complex with a large interaction area. NMR and/or neutron scattering techniques could be used if analogs of partly dissociated complexes were available. From the theoretical viewpoint, a detailed term-by-term analysis as performed by Ben-Naim for the global processes of protein folding and protein association [29] might be extended to the study of transition states of protein complex separation. Clearing up this question will be a prerequisite before proceeding further along the lines of the present work.

Acknowledgements

Our calculations were started on the vectorial processor of the IBM 30/90 computer of CNUSC, Montpellier, with a special grant under the C3NI program, and developed on the Cray C98 supercomputer of IDRIS, Orsay, with a C.N.R.S. grant. Graphism was done on the IBN RISC 6000/32 work station belonging to the local facilities of the Laboratoire de Physique Quantique in Toulouse.

References

1. J. Durup: *Actualités de Chimie Thérapeutique*, 20e série, Soc. Thérap. (XVIIIth Internat. Meetg. on Medic. Chem., Toulouse, 7–9 July 1992), p. 309 (1993).
2. P. M. Colman, W. G. Laver, J. N. Varghese, A. T. Baker, P. A. Tulloch, G. M. Air, and R. G. Webster: *Nature* **326**, 358 (1987).
3. F. Alary, J. Durup, and Y.-H. Sanejouand: *J. Phys. Chem.* **97**, 13864 (1993).
4. F. Alary: Thèse de Doctorat, Université Paul Sabatier, Toulouse, France (1994).
5. J. Durup, F. Alary, and Y.-H. Sanejouand: in *Statistical Mechanics, Protein Structure and Protein–Substrate Interaction*, S. Doniach, Ed., Plenum Press, New York, p. 339 (1994).
6. B. R. Brooks, R. E. Bruccoleri, B. D. Olafson, D. J. States, S. Swaminathan, and M. Karplus: *J. Comput. Chem.* **4**, 187 (1983).
7. T. O. Fischmann, G. A. Bentley, T. N. Bhat, G. Boulot, R. A. Mariuzza, S. E. V. Phillips, D. Tello, and R. J. Poljak: *J. Biol. Chem.* **266**, 12915 (1991).
8. T. O. Fischmann: Thèse de Doctorat, Université de Paris VII, Paris, France (1991).
9. R. A. Mariuzza and R. J. Poljak: *Curr. Opin. Immunol.* **5**, 50 (1993).
10. T. N. Bhat, G. A. Bentley, G. Boulot, M. I. Green, D. Tello, W. Dall'Acqua, H. Souchon, F. P. Schwarz, R. A. Mariuzza, and R. J. Poljak: *Proc. Natl. Acad. Sci. USA* **91**, 1089 (1994).
11. M. A. Ech-Cherif El-Kettani and J. Durup: *Biopolymers* **32**, 561 (1992).
12. C. F. Wong and J. A. McCammon: *J. Amer. Chem. Soc.* **108**, 3830 (1986).
13. P. A. Kollman: *Curr. Opin. Struct. Biol.* **4**, 240 (1994).
14. G. Ravishanker, M. Mezei, and D. L. Beveridge: *Faraday Symp. Chem. Soc.* **17**, 79 (1982).
15. D. E. Smith, L. Zhang, and A. D. J. Haymet: *J. Amer. Chem. Soc.* **114**, 5875 (1992); D. E. Smith and A. D. J. Haymet: *J. Chem. Phys.* **98**, 6445 (1993).
16. D. van Belle and S. J. Wodak: *J. Amer. Chem. Soc.* **115**, 647 (1993).
17. P. Linse: *J. Amer. Chem. Soc.* **115**, 8793 (1993).
18. D. A. McQuarrie: *Statistical Mechanics*, Harpers & Row, New York, Section 14-1 (1976).
19. J.-M. Depaepe, J.-P. Ryckaert, E. Paci, and G. Ciccotti: *Mol. Phys.* **79**, 515 (1993).
20. E. S. Ward, D. Güssow, A. D. Griffiths, P. T. Jones, and G. Winter: *Nature* **341**, 544 (1989).
21. P. Pechukas: *Dynamics of Molecular Collisions*, W. H. Miller, Ed., Plenum Press, New York, Part B, Ch. VI (1976).
22. S. H. Northrup and H. P. Erickson: *Proc. Natl. Acad. Sci. USA* **89**, 3338 (1992); S. H. Northrup: *Curr. Opin. Struct. Biol.* **4**, 269 (1994).
23. R. P. Rand, N. L. Fuller, P. Butko, G. Francis, and P. Nicholls: *Biochemistry* **32**, 5925 (1993).
24. J. D. Dunitz: *Science* **264**, 670 (1994).
25. F. Alary and J. Durup: to be published.
26. P. E. Smith and B. M. Pettitt: *J. Chem. Phys.* **95**, 8430 (1991).
27. H. Schreiber and O. Steinhauser: *Biochemistry* **31**, 5856 (1992); *Chem. Phys.* **168**, 75 (1992).
28. R. J. Loncharich and B. R. Brooks: *Proteins* **6**, 32 (1989).
29. A. Ben-Naim: *Biopolymers* **29**, 567 (1990).

Calculation of Atom-Centered Partial Charges for Heme

JOHN I. MANCHESTER, MARK D. PAULSEN, and RICK L. ORNSTEIN[*]
Environmental Molecular Sciences Laboratory, Pacific Northwest Laboratory, Richland, WA 99352, U.S.A.

Abstract. Atom-centered partial charges are calculated for the Fe-heme in cytochrome P450cam for use in molecular dynamics simulations of polar substrates bound in the active site of the enzyme. Charges are fit to the electrostatic potential produced by ab initio UHF wavefunctions for an Fe-porphine model. Basis set dependence of these charges is observed using the LANL1DZ, LANL2DZ and augmented 6–31G levels of theory. Upon geometry optimization of the enzyme, these charge sets cause varying degrees of distortion of the porphyrin from its crystallographically observed conformation. Scaling the charges calculated from the augmented 6–31G basis by 75% reduces the heme distortion while preserving reasonable interactions with a polar substrate. A comparison of the calculated charges with other published values is presented.

Key words. Cytochrome P450, forcefield, iron, porphyrin, potential-derived charges.

Introduction

Cytochromes P450 are a class of hemeproteins that catalyze a variety of oxygenation reactions by activating molecular oxygen and inserting it into organic compounds such as hormones and xenobiotics. In an anaerobic environment, these enzymes are also capable of reductively dehalogenating heavily halogenated hydrocarbons [1, 2], commonly used in organic synthesis and as solvents. The ability of P450s to catabolize these compounds has attracted attention to the possibility of using reconstituted systems, or genetically engineered organisms, to degrade halohydrocarbons that now constitute a major contaminant class at industrial waste sites. Wackett and coworkers [3] observed significant rates for dechlorination of hexachloroethane using a purified P450cam system, but observed no dechlorination of 1,1,1-trichloroethane (TCE). In a recent MD simulation of the wild-type enzyme with TCE bound in the active site [4], we observed that the substrate populated an unreactive conformation for the duration of production dynamics – a result in qualitative agreement with the experimental observation that TCE is not catabolized by P450cam.

An examination of the components of the enzyme-substrate intermolecular energies in the MD simulations indicated that it was electrostatic repulsion between the heme group and TCE which dominated enzyme-substrate interactions and determined the observed substrate orientation. Although the results of this MD simulation are in qualitative agreement with experiment, a comparison of the partial charges used on the heme cofactor with those used by others in simulations of hemeproteins suggests that the charge set we used may have exaggerated the heme-TCE repulsion. In many ways the parameterization of atom-centered partial charges is the most difficult and subjective part of any empirical forcefield. Re-

---

[*] Author for correspondence.

A. Pullman et al. (eds.), *Modelling of Biomolecular Structures and Mechanisms*, 181–188.
© 1995 *Kluwer Academic Publishers. Printed in the Netherlands.*

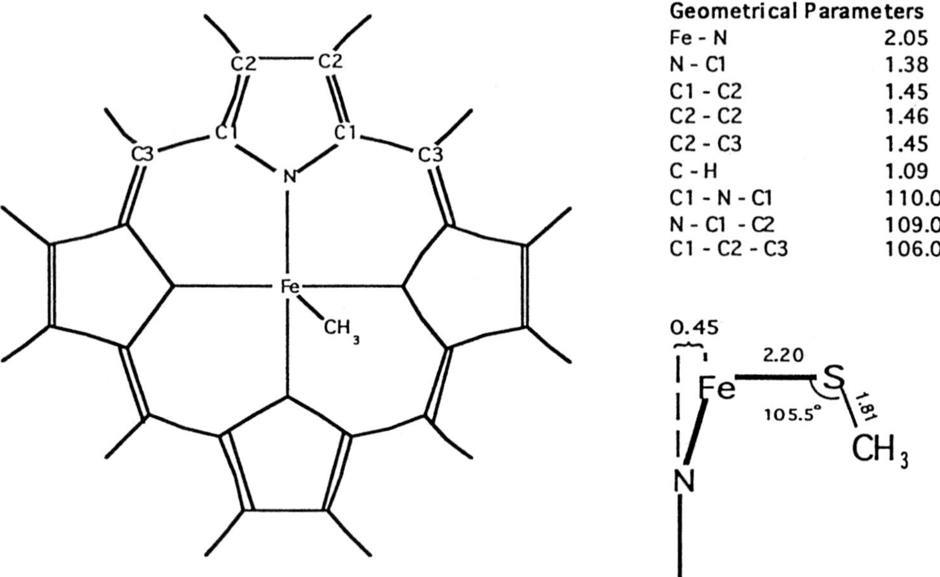

Fig 1.   *Ab initio* calculations were performed on this planar heme model. Geometrical parameters were obtained assuming four-fold symmetry and averaging internal coordinates in the crystal structure of camphor-bound P450cam [7]. Heme substituents were replaced with H atoms, and Fe lies 0.45 Å out of the heme plane. All bond lengths shown are in Å, and bond angles are in degrees. Note that the view on the left is along Fe—S bond.

cently, however, electrostatic potential-derived (PD) charges have gained widespread acceptance for use in molecular mechanics calculations. We applied this approach to reparameterize the charges used in our MD simulations, which implement the consistent-valence forcefield [5]. In this paper, we assess the effect of basis set on the PD charges for a model of the heme cofactor and evaluate the charge sets for the effect on substrate–enzyme interaction energies and on equilibrium heme geometry. A comparison with other published charge sets for the heme cofactor is also presented.

Methods

Ab initio calculations were performed using Gaussian 92 [6] on a model of the heme prosthetic group in cytochrome P450, shown in Figure 1. All substituents were removed from the heme, leaving an Fe-porphine, and the cysteine axial ligand was truncated to methyl mercaptide. Unique bond lengths and bond angles are indicated on the diagram; they were obtained by assuming four-fold symmetry and averaging the corresponding geometrical parameters from the crystal structure of camphor-bound P450cam [7]. The porphine model used was constrained to planarity, except for Fe (0.45 Å below the plane) and its axial ligand. The Fe was modeled as a +3 ion in a high-spin state ($S = 5/2$), the experimentally observed state for the binary complex of P450cam and a variety of substrates and substrate analogs. Unrestricted Hartree–Fock (UHF) calculations were performed on this system without geometry optimization, using three basis sets: LANL2DZ, LANL1DZ and an augmented 6–31G basis. The LANL2DZ [11] basis set consists of the standard 3–21G basis for all first- and second-row atoms, together with Hay and Wadt's [11] (341/311/41) valence contraction for Fe. The $1s$, $2s$ and $2p$ orbitals of Fe are represented by an effective core potential. LANL1DZ [11] consists of Dunning's double-zeta basis set [12] on first- and second-row elements, with the (21/11/41) valence contraction. The final basis set we used consists of the 6–31G basis for first- and second-row elements (polarization functions were used on all heavy atoms with the addition of diffuse functions on Fe ligands) with Wachter's [13] all-electron contraction (62111111/3312/32) on Fe. Single-point SCF convergence criteria were imposed. PD-charges were calculated using Breneman's CHELPG algorithm [8] as implemented in Gaussian 92. CHELPG employs a grid method to select points at which the electrostatic potential of the molecule is sampled. The method has the advantage that the resulting charges are relatively insensitive to internal bond rotations and the orientation of the molecule in space. Other methods [9,10] demonstrate some sensitivity to these properties [8]. Fe was assigned a van der Waals radius of 1.95 Å. For all other atoms the default atomic radii were used, and the fitting procedure was constrained to reproduce the total charge of the molecule.

Molecular mechanics interaction energies (van der Waals and Coulomb components) between the substrate and the heme cofactor were calculated using the Discover simulation package (Biosym Technologies v2.9). TCE was docked into the active site of P450cam by positioning the substituted carbon of TCE at the coordinates of atom C5 of the camphor molecule in the P450cam crystal structure. Camphor is hydroxylated at the C5 position by the wild-type enzyme. Charges on TCE had previously been calculated with CHELPG on a geometry fully optimized at 6–31G*. A few steps of geometry optimization were performed on the complex so that the substrate relaxed to nonbonded contact with the heme-Fe and (in some cases) the pyrrole nitrogens. Only the porphine model of the heme cofactor and the substrate were included in these calculations.

To assess the influence of various charge sets on the equilibrium geometry of the heme, the calculated PD-charges from different wavefunctions were assigned to the atoms of the porphine core, while charge balance (to a total of -2) was maintained on the entire porphyrin by adjusting charges on the substituents in small increments. One thousand steps of conjugate gradients minimization were performed on the crystal structure of camphor-bound cytochrome P450cam, including the crystallographic waters, using an all-hydrogen model and the consistent-valence forcefield [5]. The entire enzyme was free to move during minimization. Nonbonded interactions were evaluated using a group-based switching function between 12.0 and 15.0 Å, and the nonbonded pair list was updated at each minimization step.

Results and Discussion

To determine the sensitivity of electrostatic potential-derived (PD) charges to the basis set used in *ab initio* calculations on the heme, we performed UHF single-point calculations using three different basis sets. The resulting charges are shown

TABLE I

Atom-centered partial charges on the heme from various sources (atoms labeled as in Figure 1). See text for a description of calculations. CVFF denotes the charge set previously used in conjunction with Hagler's [5] consistent-valence forcefield in simulations on cytochrome P450

	Fe	S	N	C1	C2	C3
CVFF	1.22	0.10	−0.55	0.05	−0.10	−0.10
6–31 + G*	1.50	−0.24	−0.84	0.58	−0.28	−0.64
LANL1DZ	1.94	−1.04	−1.25	0.98	−0.35	−0.86
LANL2DZ	1.02	−0.34	−0.85	0.62	−0.20	−0.46
LANL1MB [16]	1.39	−0.91	−0.38	0.10	−0.10	−0.10
IEHT [18]	0.20	−0.47	−0.18	0.03	−0.01	0.40
INDO/S [22]	1.50	−0.65	−0.40	0.15	0.15	0.15

in Table I. The larger magnitudes of the charges calculated from LANL1DZ relative to LANL2DZ, especially on Fe and its ligands, indicate that the Fe-ligand interactions are more ionic than covalent in nature. This result is in part due to freezing the $3s$ and $3p$ electrons in the calculation. Although LANL2DZ consists of a larger basis for Fe, the standard 3–21G basis is used for first- and second-row atoms. In an attempt to achieve a more balanced basis set, we chose Wachter's all-electron double-zeta quality basis for Fe [13], and an augmented 6–31G basis set for the remaining atoms in the heme. In fitting PD charges for small organic molecules, the 6–31G* basis set is commonly used [14]. The augmented 6–31G basis produced small changes in the partial charges compared to the LANL2DZ basis set, with the exception of charges on Fe and C3 atoms (see Figure 1).

One limitation in treating this open-shell system with UHF is that spin contamination can arise due to the mixing of states of higher multiplicity. The value of S^2 for the pure high-spin state is 8.75; LANL2DZ, LANL1DZ and the augmented 6–31G calculations converged with S^2 values of 8.754, 8.927 and 10.775, respectively. It is at present not understood what effect spin contamination has on PD-charges. In an attempt to address this question, we also performed several ROHF calculations, which constrained S^2 to the proper value. The Gaussian 92 implementation of CHELPG gave inaccurate results for charges fit to the ROHF wavefunctions (results not shown). The source of this discrepancy is still under investigation.

Others have used a variety of quantum mechanical methods to determine iron-heme partial charges for use in molecular mechanics forcefields. Three of these charge sets, developed for use in conjunction with the Amber potential function [15], are also summarized in Table I. Most recently, Jones *et al.* [16] used natural bond order analysis [17] on an *ab initio* wavefunction calculated with the LANL1MB [11] basis set, which includes Hay and Wadt's effective core (through $3p$) on Fe and the minimal STO-3G basis on first- and second-row elements. The authors addressed the binding of different enantiomers of nicotine to the resting state of P450cam, the high-spin Fe (III) heme that is also under investigation in the present study. Lopez and Kollman [18] parameterized the Fe (II) heme in hemoglobin and myoglobin using iterative extended Hückel (IEHT) calculations. There is some ambiguity in directly comparing this charge set with those developed for P450 since the two hemes have different axial ligands (histidine versus cysteine for P450) and different oxidation states. In addition, the magnitudes of these charges are not well matched to the other potential types of the consistent-valence forcefield; they are up to an order of magnitude smaller than charges on roughly analogous atoms in the forcefield. For example, in the consistent-valence forcefield, a pyrrole N has a charge of -0.56, and a typical aromatic carbon has a charge of between -0.10 and -0.15. Karplus and coworkers [19] used a similar IEHT approach [20] in developing parameters for MD simulations on myoglobin using CHARMM [21]. Finally, Loew and coworkers [22] conducted molecular dynamics simulations of valproic acid-bound cytochrome P450cam, using charges obtained from INDO/S studies on Fe-porphyrin compounds [23].

To qualitatively assess the effects of these different charge sets on interactions between the heme and a polar substrate, we calculated classical interaction energies between the Fe-porphine model in Figure 1 and TCE by substituting each of the charge sets into the consistent-valence forcefield. A reactive conformation of the substrate is one in which a Cl atom of TCE is in contact with Fe; an unreactive conformation has a H contacting Fe. The difference in these interaction energies is a measure of each charge set's preference for the reactive or unreactive conformation of the substrate. This preference is plotted in Figure 2. The LANL1MB [16] and IEHT [18] charge sets yield preferences for the unreactive substrate conformation that are similar to those calculated with our PD-charges. However, the CVFF charge set gives a significantly larger preference for the unreactive conformation.

The INDO/S charge set is alone in placing positive charges on all C2 and C3 atoms, while the CVFF is the only set in which S is positively charged. Both the CVFF charges and the INDO/S charges have been used successfully in studies of camphor, camphor analogs and other nonpolar organic molecules bound in the active site of P450cam [22, 24–26]. However, TCE and other heavily halogenated compounds can have substantial dipole moments, which will make their interactions more sensitive to the partial charges used to represent the heme group.

In addition to correctly predicting the relative energetics of substrate-heme interactions, it is important for the partial charge set to be balanced with the bonded terms in the molecular mechanics forcefield so that the geometry of the minimized heme group is close to that determined experimentally. We determined the rms deviation in atomic position of the heavy atoms in the porphine core relative to the crystal structure of the heme group in P450cam for each of the PD-charge sets calculated. For LANL2DZ, LANL1DZ and the augmented 6–31G basis sets, respectively, these values are 0.179, 0.585 and 0.190 (see Table II). Charges from LANL1DZ produce large distortions of the heme upon minimization in the environment of the enzyme. Although the charges from LANL2DZ produce a slightly smaller distortion than the augmented 6–31G basis, the latter is a more complete basis and likely provides a more accurate description of the electron distribution in the heme.

The primary distortions that occur upon minimization of the heme with different charge sets are displacement of Fe toward the heme plane and an overall buckling of the porphyrin due to electrostatic attraction between the C3 atoms and Fe. Scaling these charges by a constant factor reduces the magnitudes of these 1,4-

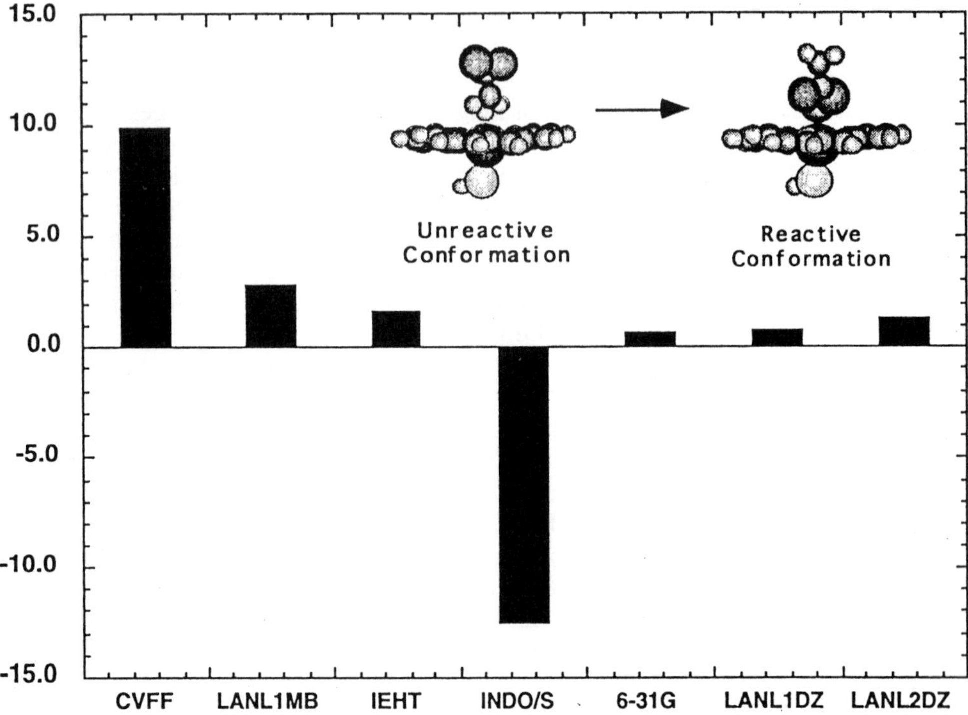

Fig. 2.   The preference for 1,1,1-trichloroethane to occupy a reactive conformation (with the chlorine atoms facing Fe) is the difference between the interaction energy in the unreactive and reactive conformations. The CVFF charges overestimate the preference for the unreactive conformation, while the INDO/S [22] charges overestimate the preference for the reactive conformation relative to other charge sets examined.

TABLE II

Nonbonded (Coulombic) interactions within the heme can cause it to distort from the equilibrium geometry, leading to artifacts in the MD simulation. Charges calculated with the augmented 6–31G basis set were scaled and assigned to the heme in the crystal structure of camphor-bound cytochrome P450cam, which was subjected to 1000 steps of energy minimization. The rms values correspond to the deviation of the porphine 'core' atoms of the heme in the minimized structure relative to the crystal structure. Preference for a reactive conformation is calculated as in Figure 2.

Charge set	RMS deviation from crystal structure (Å^2)	Preference for reactive conformation (kcal/mol)
CVFF	0.136	9.923
LANL2DZ	0.179	1.338
LANL1DZ	0.585	0.723
6–31G	0.190	0.699
75% 6–31G	0.167	0.104
50% 6–31G	0.108	0.042
25% 6–31G	0.119	−0.299
Neutral	0.125	−0.523

interactions, but it also diminishes Coulombic interactions with TCE, which are important for accurately modeling heme-substrate dynamics. Table II shows the rms deviation between the crystal structure and the structure minimized with 6–31G charges that were scaled by 75%, 50% and 25%. These calculations were also performed on a neutral heme to provide a baseline, in which only van der Waals energies are present. Cutting the 6–31G charges by half reduces the distortions upon minimization to an acceptable value; these charges are also more closely matched to the magnitudes of charges on analogous atoms in the consistent-valence forcefield. However, the preference for the unreactive conformation is also substantially reduced by this charge set. As a compromise between deviation from the crystal structure and preference for the unreactive conformation, 6–31G charges scaled by 75% are currently being used in simulations of TCE-bound P450cam.

Both the size and low-lying degeneracy of states in the open-shell iron-porphyrin present a substantial challenge to obtaining an accurate quantum mechanical description of this system. Improvements in algorithms and computational resources, however, have helped to unravel these complexities. Yamamoto and Kashiwagi [27] conducted *ab initio* RHF calculations on a model for peroxidase compound II, an iron-porphine with oxygen and histidine as axial ligands. The authors also performed complete active space (CAS) SCF calculations and demonstrated the importance of including electron correlation in the porphyrin wavefunction, and later extended these calculations with an enhanced basis set [28]. Harris and Loew [29] recently used a restricted open shell (ROHF/CI) extension of the INDO/S semiempirical method, to establish a mechanistic link between spin state and the observed spectra of the heme in P450cam. The INDO/S ROHF/CI method makes provisions for the effects of electron correlation using a limited configuration interaction approach. The present study constitutes an effort to provide a realistic model of the charge distribution for the resting state heme in P450cam. Future studies are directed at characterizing the electronic properties of the resting and intermediate states of the heme in cytochrome P450 with more rigorous methodology.

Acknowledgements

The authors express gratitude to Dave Feller, Thom Dunning, Jr., Sotiris Xantheas, Mark Thompson, Rick Kendall, Jeff Hay, Jeff Jones, and Arka Mukherjee for valuable discussions. Special thanks to Eric Glendening and Professor Robert Rein for numerous consultations and input. Pacific Northwest Laboratory is a multiprogram national laboratory, operated for the U.S. Department of Energy by Battelle Memorial Institute under Contract DE-AC06-76RLO 1830. John Manchester is a graduate student in the Dept. of Biophysics, Roswell Park Division, State University of New York at Buffalo. This work was supported by a grant (KP0402) from the Health Effects and Life Sciences Research Division within the Office of Health and Environmental Research of the Office of Energy Research of the U.S. Department of Energy (RLO). Computer resources were provided in part by the National Energy Research Supercomputer Center (Livermore, CA) of the U.S. Department of Energy.

References

1. T. L. MacDonald: *CRC Crit. Rev. Toxicol.* **11**, 85 (1983).
2. C. E. Castro, R. S. Wade, and N. O. Besler: *Environ. Tox. Chem.* **8**, 13 (1985).
3. M. S. P. Logan, L. M. Newman, C. A. Schanke, and L. P. Wackett: *Biodegradation* **4**, 39 (1993).
4. M. D. Paulsen and R. L. Ornstein: *J. Comp. Aided Molec. Design* **8**, 389 (1994).
5. A. T. Hagler: in V. J. Hruby and J. Meienhofer (Eds.), *The Peptides*, Vol. 7, Academic Press, New York, p. 213 (1985). P. Dauber-Osguthorpe, V. A. Roberts, D. J. Wolff, M. Genest, and A. T. Hagler: *Proteins: Struct. Funct. Genetics* **4**, 31 (1988).
6. Gaussian 92, Revision-D.2, Gaussian, Inc., Pittsburgh PA. M. J. Frisch, G. W. Trucks, M. Head-Gordon, P. M. W. Gill, M. W. Wong, J. B. Foresman, B. G. Johnson, H. B., Schlegel, M. A. Robb, E. S. Replogle, R. Gomperts, J. L. Andres, K. Raghavachari, J. S. Binkley, C. Gonzalez, R. L. Martin, D. J. Fox, D. J. Defrees, J. Baker, J. J. P. Stewart, and J. A. Pople (1992).
7. T. L. Poulos, B. C. Finzel, and A. J. Howard: *J. Mol. Biol.* **195**, 687 (1987).
8. C. M. Breneman and K. B. Wiberg: *J. Comp. Chem.* **11**, 361 (1990).
9. L. E. Chirlian and M. M. Francl: *J. Comp. Chem.* **8**, 894 (1987).
10. U. C. Singh and P. A. Kollman: *J. Comp. Chem.* **5**, 129 (1984).
11. P. J. Hay and W. R. Wadt: *J. Chem. Phys.* **82**, 270 (1985). P. J. Hay and W. R. Wadt: *J. Chem. Phys.* **82**, 284 (1985). P.J. Hay and W.R. Wadt: *J. Chem. Phys.* **82**, 299 (1985).
12. T. H. Dunning and P. J. Hay: Plenum, New York (1976).
13. A. J. H. Wachter: *J. Chem. Phys.* **52**, 1033 (1970).
14. H. A. Carlson, T. B. Nguyen, O. Modesto, and W. L. Jorgensen: *J. Comp. Chem.* **14**, 1240 (1993).
15. S. J. Weiner, P. A. Kollman, D. T. Nguyen, and D. A. Case: *J. Comp. Chem.* **7**, 230 (1986).
16. J. P. Jones, W. F. Trager, and T. J. Carlson: *J. Am. Chem. Soc.* **115**, 381 (1993).
17. A. E. Reed and F. J. Weinhold: *QCPE Bull.* **5**, 141 (1985). A. E. Reed, R. B. Weinstock, and F. J. Weinhold: *J. Chem. Phys.* **83**, 735 (1985).
18. M. A. Lopez and P. A. Kollman: *J. Am. Chem. Soc.* **111**, 6212 (1989).
19. K. Kuczera, J. Kuriyan, and M. Karplus: *J. Mol. Biol.* **213**, 351 (1990).
20. M. C. Zerner, M. Gouterman, and H. Kobayashi: *Theor. Chim. Acta* **6**, 363 (1966).
21. B. R. Brooks, R. E. Bruccoleri, B. D. Olafson, D. J. States, S. Swaminathan, and M. Karplus: *J. Comp. Chem.* **4**, 187 (1983).
22. J. R. Collins, D. L. Camper, and G. H. Loew: *J. Am. Chem. Soc.* **113**, 2736 (1991).
23. W. D. Edwards, B. Weiner, and M. C. Zerner: *J. Am. Chem. Soc.* **108**, 2196 (1986). W. D. Edwards and M. C. Zerner: *Theor. Chim. Acta* **72**, 347 (1987). J. Ridley and M. C. Zerner: *Theor. Chim. Acta* **32**, 111 (1973). J. Ridley and M.C. Zerner: *Theor. Chim. Acta* **42**, 223 (1976). A. D. Bacon and M. C. Zerner: *Theor. Chim. Acta* **53**, 21 (1979). M. C. Zerner, G. H. Loew, R. C. Kirchner, and U. T. Mueller-Westerhoff: *J. Am. Chem. Soc.* **102**, 589 (1980).
24. P. R. Ortiz de Montellano, J. A. Fruetel, J. R. Collins, D. L. Camper, and G. H. Loew: *J. Am. Chem. Soc.* **113**, 3195 (1991).
25. M. D. Paulsen, D. Filipovic, S. G. Sligar, and R. L. Ornstein: *Protein Science* **2**, 357 (1993).
26. M. D. Paulsen and R. L. Ornstein: *Protein Engineering* **6**, 359 (1993).
27. S. Yamamoto and H. Kashiwagi: *Chem. Phys. Lett.* **145**, 111 (1988). S. Yamamoto, J. Teraoka, and H. Kashiwagi: *J. Chem. Phys.* **88**, 303 (1988).
28. S. Yamamoto and K. Kashiwagi: *Chem. Phys. Lett.* **205**, 306 (1993).
29. D. Harris and G. H. Loew: *J. Am. Chem. Soc.* **115**, 5799 (1993).

Molecular Dynamics Simulations of Phenylimidazole Inhibitor Complexes of Cytochrome P450$_{cam}$

DAN L. HARRIS, YAN-TYNG CHANG and GILDA H. LOEW
Molecular Research Institute, 845 Page Mill Road, Palo Alto, CA 94304, U.S.A.

Abstract. Molecular dynamics simulations have been performed on three phenylimidazole inhibitor complexes of P450$_{cam}$, utilizing the X-ray structures and the AMBER suite of programs. Compared to their corresponding optimized X-ray structures, very similar features were observed for the 1-phenylimidazole (1-PI) and 2-phenylimidazole (2-PI) complexes during a 100 ps MD simulation. The 1-PI inhibitor binds as a Type II complex with the imidazole nitrogen as a ligand of the heme iron. Analysis of the inhibitor-enzyme interctions during the MD simulations reveals that electrostatic interactions of the imidazole with the heme and van der Waals interactions of the phenyl ring with nearby hydrophobic residues are dominant. By contrast, 2-PI binds as a Type I inhibitor in the substrate binding pocket, but not as a ligand of the iron. The interactions of this inhibitor are qualitatively different from that of the Type II 1-PI, being mainly electrostatic/H-bonding interactions with a bound water and polar residues. Although the third compound, 4-PI, in common with 1-PI, also binds as a Type II inhibitor, with one nitrogen of the imidazole as a ligand to the iron, the MD average binding orientation deviates significantly from the X-ray structure. The most important changes observed include: (1) the rotation of the imidazole ring of this inhibitor by about 90° to enhance electrostatic interactions of the imidazole NH group with the carbonyl group of LEU244, and (2) the rotation of the carbonyl group of ASP251 to form a H-bond with VAL254. An analysis of the H-bonding network surrounding this substrate in the optimized crystal structure revealed that there is no H-bonding partner either for the free polar NH group in the imidazole ring of 4-phenylimidazole or for the polar carbonyl group of the nearby ASP251 residue. The deviation of the dynamically averaged inhibitor-enzyme structure of the 4-PI complex from the optimized crystal structure can therefore be rationalized as a consequence of the optimization of the electrostatic interactions among the polar groups.

Key words. P450$_{cam}$, phenylimidazole inhibitor, molecular dynamics simulation

Introduction

Poulos *et al.* have obtained [1] and refined [2] the crystal structures of three inhibitor complexes of the ferric form of cytochrome P450$_{cam}$ with 1-phenylimidazole (1-PI), 2-phenylimidazole (2-PI) and 4-phenylimidazole (4-PI) shown below and found that they bind in the substrate binding site and inhibit the function of this enzyme. The free energies of binding of these inhibitors are significant with dissociation constants in the μM range [3].

A. Pullman et al. (eds.), *Modelling of Biomolecular Structures and Mechanisms*, 189–202.
© 1995 *Kluwer Academic Publishers. Printed in the Netherlands.*

Despite their similarities, comparisons of the crystal structures of these phenyl-imidazole inhibitor-P450$_{cam}$ complexes reveal a number of qualitatively different features. First, as shown in Figure 1, the binding orientations of the inhibitors are different. Specifically, the 1-PI and the 4-PI isomer both form Type II complex with a N of the imidazole ring acting as a ligand of the heme iron atom of the P450$_{cam}$ enzyme, but the orientations of the phenyl ring and the imidazole ring differ in these two inhibitors. In addition, there is no free N atom to interact with other residues in the 1-isomer while in the 4-isomer, there is one. The 2-PI isomer forms a Type I complex, binding in the substrate binding site but not as a ligand to the heme iron atom. Also, there are two bound waters in the site, one acting as a ligand to the iron, with the phenyl ring of the 2-isomer located nearby, and the other interacting with one of the two N atoms of the imidazole ring of 2-PI. This water serves as a H-bond bridge to the ASP251. The other N of the imidazole ring in the 2-PI isomer is H-bonded to the hydroxyl group of TYR96. In addition to the different orientations of the inhibitors, the surrounding residue structures, especially those in the central region of helix I, in the three complexes are perturbed to different degrees from the substrate free form, with the 1- and 4-isomer environments more similar to each other.

In this paper, we have performed molecular dynamics simulations for all three phenylimidazole complexes of P450$_{cam}$. Our goals were: (i) to use the dynamic behavior to obtain additional insight into the mode of inhibitor enzyme interactions; (ii) to identify and characterize the key inhibitor-enzyme interactions for each complex; (iii) to elucidate the origin of the observed differences in binding of the three inhibitors; and (iv) to provide knowledge for future design of more potent inhibitors for the P450$_{cam}$ enzyme.

Method

The crystal structures of the 1-phenylimidazole, 2-phenylimidazole, and 4-phenyl-imidazole complexes with cytochrome P450$_{cam}$ used were those in the Brookhaven Crystallographic Database as documented in the published work of Poulos and Howard [1,2]. The neutral form of each inhibitor was used, and for the 2-PI, the only inhibitor for which there was a choice of the two imidazole nitrogens, the proton was assigned to N3 after consideration of optimization of hydrogen bonding interactions in its crystallographic orientation. Each of the inhibitors was then fully geometry optimized using the GAUSSIAN92 programs and a 6-31G* basis set [4]. Charge parametrization for each of the inhibitors was developed from CHELPG charges fit to the molecular electrostatic potential for each of the inhibitors derived from a 6-31G*//6-31G* single point calculation. The intramolecular harmonic vibrational parametrization was taken principally from AMBER [5] and MSI/Quanta [6]. The charges for the heme were taken to be those of the resting state heme as determined from ZINDO Mulliken charges from that state [7]. Intramolecular/vibrational force constants for the heme were derived from the standard AMBER parameter set.

The P450$_{cam}$/inhibitor energy minimization and molecular dynamics simulations were performed using AMBER 4, employing the all atom representation and full protein dynamics including all crystallographic waters. Initial hydrogen positions

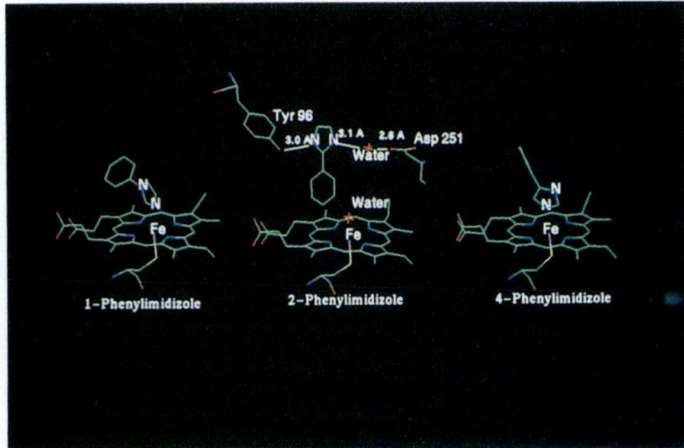

Fig. 1. X-ray orientation of phenylimidazole in the complex of P450$_{cam}$. From left to right: 1-PI, 2-PI, 4-PI.

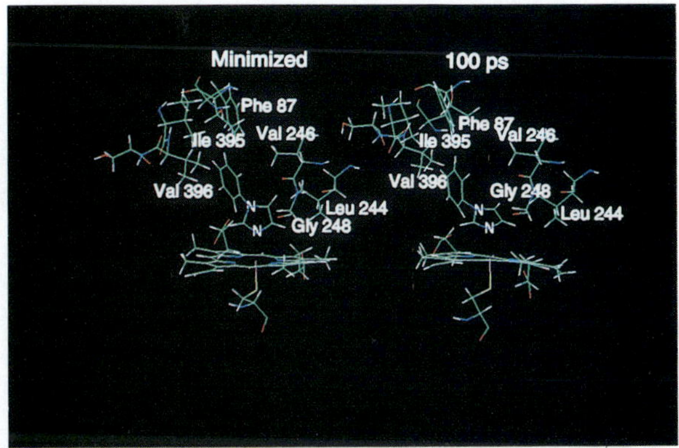

Fig. 3. The structures of P450$_{cam}$–1PI complex from the energy minimized and at 100 ps of the MD simulation.

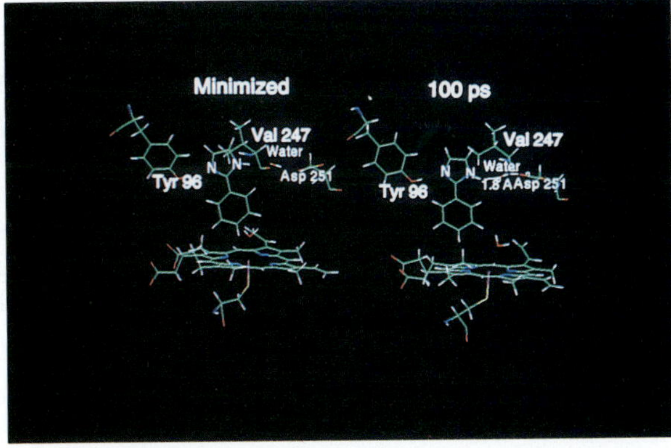

Fig. 4. The structures of P450$_{cam}$–2PI complex from the energy minimized and at 100 ps of the MD simulation.

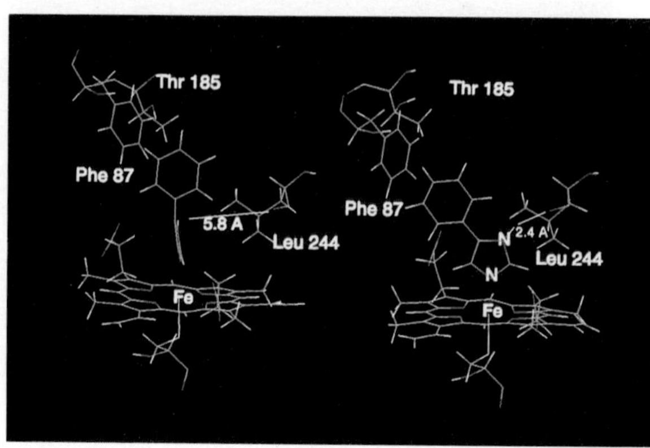

Fig. 5.   The structures of P450$_{cam}$–4PI complex from the energy minimized and at 100 ps of the MD simulation. (a) The imidazole ring rotates such that its N—H group can get close to the carbonyl group of LEU244.

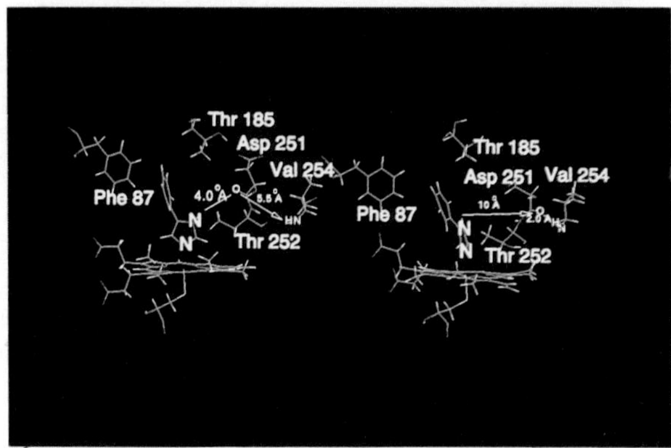

Fig. 5(b).   Viewed from a direction perpendicular to view a, the carbonyl group of ASP251 also rotates to form a H-bond with VAL254.

were initially optimized by running 200 steps of energy minimization and 2–5 ps of limited dynamics allowing just the hydrogens to reorient. Each of the structures was then completely energy minimized using 200 steps of steepest descents plus sufficient additional conjugate gradient minimization to reduce the gradient below 1 kcal/Å. Molecular dynamics simulations were then performed at constant temperature of 300 °K using a radially dependent dielectric constant ($D = r$) and employing a 15 Å nonbonded interaction truncation. Following a short equilibration period of 10 ps, an additional 100 ps of production dynamics was performed and coordinate information was saved every 0.250 ps for subsequent analysis.

Results and Discussion

For each 100 ps MD simulation of the three P450$_{cam}$–inhibitor complexes, B-factors were calculated for the inhibitor and the protein and compared with experiment. In addition, important structural features, and the interaction energy between the inhibitor and nearby residues were monitored during the 100 ps MD simulations. In the energy analysis, the van der Waals, electrostatic, and the total interaction energies were obtained. The results for each of the three complexes are described below.

I. P450$_{cam}$–1PI COMPLEX

Table I reports the B-factors and RMS coordinate differences between the energy minimized crystal structure and the MD averaged structure for both the protein and the 1-PI inhibitor. We see from this table that the mean B-factor as determined from the X-ray structure and from the MD simulation for both the protein and the inhibitor are very similar. Moreover, the overall RMS difference between the minimized crystal structure and the MD average crystal structure is 1.5 Å, and for the inhibitor is 0.9 Å. Figure 2 shows a comparison between the mean residue B-Factors as determined from the MD simulation and that derived from crystallography. The main differences are in surface exposed residues that are not near the buried binding site of this enzyme. Taken together, the results in Table I and Figure 2 indicate that the qualitative features and fluctuations of the inhibitor–enzyme complex structure are similar in the MD simulation and in the crystal structure.

Figure 3 shows the important nonbonded interactions between the inhibitor and the binding pocket of P450$_{cam}$ in the minimized crystal structure and after 100 ps in the dynamics trajectory of the 1-PI bound P450$_{cam}$. It is clear from Figure 3 that the two structures are very similar with respect to the inhibitors orientation and interactions. Comparing the two, we see that the strong interaction between the imidazole N (N3) of 1-PI and the heme iron, typical of Type II inhibitor binding, is preserved. The Fe-N3 distance is 2.1 Å in the minimized structure and 2.24 Å at 100 ps. Other important interactions of the 1-phenylimidazole with the P450$_{cam}$ binding site residues are van der Waals contact of the phenyl ring with hydrophobic sidechains, manifest by 3–4 Å heavy-atom/heavy-atom distances. These distances fluctuate about a well-defined mean indicating that no large scale reorientation occurs on the time scale of this simulation.

Table II shows the major contributions to the total interaction energy of the 1-

TABLE I

B-factors and RMS coordinate differences between the crystal structure and the MD average structure

Inhibitor	Avg. protein atom B factor[a]		Avg. inhibitor atom B factor		RMS difference[b]	
	XRAY	MD	XRAY	MD	All atoms	Inhibitor atoms
1-PI	22.8	20.5	28.8	20.8	1.5 Å	0.9 Å
2-PI	19.6	19.4	15.9	12.0	1.4 Å	0.5 Å
4-PI	24.1	19.9	26.9	28.9	1.4 Å	2.0 Å

[a]Units Å^2.
[b]Rms difference between minimized crystallographic positions and MD average crystal structure.

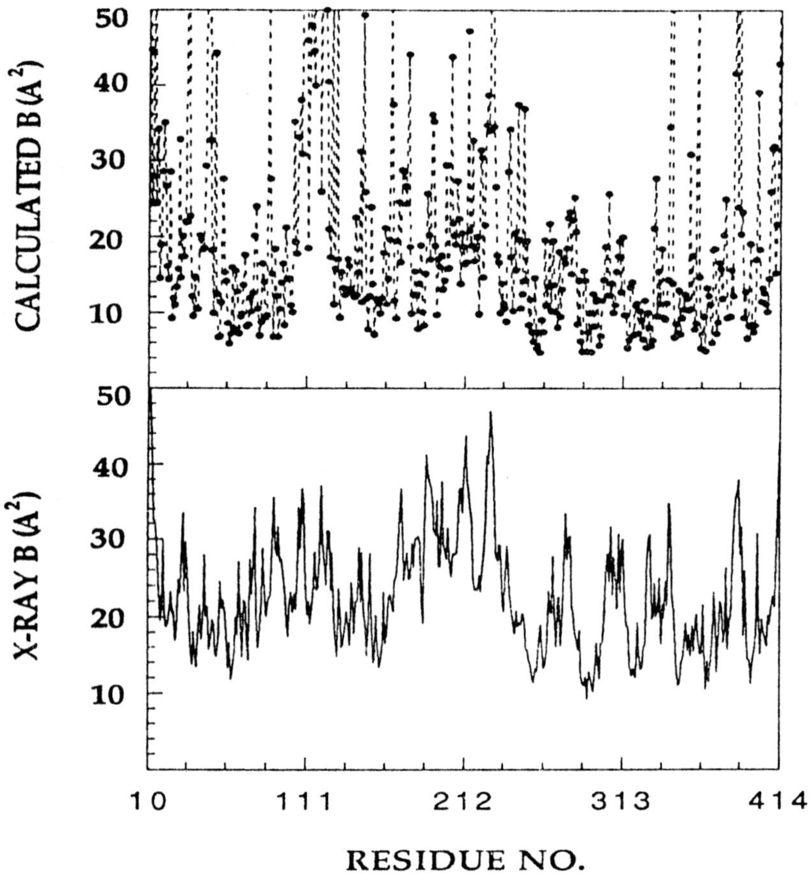

Fig. 2.   X-ray-MD calculated B factor comparison for P450$_{cam}$-1PI complex.

PI inhibitor with the enzyme. The importance of the electrostatic interaction of the 1 -phenylimidazole N (N3) with the ferric heme iron is shown by the major contribution it makes to the total interaction energy. The remainder of the interactions are spread among a number of surrounding residues in close contact with 1-PI.

II. P450$_{cam}$–2PI COMPLEX

Table I reports the B-factor and RMS coordinate differences between the energy minimized crystal structure and the MD averaged structure for both the protein and the 2-PI. We see from this table that the mean B-factor as determined from the X-ray structure and from the MD simulation for both the protein and 2-PI are very similar. Moreover, the overall RMS difference between the minimized crystal structure and the MD average crystal structure is 1.4 Å, and for the inhibitor is 0.5 Å. Thus, despite the fact that this Type I inhibitor is not anchored by the strong electrostatic interactions with the heme unit as is 1-PI, it still remains in a

TABLE II

Major contributions to the interaction energy between 1-PI inhibitor and surrounding residues in the energy minimized complex and the 100 ps MD simulation[a]

	Minimization			⟨100 ps MD⟩		
	vdw	ees	Total	⟨vdw⟩	⟨ees⟩	⟨Total⟩
HEME[b]	−5.170	−21.427	−26.597	−4.862	−20.640	−25.503
PHE87	−1.498	−0.070	−1.568	−0.751	0.099	−0.652
THR185	−0.440	0.190	−0.250	−0.238	0.095	−0.143
LEU244	−1.312	−0.278	−1.590	−1.614	−0.443	−2.057
VAL247	−1.097	0.249	−0.848	−1.192	0.200	−0.992
GLY248	−1.186	−0.027	−1.213	−1.091	−0.055	−1.146
ASP297	−0.956	−1.113	−2.069	−0.764	−1.098	−1.863
ILE395	−0.644	0.154	−0.490	−0.579	0.117	−0.462
VAL396	−1.070	−0.033	−1.104	−0.693	−0.025	−0.719
TOTAL	−20.34	−23.09	−43.43	−18.77	−22.20	−40.97

[a]Units of all energies are kcal/mol.
[b]In these simulations, the heme unit and the CYS357 are treated as a united residue. The same applies to all other energy analysis presented in this paper.

relatively stable orientation with the qualitative features and fluctuations of the inhibitor-enzyme complex structure in the MD simulation similar to that in the crystal structure.

An analysis of the interactions of this inhibitor with the binding site of P450$_{cam}$ shown in Figure 4 in the energy minimized crystal structure and the structure at 100 ps of MD clearly shows the origin of the stability of the orientations of the inhibitor and the resulting similarity between the minimized crystal structure and at the end of the simulations. As shown in this figure, TYR96 is hydrogen bonded to the imidazole N(N1). The oxygen of the internal crystallographic water is hydrogen bonded to the imidazole N3—H3 in our nomenclature and the two hydrogens of this water are themselves hydrogen bonded to the oxygen of ASP251 and the oxygen of the carbonyl of VAL247. This hydrogen bond network, as well as that of the N1 imidazole of the 2-PI inhibitor with TYR96, is persistent throughout the entire 100 ps of production dynamics as supported by an analysis of the contact distance between the atoms involved. This stable hydrogen bonded cross linked interaction of 2-PI is largely responsible for its small X-ray and computed B-factor compared to those of the other two inhibitors since all three inhibitors have many van der Waals contacts with substrate cavity hydrophobic sidechains.

Table III gives the major components of the interaction energy between the 2-PI inhibitor and key residues. In contrast to the results for 1-PI, one clearly sees that the interaction between 2-PI and the heme is greatly reduced because of the absence of the N—Fe bond found in 1-PI complex. The largest interactions are those of 2-PI with TYR96 and with the internal crystallographic water located near ASP251 and VAL247. The interactions of this water and of the water that acts as the ligand to the iron are given in Table IV. These energy analyses strongly support the conclusion that electrostatic/H-bond interactions are the cause of the observed stable binding orientation of 2-PI in P450$_{cam}$. Thus, the stabilities of the orientation of both the Type II inhibitor, 1-PI, and the Type I inhibitor, 2-PI,

TABLE III

Major contributions to the interaction energy between 2-PI inhibitor and surrounding residues in the energy minimized complex and the 100 ps MD simulation[a]

	Minimization			$\langle$100 ps MD$\rangle$		
	vdw	ees	Total	$\langle$vdw$\rangle$	$\langle$ees$\rangle$	$\langle$Total$\rangle$
HEME	−4.404	0.182	−4.223	−3.573	0.125	−3.448
TYR96	−1.218	−7.015	−8.639	−1.116	−6.022	−7.387
VAL247	−0.938	−0.844	−2.789	−1.378	−0.264	1.644
ASP251	−1.359	−2.133	−2.493	−0.511	−2.410	−2.922
H_2O(H-bond)	−1.036	−10.383	−9.398	−0.999	−10.082	−8.995
H_2O ligand	−1.542	−1.570	−2.112	−0.391	−10.082	−0.681
TOTAL	−22.15	−20.19	−42.82	−19.17	−0.290 −18.79	−38.12

[a]Units of all energies are kcal/mol.

TABLE IV

Major contributions to the interaction energy of each of the bound waters in 2-PI inhibitor complex with surrounding residues in the energy minimized complex and the 100 ps MD simulation[a]

	Minimization			$\langle$100 ps MD$\rangle$		
	vdw	ees	Total	$\langle$vdw$\rangle$	$\langle$ees$\rangle$	$\langle$Total$\rangle$
(A) H-bond H_2O						
2PI	1.036	−10.383	−0.398	0.999	−10.082	−8.995
THR185	0.090	−3.953	−385	−0.243	0.323	0.071
VAL247	0.206	6.743	−6.912	0.137	−5.460	−5.610
ASP251	1.574	−16.315	−14.193	1.641	−15.556	−13.752
(B) Ligand water						
2PI	−0.542	−1.570	−2.112	−0.391	−0.290	−0.681
HEME	5.217	−26.345	−21.127	4.682	−24.833	−20.152
THR252	−0.201	0.441	0.238	−0.096	−3.455	−3.801

[a]Units of all energies are kcal/mol.

binding in the substrate binding site but not to the heme iron are comparable, but for qualitatively different reasons.

III. P450$_{cam}$–4PI COMPLEX

As shown in Table I, both the mean protein B-factor and the mean 4-PI B-factor show qualitative agreement between the MD simulations and the crystal structure. The overall RMS difference in the minimized complex and the MD average structure is 1.4 Å, similar to the magnitude obtained for the P450$_{cam}$–1PI and P450$_{cam}$–2PI complexes. However, the 2.0 Å RMS difference in the 4-PI inhibitor itself between the minimized and MD average crystal structures is quite large compared to the RMS of 1-PI (0.9 Å) and 2-PI (0.5 Å). This result is at first surprising, given that both 1-PI and 4-PI bind as Type II inhibitors with the N of the imidazole as a ligand of the heme iron atom. However, examination of the

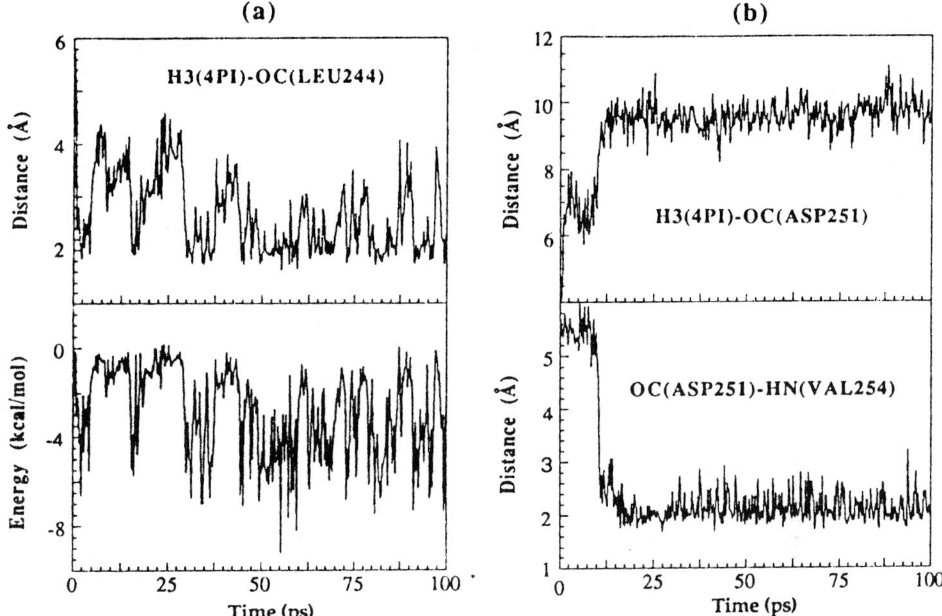

Fig. 6. (a) The correlation of the distance between H3 of 4-PI and the carbonyl oxygen of LEU244 and the electrostatic interaction energy between 4-PI and LEU244. (b) The motion of the carbonyl group of ASP251 leads to an increase of r(H3(4PI)-OC(ASP251)), and a decrease of r(OC(ASP251)-HN(VAL254)). A H-bond is formed between ASP251 and VAL254.

motion of 4-PI and of nearby residues in the binding site, provided insight into the origin of this difference. In 4-PI, but not in 1-PI, there is a second NH group of the imidazole available for electrostatic/H-bonding interactions with the binding site. However, in the X-ray and energy minimized structures, the few polar residues with electron-donor functional groups near this NH group are too far away to form a strong electrostatic interaction. For example, the hydroxyl oxygen OG of THR252 is 3.7 Å from this polar H atom of 4-PI in the energy minimized structure. Similarly, the distances between H3(4-PI) and OC(VAL247), OC(ASP251), OD1(ASP251), OD2 (ASP251) in the minimized structure are quite large, i.e. 4.0, 4.0, 7.6, and 6.4 Å, respectively. Therefore, very early in the MD simulations (before 5 ps), the imidazole ring of the inhibitor rotates to search for stronger electrostatic interactions of the N3—H3 group with its environment. As shown in the two views of Figure 5, the N—H group rotates by about 90° and forms a good interaction with the carbonyl oxygen of LEU244. Because of the requirement to simultaneously maintain the strong N1–Fe electrostatic interaction and the H3(4PI)–OC(LEU244) interaction, a translation of the entire HEME + 4-PI complex toward the LEU244 direction was also observed. The magnitude of this translation is about 1–2 Å.

Figure 6(a) shows a strong correlation between the fluctuation of the distance between the H3 atom of 4-PI and the OC atom of LEU244 and the fluctuation of

TABLE V

Major contributions to the interaction energy between 4-PI inhibitor and surrounding residues in the energy minimized complex and the 100 ps MD simulation[a]

	Minimization			$\langle$100 ps MD$\rangle$		
	vdw	ees	Total	$\langle$vdw$\rangle$	$\langle$ees$\rangle$	$\langle$Total$\rangle$
HEME	−2.925	−21.895	−24.820	−4.048	−20.394	−24.441
LEU244	−0.740	−0.110	−0.850	−1.961	−2.837	−4.963
VAL247	−1.955	0.138	−2.095	−1.795	0.610	−1.185
GLY248	−0.893	−0.362	−1.247	−1.362	0.375	−0.988
ASP251	−0.388	0.872	−1.262	−0.189	−0.093	−0.282
TOTAL	−17.94	−24.63	−42.58	−19.18	−25.01	−43.35

[a]Units of all energies are kcal/mol.

the electrostatic interaction energy between the inhibitor and LEU244 during the 100 ps simulations. The MD average of this distance is 2.66 Å, compared to the value of 6.40 Å in the minimized structure. Simultaneously, the LEU244 carbonyl group maintains a H-bond with the GLY248 HN, although the distance of this H-bond does increase slightly from 1.83 Å in the minimized structure to 2.03 Å from the average of MD simulations.

Another anomaly found in the X-ray structure is that the carbonyl group of ASP251 does not have a H-bond interaction with any of its nearby residues. In the energy minimized structure, the backbone OC of ASP251 already moves from its X-ray position toward the binding site cavity. The magnitude of this motion is about 1 Å. This movement allows a slightly better electrostatic interaction between the OC(ASP251) group and the HN group of 4PI. However, the distance between these two groups is still too far away (4.03 Å) to form a H-bond. During the MD simulations, the OC group of ASP251 rotates in the opposite direction (i.e., away from the binding site cavity) and forms a H-bond with a new partner, the HN atom of VAL254. This rotation occurs at about 10 ps of the MD simulations and results in a decrease in the distance between OC of ASP251 and HN of VAL254 from 5.50 (in the minimized structure) to 2.48 Å in the MD average and a correlated increase in the distance between H3 of 4-PI and OC of ASP251 from 4.03 in the minimized structure to 9.29 Å from the MD average. The correlation of the change of these two distances as a function of time is shown in Figure 6(b). Along with this motion, the H-bond network among nearby residues also changes.

Table V compares the van der Waals, electrostatic and total interaction energies for the energy minimized 4-PI complex and the MD average. As in the other Type II complex, with 1-PI, the N1—Fe heme electrostatic interaction remains the strongest during the simulation. However, consistent with the change in orientation observed, the change of 4-PI interactions with its surrounding residues is also revealed in the energy analysis. The most significant change comes from the increasing attraction (about 4 kcal/mol) between 4-PI and LEU244. On the other hand, the interaction energies with VAL247 and ASP251 each decrease by about 1 kcal/mol.

In contrast to the behaviors of the P450$_{cam}$–1PI and P450$_{cam}$–2PI complexes, the MD simulations of P450$_{cam}$–4P1 complex clearly show a large deviation from

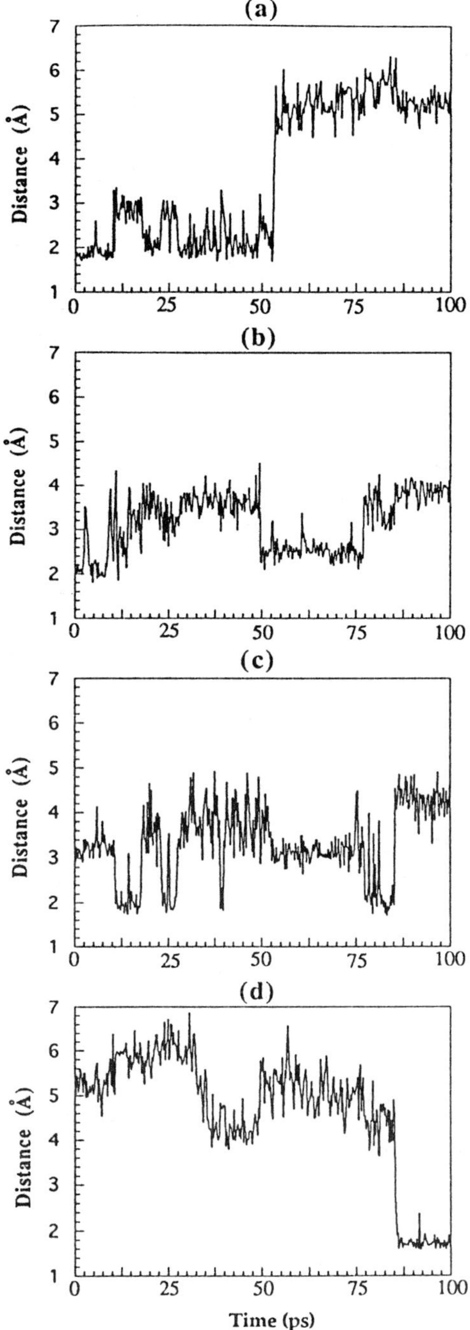

Fig. 7.   The distance between HW2 of the extra water and (a) OC(ASP251), (b) OG(THR252), (c) OC(VAL247), and (d) OD1(ASP251).

the X-ray structure. This deviation is due to the optimization of the electrostatic interactions between 4PI and P450$_{cam}$ and consequent optimization between the residues of P450$_{cam}$. In principle, a water molecule may serve as a bridge which interacts with both the NH group of 4PI and the CO group of ASP251 via H-bond network. Such a water molecule, if present, would alter the dynamics of the 4PI substrate. Therefore, we have performed another MD simulation by artificially placing an extra water molecule between 4PI and ASP251.

As expected, the results from this modified simulation show the HN group of 4PI maintains a strong H-bond with the extra water both in the energy minimized structure $(r(\text{H3—OW}) = 1.78\,\text{Å})$ and throughout the MD simulations $(\langle r(\text{H3—OW})\rangle = 1.86\,\text{Å})$. However, the interaction between the extra water and OC(ASP251) only persists for about 50 ps. OC(ASP251) then moves toward VAL254 and ASN255 as observed in the unmodified simulation. The H-bonds between the hydrogen atoms of the extra water and their surrounding residues change periodically. In the minimized structure, one of the two polar hydrogens in the extra water forms a strong interaction with OC of VAL247, while the other one interacts with OC group of ASP251. During the MD simulations, HW1 of this extra water interacts with OC(VAL247) first, then with OG(THR252), OC(VAL247), OG(THR252). The sequence of interaction parter for HW2 is OC(ASP251)-OG(THR252)-OC(VAL247)-OD1(ASP251). As an illustration, Figure 7(a)–(d) shows the distances between HW2 and OC(ASP251), OG(THR252), OC(VAL247), and OD1(ASP251), respectively.

An interesting facet of this trajectory is the rotation of the ASP carboxy side chain toward the extra water (as demonstrated by the distance between HW2 and OD1(ASDP251) shown in Figure 7(d)) at a later period ($\sim$90 to 100 ps) of this simulation. As suggested by Sligar and coworkers [8, 9], ASP251 plays a role in P450$_{cam}$ oxygen activation. One of the possible mechanisms of this activation involves the interaction of the carboxy OH group of ASP251 with a water molecule. Our simulation supports the possibility of such an interaction if a water molecule is nearby.

Conclusions

Molecular dynamics simulations using AMBER have been performed on three P450$_{cam}$–phenylimidazole complexes. The B factors for the protein and inhibitors obtained from the simulations are in reasonable agreement with the X-ray values in all three complexes. For the P450$_{cam}$–1PI complex, the strong electrostatic interaction of the 1-phenylimidazole N(N3) interacting with the ferric heme iron, coupled with the van der Waals interactions, are sufficient to maintain a bound orientation close to that found in the crystal structure. For the P450$_{cam}$–2PI complex, the electrostatic/H-bond interactions among 2-PI, TYR96, ASP251, VAL247 and the bound water, rather than interactions with the heme, are the cause of the stable binding orientation of 2-PI in P450$_{cam}$. For the P450$_{cam}$–4PI complex, the 4-PI despite binding as a ligand to the heme, exhibits a significant deviation of the MD average imidazole ring orientation and the ASP251 OC group orientation also deviates from the crystal structure. The cause of these deviations is, in both instances, the optimization of the electrostatic interaction of the two polar groups (NH of 4-PI and OC of ASP251) with their surroundings. This is verified by a modified simulation of the P450$_{cam}$–4PI complex with an extra water in which the H-bond interaction between 4PI NH group and the oxygen atom of the extra water persists.

As a first step in the understanding of the interactions of inhibitors with P450$_{cam}$, and the possible use of these insights for optimum inhibitor design, we have determined and compared the time dependent interactions of three phenylimidazoles with P450$_{cam}$ that have known inhibitor bound P450$_{cam}$ crystal structures. As a result of their differing architecture, the three phenylimidazoles differ in their ability to hydrogen bond to residues of P450$_{cam}$ which, in turn, alters their preferred binding orientations as well as the main source of their interaction energy with the enzyme. While the results clearly indicate why 1-PI and 2-PI bind with similar affinities, it is more difficult to understand the lower affinity measured for the 4-PI complex [3]. Poulos and coworkers rationalized these results by suggesting that, whereas 1-PI has no free imidazole nitrogen once bound to P450$_{cam}$, 4-PI has one "free" nitrogen with no optimal hydrogen bond partners [1]. However, the MD simulations reveal changes in both the inhibitor and residues in the binding site to alleviate this deficiency. In considering the relative free energies of binding, one must take into account not only the energy of binding of the inhibitors to the enzyme but the relative solvation energy of the inhibitors since they transfer from solution to the protein interior. Thus, it is possible that the low affinity of 4-PI is related to its higher desolvation energy relative to the other two inhibitors. The molecular dynamics results presented here provide a good foundation for further exploration of relative thermodynamics of desolvation of 1-PI and 4-PI in binding to P450$_{cam}$.

Acknowledgments

Support by National Institutes of Health Grant GM 27943 (D.L.H. and G.H.L.) and Environmental Protection Agency Grant No. CR818677-01-0 (Y.T.C.) for this work is gratefully acknowledged. We also thank Dr. T. Poulos for his encouragement and helpful discussions and gratefully acknowledge a grant to perform these calculations on the CRAY C90 at the Pittsburgh Supercomputing Center sponsored by the National Science Foundation.

References

1. T. L. Poulos and A. J. Howard: *Biochemistry* **26**, 8165 (1987).
2. T. L. Poulos and A. J. Howard: Brookhaven Crystallographic Database (1993).
3. J. D. Lipscomb: *Biochemistry* **19**, 3590 (1980).
4. Gaussian 92, Revision A, M. J. Frisch, G. W. Trucks, M. Head-Gordon, P. M. W. Gill, M. W. Wong, J. B. Foresman, B. G. Johnson, H. B. Schlegel, M. A. Robb, E. S. Replogle, R. Gomperts, J. L. Andres, K. Raghavachari, J. S. Binkley, C. Gonzalez, R. L. Martin, D. J. Fox, D. J. Defrees, J. Baker, J. J. P. Stewart, and J. Pople: Gaussian, Inc., Pittsburgh PA (1992).
5. D. A. Pearlman, D. A. Case, J. C. Caldwell, G. L. Seibel, U. C. Singh, P. Weiner, and P. A. Kollman: AMBER 4.0, Department of Pharmaceutical Chemistry, University of California, San Francisco (1991).

6. QUANTA: Version 3.3, Molecular Simulations Inc. (1992).
7. D. L. Harris (unpublished work).
8. J. Aikens and S. G. Sligar: *J. Am. Chem. Soc.* **116**, 1143 (1994).
9. N. C. Gerber and S. F. Sligar: *J. Biol. Chem.* **269**, 4260 (1994).

The Effect of Hydrostatic Pressure on Protein Crystals Investigated by Molecular Simulation

DARRIN M. YORK,[1,2]* TOM A. DARDEN[4] and LEE G. PEDERSEN[3,4]
[1]*Department of Chemistry, Duke University, Durham, NC 27706, U.S.A.*
[2]*North Carolina Supercomputing Center, a Division of MCNC, RTP, NC 27709-12889, U.S.A.*
[3]*Department of Chemistry, University of North Carolina, Chapel Hill, NC 27599-3290, U.S.A.*
[4]*National Institute of Environmental Health Sciences, Research Triangle Park, NC 27709, U.S.A.*

Abstract. The effect of hydrostatic pressure on protein crystal structures is examined with molecular simulation. Four 1 ns molecular dynamics simulations of bovine pancreatic trypsin inhibitor in a crystal unit cell have been performed at solvent densities corresponding to 32%, 36%, 40%, and 44% solvent. Electrostatic interactions in the crystalline environment were treated rigorously with Ewald sums. The effect of varying the solvent density at constant unit cell volume is analyzed with respect to changes in protein structure, atomic fluctuations, solvation, and crystal packing. The results indicate the solvent density range 36–40% gives excellent overall agreement with high resolution crystallographic data ($\sim$0.3 Å rms backbone deviation). The low density (32%) and high density (44%) simulations have larger deviations.

Key words: Crystal, Ewald, molecular dynamics, pressure, protein, simulation.

Introduction

The majority of experimental structural information for biological macromolecules to date has been provided by X-ray crystallography. There is considerable evidence that protein structure generally is not significantly altered upon crystallization [1,2]. Nonetheless, the structure of a protein in a crystalline environment and in solution may be expected to differ [3]. The degree to which this occurs depends on the presence of crystal packing contacts, and changes in solvation. Typically, protein crystals contain between 30% and 70% solvent by volume [1]. However, solvent contents may vary, being as high as 90% or as low as 25% [2]. It is therefore of interest to understand the effect of solvent content on the crystal packing and solvation of protein crystal structures.

Crystal packing and solvation may be attenuated by external variables such as ion concentration, pH, temperature, and pressure. Here, we focus on the latter. Pressure induced effects on protein structure and solvation have been studied by a variety of experimental techniques [4]. It is known, for example, that pressure can induce denaturation [5] and cause changes in secondary structure [6]. Recently, the effect of hydrostatic pressure on the structure [7] and solvation [8] of hen egg-white lysozyme at 1 atm and 1000 atm has been studied by X-ray crystallography.

Molecular dynamics (MD) simulations provide a useful tool for probing the effect of environment on protein structure. Several studies have been reported that examine the effect of solvation on protein structure by directly comparing

*Author for correspondence.

A. Pullman et al. (eds.), *Modelling of Biomolecular Structures and Mechanisms*, 203–215.

molecular dynamics simulations in a crystalline environment to those in solution [9–12]. There are several compelling reasons why crystal simulations are of interest. The abundance of X-ray crystal structures provides a wealth of experimental structural data that frequently are used as starting points for theoretical investigations. Consequently, crystal simulations can be used to assess the reliability of simulation force fields and methodologies by direct comparison to accurate experimental data. Moreover, unlike solution simulations, the boundary conditions of a crystal are well defined. Recent simulation studies of protein crystals indicate accuracy comparable to that observed between different crystallographic forms can be obtained by proper treatment of long-range electrostatic interactions [York *et al.* (1994) *Proc. Natl. Acad. Sci. USA* **91**, 8715].

In this study, we examine the effect of variations in pressure on the structure of bovine pancreatic trypsin inhibitor (BPTI) in a crystalline environment using molecular dynamics. The form I crystal of BPTI has been determined by X-ray crystallography [13], and subsequently this protein has been the focus of a number of theoretical investigations (for review see Reference 14). Recently two studies have investigated the effect of pressure on the structure of BPTI in solution [15,16]. However, to our knowledge, the effect of pressure on a protein crystal has not been studied by molecular simulation. Herein we examine the effect of variations in solvent density (and hence pressure) at constant unit cell volume on the structure, atomic fluctuations, solvation, and crystal packing of BPTI. We demonstrate that with appropriate solvent density, excellent agreement with high resolution crystallographic data can be obtalned ($\sim$0.3 Å deviation), whereas elevated or reduced density leads to greater deviation.

Methods

We have performed four 1-ns molecular dynamics simulations of BPTI in a crystal unit cell with variable solvent density. The crystalline environment modeled in this study was that of the form I crystal reported by Diesenhofer and Steigman [13] (space group $P2_12_12_1$, $a = 43.1$ Å, $b = 22.9$ Å, $c = 48.6$ Å). The unit cell contains four protein molecules and 240 crystallographic waters (60 per protein molecule). The net charge of each protein molecule was taken to be +6 corresponding to the normal protonation state of the component amino acids at neutral pH. Twenty four chloride counterions were added to neutralize the system. Water molecules were packed in the interstitial space between protein molecules at four different densities. The resulting crystalline systems contained a total of 492, 552, 612, and 672 water molecules, corresponding to approximately 32%, 36%, 40%, and 44% solvent by volume. The form I crystals have been estimated experimentally to be approximately 36% solvent [17]; however, no density measurements have been reported.

Molecular mechanics and dynamics calculations were performed using a modified version of the AMBER3.0 software package. The all-atom force field [18] was employed for protein molecules, and water was treated using the TIP3P model [19]. Chloride ion parameters were taken from Lybrand *et al.* [20] ($r_0 = 2.495$ Å, $\epsilon = 0.107$ kcal/mol). Covalent bonds involving hydrogen were constrained using a modified SHAKE algorithm [21].

Electrostatic forces in the crystal were treated using the Ewald summation convention calculated using the Particle-Mesh Ewald technique [22]. The external dielectric was taken to be infinite so that the surface term in the Ewald expression vanishes. Rigorous treatment of long-range electrostatic forces in this way has been demonstrated to be necessary for proper behavior in simulations of large protein crystals [12,23].

The crystallographic structure provided the unrefined geometry of the protein heavy atoms and structural waters. Hydrogens were added using the internal geometries in the AMBER database, followed by conjugate gradient energy minimization keeping the heavy atom positions fixed. Bulk water molecules were initially packed in the unit cell to low density (32%), energy minimized, and equilibrated with 100 ps of MD keeping the solute (heavy atoms and minimized hydrogens) fixed. Chloride ions were added as follows. The potential at each bulk water oxygen position was evaluated, and the 24 water molecules with the most favorable electrostatic potentials were replaced by chloride ions. The resulting (low density) system was then re-equilibrated with constrained energy minimization and molecular dynamics. Higher density systems were constructed by periodically checking the solvent bath for cavities (1.4 Å radius) during equilibration, and packing the cavities with additional waters. Once fully packed, each higher density system (36%, 40%, 44%) was equilibrated in the same manner as the low density (32%) system. The equilibrated systems were relaxed with 200 steps unconstrained energy minimization to arrive at the starting configurations for full MD. Initial velocities were obtained from a Maxwellian distribution at 1 K, and integration was performed using a 1 fs time step. Systems were initially heated to 298 K over 10 ps by coupling to a thermal bath (temperature relaxation time 0.4 ps). Molecular dynamics was carried out at constant temperature and volume to 1 ns with coordinate files output every 0.5 ps.

Results and Discussion

Here we give a preliminary account of the overall changes in structure and atomic motions predicted for a protein crystal at variable solvent densities. A more thorough study, particularly of the details of the protein-water interactions, is forthcoming.

(i) ROOT-MEAN-SQUARE POSITIONAL DEVIATIONS

The deviation in structure of the simulated protein molecules from the crystallographic structure can be monitored by the root-mean-square positional deviation (rmsPD). The rmsPD of one structure relative to another is obtained by optimal superposition (in a least squares sense) of a set of topologically equivalent atoms [24]. Typically, atom sets consisting of backbone atoms —(N—C_α—C)—, or all heavy (non-hydrogen) atoms, are chosen for the least square fit. Recently, the use of atom sets consisting of heavy atoms with relatively low crystallographic temperature factors ($<20\,\text{Å}^2$) has been recommended for fitting procedures [25].

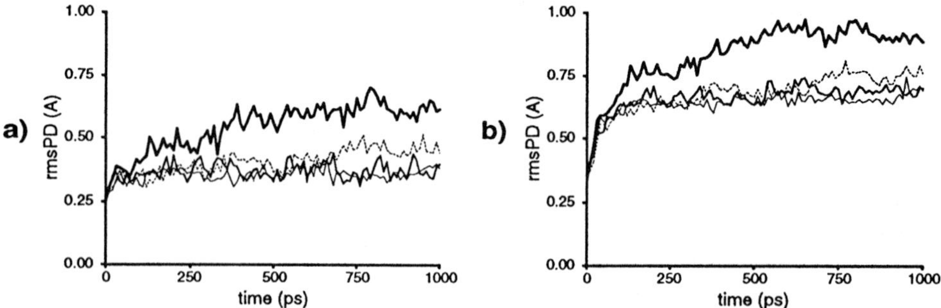

Fig. 1.   Time evolution of the root-mean-square positional deviation (rmsPD) of the instantaneous unit cell average (average of the 4 protein monomers at each time point) for (a) backbone (N—C_α—C) atoms, and (b) heavy atoms with crystallographic temperature factors <20 Å^2. Different simulations are indicated by different line styles: 32% (thick solid line), 36% (medium line), 40% (thin solid line), 44% (thin dotted line).

We include the latter convention in this section for comparison. Figure 1(a,b) shows the time evolution of the rmsPD values of the simulated structures relative to the crystallographic structure for each simulation. Equilibration of the rmsPD occurs by 250 ps for the higher density simulations (36%, 40%, and 44%). The low density simulation (32%) requires slightly longer (~400 ps). The asymptotic values of the rmsPD for backbone atoms (32%, 36%, 40%, 44%: 0.61, 0.38, 0.36, 0.42 Å, respectively) and heavy atoms (0.92, 0.68, 0.65, 0.71 Å, respectively) are quite small relative to earlier MD simulations of proteins of comparable size [25]. The stability of the rmsPD values over the relatively long simulation period (1 ns) suggests the structures are equilibrated within a local region of configuration space.

(ii) AVERAGE STRUCTURES

Crystallographic structures represent a time average of an ensemble of protein molecules in a crystal lattice. The fundamental (periodic) repeating unit of the crystal is the unit cell. However, within the unit cell, there can be several structurally equivalent asymmetric units that are related by crystallographic symmetry. In the case of form I crystals of BPTI, the asymmetric unit consists of a single protein molecule. The simulation average structure should therefore include contributions from all of the protein molecules at each time point. The average structure can be obtained by transforming each of the protein molecules to a common local reference frame by applying the inverse $P2_12_12_1$ symmetry operations at each time point. The sampling distribution used to obtain the average involves the 'ensemble' of $M = 4$ transformed protein structures at each time point. The total number of structures in the distribution is $M \times N$ where N is the number of time points and M is the number of protein molecules (asymmetric units) in the unit cell. We denote quantities derived from this distribution by the prefix *cell-* (e.g. *cell*-average, *cell*-variance). Alternately, if we are interested in the statistical behavior of individual molecules, a distribution for each molecule can be constructed involving

TABLE I

Comparison of the rmsPD (Å) of the *cell*-average and *time*-average structures (0.25 to 1 ns) from the X-ray crystal structure

Structure/atom set	Simulation			
	32%	36%	40%	44%
cell-average structures				
backbone[a]	0.54	0.33	0.31	0.39
heavy atom[b]	0.84	0.63	0.60	0.67
time-average structures[c]				
backbone[a]				
M_1	1.09	0.46	0.31	0.45
M_2	0.41	0.36	0.35	0.37
M_3	1.16	0.33	0.36	0.40
M_4	0.38	0.35	0.39	0.42
heavy atom[b]				
M_1	1.40	0.77	0.64	0.76
M_2	0.72	0.68	0.69	0.67
M_3	1.45	0.62	0.64	0.71
M_4	0.75	0.69	0.67	0.73

[a]Main chain N—Cα—C atoms (174 total).
[b]All heavy (non-hydrogen) atoms (454 total).
[c]Monomers 1–4 are designated $M_1, \ldots, M_4$.

only sampling in time, denoted by the prefix *time*- (e.g. *time*-average, *time*-variance).

Table I compares the rmsPD of the simulation *cell*-average and *time*-average structures with respect to the crystallographic structure. Atomic positions from the trajectories were sampled every 0.5 ps from 0.25 to 1 ns. The average structure from the 40% simulation has the lowest rmsPD (0.31 Å backbone), whereas the average structure from the 32% simulation has the largest rmsPD (0.54 Å backbone). The 36% and 40% structures are remarkably similar to each other [rmsPD 0.12 Å (backbone) and 0.23 Å (all heavy atoms)]. This is consistent with observations that the 1 atm and 1000 atm crystal structures of egg-white lysozyme have rmsPD values of 0.12 Å (main-chain N, C_α, C_β, C, O atoms) and 0.20 Å (all heavy-atoms) [7]. Figure 2 shows the rmsPD values of the *cell*-average structures as a function of alpha carbon. The main deviations in the 32% structure occur at the protein termini, and at residues 12 and 36. The latter are flexible glycine residues that are not involved in secondary structure. Peaks in the alpha carbon rmsPD observed in all the *cell*-average structures occur at residues Ala[25] and Gly[28]. These residues correspond to the first and last residues of a four-residue hydrogen bonded turn separating β-strands in an antiparallel sheet (see below).

(iii) GLOBAL STRUCTURAL PROPERTIES

Changes in the overall shape of the protein structures are reflected by global structural properties such as the radius of gyration and solvent accessible surface area (Table II). The radius of gyration shows a slight monotonic decrease with

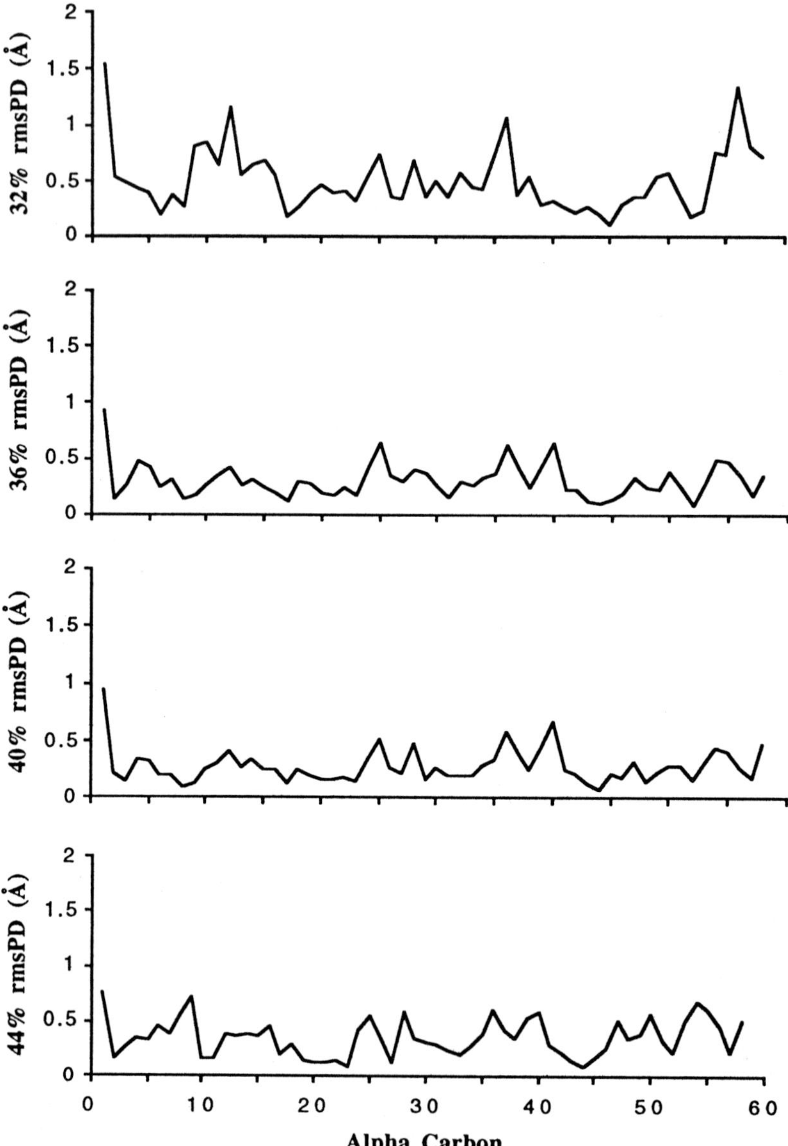

Fig. 2.   Comparison of the rmsPD of the *cell*-average structures (0.25–1 ns) from the X-ray structure
as a function of alpha carbon.

increasing solvent density (10.93–10.85). The values are all slightly smaller than
for the crystallographic structure by about 1 percent. The radius of gyration can
be represented as a vector quantity by its principal components (Rg_x, Rg_y, Rg_z).
The deviation in the shape of the simulated *cell*-average structures from the
crystallographic structure can be measured by taking the norm of the difference
of the principle components (*dev. Rg* in Table II). The deviation of the radius of

TABLE II

Radius of gyration and solvent accessible surface area of the *cell*-average structures $\langle \cdots \rangle_{cell}$ and X-ray crystal structure

Property	Structure				
	$\langle 32\% \rangle_{cell}$	$\langle 36\% \rangle_{cell}$	$\langle 40\% \rangle_{cell}$	$\langle 44\% \rangle_{cell}$	X-ray[a]
Rg[b]	10.93	10.91	10.89	10.85	11.0
Rg$_x$	8.39	8.42	8.43	8.39	8.59
Rg$_y$	5.41	5.32	5.29	5.30	5.24
Rg$_z$	4.46	4.44	4.43	4.38	4.44
dev. Rg[c]	0.26	0.19	0.17	0.22	
Surface area[d]	4151	4178	4214	4184	4279

[a]X-ray structure with minimized AMBER hydrogens.
[b]Radius of gyration Rg (Å); the three principle components are designated (Rg$_x$, Rg$_y$, Rg$_z$).
[c]Deviation in the radius of gyration of the simulation average structures relative to the X-ray structure computed as the norm of the difference of the principle components.
[d]Solvent accessible surface area (Å^2) calculated using the program DSSP [24].

gyration is smallest for the 40% *cell*-average structure (0.17 Å) indicating the globular shape is more similar to the crystallographic structure than that of the other *cell*-average structures. Interestingly, the solvent accessible surface area does not show the same monotonic behavior with solvent density as the radius of gyration, but has a maximum at the 40% structure (4214 Å^2) where it is closest to the crystallographic value (4279 Å^2).

(iv) ATOMIC FLUCTUATIONS

Atomic fluctuations give information about the motion of atoms vibrating in a local potential of mean force. The root-mean-square positional fluctuations (rmsPF) can be measured directly from the MD simulations using either the *cell*-variance or the *time*-variances (Figure 3). Atomic fluctuations can be estimated from the crystallographic B values through the equation $\langle \Delta r_i^2 \rangle^{1/2} = (3B_i/8\pi^2)^{1/2}$, where $\langle \Delta r_i^2 \rangle^{1/2}$ is the rmsPF of atom i, and B_i is the corresponding B-value [27]. The absolute values of the fluctuations derived from the crystallographic data contain additional contributions such as lattice disorder that are not present in the MD simulations. Consequently, the absolute values of the experimentally derived fluctuations are less reliable than are their relative values [28]. In way of comparison, it is useful to compute the Pearson's correlation coefficient between the simulated and experimentally derived fluctuations. Table III shows the average backbone and side-chain heavy atom atomic fluctuations determined from the simulation *time*-variances and *cell*-variance, and estimated from the crystallographic B values. In all cases, the correlation coefficient was in the range 0.5–0.7. The average fluctuations decrease considerably from the 32% to the 44% simulation. This is consistent with the intuitive argument that, within some range, the pressure effect would tend to damp atomic fluctuations in the protein. In contrast, the fluctuations from the 36% and 40% simulations are almost identical. It is probable that the 36–40% solvent density range closely resembles that of the native crystal, and

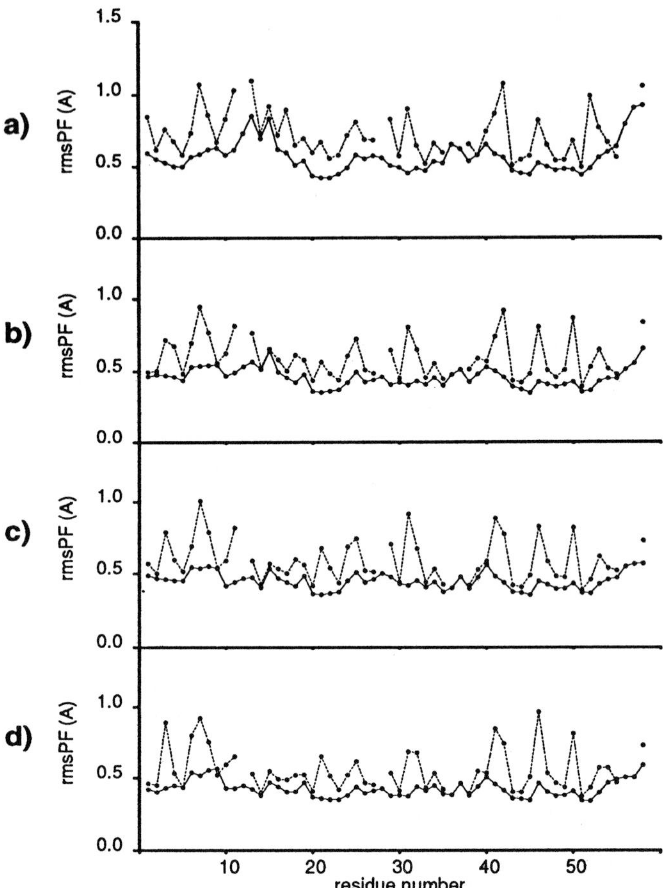

Fig. 3.   Comparison of the rmsPF (0.25–1 ns) for main-chain heavy atoms (N—Cα—CO) (solid lines) and side-chain heavy atoms (dotted line) for the simulations: (a) 32%, (b) 36%, (c) 40%, and (d) 44%.

hence one might expect the structure and fluctuations to be stable in this range. Similarly, a minor decrease in the average experimental B-factors (1 Å^2) was observed in the 1000 atm crystal structure of lysozyme, corresponding to a difference of approximately 0.04 Å rmsPF [7].

(v) SECONDARY STRUCTURE

Secondary structure reflects the nature of intramolecular hydrogen bonding. Remarkably, the secondary structural assignments for the 36%, 40%, and 44% *cell*-average structures are identical to those of the crystallographic structure (Table IV). In contrast, the 32% *cell*-average structure has deviations in secondary structure at residues 3–6, 43, and 56. The latter two involve residues in loops or turns. The most striking feature of the 32% *cell*-average structure is that residues 3–6

TABLE III

Simulated and experimentally derived root-mean-square positional fluctuations (Å) in the BPTI form I crystal. Average simulated fluctuations derived from the *time*-variances and *cell*-variance (see text) are shown. The correlation coefficients for the simulation results with the experimental results are given in parentheses

distribution/atom set	Simulation				X-ray
	32%	36%	40%	44%	
cell-distribution					
backbone[a]	0.57 (0.7)	0.45 (0.6)	0.45 (0.6)	0.42 (0.5)	0.68
sidechain[b]	0.73 (0.7)	0.60 (0.7)	0.60 (0.7)	0.57 (0.6)	0.78
time-distribution[c]					
backbone[a]	0.95 (0.5)	0.55 (0.5)	0.55 (0.5)	0.46 (0.5)	0.68
sidechain[b]	1.15 (0.6)	0.72 (0.7)	0.71 (0.7)	0.64 (0.7)	0.78

[a]Main-chain N—Cα—CO atoms (232 total).

[b]Heavy (non-hydrogen) side-chain atoms (222 total).

[c]Atomic fluctuations were computed from the *time*-distributions for each protein monomer and then averaged over the four monomers.

TABLE IV

Secondary structural assignments for the simulation *cell*-average structures and crystallographic structure in the BPTI form I crystal

Structure Secondary Structural Assignment[a]

RPD FC LEPPYTGPCKAR I IRYFYNAKAGLCQTVYGGCRA KR NNFKSAEDCMRT CGGA[b]

```
                    1             2             3             4             5
       1 2 3 4 5 6 7 8 9 0 1 2 3 4 5 6 7  8 9 0 1 2 3 4 5 6 7 8 9 0 1 2 3 4 5 6 7 8 9 0 1 2 3 4 5 6 7 8 0 1 2 3 4 5 6 7 8
```

Structure					
X-ray	GGGGS	S	EEEEEEETTTTEEEEEEE	SSS	SS	BSSHHHHHHHS
32%	HHHHS	S	EEEEEEETTTTEEEEEEE	SSS	S	BSSHHHHHHHT
36%	GGGGS	S	EEEEEEETTTTEEEEEEE	SSS	SS	BSSHHHHHHHS
40%	GGGGS	S	EEEEEEETTTTEEEEEEE	SSS	SS	BSSHHHHHHHS
44%	GGGGS	S	EEEEEEETTTTEEEEEEE	SSS	SS	BSSHHHHHHHS

[a]Secondary structural assignments were made using the program DSSP [24]: H = α-helix, G = 3_{10}-helix, E = β-sheet, T = hydrogen bonded turn, S = bend, B = isolated β-bridge.

[b]BPTI amino acid sequence.

have converted from the 3_{10}-helix to an α-helix. This is consistent with theoretical studies that suggest a relatively low barrier for the $3_{10} \leftrightarrow$ α-helix transition (0.2 kcal/mol), with the α-helical form favored by aqueous conditions [29].

BPTI contains an interesting aromatic hydrogen bonding interaction between residues Gly[37]----Tyr[35]----Asn[44]. The backbone NH of Gly[37] and the side-chain NH2 of Asn[44] are hydrogen bond donors on opposite sides of the aromatic Tyr[35] ring. This delicate interaction is necessary for preserving the native structure since the Y35G mutant crystal structure has significant deviation in this region [32]. The aromatic hydrogen bond interaction is maintained in the 36%, 40%, and 44% simulations; however, in the 32% simulation it is significantly weakened. An

TABLE V
Intermolecular hydrogen bond (heavy atom–heavy atom) distances in BPTI crystal form I[a]

| H-bond X—H····Y | | X–Y distance (Å) | | | | |
donor (X)	acceptor (Y)	X-ray	$\langle 32\% \rangle_{cell}$	$\langle 36\% \rangle_{cell}$	$\langle 40\% \rangle_{cell}$	$\langle 44 \rangle_{cell}$
Asp[3] N	Glu[49] OE1	3.07	**3.65**	3.38	3.16	3.04
Arg[17] NE	Lys[26] O	2.97	3.12	2.76	2.74	2.74
Arg[17] NH-1	Ala[58] O	–	3.33	2.76	2.65	2.70
Arg[17] NH-2	Ala[58] O	2.95	3.35	3.13	3.42	3.25
Arg[17] NH-1	Ala[58] OXT	–	–	3.13	3.37	3.17
Arg[17] NH-2	Ala[58] OXT	2.67	**3.42**	2.73	2.66	2.70
Arg[39] NH-1	Tyr[21] OH	2.70	**3.45**	3.30	**3.76**	**3.84**
Arg[39] NE	Glu[49] OE-1	3.04	2.66	2.69	2.71	2.73
Arg[39] NH-2	Glu[49] OE-2	2.80	3.00	2.69	2.71	2.70
Arg[42] NH-1	Tyr[10] OH	3.26	–	**3.88**	3.50	**3.98**
Arg[42] NH-1	Arg[39] O	3.28	–	–	–	–
Ala[48] N	Arg[39] NH-2	3.34	3.70	3.67	3.65	3.74

[a]Donor–acceptor heavy-atom–heavy-atom distances less 4 Å are shown. Distances shown in **bold** differ from the crystallographic values by more than 0.5 Å.

extensive discussion of intramolecular hydrogen bonding in BPTI is given in References 17, 30, and 31.

(vi) INTERMOLECULAR HYDROGEN BONDING

The crystal structure of BPTI (form I) contains several intermolecular hydrogen bonds that stabilize the protein molecules in the crystal lattice [13]. An interesting question arises as to whether the crystal packing environment is disrupted by variations in hydrostatic pressure. Table V compares the intermolecular hydrogen bond donor–acceptor distances for the crystallographic and *cell*-average structures. There are ten intermolecular hydrogen bonds present in the crystallographic structure. Of these ten, five are preserved in the 32% simulation (where 'preserved' is defined as having a distance deviation of less than 0.5 Å from the crystallographic result), eight are preserved in the 36% and 40% simulations, and seven are preserved in the 44% simulation. In all of the simulations the interaction between Arg^{39} and Arg^{42} was absent, and the interaction between Arg^{39} and Tyr^{21} was significantly reduced. We conclude that the crystal packing environment is sensitive to pressure. For an extensive discussion of intermolecular hydrogen bonding in BPTI crystal structures see Reference 30.

(vii) SOLVENT MOBILITY

Protein crystals generally contain solvent molecules (water and ions) that are structurally 'bound' to the protein molecules in addition to bulk-like solvent in the interstitial spaces. The mobility of the solvent might be expected to change

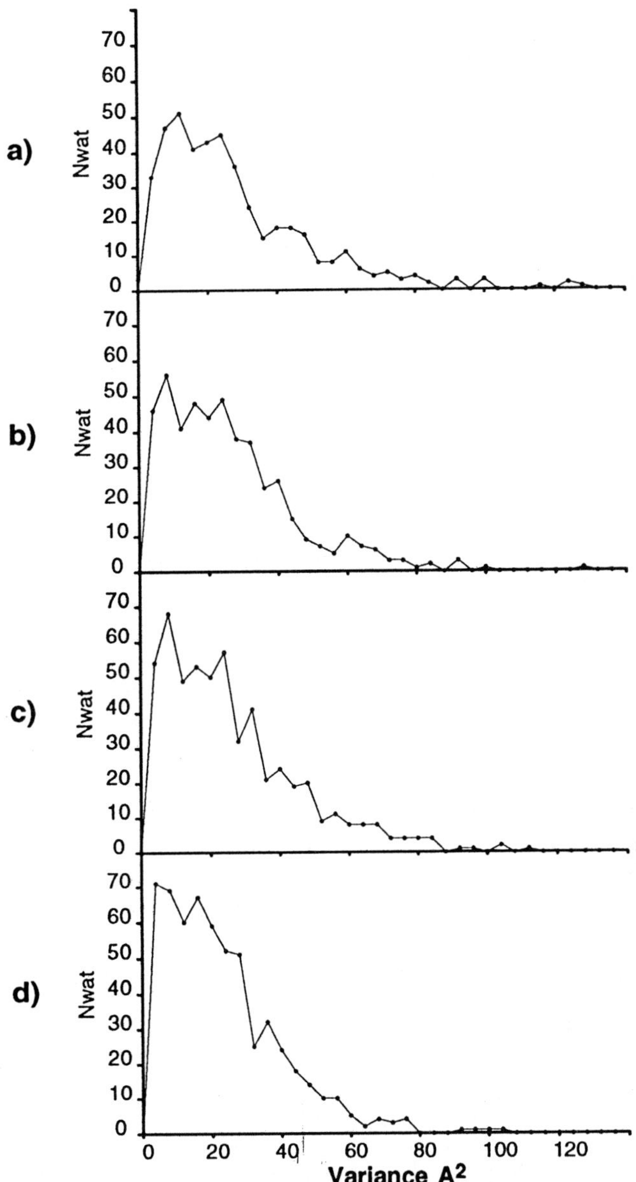

Fig. 4.   Histograms of the water molecule variance (0.25–1 ns) in the simulations: (a) 32%, (b) 36%, (c) 40%, and (d) 44%.

with variation in pressure. The average water and chloride ion variance (0.25–1 ns) is largest in the low density (32%) simulation (25.7, 9.4 Å^2, respectively) and smallest in the high density (44%) simulation (20.2, 6.5 Å^2). Surprisingly, the 36% simulation has slightly smaller water and chloride ion variances (22.7, 6.6 Å^2) than the 40% simulation (23.9, 7.1 Å^2). Hence, there appears to be an inflexion in the average solvent mobility as a function of pressure in the density range 36–40%. This compares with the results of the 1 atm and 1000 atm crystal structures of egg-white lysozyme that indicate fluctuations in the structural solvent are relatively insensitive to pressure as indicated by the B factors [8]. In all of the simulations, the variance of the water molecules was approximately three times greater than the variance of the chloride ions. The relatively large variance (related to the diffusion constant) for the TIP3P water has been previously noted [14]. Histograms showing the distribution of water molecule variance in the simulations are shown in Figure 4. The general trend as the solvent density increases is that of a slight shift in the distribution towards lower variances.

Conclusion

Simulations of BPTI in a crystal unit cell (form I) have been carried out to 1 ns at solvent densities corresponding to approximately 32%, 36%, 40%, and 44% solvent by volume. The solvent content estimated for the native crystal structure is in the range of 36–40%. Results of the 36% and 40% simulations demonstrate that the protein structure, atomic fluctuations, and crystal packing environment are stable in this range, and agree well with crystallographic data (average structures give rms backbone deviations of ~0.3 Å from the native crystal structure). The 40% simulation gives the best overall agreement with experimental data. The 32% and 44% simulations show more substantial deviations in structure, particularly the low density (32%) simulation. Atomic fluctuations are considerably exaggerated in the 32% simulation, and damped in the 44% simulation relative to the stable fluctuations observed for the 36% and 40% simulations. Crystal packing is also disrupted in the 32% and 44% simulations. The average variance of the solvent and counterions as a function of solvent density (pressure) has an inflexion in the 36–40% solvent density range. The results reported here represent the first study of the effects of hydrostatic pressure on large protein crystals using molecular simulation.

Acknowledgement

DY acknowledges the National Science Foundation for a postdoctoral fellowship jointly funded by the North Carolina Supercomputing Center, MCNC, RTP, NC. This work was supported by grants from NIEHS (LP,TD). Supercomputing time was provided by the Biomedical Supercomputing Center, National Cancer Institute, FCRDC, the Pittsburgh Supercomputing Center, Pittsburgh, PA, and the North Carolina Supercomputing Center, RTP, NC.

References

1. B. Matthews: in H. Neurath and R. Hill, Eds., *The Proteins*, *3rd ed.*, *Vol. 3*, Academic Press, New York, p. 403 (1977).
2. T. Creighton: in *Proteins*, *2nd ed.*, W. H. Freeman & Co., New York (1993).
3. A. Bax: *Ann. Rev. Biochem.* **58**, 223 (1989).

4. G. Weber: in R. van Eldik and J. Jonas, Eds.: *High Pressure Chemistry and Biochemistry*, D. Reidel Publishing Company, Boston (1986).

5. P. Wong and K. Heremans: *Biochim. Biophys. Acta.* **956**, 1 (1988).

6. D. Carrier, H. Mantsch, and P. Wong: *Biopolymers* **29**, 837 (1990).

7. C. Kundrot and F. Richards: *J. Mol. Biol.* **193**, 157 (1987).

8. C. Kundrot and F. Richards: *J. Mol. Biol.* **200**, 401 (1988).

9. W. F. van Gunsteren and M. Karplus: *Biochemistry* **21**, 2259 (1982).

10. W. F. van Gunsteren and H. J. C. Berendsen: *J. Mol. Biol.* **176**, 559 (1984).

11. P. Krüger, W. Strassburger, A. Woller, and W. F. van Gunsteren: *Eur. Biophys. J.* **13**, 77 (1985).

12. D. York, T. Darden, L. Pedersen, and M. Anderson: *Biochemistry* **32**, 1443 (1993).

13. J. Deisenhofer and W. Steigmann: *Acta Crystallogr. B* **31**, 238 (1975).

14. V. Daggett and M. Levitt: *Ann. Rev. Biophys. Biomol. Struct.* **22**, 353 (1993).

15. D. Kitchen, L. Reed, and R. Levy: *Biochemistry* **31**, 10083 (1992).

16. R. Brunne and W. van Gunsteren: *FEBS Lett.* **323**, 215 (1993).

17. A. Wlodawer, J. Walter, R. Huber, and L. Sjölin: *J. Mol. Biol.* **180**, 301 (1984).

18. S. J. Weiner and P. A. Kollman: *J. Comp. Chem.* **7**, 230 (1986).

19. W. L. Jorgensen, J. Chandrasekhar, J. D. Madura, R. W. Impey, and M. Klein: *J. Chem. Phys.* **79**, 926 (1983).

20. T. Lybrand, J. McCammon, and G. Wipff: *Proc. Natl. Acad. Sci. USA* **83**, 833 (1986).

21. J. Ryckaert, G. Ciccotti, and H. Berendsen: *J. Comput. Phys.* **23**, 327 (1977).

22. T. Darden, D. York, and L. Pedersen: *J. Chem. Phys.* **98**, 10089 (1993).

23. D. York, T. Darden, and L. Pedersen: *J. Chem. Phys.* **99** 8345 (1993).

24. A. D. McLachlan: *J. Mol. Biol.* **128**, 49 (1979).

25. D. Kitson, F. Avbelj, J. Moult, D. Nguyen, J. Mertz, D. Hadzi, and A. Hagler: *Proc. Natl. Acad. Sci. USA* **90**, 8920 (1993).

26. W. Kabsch and C. Sander: *Biopolymers* **22**, 2577 (1983).

27. M. Karplus and G. Petsko: *Nature* **347**, 631 (1990).

28. G. Petsko and D. Ringe: *Ann. Rev. Biophys. Bioeng.* **13**, 331 (1984).

29. M. Smythe, S. Huston, and G. Marshall: *J. Am. Chem. Soc.* **115**, 11594 (1993).

30. A. Wlodawer, J. Deisenhofer, and R. Huber: *J. Mol. Biol.* **193**, 145 (1987).

31. K. Berndt, P. Güntert, L. Orbons, and K. Wüthrich: *J. Mol. Biol.* **227**, 757 (1992).

32. D. Housset, K-S. Kim, J. Fuchs, C. Woodward, and A. Wlodawer: *J. Mol. Biol.* **220**, 757 (1993).

Twists and Turns in DNA: Predicting Base Sequence Effects on the Conformation of the Double Helix

RICHARD LAVERY
Laboratoire de Biochimie Théorique, CNRS URA 77, Institut de Biologie Physico-Chimique, 13, rue Pierre et Marie Curie, Paris 75005, France.

Abstract. The present study presents the results of molecular modelling of base sequence effects on the structure of the DNA double helix. An appropriate choice of the modelling algorithm and the use of helical symmetry has enabled a very thorough investigation of the conformational space of all unique tetranucleotide repeating sequences. The results indicate that each sequence is associated with a considerable number of stable sub-states characterised by their sugar puckers. Using this data it is possible to envisage predicting the conformation of much longer pieces of DNA using the superposition of tetranucleotide fragments.

Key words. Molecular modelling, nucleic acid sub-states, sugar pucker.

Introduction

Today we are facing an explosion of information concerning nucleic acid sequences. This information, resulting in part from the Human Genome Project, opens many perspectives for understanding both the functioning and the malfunctioning of living organisms. However, it is not sufficient to stop simply with the sequence of bases which constitutes the genetic library. Biological activity requires that this information first be translated from a one dimensional code into a three dimension code, that is to say, from sequence into the local structure and dynamic behaviour of the DNA double helix [1]. Only then can the code be exploited and, through a series of exquisitely precise protein-nucleic acid interactions, play its role in instructing cellular function.

Understanding the mechanics and the energetics which govern the fine structure of DNA as a function of its sequence will certainly require a large structural database [2]. Unfortunately, such a database is not currently available. Despite important progress in experimental techniques for determining macromolecular structure, both X-ray crystallography and NMR spectroscopy suffer from certain limitations in the case of oligonucleotide studies. In the case of crystallography it is often difficult to obtain high quality crystals of B-DNA and, in addition, distinguishing between intrinsic sequence effects and lattice induced deformations is not easy. In the case of NMR studies, the linear nature of the DNA molecule makes it more difficult to determine fine structure from short NOESY interactions, while COSY data only gives access to a limited number of backbone angles. Confronted with these difficulties, we have attempted to develop modelling techniques which would allow a larger data base to be built up quickly and would perhaps lead to some understanding of the underlying mechanics by which base sequence becomes 3D structure. Work carried out over the last few years has enabled us to show

A. Pullman et al. (eds.), Modelling of Biomolecular Structures and Mechanisms, 217–230.

that our approach apparently correlates well with available experimental data. It has also brought to light the importance of sugar pucker in controlling DNA conformation which has, in turn, led to methods for building a large structural database of DNA and the possibility of predicting the conformation of long DNA fragments. These developments are briefly described below.

Methodology

In attempting to model nucleic acid fine structure, we began with the decision to build a simulation algorithm which would explicitly take into account the stereochemistry of the nucleic acids and their common helical structure. We also chose to simplify our representation in order to reduce as far as possible the number of variables necessary for modelling a given fragment of DNA. These choices led to the Jumna (JUnction Minimisation of Nucleic Acids) algorithm [3] which combines the direct use of helicoidal coordinates for positioning individual nucleotides with respect to a common helical axis system with internal coordinates (dihedral angles and, inside the sugar rings, valence angles) for describing the internal flexibility of each nucleotide. Junctions between successive 3'-mono-phosphate nucleotides are maintained using harmonic distance and valence angle constraints. In this way, we achieve a ten-fold reduction in the number of degrees of freedom representing the system compared to a classical molecular mechanics calculation in Cartesian coordinates. Although we necessarily loose some fine detail with our model, this is offset by improved possibilities for searching conformational space.

Jumna has another important advantage, namely, the possibility of controlling chosen features of nucleic acid conformation during the simulation. The first, and most important aspect, of this control is the possibility of introducing helical symmetry by simply grouping together sets of helically equivalent variables. This can lead to a further order of magnitude reduction in the number of variables representing the system and opens the way for treating infinite regular polymers very efficiently. The second aspect of control is the possibility of introducing a wide variety of constraints (inter-atomic distances, dihedral angles, helical parameters, sugar puckers, axis curvature, etc.). Recent extensions also allow the automatic calculation of 1D and 2D energy maps in order to follow conformational transitions. Using Jumna we have been able to study the conformation and flexibility of both oligomers and polymers of DNA and RNA [4–9], and to investigate the transition pathways which link stable conformers [3,10,11].

Today Jumna contains some 15 000 lines of code. Its force field FLEX [12–15] is based on Lennard–Jones terms [12] and atomic charges [13] specifically developed for treating nucleic acids. Dihedral angle barriers and valence angles have been largely parameterized on the basis of quantum chemical calculations. Water and counterion effects are currently reduced to a simple electrostatic screening using a sigmoidal distance dependent dielectric function [3,16] and reduced net charges on each phosphate group $(-0.5e)$. In order to analyse and compare nucleic acid structures we use the Curves algorithm [17,18] which calculates both the ideal axis representing a given nucleic acid fragment and a full set of helicoidal parameters. These parameters naturally obey the Cambridge convention [19].

TABLE I

Energy per dinucleotide repeat unit for the stable sub-states of the regular DNA polymers studied (kcal/mol)

Sequence	Energy					
$(AA)_n$	−82.5*	−82.4*	−81.9*			
$(GG)_n$	−118.1	−115.2	−115.1*	−114.2*		
$(CG)_n$	−119.4	−118.7	−118.0	−114.0[#]		
$(TA)_n$	−81.6	−80.8	−80.2	−80.0[#]		
$(CA)_n$	−100.9	−100.8	−100.6	−100.4	−100.3	−100.2
	−99.6	−99.6	−99.2	−97.2[#]	−97.1*	−95.2[#]
	−95.0[#]					
$(GA)_n$	−99.5	−98.4[#]	−98.3	−97.6	−97.0	−96.7[#]

Note. Sub-states having purines with O1′-endo sugars are indicated by the symbol #. Sub-states of homopolymers showing dinucleotide symmetry are indicated by the symbol *.

Results

The modelling techniques described above have been put to use over the last few years in a systematic attempt to understand the origins of base sequence effects on DNA. Our first studies were limited to the 10 unique dinucleotide steps, which were studied with the help of 6 DNA polymers having regular repeating base sequences [4]:

$(AA)_n$	AA	TT*	
$(GG)_n$	GG	CC*	
$(CG)_n$	CG	GC	
$(TA)_n$	TA	AT	
$(CA)_n$	CA	AC	TG* GT*
$(GA)_n$	GA	AG	TC* CT*

(Note that although the four nucleic acid bases can be combined in 16 different ways to make a two-base sequence, there are in fact only 10 unique dinucleotide junctions. Six combinations, marked above with asterisks, are simply the complementary strands of another dinucleotide occuring on the same line).

These polymers were subjected to helical symmetry constraints and energy minimisation was used to find the most stable conformations of each sequence. End effects were avoided by minimising the energy per repeating unit of each polymer rather than its total energy. Seven nucleotides were taken into account on either side of the central dinucleotide unit. The results obtained can therefore be considered to apply to infinite regular polymers. The conformation space belonging to the B-DNA family of structures was studied as thoroughly as possible by using many starting points for each minimisation and by deforming and then relaxing the conformations obtained to overcome conformational energy barriers [4,20]. The results of this study led to the finding that each base sequence was associated with a number of stable sub-states having similar energies (Table I). Although the mean values obtained for both backbone and helicoidal parameters were close to experimental values, it was found that the various sub-states of each

TABLE II

Range of backbone and helical parameters for the stable sub-states of the regular DNA polymers studied. For comparison both fibre and crystallographic data on B-DNA are shown

Parameter	Jumna			Experimental	
	Minimum	Maximum	Mean	Fibre	Crystals
α	−72	−58	−65	−41	−64
β	159	188	175	135	166
γ	50	67	58	37	50
δ	91	150	133	139	132
ϵ	−177	−161	−172	−134	−179
ζ	−145	−90	−110	−102	−94
χ	−149	−94	−118	−102	−104
Phase	74	184	150	154	150
Amplitude	28	46	38		39
Xdisp	−3.5	−0.2	−1.6	0.0	0.5
Inclin	−18.0	12.0	−0.3	1.5	−1.2
Propeller	−22.0	4.0	−6.4	−13.3	−11.0
Buckle	−8.0	21.0	−1.8	0.0	0.9
Rise	2.8	4.0	3.3	3.4	3.4
Twist	28.0	45.0	35.5	36.0	36.2

Note. The experimental data shown refer to the mean structures of 8 high resolution DNA decamer crystals [21] and to the best available fibre diffraction data [22].

TABLE III

Sub-states classified according to their sugar puckering (S: C2′-endo, low amplitude, X: C2′-endo, high amplitude, E: O1′-endo). The sugars in each strand are given in the 5′ → 3′ sense and follow the order of the bases given in the first column of the table

Sequence	1	2	3	4	5	6
$(AA)_n$	SX:SS*	SX:ES*	SX:SE*			
$(GG)_n$	XX:XX	SS:XX	XS:SX*	SX:ES*		
$(CG)_n$	SX:SX	XS:XS	ES:ES	SE:SE$^{\#}$		
$(TA)_n$	SX:SX	XS:XS	ES:ES	XE:XE$^{\#}$		
$(CA)_n$	ES:XS	XS:SX	XS:XS	ES:SX	SX:XS	ES:ES
	XS:ES	SX:SX	SX:ES	SE:XS$^{\#}$	SE:XX$^{\#}$	SE:SE$^{\#}$
	SE:ES$^{\#}$					
$(GA)_n$	XS:SX	ES:SX$^{\#}$	SX:ES	XS:SE	SX:SE	ES:SE$^{\#}$

Note. Sub-states having purines with O1′-endo sugars are indicated by the symbol #. Sub-states of homopolymers showing dinucleotide symmetry are indicated by the symbol *.

sequence often had very different conformations, leading to a wide range for most of the structural parameters (Table II).

Further investigation of the sub-state conformations showed that the set of substates associated with each base sequence could be distinguished on the basis of their sugar puckers alone (Table III). The sugars were found to lie within three

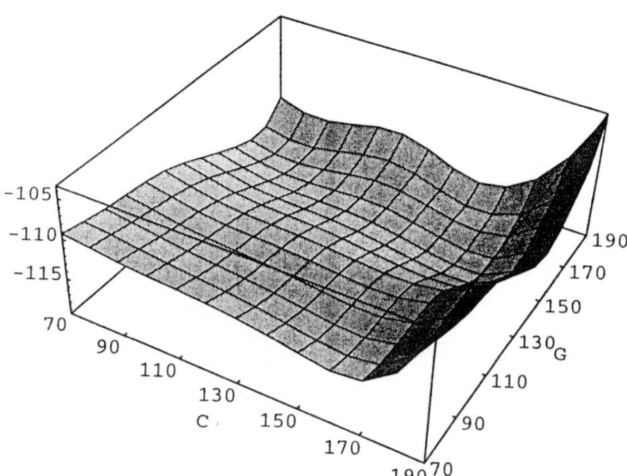

Fig. 1.   Energetic surface of the (CG)$_n$ polymer as a function of its sugar phase angles (both dinucleotide and dyadic symmetry are imposed on the polymer, leading to the presence of only two symmetry distinct sugars).

pseudorotational families which we have named S (C2'-endo puckers with low amplitude), X (C2'-endo puckers with high amplitude) and E (O1'-endo puckers, associated commonly, although not exclusively, with pyrimidine nucleotides). These sugar families were found to be correlated with a number of other structural parameters and thus to have a major role in determining the local conformation of the double helix. These findings were subsequently confirmed by modifying Jumna to automatically calculate energy maps as a function of sugar pucker. All the sub-state minima previously found could be located on these maps, which, in addition, allowed us to visualise the transition pathways and the energy barriers separating the different sub-states. In general it was found that sub-states containing only S and X sugars were separated by very small energy barriers below 1 kcal/mole, while transitions from these sub-states to those containing E sugars required passing barriers of roughly 2 kcal/mole (see example in Figure 1). This study also allowed us to demonstrate the direct coupling of sugar pucker to a variety of other parameters including both helicoidal variables and backbone dihedrals (Figure 2).

Although these results led to some interesting data on the nature of base sequence effects and, notably, on the role of sugar pucker, it was also clear that explaining available experimental data would require going beyond the nearest neighbour dinucleotide model [2]. If we are concerned mainly with internucleotide parameters such as rise, twist and wedge angles, then the next step beyond the dinucleotides involves studying tetranucleotide sequences. This implies the passage to a next nearest neighbour model where each unique dinucleotide step will now be studied in the context of all possible 5' and 3' neighbouring base pair steps. This considerably complicates the study to be undertaken, since a rapid calculation shows that their are a total of 136 unique tetranucleotide sequences. Although these sequences can again be grouped into a smaller number of regular polymers

(a) X disp

(b) Twist (CpG)

(c) Glycosidic angle (G)

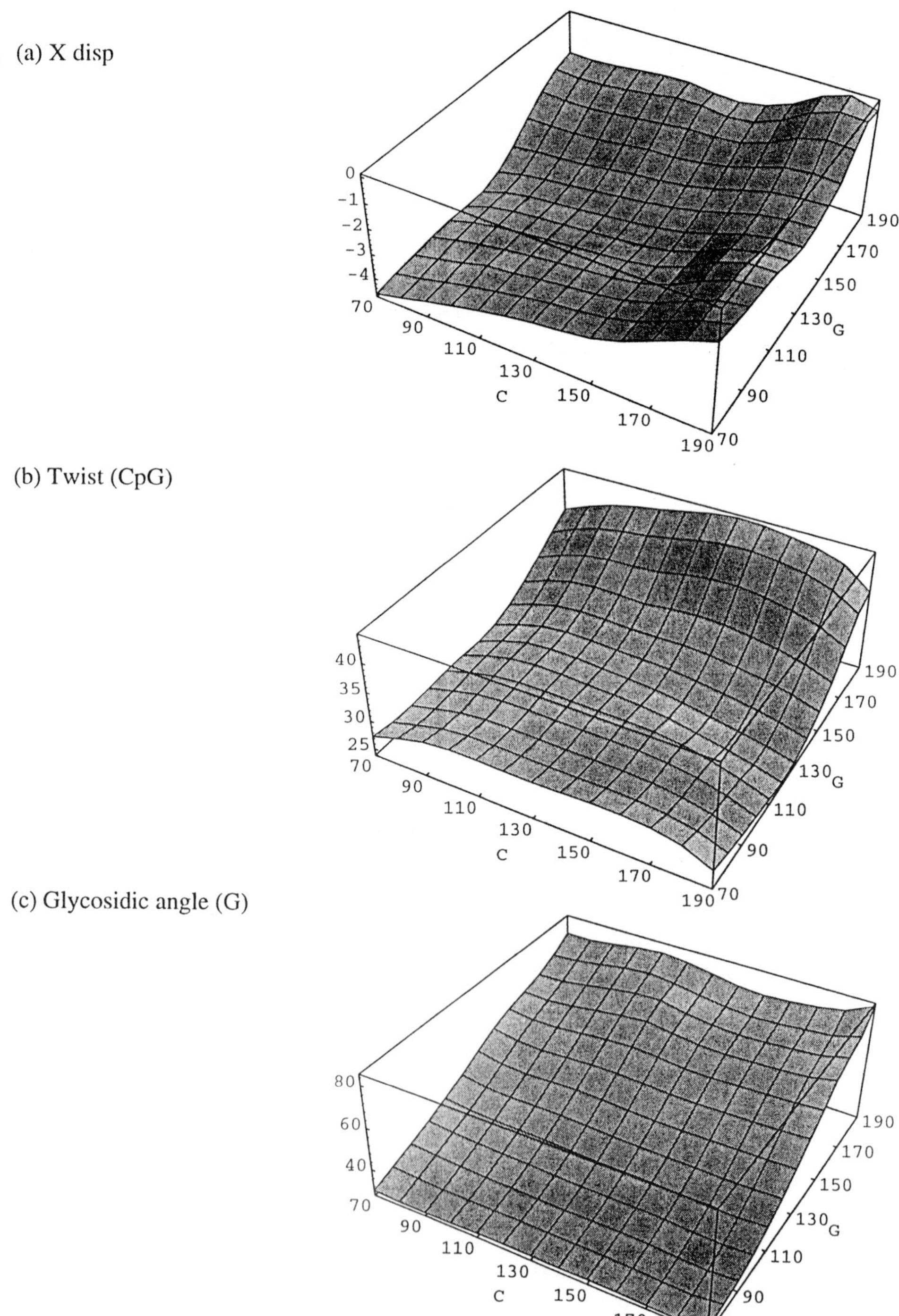

Fig. 2. Variation of certain structural parameters of the $(CG)_n$ polymer as a function of its sugar phase angles.

TABLE IV

Thirty-nine regular polymer sequences containing the 136 unique tetranucleotides

$(GGUU)_n$	GGGG			
$(AAAA)_n$	AAAA			
$(CGCG)_n$	CGCG	GCGC		
$(TATA)_n$	TATA	ATAT		
$(GTGT)_n$	GTGT	TGTG		
$(GAGA)_n$	GAGA	AGAG		
$(CCGG)_n$	CCGG	CGGC	GGCC	
$(TTAA)_n$	TTAA	TAAT	AATT	
$(TCGA)_n$	TCGA	CGAT	GATC	
$(TGCA)_n$	TGCA	ATGC	CATG	
$(ACGT)_n$	ACGT	CGTA	GTAC	
$(AGCT)_n$	AGCT	TAGC	CTAG	
$(AGGT)_n$	AGGT	GGTA	GTAG	TAGG
$(AGGA)_n$	AGGA	GGAA	GAAG	AAGG
$(TGGT)_n$	TGGT	GGTT	CAAC	TTGG
$(GCGG)_n$	GCGG	CGGG	GGGC	GGCG
$(ACGA)_n$	ACGA	CGAA	GAAC	CGTT
$(AGCA)_n$	AGCA	TTGC	CAAG	AAGC
$(AGCG)_n$	AGCG	GCGA	CGAG	GAGC
$(ATAA)_n$	ATAA	TAAA	AAAT	AATA
$(ATAC)_n$	ATAC	TGTA	ATGT	CATA
$(GATG)_n$	GATG	ATGG	TGGA	GGAT
$(GATA)_n$	GATA	ATAG	TAGA	AGAT
$(GGAG)_n$	GGAG	GAUG	AGUG	GGGA
$(AGAA)_n$	AGAA	GAAA	AAAG	AAGA
$(CGAC)_n$	CGAC	GGTC	CGGT	CCGA
$(TGAT)_n$	TGAT	AATC	CAAT	TTGA
$(CAGC)_n$	CAGC	AGCC	TGGC	CTGG
$(TAGT)_n$	TAGT	AGTT	TAAC	CTAA
$(CAGA)_n$	CAGA	AGAC	TGTC	CTGT
$(CAGT)_n$	CAGT	AGTC	TGAC	CTGA
$(GGTG)_n$	GGTG	GTGG	TGGG	GGGT
$(AGTA)_n$	AGTA	GTAA	TAAG	AAGT
$(CGTC)_n$	CGTC	GGAC	CGGA	ACGG
$(TGTT)_n$	TGTT	AAAC	CAAA	TTGT
$(CTGC)_n$	CTGC	GGCA	AGGC	CAGG
$(ATGA)_n$	ATGA	TGAA	GAAT	AATG
$(GTGC)_n$	GTGC	CGCA	ACGC	CGTG
$(GTGA)_n$	GTGA	TGAG	GAGT	AGTG

(Table IV), there are nevertheless 39 polymers to be studied, compared to the six
necessary for the dinucleotides. In addition, each polymer must be treated using
tetranucleotide rather than dinucleotide symmetry constraints leading to 8 unique
sugars per repeating unit.

The size of the problem confronting us necessitated the development of auto-
mated procedures for finding sub-state energy minima. The procedure we have
developed is based on the fact that all sub-states found for each dinucleotide

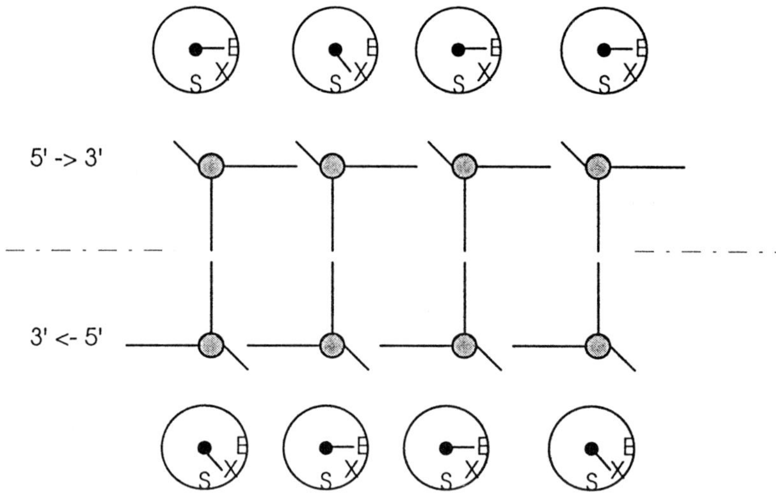

Fig. 3.   Combinatorial searching on the basis of the sugar puckerin gof each nucleotide in the repeat
unit of a regular DNA polymer.

sequence could be uniquely labelled by their sugar puckers. This led to the idea
of a combinatorial search based on puckers of each sugar in the repeating unit as
illustrated in Figure 3. For each step of this search the symmetry unique sugars
were first constrained to a given set of puckers and an energy minimisation was
performed. The constraints were then released and a second minimisation was
carried out.

The results of this study (see the example in Table V for the case of the
$(ACGT)_n$ polymer) showed that there are on average 20 sub-states per polymer.
This number however varies from a minimum of 5 to a maximum of 60. In
addition, since many sequences possessing mononucleotide, dinucleotide or inversion symmetry actually have sub-states which do not express this symmetry
[8,20,23], the total number of sub-states is increased by multiple copies having
identical energies and related by the symmetry operations applicable to the polymer in question. We can also note that although only S, X and E sugar puckers
were used as starting points in the combinatorial search, certain sub-states are
generated with C3'-endo sugars, usually characteristic of A-DNA, while other
sub-states contain unusual backbone geometries involving the ϵ/ζ or α/γ dihedral
angles.

From these results it is possible not only to understand more clearly the extent
and the nature of base sequence effects, but also to attempt to construct longer
sequences of DNA by tetranucleotide superposition. If we assume that next nearest
neighbour interactions will be generally sufficient to describe sequence effects, then
the tetranucleotide database we have constructed already contains the information
needed to assemble a DNA fragment with any chosen sequence. The technique
we have developed for this construction is illustrated schematically in, Figure 4
for the sequence CATGACGTCATG containing the Cre site [24] whose structure

TABLE V

Sub-states of the $(ACGT)_n$ polymer (energies per tetranucleotide repeat unit in kcal/mol)

I	ACGTTGCA	Energy	Sym.	$\alpha\gamma$
1	XSXSSXSX	−199.542	*	.
2	XXSSSSXX	−199.515	*	.
3	XSXSSSXX	−199.470	.	.
4	XESSSSXX	−199.232	.	.
5	XXSSXSES	−199.221	.	.
6	SESXXSXS	−199.054	.	.
7	SESXXSES	−199.037	*	.
8	XSXSXSXS	−198.961	.	.
9	XESSSSEX	−198.832	*	.
10	XXSSESXX	−198.678	.	.
11	SESXSSEX	−198.663	.	.
12	XXSSXSXS	−198.663	.	.
13	SXSXXSXS	−198.419	*	.
14	XSXSXXSS	−198.294	.	.
15	XXSEESXX	−198.154	*	.
16	XXSSXXSS	−198.043	.	*
17	SESESSXX	−197.762	.	.
18	SESEXSES	−197.529	.	.
19	SESEXSXS	−197.359	.	.
20	SXSEXSXS	−197.356	.	.
21	XESESXSX	−197.213	.	.
22	XXXSXSSS	−197.104	.	*
23	SESESSEX	−196.949	.	.
24	SESEEESES	−196.516	*	.
25	SSXXESES	−196.352	.	.
26	XSXSXXSS	−196.201	.	*

Note. Only the 19 symmetry unique minima are shown, if symmetry equivalent copies are included there are a total of 45 minima. The last two columns of the table indicate which minima exhibit the inversion symmetry of the base sequence and which minima contain unusual α/γ backbone conformations.

has already been obtained using NMR data [7]. Firstly, we decompose the sequence into its constituent tetranucleotides which consequently overlap by trinucleotide sequences. Next we search the database for tetranucleotide sub-states which have the same structure as their neighbours for the six overlapping nucleotides. As indicated in Figure 4, each overlap is tested using an angular RMS value which takes into account the phosphodiester backbone angles (α, β, γ, ϵ, ζ), the sugar conformation (phase, amplitude) and the glycosidic angle (χ) of each nucleotide. Excluding backbone linkages outside the trinucleotide overlapping fragment leads to a total of 38 parameters which contribute to the RMS value calculated. It might be suspected that since neighbouring tetranucleotides within an irregular sequence generally come from different initial polymer studies that there would be little chance of obtaining good superposition. However, the tests we have carried out

(a) Tetranucleotides composing the Cre sequence CATGACGTCATG

C	A	T	G	A	C	G	T	C	A	T	G
C	A	T	G								
	A	T	G	A							
		T	G	A	C						
			G	A	C	G					
				A	C	G	T				
					C	G	T	C			
						G	T	C	A		
							T	C	A	T	
								C	A	T	G

(b) Variables taken into account during the superposition tests (indicated in the schematic diagram by bold lines). After successful superposition the two nucleotide indicated by the black circles are added to the polymer under construction.

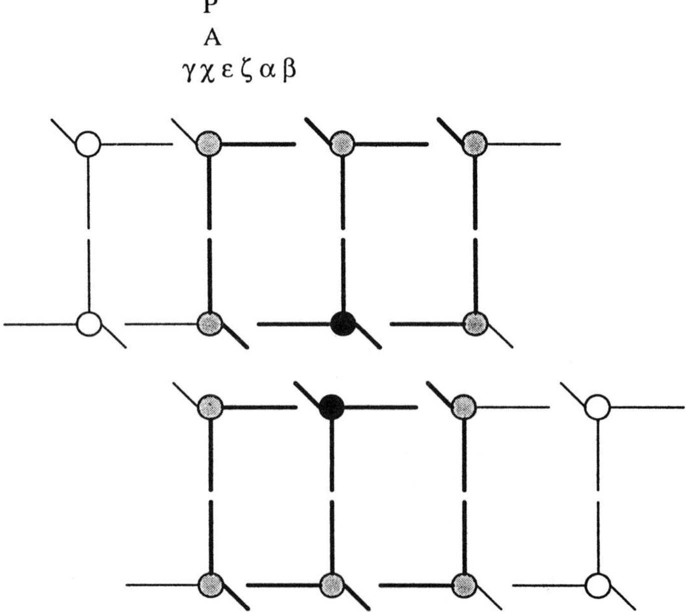

Fig. 4.   Superposition technique for building long sequences from the tetranucleotide database.

TABLE VI

Successive sub-state elimination during the construction of the
Cre sequence CATGACGTCATG (the subscript i indicates that
two sub-states n and ni are related by inversion symmetry)

A	C	G	T
	1 → 45		

G	A	C	G	T	C
	1	1	1		
	2	9	2		
	5	9i	5		
	6	10	6		
	7	16	7		
	8	16i	8		
	11	19	11		
	12	25	12		
	13	25i	13		
	16	27	16		
	17	27i	17		

T	G	A	C	G	T	C	A
	8	2	1	2	8		
	12	5	10	5	12		
	17	11	16	11	17		
			16i				

A	T	G	A	C	G	T	C	A	T
	3	8	2	1	2	8	3		
	6						6		

C	A	T	G	A	C	G	T	C	A	T	G
	1	3	8	2	1	2	8	3	1i		
	3	6						6	3		
	7i								7		
	16i								16		
	21i								21		
	30								30i		

so far suggest that this is not the case and successful superpositions can generally
be made with a remarkable quality of around $3° \rightarrow 4°$ RMS.

The second result of our studies concerns the number of conformations which
can be constructed for a given sequence. In the case of the Cre sequence, the
number of sub-states in the database for each of its constituent tetranucleotides
enables us to calculate that there are roughly 3×10^{12} possible structural combinations which could be made. However, once superposition is taken into account
a very different picture emerges. As illustrated in Table VI, fitting together tetranucleotides within an irregular sequence leads to a dramatic decrease in structural
multiplicity and, in general, it is found that the final DNA fragment is compatible
with only a small number of global conformations. For Cre, the central tetranucleotide ACGT begins with 45 stable sub-states. Adding the two neighbouring tetranucleotides (GACG on the 5′ side and CGTC on the 3′ side) and searching for fits

TABLE VII

RMS fit between the structure of the Cre dodecamer CATGACGTCATG obtained by minimisation (conformation Ib in Ref. 5) and that obtained by construction from the tetranucleotide database (values are in degrees, 8 angular variables are compared for each nucleotide)

Res	γ	ϵ	ζ	α	β	χ	Pha	Amp	Rms
C1	–	4.1	39.8	2.4	7.7	32.5	77.3	6.0	35.3
A2	3.3	0.3	0.1	1.5	2.3	6.9	1.5	0.3	2.93
T3	1.3	0.7	0.2	1.0	1.2	2.8	0.1	0.9	1.29
G4	1.6	0.9	2.6	1.2	1.1	5.8	8.2	1.4	3.82
AS	1.4	1.7	3.4	1.1	0.7	4.4	0.7	0.1	2.16
C6	0.3	0.7	2.8	1.3	0.7	1.4	2.5	0.1	1.54
G7	0.2	0.9	1.8	0.9	3.7	3.5	3.5	1.0	2.35
T8	1.4	0.1	7.4	2.0	0.2	1.9	0.3	1.5	2.88
C9	1.5	0.1	2.0	0.1	2.4	0.9	6.8	2.1	2.81
A10	0.1	0.0	1.7	0.0	2.7	3.5	2.3	0.7	1.87
T11	1.0	0.2	0.6	0.9	6.3	1.0	7.4	1.3	3.52
G12	2.5	–	–	–	–	8.8	12.0	5.8	8.08
G13	2.6	–	–	–	–	8.7	12.0	5.7	8.08
T14	1.0	0.2	0.7	0.8	6.2	1.0	7.3	1.3	3.49
A15	0.2	0.0	1.7	0.1	2.6	3.6	2.2	0.7	1.87
C16	1.5	0.1	2.0	0.1	2.5	0.8	6.9	2.1	2.84
T17	1.4	0.1	7.4	2.1	0.1	2.0	0.4	1.5	2.90
G18	0.2	0.9	1.8	0.9	3.8	3.5	3.6	1.1	2.38
C19	0.4	0.7	2.8	1.2	0.6	1.4	2.5	0.1	1.53
A20	1.4	1.7	3.5	1.1	0.6	4.3	0.7	0.1	2.17
G21	1.5	0.9	2.7	1.3	1.1	5.8	8.3	1.4	3.83
T22	1.3	0.7	0.2	1.0	1.2	2.8	0.1	0.9	1.28
A23	3.2	0.2	0.1	1.4	2.4	7.0	1.5	0.3	2.94
C24	–	4.0	39.8	2.3	7.7	32.5	77.3	6.0	35.3

of better than 4° RMS leads to elimination of all but 11 of the ACGT sub-states and simultaneously to elimination of half of the 22 sub-states of the adjoining tetranucleotides. If we now continue the construction by adding further tetranucleotides at either end of the central fragment, the number of acceptable ACGT sub-states is reduced successively to 11 possibilities, 4 possibilities and, finally, to a single sub-state. At the end of the construction all of the central part of the oligomer has a unique structure and only its ends maintain a number of possible conformations. It should also be noted that although we only use tetranucleotide fragments for this construction, this does not exclude the appearance of sequence effects well beyond the next nearest neighbour range. As shown in the case discussed, the conformation of the central nucleotides of the Cre sequence is chosen by fixing the base sequence 5 nucleotides away from the centre of the oligomer.

Having built a conformation for our oligomer, we must now determine whether this conformation corresponds to that obtained by direct minimisation of the fragment. The answer is contained in Table VII which shows that, with the exception of the terminal nucleotides, the database constructed fragment coincides almost perfectly with one of the two conformations found for the Cre dodecamer in our earlier study [7]. (Note that this comparison was made with an arbitrary choice for the terminal nucleotides of the fragment which, as shown in Table VI, have a number of possible conformations. It should be remarked that one should not in any case expect a perfect agreement at the ends of an oligomer since the database conformations all belong to infinite regular polymeric sequences. Consequently, the explicit end-effects, present in the minimised oligomer, certainly cannot be reproduced).

These results suggest that it should indeed be possible to construct longer fragments of DNA using the tetranucleotide database. This approach will enable us to overcome the computational barriers which would normally hinder the study of long fragments. One important application in view is certainly the problem of sequence dependent DNA curvature. In this area it should be stressed that all present prediction techniques are based on a nearest neighbour model and thus only distinguish between the 10 unique dinucleotide junctions [25–27]. In contrast, our database construction is based on a next nearest neighbour model and can distinguish between the 136 unique tetranucleotides sequences.

Conclusions

Our studies of base sequence effects on DNA conformation suggest that all repetitive sequences are associated with a number of stable sub-states which are close in energy, but can vary widely in conformation. The passage from dinucleotide to tetranucleotide symmetry considerably increases the number of sub-states associated with each sequence. It is also found that all the sub-states of a given sequence can be reliably distinguished from one another on the basis of their sugar puckering.

Calculations on a total of 39 DNA polymers with repetitive sequences have enabled us to build up a structural database containing the stable sub-states of the 136 unique tetranucleotides. This data can be used, in turn, to build up longer DNA sequences by the superposition of tetranucleotides. From the tests carried out so far, it appears that building an irregular sequence in this way results in a dramatic reduction in structural multiplicity, since only few sub-states of each constituent tetranucleotide will fit within their new sequence environment. In this way, the construction of even relatively short irregular DNA fragments can lead to a unique conformation.

These findings suggest a new explanation for base sequence effects within the DNA double helix. In particular, they can explain how long range effects may occur without explicitly taking into account base-base interactions beyond the next nearest neighbour model. It may be hoped that future studies using the tetranucleotide database will enable the prediction of DNA conformation for fragments of a length which would normally be well beyond the possibilities of conventional modelling techniques.

Acknowledgement

We wish to thank the Association for International Cancer Research (St. Andrews University, UK) for their generous support of this research project and the Institut du Développement et des resources en Informatique Scientifique (Paris) for allocating supercomputer time.

References

1. R. E. Dickerson: in *Unusual DNA Structures*, R. D. Wells and S. C. Harvey (Eds.), Springer-Verlag, p. 287 (1988).
2. K. Yanagi, G. G. Privé and R. E. Dickerson: *J. Mol. Biol.* **217**, 201 (1991).
3. R. Lavery: in *Structure and Expression*, Vol. 3, *DNA Bending and Curvature*, W. K. Olson, R. H. Sarma, M. H. Sarma and M. Sundaralingam (Eds.), Adenine Press, New York, p. 191 (1988).
4. M. Poncin, B. Hartmann and R. Lavery: *J. Mol. Biol.* **226**, 775 (1992).
5. M. Poncin, D. Piazzola and R. Lavery: *Biopolymers* **32**, 1077 (1992).
6. K. Zakrzewska: *J. Biomol. Struct. Dynam.* **9**, 681 (1992).
7. O. Mauffret, B. Hartmann, O. Convert, R. Lavery and S. Fermandjian: *J. Mol. Biol.* **227**, 852 (1992).
8. K. Zakrzewska, V. I. Poltev, C. Oguey and R. Lavery: *J. Mol. Struct.* (*Theochem*) **286**, 219 (1993).
9. M. Ouali, R. Letellier, F. Adnet, J. Liquier, J. S. Sun, R. Lavery and E. Taillandier: *Biochemistry* **32**, 2098 (1993).
10. B. Hartmann, D. Piazzola and R. Lavery: *Nucleic Acids Res.* **21**, 561 (1993).
11. F. Briki, J. Ramstein, R. Lavery and D. Genest: *J. Am. Chem. Soc.* **113**, 2490 (1991).
12. V. B. Zhurkin, V. I. Poltev and V. L. Florent'ev: *Mol. Biol.* **14**, 1116 (1980).
13. R. Lavery, K. Zakrzewska and A. Pullman: *J. Comp. Chem.* **5**, 363 (1984).
14. R. Lavery, H. Sklenar, K. Zakrzewska and B. Pullman: *J. Biomol. Struct. Dynam.* **3**, 989 (1986).
15. R. Lavery, I. Parker and J. Kendrick: *J. Biomol. Struct. Dynam.* **4**, 443 (1986).
16. B. Hingerty, R. H. Richie, T. L. Ferrel and T. E. Turner: *Biopolymers* **24**, 427 (1985).
17. R. Lavery and H. Sklenar: *J. Biomol. Struct. Dynam.* **6**, 655 (1989).
18. R. Lavery and H. Sklenar: in *Structure and Methods*, Vol. 2, *DNA Protein Complexes and Proteins*, R. H. Sarma and M. H. Sarma (Eds.), Adenine Press, p. 215 (1990).
19. R. E. Dickerson, M. Bansal, C. R. Calladine, S. Diekmann, W. N. Hunter, O. Kennard, R. Lavery, H. C. M. Nelson, W. K. Olson, W. Saenger, Z. Shakked, H. Sklenar, D. M. Soumpasis, C.-S. Tung, E. von Kitzing, A. H.-J. Wang and V. B. Zhurkin: *J. Mol. Biol.* **205**, 787 (1989).
20. R. Lavery and B. Hartmann: *Biophys. Chem.* **50**, 33 (1994).
21. A. Lipanov, M. L. Kopka, M. Kaczor-Grzeskowiak, J. Quintana and R. E. Dickerson: *Biochemistry* **32**, 1373 (1993).
22. S. Arnott, R. Chandrasekaran, D. L. Birdsall, A. G. W. Leslie and R. L. Ratliff: *Nature* **283**, 743 (1980) and coordinates communicated to our laboratory by S. Arnott.
23. R. Lavery: in *Advances in Computational Biology*, Vol. 1, H. O. Villar (Ed.), JAI Press, Connecticut, p. 69 (1994).
24. M. R. Montminy, K. A. Sevarino, J. A. Wagner, G. Mandel and R. H. Goodman: *Proc. Natl. Acad. Sci. USA*, **83**, 6682 (1988).
25. A. Bolshoy, P. McNamara, R. E. Harrington and E. N. Trifonov: *Proc. Natl. Acad. Sci. USA* **88**, 2312 (1991).
26. S. Cacchione, P. De Santis, D. Foti, A. Palleschi and M. Savino: *Biochemistry* **28**, 8076 (1989).
27. R. K.-Z. Tan and S. C. Harvey: *J. Biomol. Struct. Dynam.* **5**, 497 (1987).

Rotational Motions of Bases in DNA

FATMA BRIKI,[1] JEAN RAMSTEIN,[1] RICHARD LAVERY[2] and DANIEL
GENEST[1]
[1]CBM-CNRS – 1A Av. de la Recherche Scientifique, 45071 Orleans cedex 02, France
[2]IBPC – Lab. de Biochimie théorique, 13 rue P. et M. Curie, 75005 Paris, France

Abstract. Two different kinds of dynamics simulation have been performed to investigate the rotation
of the bases in DNA. A Brownian dynamics simulation using data from fluorescence experiments
establishes a link between the rotation observed at the nanosecond time scale and the base pair opening
reaction which is a process detected at the millisecond time scale. A 200 ps molecular dynamics
simulation of an octanucleotide shows that rotational motions with frequencies lower than $1\,ps^{-1}$ and
characterized by a relaxation time of 17 ps are responsible for the fluorescence anisotropy decay
observed in the supra nanosecond time domain. They correspond to collective motions of groups of
atoms. Both simulations show that a continuity of the base rotation exists from ps to ms.

Key words. Base pair opening, Browian dynamics, DNA, molecular dynamics, rotational motions,
simulation.

Introduction

The base pair opening process in DNA consists of the temporary disruption of
the hydrogen bonds of a single base pair at temperatures far below the melting
point. It is related to the rotation of the bases about the helix axis. This reaction
has been studied by hydrogen exchange experiments. It is an important process
since, during the open state, the bases are accessible to the solvent. Furthermore
it has been suggested that it could play an important role in the propagation of
the replication fork [1].

Quantitative data have been reported from ^{1}H exchange NMR experiments on
oligonucleotides [2,3]. They concern the activation free energy, which is about
15–30 kcal/mol, and the average time between successive openings, which is found
to be in the millisecond to second range according to the length and the sequence
of the DNA and the nature of the bases.

On the other hand, rotational motions of bases have been detected in the 10^{-9} to
10^{-7} s time range by studying fluorescence anisotropy decay of ethidium bromide
intercalated between base pairs [4]. These motions were interpreted as Brownian
rotations of bases about the helix axis. From these experiments, different parameters were deduced: amplitudes of fluctuations, rotational diffusion constant
and restoring torque constant.

Therefore two different experimental techniques, which are sensitive to very
different time scales, detect phenomena which are related to rotational motions
in DNA. In fluorescence, rotational motions are detected but not the open state
while in NMR, the open state is observed but the motions are not directly observed. Thus the question arises whether these experimental observations analyze
the same phenomenon.

In order to answer to this question we have calculated several ms trajectories

A. Pullman et al. (eds.), Modelling of Biomolecular Structures and Mechanisms, 231–239.

by Brownian dynamics simulations (BD) of the rotation of a base, in which information given by fluorescence is introduced as parameters. The data which are obtained are then compared to those given by NMR. A good agreement indicates that the observations from fluorescence and NMR concern the same phenomenon. Therefore a link between the nanosecond and millisecond time scale may be established by numerical simulation.

We have also investigated the possibility of linking fluorescence data to motions occurring in the picosecond time range. From a 200 ps molecular dynamics simulation (MD) of the double stranded self complementary nucleotide d(CTGAT-CAG) in the presence of water molecules and Na^+ ions, the orientation trajectory of a base has been studied and different quantities have been calculated and compared to fluorescence data. The MD simulation shows essentially two kinds of rotational motions: (i) sub-picosecond vibrational modes split into four well separated lines in the high frequency part of the spectrum ($>10\,ps^{-1}$); and (ii) slow diffusive motions in the low frequency part ($<1\,ps^{-1}$). A good agreement between fluorescence and MD is found if only the diffusive motions are considered. The fast rotational motions do not seem to be related to the same phenomenon. The origin of both classes of motions is elucidated.

Methods

1. THE LANGEVIN OSCILLATOR

The Langevin equation for a rotating body is:

$$I\,d^2\theta/dt^2 = -\gamma\,d\theta/dt + G(t) + R(t) \tag{1}$$

where θ is the angular displacement, I the inertial momentum, γ the friction coefficient, $G(t)$ an external torque derived from a mean force potential, and $R(t)$ a Gaussian random torque with:

$$\langle R \rangle = 0$$
$$\langle R(0)R(t) \rangle = 2KTI\gamma\delta(t)$$

where K is the Boltzmann constant, T the temperature and $\delta(t)$ the Dirac function.

If $G(t)$ is derived from a harmonic potential, Equation (1) becomes:

$$I\,d^2\theta/dt^2 = -\gamma\,d\theta/dt - C\theta + R(t) \tag{2}$$

where C is the restoring torque constant. Relationships exist between the different parameters of Equations (1) and (2):

$$\beta = \gamma/I\omega^2$$
$$C = KT/rms^2 \tag{3}$$

where rms is the root mean square of θ, β is the relaxation time of the autocorrelation function of the angular displacement and $\omega^2 = C/I$. Introducing the Einstein equation between γ and the rotational diffusion constant $D = KT/\gamma$, one finally gets:

$$D = rms^2/\beta \tag{4}$$

The quantities C and D have been deduced from fluorescence anisotropy experiments under the approximation of Equation (2) [5–7]. Therefore, under the same approximation, the same quantities may be obtained from MD using Equations (3) and (4).

2. BROWNIAN DYNAMICS SIMULATION

The simulation is based on Equation (1) in the case of a diffusive motion, i.e. the inertial term may be neglected. The numerical solution of Ermack and McCammon [8] was used in which we did not consider hydrodynamics interactions:

$$\Delta\theta = DG\,\Delta t/KT + X \tag{5}$$

where $\Delta\theta$ is the angular increment during Δt and X is a Gaussian random number with 0 mean and $2D\,\Delta t$ variance. For the diffusive approximation to be valid Δt must be sufficiently large that the angular velocity has lost memory. Furthermore, Equation (5) assumes $G(t)$ to be constant during Δt. The diffusive character of the rotation is shown by fluorescence experiments since these motions have characteristic times in the range 1–30 ns.

The rotational diffusion constant D is obtained from the value of the relaxation time $\rho = 14$ ns at 300 °K of the fluorescence anisotropy decay of ethidium bromide intercalated in DNA [4]. ρ is supposed to represent the rotation of a cylinder about the helix axis. The cylinder is composed of an ethidium bromide molecule with the two adjacent base pairs. It therefore has a height of 10.2 Å. Its diffusion constant D_c is related to ρ by $D_c = 1/4\rho$ so that $D_c = 1.8\ 10^7\ \mathrm{s}^{-1}$ which gives for a single base $D = 2 \times 10^8\ \mathrm{s}^{-1}$. No sequence specificity was experimentally detected.

The external torque $G(t)$ cannot be obtained from fluorescence experiments without a drastic approximation of the harmonic oscillator. This approximation which could be valid in the nanosecond time range, is certainly not in the millisecond time domain where higher angular fluctuations are expected. We have preferred to use the results of Ramstein and Lavery [9] who calculated the adiabatic potential corresponding to the transition from the equilibrium state to the open state of the central thymine in the double stranded dA5-dT5 oligonucleotide. This potential is approximately linear in the angular range 0–15°. For $\theta = 15°$ a kink is observed which is interpreted as the occurrence of the open state. We have completed this potential by imposing that if $\theta \leqslant 0°$ at a given step of the simulation, its value is set to 0° preventing from opening toward the minor groove and the external torque is suppressed since this corresponds to the equilibrium position.

The time increment was chosen to be $\Delta t = 2$ ps so that $\langle X^2 \rangle \leqslant 1.5°$ on one hand, and the angular velocity has relaxed (see MD results) on the other hand. 6×10^9 steps of simulation corresponding to 12 ms were performed at different temperatures [10]. For the different temperatures the value of D was readjusted to correct for temperature and viscosity changes.

3. MOLECULAR DYNAMICS SIMULATION

A 200 ps MD simulation of the double stranded self complementary octanucleotide d(CTGATCAG) in the B form has been performed with the GROMOS package

TABLE I
Comparison between BD and ^{1}H NMR data

Quantity	BD	NMR
Base pair lifetime	180 ms at 15 °C	60–120 ms at 15 °C [1]
Activation free energy	20 kcal M^{-1}	22 kcal M^{-1} [2]

[1]The experimental value depends on the sequence (from Ref. [13]).
[2]For poly A-U.

in the presence of 1564 water molecules and 14 Na$^+$ ions [11]. All bond lengths were kept constant with the SHAKE algorithm. The analysis focused on the orientation of the most central thymine, and particularly on the angular fluctuations of the glycosidic bond about the helix axis. Thus, the angle θ represents the orientation of the transversal component of C1$'$N1.

The quantities which have been analyzed are the rms, the frequency spectrum of the trajectory, the displacement and velocity autocorrelation functions and the mean square displacement. In the case of a multiexponential decaying correlation function $f(t) = \Sigma a_i \exp(-t/\beta_i)$ an average relaxation time is defined as:

$$\langle \beta \rangle = \sum (a_i \beta_i)/\sum a_i \tag{6}$$

Results

1. FROM NANOSECOND TO MILLISECOND [10]

During the trajectory, the opening angle was always found to be within the limits 0°–13°. The average time τ between two successive events for which a given value of θ is reached has been computed as a function of θ. For fluctuations larger than 3° a linear variation is obtained between $\log(\tau)$ and θ. Extrapolation to $\theta = 15°$ gives the average time τ_0 separating the occurrence of two successive open states, i.e. the life time of the central A-T base pair of dA5-dT5. We have found a value of 15 ms at 300 K.

Similar computations have been done at different temperatures, from which the activation free energy of the opening process is determined. A value of 20 kcal M^{-1} is obtained.

Table I shows the comparison between these values and the values measured by ^{1}H exchange NMR. A satisfying agreement is observed and our results suggest that the origin of the open state is in the rotation of the bases about the helix axis, which can be quantitatively analyzed in the nanosecond time range.

2. FROM PICOSECOND TO NANOSECOND [11]

The frequency spectrum of the trajectory, shown on Figure 1, exhibits several groups of well-separated lines: a group in the low frequency part (between 0 and 1 ps^{-1}) and a few lines in the high frequency zone (higher than 8 ps^{-1}). The total rms of the angular fluctuations is 8.3°. The normalized autocorrelation function

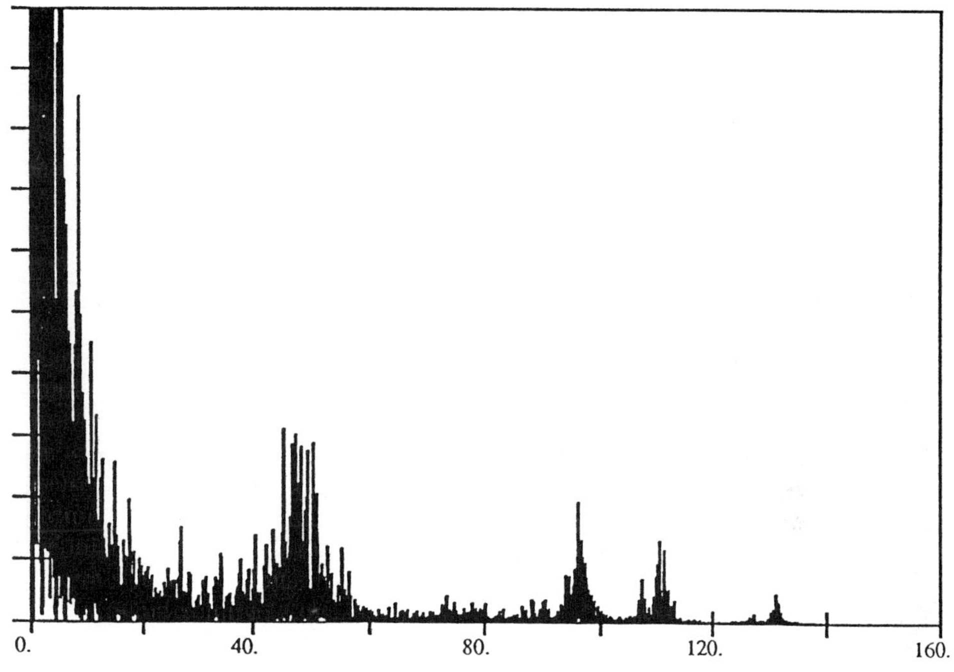

Fig. 1.   Frequency spectrum of the 200 ps angular trajectory calculated from the MD simulation. Frequencies are in radians ps^{-1}. Taken from Ref. [11] with kind permission of Adenine Press.

of the displacement can be satisfactorily fitted by a sum of two exponential terms using the Marquardt algorithm:

$$f(t) = a_1 \exp(-t/\beta_1) + (1 - a_1) \exp(-t/\beta_2) \tag{7}$$

with a_1, β_1 and β_2 equal to 0.6, 0.5 ps and 17.4 ps, respectively. The average relaxation time of $f(t)$ is calculated from Equation (6) which gives $\langle \beta \rangle = 7.26$ ps. Fitting the autocorrelation function with 2, 3, 4 or 5 exponential always gives a long relaxation time of about 17 ps with 40% contribution and short relaxation times ranging from 0.01 to 0.7 ps with an average value around 0.5 ps. Therefore the long relaxation time of 17.4 ps does not depend on the number of exponential used to analyze $f(t)$ and is really significant.

Under the same approximation as that used in fluorescence (Equation 2), it is possible to calculate the restoring torque constant C and the rotational diffusion constant D for the total angular fluctuations by introducing the values of the total rms (8.3°) and $\langle \beta \rangle$ in Equations (3) and (4). The results are given in Table II and it can be seen that no agreement with fluorescence experiments is obtained.

At this stage we make two assumptions: (1) the long and short relaxation times are assigned to the low and high frequency motions respectively; and (2) the low frequency motions are not correlated to the high frequency motions. The first

TABLE II
Comparison between MD and fluorescence data

Quantity	Full motion	Fast component	Slow component	Experiment
rms	8.3°	6.4°	5.1°	10° [1]
β	7.26×10^{-12}	0.6×10^{-12}	17.4×10^{-12}	–
C	0.20×10^{-18}	0.33×10^{-18}	0.50×10^{-18}	0.38×10^{-18} [2]
				0.46×10^{-18} [3]
D	28.6×10^{8}	206.6×10^{8}	4.7×10^{8}	2×10^{8}

Values are given in SI units.
[1] This value corresponds to the amplitude of sub-nanosecond motions detected in fluorescence (from Ref. [4]).
[2] Taken from Ref. [5].
[3] Take from Ref. [7].

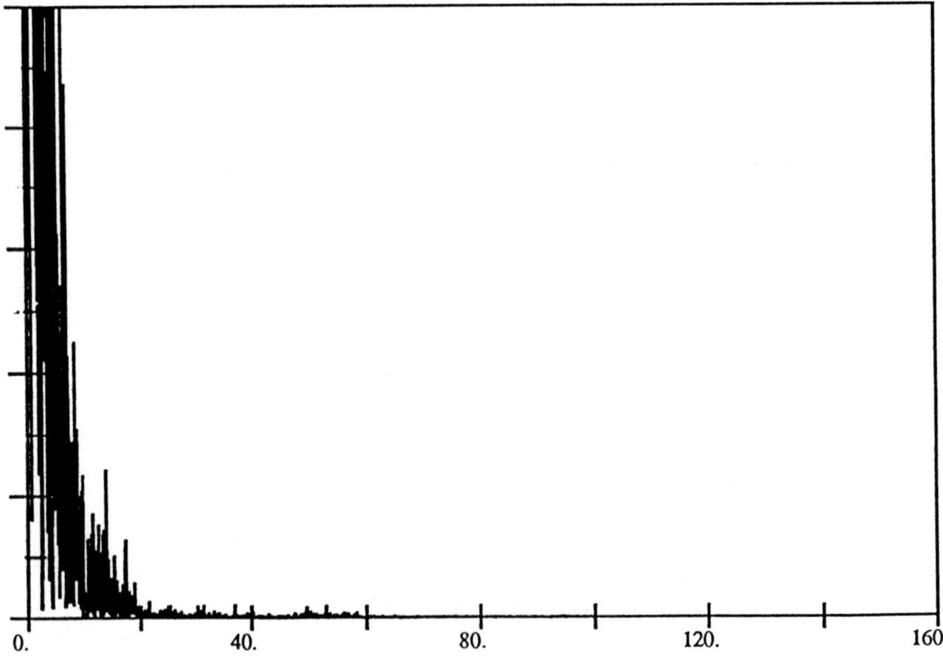

Fig. 2.   Frequency spectrum of the x-component of the center of mass of furanose ring corresponding to the central thymine. Frequencies in radian ps^{-1}.

hypothesis has been shown to be valid [11]. With these assumptions the preexponential terms in Equation (7) are given by:

$$a_1 = g^2/\text{rms}^2 \quad \text{and} \quad (1 - a_1) = h^2/\text{rms}^2$$

where g and h are the rms's corresponding to the fast and slow motions, respectively, which can therefore be determined. The high frequency motions are responsible for an rms of 6.4° and the low frequency motions for an rms of 5.2°.

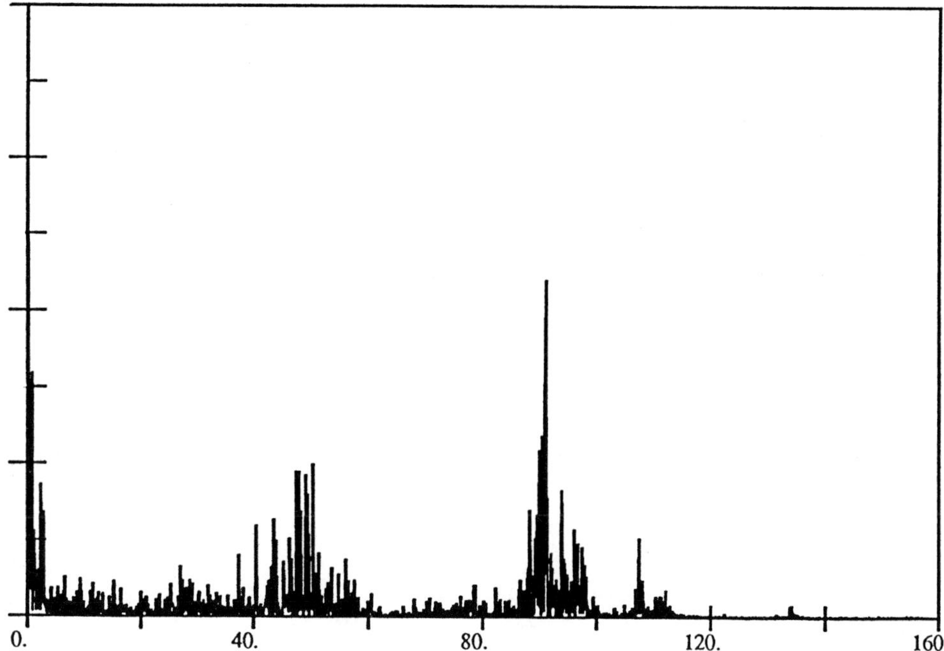

Fig. 3.   Frequency spectrum of the x-component of C1$'$ corresponding to the internal deformation of the same furanose ring as in Figure 2. Frequencies in radian ps^{-1}.

Using these values and their corresponding relaxation times, we are in position to calculate C and D for each of these two kinds of motions. Results are reported in Table II and it can be seen that the low frequency motions describe well the experimental observations. On the contrary there is no agreement between experiment and data derived from the fast motions.

We have recently analyzed the dynamics of the oligonucleotide in terms of sub-unit motions [12]. The motion of each sub-unit was decomposed on translation of its center of mass, rotation about its center of mass and internal deformation of the sub-unit. The corresponding coordinate trajectories were calculated. Figure 2 shows the frequency spectrum of a component of the center of mass corresponding to the central thymine sugar ring, considered as a sub-unit. It is obvious that only the low frequencies are present.

On the contrary, Figure 3 shows the frequency spectrum of a component of the C1$'$ atom calculated from the trajectory corresponding to internal deformation of the same subunit. It can be seen that the spectrum contains only the high frequecies. Identical results are obtained for the N1 atom belonging to the thymine base sub-unit. Thus the slow motions are due to concerted motions of groups of atoms, namely the furanose ring and the base, while the deformation of these sub-units are responsible for the fast motions.

The normalized autocorrelation function for the angular velocity is shown on Figure 4. It is obvious that the velocity has completely relaxed for time greater than

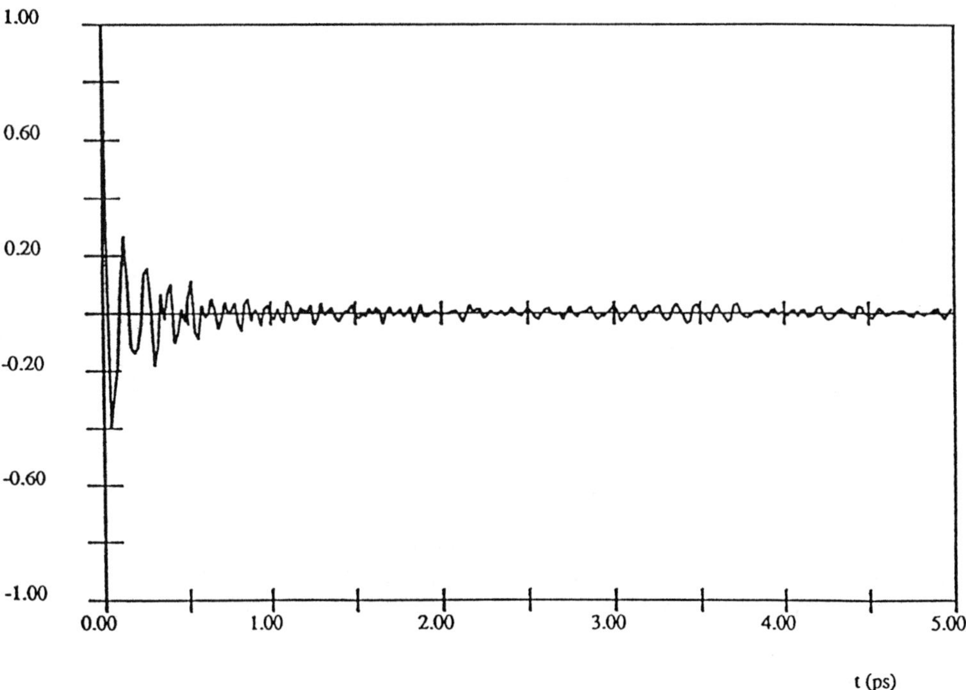

Fig. 4.   Autocorrelation function of the angular velocity calculated from the MD simulation. Taken
from Ref. [11] with kind permission of Adenine Press.

0.8 ps. During the same time the angular displacement autocorrelation function of
the slow component of the motion has decayed only by less 4.5%. These observations are characteristic of a diffusive Brownian motion and show that the value
of Δt in Equation (5) must be greater than 0.8 ps. This was effectively the case in
the Brownian dynamics simulation.

Discussion and Conclusion

We have presented two kinds of numerical simulations, in order to link observations concerning the base rotation in DNA from the picosecond to the millisecond time domain. Picosecond observations are extracted from a molecular dynamics simulation and millisecond observations are relative to the opening base pair
process studied by ^{1}H-NMR. A relay between these extremes is brought by fluorescence anisotropy decay experiments of intercalated dyes which give information
in the range 10^{-9}–10^{-7} s.

The Brownian dynamics simulation leads to the expected values for the base
pair life time and for the activation free energy of the opening process. The
parameters used in the simulation are obtained from fluorescence experiments,
namely the diffusion constant and the validity of the diffusive motion. We have
also used a potential of mean force which was not deduced from fluorescence but

calculated by Ramstein and Lavery [9]. The agreement between the simulation and NMR data shows that the opening process is really due to rotational motions of the bases which are observed in the supra-nanosecond time scale.

The molecular dynamics simulation was performed over a sub-nanosecond time scale. Data from simulation are not directly consistent with data from fluorescence experiments. However, simulation reveals two kinds of rotational motions. The fast sub-picosecond motions do not seem to be related to fluorescence measurements. On the contrary the slow suprapicosecond motions give a satisfying description of experimental results. We think the reason for these observations is that the angle which is analyzed in the simulation, namely the transversal fluctuation of C1'N1, does not only represent the base rotation related to the opening process. It also contains contributions from out-of-plane motions, sugar ring deformation and base deformation. Our study shows that the sugar ring deformation, which contains only high frequency motions, does not contribute to the fluorescence anisotropy decay. On the contrary, the overall motions of the furanose ring and of the base are at the origin of the fluorescence results and therefore of the opening process.

In conclusion, numerical simulations have allowed to establish a bridge between different techniques and have shown that a particular phenomenon, the opening base process in DNA, can be observed from picosecond to millisecond.

References

1. Y. Z. Chen, W. Zhuang, and E. W. Prohofsky: *J. Molec. Struct. Dynam.* **10**, 415 (1992).
2. J. G. Moe and I. M. Russu: *Biochemistry* **31**, 8421 (1992).
3. M. Gueron, M. Kochoyan, and J. L. Leroy: *Nature* **382**, 89 (1987).
4. D. Genest and P. Wahl: *Biochim. Biophys. Acta* **52**, 502 (1978).
5. J. Thomas, S. Allison, C. Appeleof, and M. Schurr: *Biophys. Chem.* **12**, 177 (1980).
6. D. Genest, P. Wahl, M. Erard, M. Champagne, and M. Daune: *Biochimie* **64**, 419 (1982).
7. M. Collini, G. Chirico, and G. Baldini: *Biopolymers* **32**, 3447 (1992).
8. D. L. Ermack and J. A. McCammon: *J. Chem. Phys.* **69**, 1352 (1978).
9. J. Ramstein and R. Lavery: *Proc. Natl. Acad. Sci. USA* **85**, 7231 (1988).
10. F. Briki, J. Ramstein, R. Lavery, and D. Genest: *J. Am. Chem. Soc.* **113**, 2490 (1991).
11. F. Briki and D. Genest: *J. Molec. Struct. Dynam.* **11**, 43 (1993).
12. F. Briki and D. Genest: *J. Molec. Struct. Dynam.* in press.
13. J. L. Leroy: *Regard sur la Biochimie* **5**, 57 (1990).

MOIL-View – A Program for Visualization of Structure and Dynamics of Biomolecules and STO – A Program for Computing Stochastic Paths

CARLOS SIMMERLING,[1] RON ELBER,[2] and JING ZHANG[1]
[1]*Department of Chemistry, University of Illinois at Chicago, 829 West Taylor, Chicago IL 60607, U.S.A.*
[2]*Department of Physical Chemistry and the Israel Center for Structural Biology, The Hebrew University, Jerusalem, 91904 Israel*

Abstract. The structure and implementation of two computer programs are described. MOIL-view is a user friendly and flexible program for visualization of molecular structure and dynamics. STO is a program to compute most probable stochastic trajectories between fixed points. STO is useful to compute reaction paths in complex thermal systems.

Key words. MOIL-View, biomolecules, STO.

I. Introduction

MOIL-View is a public domain program, designed to allow the user to view molecular structures on Silicon Graphics or IBM RS/6000 workstations. While many of the features of MOIL-View are most useful for biomolecules, the program can also be used for a wide variety of other types of molecules. The program is written in FORTRAN, using the Silicon Graphics GL graphics language. The source code is made available in order to allow users to customize the program for their specific needs. MOIL-View is well interfaced with the program MOIL [1], but can also be used as a standalone visualization program, supporting the PDB [2] file format.

MOIL-View allows the user to load a file containing the coordinates of atoms in the molecule. Bonds between particles are then determined either by a distance algorithm, or a MOIL-style connectivity file [1] that contains an explicit list of bonds. These coordinates are then displayed as a stick model of the molecule(s). This model can be rotated, translated and scaled, the colors of particles can be changed, and the program can find and display hydrogen bonds. The program can display distances between selected particles as well as angle and torsion values. A second molecule can be loaded, and the two can be overlapped and compared. The modified coordinates of the molecules can finally be saved back to a data file.

In addition to stick models, MOIL-View can also display molecules using spheres and protein ribbons, or mixtures of the three. Spheres can be solid, or have surfaces represented by points or lines. The radius of the sphere is based on the VDW radius of the atom, and can be adjusted by the user. Hardcopy of these structures is available through postscript line drawings or bitmap images. MOIL-View can also build contact maps for the molecule, calculating a matrix of monomer contacts which then can be printed in a graphical format.

A. Pullman et al. (eds.), Modelling of Biomolecular Structures and Mechanisms, 241–265.
© 1995 *Kluwer Academic Publishers. Printed in the Netherlands.*

The user has control over many of the parameters used to display the structure. These include particle colors, sphere (luality level, cutoff distances for hydrogen bonds, lighting, shading, VDW radius of particles, background color, centering, ribbon colors, and more. A simple command interpreter, adapted from MOIL [1], allows the user to select particles according to logical commands. Selected particles can then be the subject of many functions, such as changing colors or creation of ribbons.

Finally, movies of dynamics trajectories can be viewed. MOIL-View supports CHARMm [3] format DCD type trajectory files. A compression algorithm is provided for faster playback of detailed structures on slower machines. Support is also provided for NTSC output and recording to videotape.

In MOIL-View, there are two main interfaces between the user and the program. These are the toolbox, a window that controls transformation of structures, and the menu, which contains option choices that control how the molecule is displayed, and access to the various modules of the program. The menus in MOIL-View are obtained by pressing and holding the right mouse button at any time during the execution of the main program. The mouse button is held down while the menu choice is made, and finally released when the desired choice is highlighted. The full menu structure is given in Appendix 1. In this text, locations in the menu hierarchy will be referred to using a path format, for example Options/Change Parameters/Change VDW radius would describe the menu location for changing the VDW radius of a particle.

More information on these options will be given below, along with more specific details on the programming aspects of each.

II. Details

MOIL-View is written in FORTRAN, using Silicon Graphics GL graphics library routines. While it is not possible to present all of the details of such a program, we present a general overview of the programming techniques used, followed by a specific example of the use of MOIL-View. Detailed descriptions of each option in MOIL-View are given in the user's manual, available through anonymous ftp to 128.248.186.70. In this section we describe the 3D world in which the objects are designed, the transformation of this 3D space to the 2D screen, the different options available for graphically representing the molecule, and finally how the user controls these options and the transformation of the structure.

Some details concerning the graphical objects that are available using GL will facilitate the understanding of how MOIL-View creates molecular structures. GL contains routines for drawing points, lines, polygons, triangle meshes and quadrilateral strips. Meshes and strips are simply strings of 3 or 4 sided polygons that share edges (Figure 1). These primitives can then be combined to create more complicated objects. Each GL 'object' is a set of drawing commands given between begin and end object definition commands. This object has a unique identifier, and the object is drawn by simply calling a GL routine using this number. In this manner complicated objects need only be defined a single time, and can later be redrawn and transformed without redefining the entire structure. In addition, the SIlicon Graphics GL contains a sphere library, which approximates a sphere as a

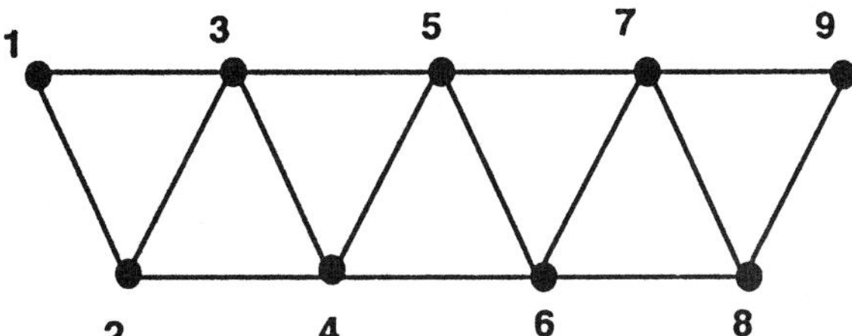

Fig, 1a.   An example of a triangle mesh graphic object. Each vertex placed after the initial two vertices adds an additional triangle to the mesh. The triangles can vary in size and the strip need not be planar.

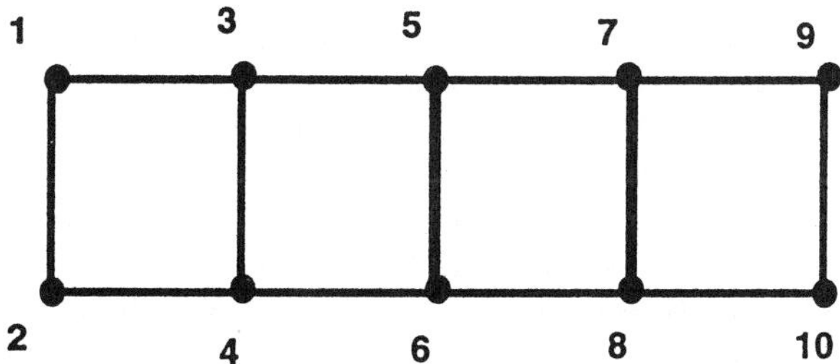

Fig. 1b.   An example of a quadrilateral strip graphic object. Each two additional vertices result in the addition of a quadrilateral segment to the strip. Quadrilaterals are more suitable than triangle meshes for ribbons, which are inherently quadrilateral in nature.

set of polygons. The number of polygons that are used to define the sphere can be specified, and affect both the quality and speed of drawing the sphere. For each graphical object in MOIL-View we describe which GL primitives are used, and how the objects are constructed.

The first function of the MOIL-View program is to define, using GL, the three dimensional space in which the molecule will be located and how this space will be represented on the screen. Information about the graphics hardware of the system, such as number of bitplanes available for display, is obtained using system calls. The main window of the program is then opened, using as many bitplanes as possible for double buffer RGB mode. Double buffering allows the program to draw on a 'hidden' screen, and then swap the buffers after drawing is complete. This allows for smooth animation. RGB mode allows the program to specify the red, green and blue components of each pixel on the screen. The size and initial position of the window are then specified to a GL routine that creates the window.

Next, calls are made to the GL that define how the graphics hardware should project this three dimensional object space onto the two dimensional window. An orthographic projection is used in MOIL-View, where objects are projected from a rectangular viewing space directly to the screen. Perspective is not used. If a Z-buffer is available, it is activated. The Z-buffer provides the proper display of overlapping pixels on the screen. Each pixel has three coordinates: the x and y of the screen, and a z value that determines the distance from the viewer. If a pixel is to be drawn at a location where another pixel has already been drawn, the Z value is examined. In MOIL-View, the new pixel is drawn only if it is closer to the view than the previous pixel. In this manner, proper overlap of objects is obtained. It is not possible to have such overlap on systems without a Z-buffer. Although drawing objects in order in decreasing distance from the viewer solves some of the problems, some objects, such as spheres, interpenetrate and cannot be properly displayed without a Z-buffer. Hence, a Z-buffer is strongly recommended for the use of MOIL-View.

The next step is to use the GL to define the lighting system that will be used. Line objects are not lit, but all solid objects, such as spheres and ribbons, can be lit and shaded. This is accomplished by defining a lighting model and material surfaces. In MOIL-View, a single light is placed in a position that is effectively at an infinite distance on the negative Z axis, that is, behind the viewer. The light is white in color. Next, materials for the surfaces are defined. The GL allows specification of the transparency, shininess, color and reflectance level of materials. A standard shininess and reflectance are used, and the colors and transparency are defined separately for each type of particle. Both of these parameters can be adjusted by the user. Lighting is accomplished automatically by the GL once these parameters are defined. In order for lighting calculations to be performed, each vertex of an object must have a surface normal vector defined. The GL calculates the effect of the lighting on the color of the vertex based on the angle between the light source, the vertex, and the surface normal vector. In this manner the GL calculates the color of an object at each vertex. In MOIL-View, Gouraud shading is used to linearly interpolate color between vertices of polygons, resulting in smooth shading across the polygon. More details concerning the lighting of objects will be given in a later section when discussing their construction.

Next, the menu system is constructed using GL calls. For each menu or submenu, a list of menu entries is given, along with an identifier denoting whether the entry will have a submenu attached to it. Starting with the main menu, each submenu is then given a number. These numbers are used to construct the full menu structure (Appendix 1). Each final menu choice is given a result code. This code is returned to the program by the GL routine that displays the menu, and is used to determine which choice was made by the user. After calling the menu routine, the program executes a series of 'if' statements to run the appropriate module.

Next, an event queue is set up, through GL calls, that alerts the main program whenever events of interest occur. Events that are queued during the program include mouse button presses and window movements. The main loop of the program waits to be notified that an event has occurred, and then performs the appropriate action.

In MOIL-View, all of the coordinate transformations of the molecule are performed through the toolbox, a small, movable window with control buttons used to transform the structure (Figure 2). The toolbox choices give the user control of rotation, translation and scaling of the molecule, width of the lines for stick models, and a Reset View button to return to the original view, discarding all transformations. It is important to note that the translation and rotation axes are in the screen-fixed system, and the coordinate axes displayed in the upper left

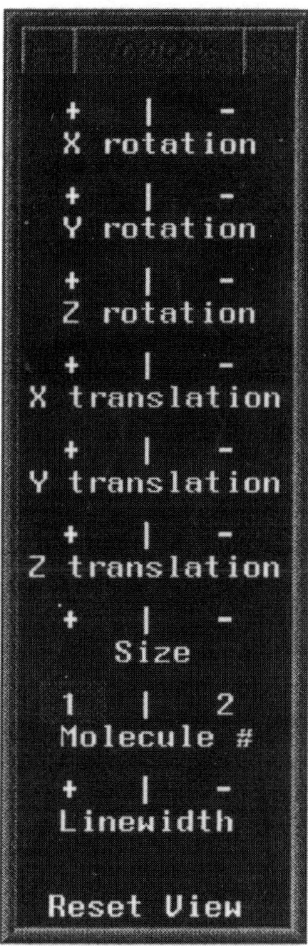

Fig. 2.   The toolbox of MOIL-View. This is the user control interface for the orientation of the molecule on the screen. Pressing the left mouse button over the labeled areas results in transformation of the structure.

corner reflect those in the original coordinate file. The toolbox is activated any time that the LEFT mouse button is pressed. If the mouse pointer is over the toolbox at the time of pressing the button, the structure will be transformed according to the choice made. Continuing to hold the button results in further changes to the structure. Since two structures can be displayed simultaneously, the toolbox allows for control of each molecule separately, or both molecules together.

The toolbox is constructed in a similar fashion as the main window: GL calls are made to initialize the window at a given size and position. The image shown in the toolbox (the rotation buttons, etc) is drawn. A separate routine is used to determine which function the user desires, and applies it to the molecule. This routine is called whenever the event queue notifies the program that the user has pressed the right mouse button. A GL call is made to determine the coordinates of the mouse cursor, and another GL call returns the location of the lower left corner of the Toolbox window. The program determines which transformation to perform by comparing the position of the mouse pointer relative to the Toolbox window. A series of 'if' statements is executed until the proper function is found. The cursor position is then used to determine the magnitude of the transformation – presses further away from the center of the toolbox results in larger steps of motion. These transformation choices are then passed to the main display routine. This routine maintains a rotation matrix that is used to transform all of the coordinates in the molecule for each rotation. Translations and scaling are done through GL calls. While GL calls can also be used for rotation, they result in rotation in a coordinate frame fixed on the molecule. For this program it was desirable to have the coordinate frame in a screen-fixed system, therefore the GL rotation routines are replaced by explicit modification of the atomic coordinates. Further details concerning the specific methods of transforming the molecule for toolbox actions, such as rotations, will be given below when the structure of the graphical objects is described.

At this point the program has constructed a three dimensional space, defined the transformation to two dimensional screen coordinates, specified the lighting model and materials that will be used, created the two user interfaces, the menu and toolbox, and activated the event queue. If a session file is found in the current directory, the session is restored. The program now waits for user input.

The first option that must be chosen by the user is to load a coordinate file. No other menu choices are active until a coordinate set is loaded. In this context we describe the file requester implemented in MOIL-View. A window is opened, similar the main program window. The program uses a system call to obtain a list of files in the current directory. These files are then listed in the new window (Figure 3). Buttons are also created for moving up the directory path, and for accepting or cancelling the selection made. The user presses the right mouse button over the desired choice. By calculating the position of the cursor relative to the window origin, the program determines which option was selected. If a new directory was chosen, the program obtains the appropriate file list and displays it in the window. If a file was chosen, the program displays it at the top of the window, and highlights it in the list of filenames. When the user presses the mouse button over the box labeled 'OK', the selected filename is returned to the main program. The program saves the last filename used for each of the different types of files: one for each coordinate set, one for each connectivity, the filename for the dynamics trajectory, and so on. When the user requests that the program open this type of file again, this path and filename becomes the default in the file requester.

Now that the user has selected a file to open, the filename is passed to the appropriate routine. Two file formats are supported, PDB [1] and CHARMm [3] CRD. The subroutine reads the list of particle names and three dimensional coordinates. These are stored using the same method as in MOIL, allowing for the possibility of interface between the programs. A vector of the length of the number of particles is created for x, y and z coordinates. Another vector contains the character names of the particles. All of the particles are divided into 'monomers'. A pointer is constructed, of the length of number of particles, that tells to which monomer a particle belongs. Finally, two vectors of length of the number

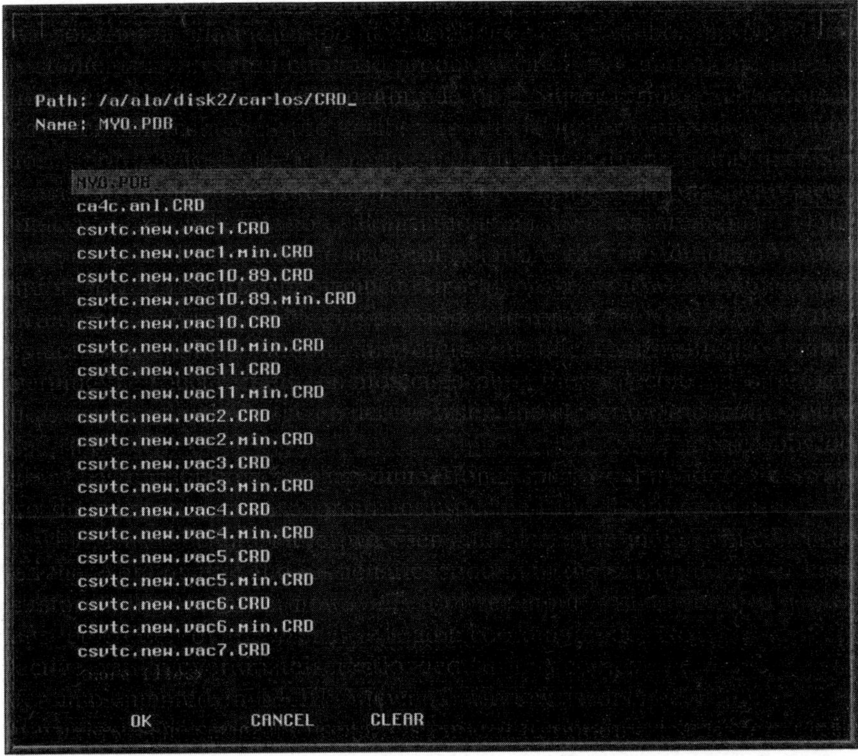

Fig. 3. The file browser of MOIL-View. Files in the current directory are displayed in the window. The current path and filename are displayed at the top of the window. The user can use the mouse to move through the directory tree and select files.

of monomers contain the names of the monomers, and a pointer to the last particle in a monomer. The coordinates and names are read from the file and the pointers are constructed by the program.

After the particle list is read, the program sets all of the default parameters for each particle. These include the red, green and blue components of the particle color, the VDW radius, the covalent radius and the particle mass (used to calculate the center of mass). These are each stored in vectors of the length of the number of particles. After reading the coordinate file, program flow passes to the default parameter routine. A series of 'if' statements compares the first few characters of the particle name to a list of standard names. If a match is found, the parameters are applied to this particle. If no match is found, carbon is selected as the default. The color and VDW radius parameters can be modified by the user, as will be described below.

Now that the program has all of the information about the particles in the system, it needs a description of how these atoms are bonded. This is available in MOIL-View using two different options. The first possibility is to load a MOIL [1] connectivity file that contains an explicit list of bonded particle pairs. This method is rapid and also most accurate. The alternate method is to calculate distances between particle pairs and compare them to the sum of the covalent radii of the two atoms. If the distance is less than this sum, the particles are added to the bond list, and the next pair is considered. Since this calculation would involve N^2 distance calculations, it becomes very time consuming for large molecules. An approximate method is employed that is similar to those used for calculation of neighbor lists in molecular dynamics calculations. A monomer list is first created, containing all monomer pairs that are within a cutoff distance. In MOIL-View, the default value for this distance is 9 angstroms. Next, a loop is performed over all particle pairs in each monomer pair. For these pairs the actual distance is calculated and compared to the covalent sum. This method results in a significant increase in the speed of determining the covalent structure of the molecule.

The list of bonds is stored in the same manner as the program MOIL [1]. There are two vectors, each of length of the number of bonds present. The first vector contains the number of the first particle in the bond, and the second contains the next particle number. This list of bonds is used to create the stick model of the molecule. The stick model is created in color: each bond between a pair of atoms is represented by a line between the two atoms. This line has two colors, with the dividing point in the middle. The color of each bond half is determined by the color of the particle on that side of the bond. All of the bonds are then combined to create the final structure.

Although the screen is two dimensional, several techniques can be used to give the illusion of three dimensions. One effective method is depth-cueing, in which the color of objects gradually fades into the background color with increasing distance from the viewer. Although this feature is available in GL, in order to implement it each graphical object must be a single color. We therefore make a temporary list of vertices for the lines that define our molecule, sorted by color. A loop is then made over the separate colors, constructing a single GL object, containing a set of line segments, for each color. Lines are defined using GL routines, specifying the three dimensional coordinates of the vertices as well as the color and linewidth. Once these objects are created for all of the particles, the program returns to the main loop to display the structure on the screen. This routine will be described in more detail below.

A single routine is used to update the screen display regardless of the function that was performed. Each routine that modifies any aspect of the molecule calls this routine to display the results. As described above, all drawing is done to the back screen buffer, while only the first buffer is displayed. First, the buffer is cleared to the current background color. Next, the program checks a flag to see if the coordinates need to be updated for rotations. If so, it performs a matrix multiplication on each of the coordinates, using the axis and angle passed from the toolbox. Each of the line objects is then updated for the new coordinates, and displayed using GL calls. If hydrogen bonds, ribbons or spheres are requested (see below), they are also displayed using GL calls. Finally, information about the selectd particles such as distances, angles and torsions are written to the bottom corner of the screen. After all drawing is complete, the front and back buffers are switched, displaying the new object on the screen.

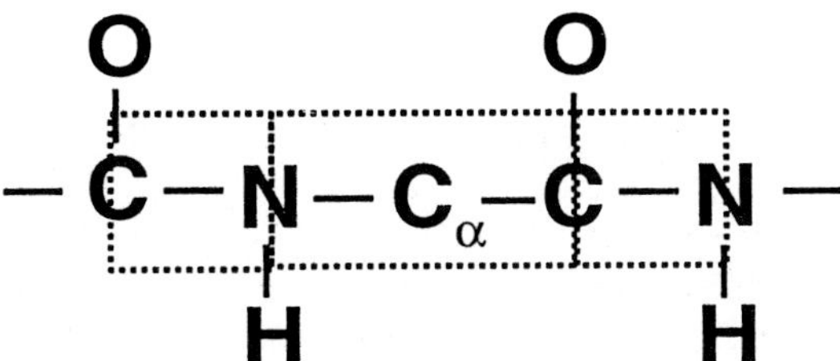

Fig. 4.   An example of a quadrilateral strip used to construct a ribbon object. Two vertices of each segment are placed along the C—O bond axis, and the other two along the N—H bond axis. Each vertex on this axis is 1/2 the bond length away from the C or N atom.

There are several other options available to the user of MOIL-View. One of these is the calculation and display of hydrogen bonds. A simple algorithm is used. Distances are calculated (similarly to the bond calculation described above) between all hydrogen atoms and potential hydrogen bond partners. Each pair is checked to make sure it does not correspond to an actual bond. If the distance is less than the defined hydrogen bond cutoff (user adjustable), then the angle between the three atoms involved is compared to an acceptable range. If both the distance and the angle are accepted, then the pair is added to the hydrogen bond list. This list is treated in a similar fashion as the normal bond list. Finally, a GL object is created for the new bonds, similarly to normal bonds with the exception of the use of dashed lines, defined through calls to GL routines.

Another option available is the display of protein 'ribbon' structures. This routine operates by searching the bond list for consecutively bonded H, N, C and O atoms that define the amide plane in proteins. A quadrilateral strip is constructed, using GL routines, with vertices of consecutive segments of the ribbon at the midpoint of the H—N and O—C bonds. The other two vertices of the quadrilateral vertices are located along the H—N and O—C vectors, but in the 'opposite direction of the first set of vertices (Figure 4). The identity of these particles is saved in a set of pointers, such that the coordinates of the ribbon can be updated later without repeating the search for the particles that define the ribbon.

In order for the ribbon surface to be properly lighted and shaded, each vertex must have a normal vector defined. This vector is determined by the following procedure: each vertex is located at the corner of two quadrilaterals. For each of these, a vector normal to the plane of the quadrilateral is found using the vector product of two sides. Since each vertex is shared by two neighboring quadrilaterals, the normal vector for a particular vertex is defined as the average of the normals of each of the two quadrilaterals. Since the lighting model only provides color for vertices, and the surfaces are smoothly shaded between vertices, this average is necessary to ensure that neighboring quadrilateral strips have continuous shaded colors. This list of vertices and normal vectors is then used to construct the ribbon structure via GL quadrilateral strip routines. The ribbon is displayed at the same time as the other graphical objects as described above. The IBM

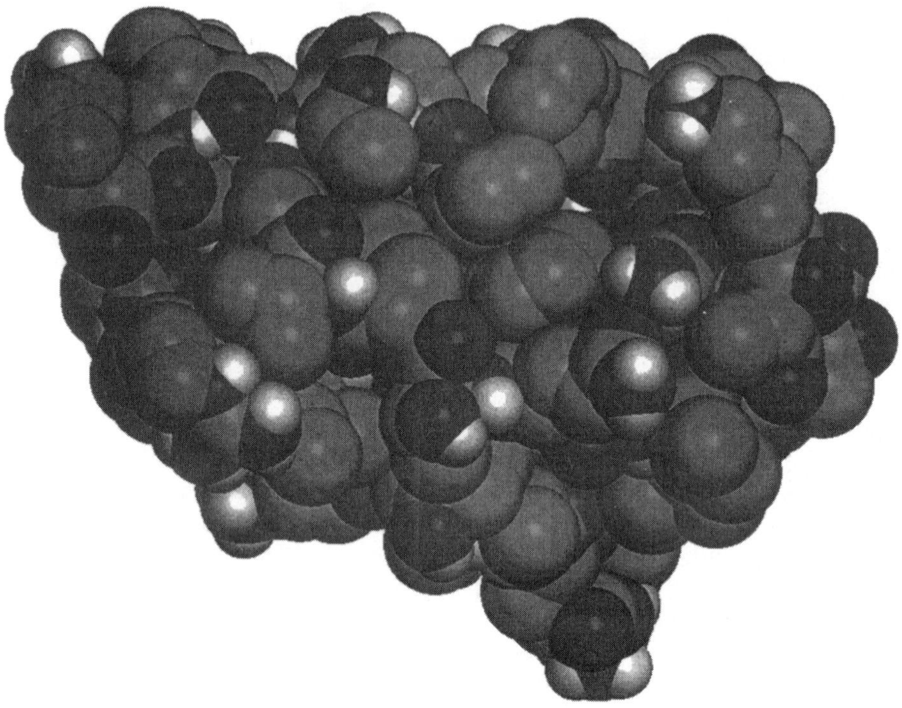

Fig. 5.   An example of a bitmap image postscript output from MOIL-View. The molecule shown is
BPTI, with all atoms displayed as spheres.

version of GL does not support such strips, therefore triangle meshes are used in a manner very similar to that described above.

Another useful graphical tool is the display of spheres to provide a space-filling model of the molecule. Spheres are available using GL calls on the SGI machines, and by a MOIL-View routine on the IBM RS/6000. A single sphere object is generated with a unit radius and origin at [0,0,0] and given an object identifier. Spheres are approximated as octahedra, with a variable level of tessellation, or subdivision of the original polygons into smaller ones for smoother surfaces. The user specifies, through a MOIL-style 'pick' command [1], a set of particles that should be represented as spheres. A vector list is created of the particle numbers that will be spheres. In the main display routine, a loop is performed over this list, and the single sphere object is scaled to the VDW radius of the particle, translated to its location, given the particle's color, and then drawn. This process is repeated for all spheres in the list. The Z-buffer ensures that the spheres overlap properly with each other and all other objects, whether drawn previously or subsequently. A sphere model of BPTI is shown in Figure 5.

It is often useful to be able to identify a particle on the screen, and obtain information such as the distance to other particles. This is accomplished in MOIL-View through the use of a GL routine that takes as input the x and y screen coordinates of the mouse cursor and returns two three dimensional points that define the line (in 3D object space) that corresponds to the selected spot on the screen. A loop is then performed over all of the particles in the system. The distance of each particle to the line is calculated as follows: the vertices returned by the GL routine are A and B. The particle is at C. We calculate the angle $A - B - C$ thorough use of the scalar product of the vectors. The distance of C to the line $A - B$ is then the sine of this angle multiplied by the magnitude of the vector $B - C$. The particle that is closest to the line is then selected. A circle is drawn around it on the screen. The program also saves the identity of the previous three particle selections. If enough prior selections exist, the program calculates distance, angle and torsion values for the new set and displays the values, along with the respective particle names, after the next screen update (Figure 8).

A useful feature of the MOIL-View program is the ability to show movies of dynamics trajectory files. MOIL-View can read files in the CHARMM[3] DCD format for this purpose. The file name is obtained as described above for coordinate files. The user is asked to input the start and end frame number, how many (if any) frames to skip before reading the next coordinate set, how long to delay between each new display, and finally whether to cycle back to the beginning of the movie after the last frame. After obtaining this information and opening the file, the program enters a loop over the number of frames. A new coordinate set is read from the file and transformed according to the current rotation matrix. The line objects are updated based on the new coordinates, as described above. Next, the pointers to the particles defining the ribbon structure are used to update the coordinates of the ribbon vertices and surface normals. If desired by the user, the hydrogen bond list is recalculated. Finally, the list of selected particles is used to update distances, angles and torsions. The event queue is polled during this time to ensure that the toolbox is active during dynamics. The screen is updated using the method described above, and the program returns to the loop for the next frame of the movie.

For large molecules, using spheres can be very slow. There are several options in MOIL-View that are designed to minimize this difficulty. First, MOIL-View uses a GL routine that disables drawing of polygons that face away from the user. For spheres, these polygons will never be seen and time should not be wasted on producing them. Another option allows users to adjust the tessellation level of the spheres, making a compromise between speed and quality. For very large molecules this may still be unacceptable, especially for viewing dynamics movies and recording to videotape. A compression algorithm was implemented to solve this problem. This 'movie' option performs similarly to normal dynamics, as described above, but after each frame is rendered the contents of the screen buffer are saved to disk. For each subsequent frame, only those pixels that changed in value from the previous frame are saved. When the rendering is complete, the file can be read from disk and played back at a speed nearly independent of the complexity of the molecule.

In addition to a screen display, it is often necessary to have a printed copy of the molecular structure. MOIL-View can create output of molecular structure suitable for printing on a postscript printer. There are two basic option available for generating such files: the structure can be redrawn using the postscript language, or the contents of the screen (as drawn by GL) can be converted to a postscript bitmap image directly. Each method has advantages and drawbacks. Bitmap files produce accurate recreations of the screen image, but are very large and print very slowly. Postscript language drawing results in smaller files that print quickly, but are not able to take advantage of the work already done by the GL and must create the proper image using only postscript tools.

Bitmap images are produced in MOIL-View using simple GL and postscript tools. The contents of the screen buffer (in the MOIL-View window area) are read, using GL routines, and written directly to the output file. A short postscript routine is also written to the file, describing the format and providing the print commands. Such files can be written in gray scale or color, and range in size from 2 to 6 megabytes. An example the bitmap postscript output for the protein BPTl is shown in Figure 5.

Drawing using only postscript tools is more complicated, since the screen image must be recreated from only the particle coordinates. Two options are available: the file can be created using the colors of the individual particles, or simply using black for all particles. The other option determines whether depth cueing (described above) should be used. These two factors influence only the actual color values used in the postscript drawing commands; the rest of the drawing procedure is identical.

First, the list of particles is sorted according to the Z value. Postscript does not have a Z-buffer, and newly drawn objects overlap those already produced. For proper overlap, drawing must occur such that nearer objects are drawn over those farther away (from the viewer). The range of these Z values is used to produce a depth-cueing scaling factor. This factor is used to scale the intensity of the particle colors, such that those with larger Z values become progressively lighter and fade into the page. Next, the X and Y coordinate ranges are determined, and a scaling factor is calculated as a conversion between atomic and page coordinates. This factor ensures that the molecule will fit properly on the page. Finally, a loop is performed over the list of sorted (by Z value) particle numbers. For each particle, loops over bonds, spheres, ribbons, and hydrogen bonds are performed to determine if an object should be drawn. Postscript lines, polygons and circles are used to create the image, with particle colors and depth cueing applied when appropriate. An example of output using postscript drawing is shown in Figure 6.

MOIL-View provides the ability to load and display a second coordinate set. Reading of coordinates, generation of bonds, and construction of GL objects are performed in the same manner as described above for the first molecule. Both molecules can be displayed at one time, and the toolbox allows for combined or separate control of the two structures. The main advantage to the display of two molecular structures is for purposes of comparison. To facilitate such comparison, MOIL-View provides the ability to automatically overlap the two structures. A range of particles can be selected for each molecule, and the selected particles can be different for the two molecules. The overlap algorithm of Kabsch [4] is used to modify the coordinates such that the root mean square deviation (RMSD) of the overlapped selection is minimized. The RMSD is printed on the screen, and the display objects are updated for the new coordinates.

Finally, a short description of the variety of routines in MOIL-View that handle

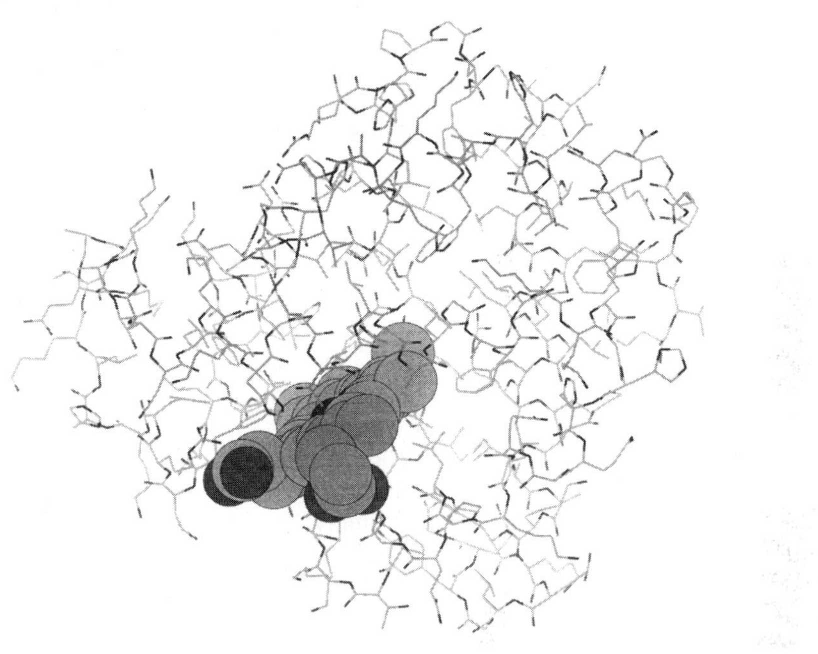

Fig. 6.   An example of the postscript drawing output of MOIL-View. The molecule is myoglobin, with sphere objects employed only for the heme. Depth cueing is on and particle colors were used.

i/o and provide interaction with the user is necessary. Since GL itself does not provide a library of pre-defined gadgets for this purpose (as exist for X windows programs), these need to be designed from scratch. These routines provide functions such as requesting the user to wait until a lengthy calculation has completed, providing an interface for input of color data and selection options, and many more. Each will not be described in detail, but the general methods employed for them will be discussed.

Each window has a backgrnund 'template' that is drawn after the window is opened. This includes text messages, areas for user input and button boxes. The methods used for the mouse interface were described above: the mouse cursor

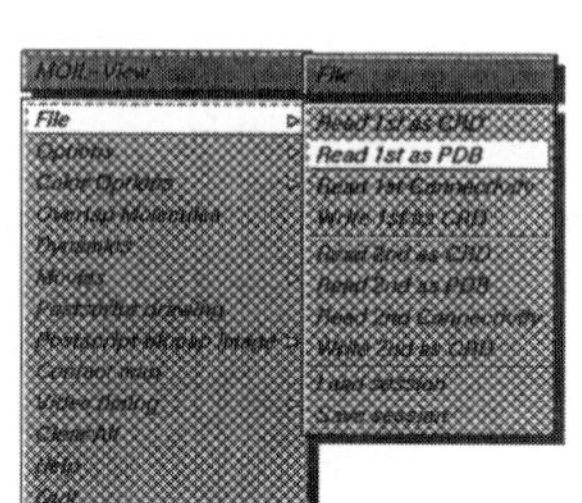
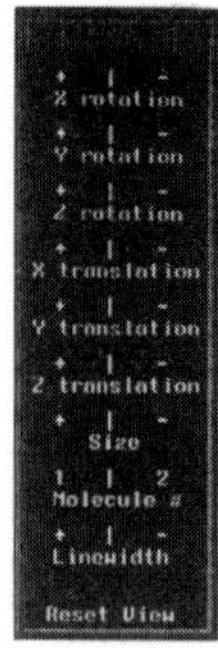

Fig. 7. A screen capture of the SGI version of MOIL-View, showing both the main window of the program and the toolbox. The user has just started the program, and the menu choice for loading a coordinate file in PDB format is shown.

coordinates are compared to the locations of the various buttons until a match is found. Text input is handled in a somewhat different fashion: a pointer to the current input line is kept, with character strings representing each line of input. As the user types characters, the keyboard input is appended to the appropriate string. The enter key switches between active lines of input, and the window screen is redrawn after each change. Finally, the character input is converted to numerical data when necessary, and the values are returned to the main program. Checks are made at each step to ensure that only valid data is input, and that string lengths are not exceeded. This provides a simple but intuitive means of providing user input through the graphical interface.

III. Example: Displaying and Printing a Ribbon Model of Myoglobin

In this section we provide an example of the possible uses of MOIL-View. We explain the process of loading a file, generating bonds, displaying and rotating a stick model, generating ribbon objects, and finally creating a postscript file. We begin the tutorial from the point where the user has just executed the program. The MOIL-View main window is blank and the toolbox window appears on the right.

Pressing and holding the right mouse button brings up the main menu. By moving the pointer to 'File', the file submenu appears.Moving the pointer to 'Read 1st as PDB' (Figure 7), and releasing the button, a new window opens on the screen. The title bar of this new window reads 'File Browser', and a list of files in the current directory is displayed inside the window. We point to the line containing MYO.PDB and press the left mouse button. This line becomes highlighted. Now we point to the box labeled 'OK' and again press the left mouse button. A small window opens and asks us to 'Wait!'. After a short delay (while the bonds are calculated), a stick model of myoglobin fills the screen. Moving the

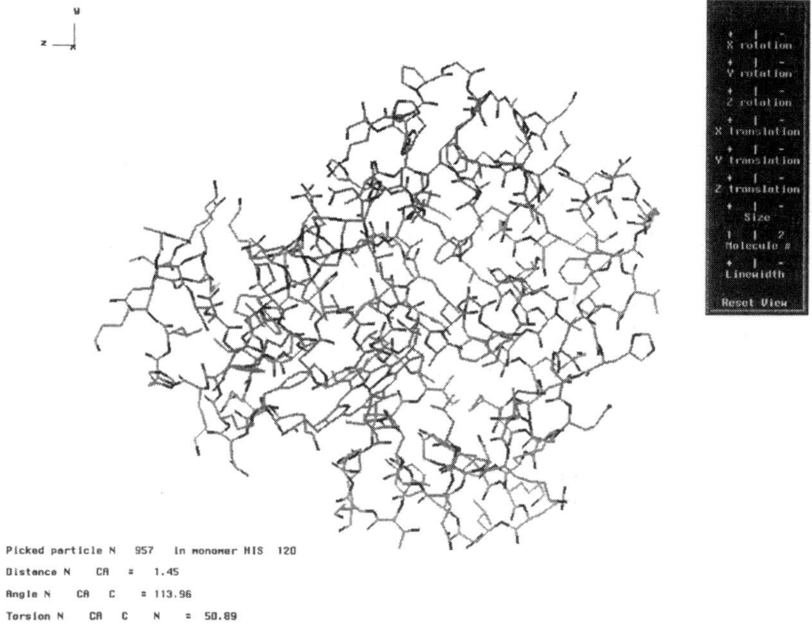

Fig. 8.   The MOIL-View screen showing a stick model of myoglobin. This screen is taken from the example session given in the text.

pointer to the toolbox window and pressing the left button will transform the object according to the item selected. Pointing to particles and pressing the middle mouse button will display distances, angles and torsions. We can make the size smaller and rotate around the Y axis to obtain the view shown in Figure 8.

Next, we generate a ribbon model for myoglobin. Pressing and holding the right mouse button brings back the menu. Under the Options/Objects Are/Ribbons sub-menu, we select 'add'. A new window opens, asking the user to 'Enter your pick command'. This command determines which part of the molecule will have a ribbon structure. Details of the pick syntax can be found the documentation for the MOIL[1] and MOIL-View programs. We type

pick chem mono * != chem mono HEME done

which means we want to pick all monomers, with the exception of the heme group. Pressing the left button on the box labeled 'OK' closes this window and opens a new one. This window allows the user to choose the color of the ribbon by mixing red, green and blue colors (without getting your hands dirty). Type 255 after Red Value and press the left button on the OK box. A red ribbon model is ADDED to the stick model that was already present. To remove the stick model (except for the heme group) we select (from the 'menu) Color Options/User Defined. The color options menu appears, and is similar. to the ribbon color window described above. Press return until the cursor is on the PICK: line. Type pick chem mono * != chem mono HEME done

just as for the ribbons. This will make all of the stick model, except the heme, have no color (which in MOIL-View means 'this portion should not be displayed'). Press OK to continue.

Now we want to save a postscript file of this structure to send to the printer. From the menu, select Postscript Drawing. The file browser will ask for a name for the new file. Press CLEAR to clear the old name, and then type ribbon.ps (or something similar). A window will ask if depth-cueing is desired (as described above). Select YES. Another window$_4$ ask if we want a color postscript file. Select NO. Now we can quit the program and print the file. In Figure 9 we show myoglobin with a ribbon structure generated through this process.

The description above is intended to briefly outline the features of the MOIL-View program, and provide a general feeling for the programming techniques that are required for a program of this type. In such a short space it is not possible, of course, to provide details for all of the various techniques and algorithms that are required in such a large program. What should be apparent is that it is possible to create a program, in FORTRAN, that provides for the viewing and manipulation of molecular structures using modern graphics workstations. The MOIL-View program was created to be of general use, and released to the public domain with the intention that users can obtain a functional program free of charge and tailor it to suit their specific needs.

The full source code to the program is available through anonymous ftp to 128.248.186.70. Separate versions are available for SGI and IBM RS/6000 machines. Complete documentation is included, describing the process of compiling the program as well as providing detailed descriptions of each of the program options.

Stochastic Path

I. INTRODUCTION

The formal definition and the search for a reaction coordinate in macromolecules is far from trivial. This is to be contrasted with the significant success made in studies of reaction paths in small molecular systems [5].

Consider first a reaction which is highly localized in space and with high energy barriers. In this case, approaches to compute reaction paths adapted from studies of small molecules are useful. However, other cases are common. In macromolecules we often find a tremendous number of local energy minima and steepest descent paths. It is therefore difficult to justify the use of the same approaches as in small molecules.

An attempt to study a complex and flexible system using a single path is similar to model a ball of badly cooked and sticky macaronis by a single macaroni. In this example each of the macaronis represents a path and each sticky contact between the macaronis corresponds to a possible transition between the paths.

Here we discuss a computational approach for the determination of a reaction coordinate. This approach is not 'sensitive' to flexible, multi-minima systems.

Fig. 9.   The output from the postscript drawing file created in the example given in the text. A ribbon
model of myoglobin, with the heme represented as a stick figure.

Thus, it is expected to work well for the last systems too. The computational approach is based on the formalism of Onsager and Machlup [6]. In the next section we present part of the results of Onsager and Machlup with some twists appropriate for our needs. In Section III we describe the algorithm and an implementation on a distributed computing system.

We thank Professor P. G. Wolynes for many stimulating discussions and directions for many important references. P.G. Wolynes analyzed analytical properties of the Onsager action for those who seek reaction paths [7] and he directed us in the present (hopefully correct) direction.

II. THE ACTION OF ONSAGER AND MACHLUP [6]

Consider a molecular system that obeys the Langevin equation of motion, that is:

$$m \, d^2\mathbf{R}/dt^2 + \gamma \, d\mathbf{R}/dt + \partial U/\partial \mathbf{R} + F = 0 \tag{1}$$

where m is the mass (matrix), $\mathbf{R}$ the coordinate vector of all the atoms in the system, γ is the friction constant, U is the systematic and the microscopic potential and F is the random force. The random force is sampled from a Gaussian distribution such that $\langle F \rangle = 0$ and $\langle F(t')F(t) \rangle = 2k_B T \gamma \, \delta(t - t')$. *The above formula provides trajectories at a given temperature T and is a useful starting point to study the dynamics of many stochastic processes. As will become clearer later it is not necessary to use the phenomenological stochastic arguments. However, they are useful for the derivation and from a conceptual point of view.*

Consider the following question: We are provided with a set of initial coordinates that represent reactants – $\mathbf{R}_r$ – and a set of final coordinates – $\mathbf{R}_f$. What is the probability of observing a specific trajectory $\mathbf{R}(t)$ between the reactants and the products? It is useful to consider first a solution for a short time interval, dt. We employ a finite difference equation in which the time is denoted by a discrete subscript. The coordinate vector at time i is $\mathbf{R}_i$. It is also convenient to forget the random force for the moment. Without the random force we can obtain an explicit solution to the next position in coordinates and velocities. We denote the coordinates and the velocities by a single vector $\mathbf{X}_i$ and the solution for the next (small) step by $\mathbf{f}(\mathbf{X}_i) = \mathbf{X}_{i+1}$, where $\mathbf{f}(\mathbf{Y})$ is the 'rule' of how to generate the next deterministic step. The existence of an explicit solution is equivalent to the statement that the conditional probability density is a delta function: $P(\mathbf{X}_{i+1}|\mathbf{X}_i; dt) = \delta(\mathbf{f}(\mathbf{X}_i) - \mathbf{X}_{i+1})$.

It is also clear what the random force is doing to the probability density. The random force adds a width to the final position of $\mathbf{X}_{i+1}$. This is since $\mathbf{f}(\mathbf{X}_i)$ is now replaced by $\mathbf{f}(\mathbf{X}_i) + \underline{F}$ ($\underline{F} = (dt^2/2m)F$). We obtain $P(\mathbf{X}_{i+1}|\mathbf{X}_i; dt) \propto \exp[-(\mathbf{f}(\mathbf{X}_i) - \mathbf{X}_{i+1})^2/\langle \underline{F}^2 \rangle]$ which is a solution for short times. To obtain a long time trajectory we accumulate many small steps:

$$P(\mathbf{X}_{i+N}|\mathbf{X}_i; t) = \int (\Pi_j \, d\mathbf{X}_j) \exp[-1/\langle \underline{F}^2 \rangle \sum_k (\mathbf{f}(\mathbf{X}_k) - \mathbf{X}_{k+1})^2] \tag{2}$$

The difference $\mathbf{f}(\mathbf{X}_k) - \mathbf{X}_{k+1}$ is $\underline{F}_k$; however, according to Equation (1) it is also equal to the deterministic part of the equation (up to a factor of $dt^2/2m$), we therefore have

$$P(\mathbf{X}_{i+N}|\mathbf{X}_i; t) = \int (\Pi_j \, d\mathbf{X}_j) \times$$

$$\times \exp[-(dt^2/2m\langle \underline{F}^2 \rangle) \sum_k (md^2\mathbf{R}_k/dt^2 + \gamma \, d\mathbf{R}_k/dt + \partial U/\partial \mathbf{R}_k)^2] \tag{3}$$

Equation (3) is a typical starting point to obtain a path integral. All that is required is to take the number of points N to be infinite and dt to be infinitesimal. There is no problem in taking dt as small as we please; after algebraic cancellations there is only one dt remaining that is used in the resulting action integral, giving

$$P(\mathbf{R}_f|\mathbf{R}_r; t) = \int D\mathbf{R}(t)\, \exp[-1/\langle F^2\rangle \int dt(md^2\mathbf{R}/dt^2 + \gamma\, d\mathbf{R}/dt + \partial U/\partial \mathbf{R})^2] \qquad (4)$$

where we return to $\mathbf{R}$ from $\mathbf{X}$ and D denotes a path integral. This is what Onsager and Machlup did [6a,b]. Here for computational preference we stay with the form (3). We further approximate the time derivatives by finite differences and as in (4) replace by $\mathbf{X}$ by $\mathbf{R}$.

$$P(\mathbf{R}_{i+N}|\mathbf{R}_i; t) = \int (\Pi_j\, d\mathbf{R}_j)\, \exp[-(dt^2/2,\langle \underline{F}^2\rangle)S] \qquad (5a)$$

$$S = \sum_k (m(\mathbf{R}_{k+1} + \mathbf{R}_{k-1} - 2\mathbf{R}_k)/(dt)^2 + \gamma(\mathbf{R}_{k+1} - \mathbf{R}_{k-1})/2dt+ \qquad (5b)$$

$$+ \partial U/\partial \mathbf{R}_k)^2$$

This equation adds all the stochastic trajectories that start at $\mathbf{R}_i$ and end at $\mathbf{R}_{i+N}$ after time t. This provides a numerical value for the conditional probability.

Computationally, the above expression has a number of important advantages.

First, the integrand is always positive, which is convenient for path integral calculations using sampling methods such as molecular dynamics or Monte Carlo.

Second, the 'action' S is always non-negative. It therefore has a clear minimum as a function of the set of structures $\{\mathbf{R}_k\}$ that represents the path. Thus, the minimum of S provides the most probable trajectory that is going from reactants to products. It is therefore a suggestive definition for a reaction coordinate in the condensed phase.

Third, since computing the probable path is equivalent to an 'energy' minimization the calculations are very stable and the stability is independent on the size of the time step – dt. This should be contrasted with solutions of differential equations, which are very sensitive to the size of time step. Clearly the resulting trajectories are approximate, depending on the level of coarse grain in time that was used. However, for motions on relatively flat energy surfaces in which the system is diffusive, the above approach is quantitative. For systems with large energy barriers, the above approach can provide qualitative information. Quantitative information is more difficult to obtain since the short residence time in the neighborhood of barriers makes it difficult to sample these events if a large time step is used.

Fourth, path integrals are very easy to parallelize. The parallelization can be done in a general way, independent of the specific form of the potential. Only minimum communication between the processors is required (the kth processor sends and accepts messages from $k \pm 1$, $k \pm 2$, a total of four neighbors independent of the total number of processors). We implemented parallel computation of the above action on a cluster of workstations. The algorithm will be presented in the next section. This type of parallelism is impossible to pursue using ordinary iteration in time of the equations of motion.

Fifth, the use of the new formalism is more natural than minimum energy paths. The effect of the temperature (a useful thing to have in condensed phase calculations) is already built in. The number of paths extracted from the Onsager–Machlup action is much smaller than the number of minimum energy paths in flexible systems. Flexible systems may put the whole concept of minimum energy paths in doubts.

The Onsager and Machlup action also has disadvantages.

First, it requires storage of the complete trajectory on the computer (or computers). This is because the whole of the trajectory is optimized at once. This can be a significant burden. In iterative algorithms only the current and perhaps a few previous steps are kept.

Second, the analytical derivatives of the action S requires second derivatives of the potential energy $- U$. These derivatives require even more storage and significant computer power to handle them effectively.

Nevertheless, the possibility of pursuing trajectories from reactants to products with time steps of 100 ps and more is exciting and may open new possibilities for atomic detail simulations on very extended time scales.

We proceed with the description of the computational implementation on a cluster of workstations.

III. COMPUTATIONAL IMPLEMENTATION OF THE ONSAGER AND
     MACHLUP ACTION

Let the number of structures be N and the number of processors (or workstations) P. The reactant and the product positions are fixed and do not require optimization. We therefore divide the $N - 2$ 'active' structures between P processors. Each processor is responsible for the optimization of $(N - 2)/P$ structures. The minimization is pursued for a given number of optimization steps $- L$. We call the $(N - 2)/P$ structures that are assigned to a given processor the 'local path segment'. The optimization of the structures is done in a standard way, either by a conjugated gradient algorithm or by simulated annealing.

To compute the energy terms that contribute to the forces at the k-th time, we need to study the sum in Equation (5b):

$$\Sigma_k[m(\mathbf{R}_{k+1} + \mathbf{R}_{k-1} - 2\mathbf{R}_k)/(\mathrm{d}t)^2 + \gamma(\mathbf{R}_{k+1} - \mathbf{R}_{k-1})/2\mathrm{d}t + \partial U/\partial \mathbf{R}_k]^2.$$

For simplicity we denote the sum by $\Sigma_k W_k^2$. It is evident that the terms $W_{k\pm1}$, $W_{k\pm2}$ and W_k depend on $\mathbf{R}_k$. Therefore when we provide a processor with a path segment of size $(N - 2)/P$ we also provide four additional structures so that the information of how to calculate the first and the last structures (on that workstation) will be complete: Figure 10. These structures are not modified by the processor that requests them only for 'energy' calculations at the edges of the local segment. The structures are modified only by the neighboring workstations. That is, only one of the computers can modify a structure. Note that the workstations that contain the first or the last segments of the set of structures do not require four structures but only two.

The calculation is pursued as follows: A UNIX shell script initiates FORTRAN programs on each of the workstations. Each workstation is taking a responsibility on a path segment (fixed in size) and is given a name of an upper and lower neighbor (another workstation) to which it will communicate later. Of course the processor at the beginning of the path is provided with only a single name of an upper neighbor and the last workstation is provided only with a lower neighbor.

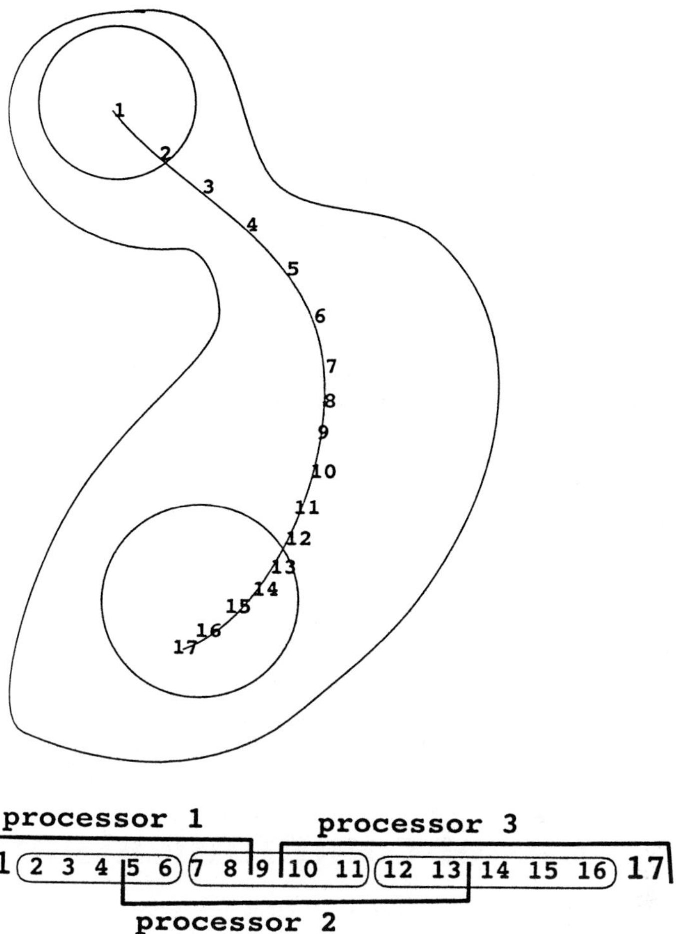

Fig. 10.   A schematic representation of the 17 structures that represent the Onsager–Machlup and their partition between three processors. Open curves are used to mark the configurations required to compute the energy assigned to a given processor. For example, processor 2 requires structures 5–13. In closed curves we marked the structures that each processor optimizes. For example, processor 2 optimizes five structures. Note that structure 1 and structure 17 are fixed.

We do not allow a workstation to optimize less than two structures and therefore there is no need for more than one neighboring workstation either up or down.

After all the workstations initiated their path segment (either by reading a pre-prepared file or by an internal calculation of a guess) they continue to optimize their set of structures for L steps, independent of the other workstations/processors. Once the L steps are completed there is a need to update the edges of the paths between the structures.

The communication is done as follows. We employ UNIX sockets for communication, that are initiated each time a communication is requested. The use of UNIX sockets makes our code very general and probably applicable to most

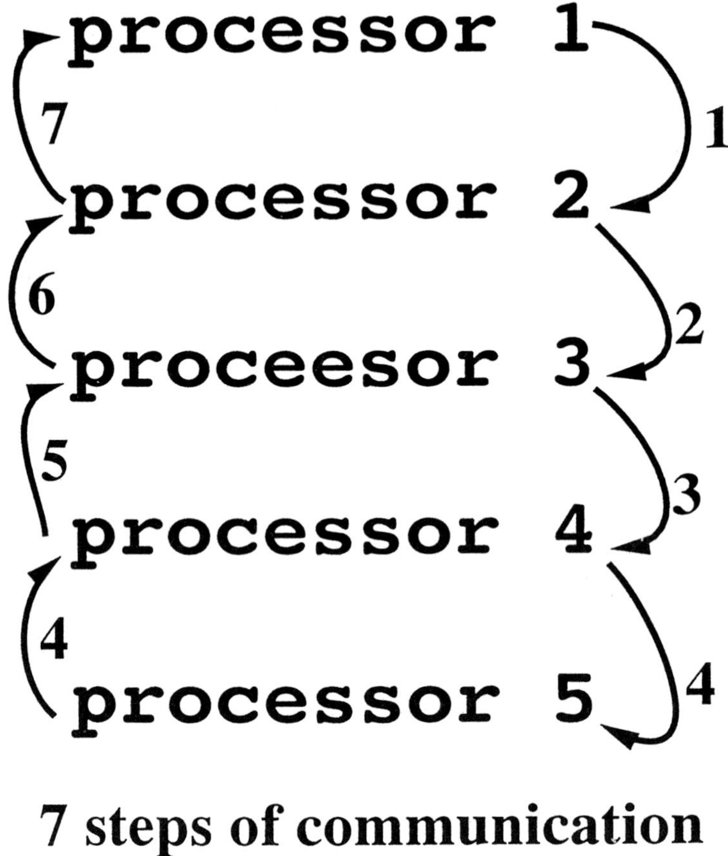

7 steps of communication

Fig. 11. The 'ring' communication between the different processors that compute the Onsager–Machlup action. The example includes five processors and the numbers near the arrows indicate the communication sequence.

UNIX workstations. However there is a penalty for generality which in our case is the speed of communication.

The communication is pursued in a ring structure (Figure 11). After the first processor completes its L minimization steps it sends its last two active structures to the second processor and waits for the second processor's reply. The second processor first accepts the two structures and then proceeds. It sends the coordinates to the third processor. The third get the coordinates and sends the required set to the fourth until the last structure is reached. The last processor sends its first structure to the one before the last and we start the ring in the backward direction. Clearly this approach is not optimal, since only two processors communicate at a time. Ideally all pairs could 'taIk' at the same time, providing us with a communication scheme that is independent of the number of processors. However, this is not allowed in the most common protocol of TCP/IP and to keep the generality of the code a ring structure was adapted.

The minimization and the communication is continued until convergence criteria are reached. The criteria include norm of the gradient and the total number of steps.

Our code is working at present on Silicon Graphics, HP 7xx and IBM R/S 6000 workstations and is available in an anonymous ftp site 128.248.186.70 and 132.64.96.20.

Appendix 1. The MOIL-View Menu Structure

1. File

    A. Read 1st as CRD
    B. Read 1st as PDB
    C. Read First Connectivity
    D. Write 1st as CRD
    E. Read 2nd as CRD
    F. Read 2nd as PDB
    G. Read 2nd Connectivity
    H. Write 2nd as CRD
    I. Load Session
    J. Save Session

2. Options

    A. Second Molecule
      1. On
      2. Off
      3. Remove

    B. Tools
      1. On
      2. Off

    C. Bond Algorithm
      1. Distance
      2. Connectivity

    D. Center Object
      1. On Pick
      2. On CM
      3. Don't Center

    E. Selection Info
      1. On
      2. Off

    F. Objects Are
      1. Spheres
      2. Lines
      3. Pick Spheres

    4. Protein Ribbons
       a. Add
       b. Remove

  G. Change Parameters
    1. Sphere Quality Level
    2. Hbond Cutoff Distance
    3. Toggle Shading
    4. Toggle Lighting
    5. Spheres are:
       a. Polygons
       b. Lines
       c. Points
       d. Opaque
       e. Transparent
    6. Toggle Z-buffer
    7. Contact Map Cutoff
    8. Show Unbonded Particles
    9. Change VDW radius

  H. Hydrogen Bonds
    1. Find
    2. Turn Off
    3. Pick HB color

  I. Solvate Molecule
    1. Shell
    2. Cubic Box

3. Color Options
  A. User Defined
  B. backbone Only
  C. Restore Default
  D. Transparency
  F. Background Color
    1. Black
    2. White
  F. Close Waters Only

4. Overlap Molecules

5. Dynamics

6. Movies
  A. Generate
  B. Watch

7. Postscript Drawing

8. Postscript Bitmap Image
   A. Color
   B. Gray Scale

9. Create Contact Map

10. Video Timing
    A. Toggle NTSC/Default

11. Clear All

12. Help

13. Quit

References

1. R. Elber, A. Roitberg, C. Simmerling, R. Goldstein, G. Verkhivker, and H. Li: 'MOIL: A molecular dynamics program with emphasis on conformational searches and reaction path calculations in large biological molecules', *Proceedings of the NATO Workshop 'Statistical Mechanics, Protein Structure and Protein–Substrate Interactions'*, Corsica, June 1993, Plenum, New York.
2. F. C. Bernstein, T. F. Koetzle, G. J. B. Williams, E. F. Meyer, M. D. Brice, J. R. Rogers, O. Kennard, T. Shimanouchi, and M. Tasumi: *J. Mol. Biol.* **112**, 535 (1977).
3. B. R. Brooks, R. E. Bruccoleri, B. D. Olafson, D. J. States, S. Swaminathan and M. Karplus: *J. Comput. Chem.* **4**, 187 (1983).
4. W. Kabsch: *Acta. Cryst.* **A34**, 827 (1978).
5. For a recent study (and a review) of calculations of reaction path in small (but flexible) molecules see: R. Czerminski and R. Elber: *J. Chem. Phys.* **92**, 5580 (1990).
6. (a) L. Onsager and S. Machlup: *Phys. Rev.* **91**, 1505 (1953). (b) L. Onsager and S. Machlup: *Phys. Rev.* **91**, 1512 (1953).
7. P. G. Wolynes: 'Chemical Reaction Dynamics in Complex Molecular Systems', in *Complex Systems, SFI studies in the Sciences of Complexity*, D. Stein, Ed., Addison Wesley (1989).

Rational Design of Switched Triple Helix-Forming Oligonucleotides: Extension of Sequences for Triple Helix Formation

JIAN-SHENG SUN

Laboratoire de Biophysique, Muséum National d'Histoire Naturelle, INSERM U 201, CNRS URA481, 43, rue Cuvier 75231 Paris Cedex 05, France.

Abstract. A rational design by means of molecular mechanics has been carried out in an effort to extend the range of double-helical DNA sequences that could be recognized by triple helix-forming oligonucleotides. The DNA target is composed of alternating, adjacent fragments of oligopurine·oligopyrimidine sequences, instead of a long stretch of polypurine·polypyrimidine sequence used for canonical triple helix formation. Based on the combination of different triple helix motifs in either *Hoogsteen* or *reverse Hoogsteen* configuration, mini-triple helices can be formed at each oligopurine·oligopyrimidine part of the target sequence with either parallel or antiparallel orientation with respect to the purine strand. As the adjacent purine target sequences are located in the complementary strands, the third strand oligonucleotides can be joined together through a natural phosphodiester backbone at the junctions in either a 5'-3' or a 3'-5' polarity. There are six distinct junction steps. Molecular modeling was aimed at optimizing the cooperative binding of the so-called switched triple helix-forming oligonucleotides by choosing appropriate nucleotide(s) at the junction between two adjacent mini-triple helices. A comprehensive *switch code* describing the rules for forming switched triple helices has been established. Its practical applications in extending DNA recognition by this new generation of tailor-made triple helix-forming oligonucleotides are discussed.

Key words: Design of triple helix-forming oligonucleotides, switched triple helix, switch code, antigene strategy

Introduction

Although the formation of triple helices with polynucleotides was described as early as 1957 [1], renewed interest in triple helix formation by oligonucleotides since 1987 [2,3] has been aroused by their potential biological and therapeutical applications for controlling gene expression at both transcription and replication levels [4–7]. During the last five years, many structural motifs of triple helices have been developed (see a recent review paper [8]). These developments were mainly focused on overcoming the requirement for protonation at the N3 position of cytosines in order to give rise to the C·GxC+ base triplet, along with the T·AxT base triplet, to form a (C,T)-motif canonical triple helix. In this paper, base triplets will be referred to as X·YxZ, where the symbols · and x indicate *Watson–Crick* and *Hoogsteen*-like (or *reverse Hoogsteen*-like) hydrogen-bonding interactions, respectively.

It is well known that an oligopyrimidine can form a (C,T)-motif triple helix with parallel orientation of the third strand with respect to the polypurine strand. However, an oligopurine containing guanines and adenines can recognize polypurines of Watson–Crick base pairs to give rise to a so-called (G,A)-motif triple helix [9,10]. In this (G,A)-motif, the third strand binds in an antiparallel orientation

A. Pullman et al. (eds.), Modelling of Biomolecular Structures and Mechanisms, 267–288.
© 1995 *Kluwer Academic Publishers. Printed in the Netherlands.*

with respect to the target polypurine sequence with both C·GxG and T·AxA base triplets in *reverse Hoogsteen*-like configuration. Furthermore, an oligonucleotide containing guanines and thymines can also form a (G,T)-motif triple helix [4]. Band-shift and footprint assays revealed that the orientation of such a third strand is antiparallel to the purine-rich strand of the target double-stranded DNA, rather than parallel as originally suggested [11]. However, shortly later, it was shown that the orientation of the third strand depends on its sequence [12] (see below).

In all motifs where the constituent base triplets are engaged in *Hoogsteen*-like hydrogen-bonding interactions with the nucleotides in *anti* glycosidic conformation, the orientation of the third strands is parallel with respect to the target purine strand. Conversely, in all motifs involving *reverse Hoogsteen*-like hydrogen-bonding interactions, the third strands run antiparallel to the purine strand, since *reverse Hoogsteen* configuration implies a 180° flip of the nucleotides in the third strand with the *anti* glycosidic conformation being maintained (see Figure 1).

Oligonucleotides belonging to one of the above three described motifs of triple helices contain only natural nucleotide units. Figure 1 shows those base triplets formed by natural bases and Watson–Crick base pairs. All three motifs involve *Hoogsteen* or *reverse Hoogsteen*-like hydrogen-bonding interactions between the triple helix-forming oligonucleotide and the purines of Watson–Crick base pairs. However, they differ not only according to the nucleotide composition of the triple helix-forming oligonucleotide which binds to a polypurine·polypyrimidine sequence of DNA double helix, but also by the isomorphism of base triplets.

A motif is considered as isomorphous when upon superimposition of the corresponding C1′ atoms of Watson–Crick base pairs ($X_1 \cdot Y_1$ and $X_2 \cdot Y_2$) of two constituent base triplets, $X_1 \cdot Y_1 x Z_1$ and $X_2 \cdot Y_2 x Z_2$, the position of the C1′ atoms of the third nucleotides (Z_1 and Z_2) are coincident. This isomorphism ensures a regular backbone of the third strand along the major groove of the underlying double-helical DNA regardless of its sequence. Otherwise, the third strand will zigzag along the major groove of DNA, reflecting the backbone distortion due to the non-isomorphism of base triplets. This is detrimental to the stability of the triple helix.

In the (C,T)-motif, two canonical base triplets, C·GxC+ and T·AxT, are isomorphous. But in the (G,A)-motif, the base triplets, C·GxG and T·AxA in *reverse Hoogsteen*-like configuration, are not isomorphous (Figure 1, lower part). Therefore, backbone distortion of the (G,A)-containing third strand is expected to occur at each step between a C·GxG and a T·AxA base triplet.

In the (G,T)-motif, both C·GxG and T·AxT base triplets can adopt either *Hoogsteen* or *reverse Hoogsteen*-like configurations (see Figure 1). In both cases, these two base triplets are not isomorphous, but the extent of non-isomorphism is greater in the *Hoogsteen* than in the *reverse Hoogsteen*-like configuration (Figure 1, lower part). Hence an expected backbone distortion of the third strand will be more pronounced in the *Hoogsteen* configuration. Energy-minimization studies suggest that *Hoogsteen* configuration is energetically preferred in both $(T \cdot AxT)_n$ and $(C \cdot GxG)_n$ homopolymeric triple helices, in agreement with spectroscopic studies [13,14]. However, the energy penalty associated with backbone distortion for mixed (G,T)-containing third strand is higher for *Hoogsteen* than for *reverse Hoogsteen*-like configurations. Therefore, the orientation of the (G,T)-containing

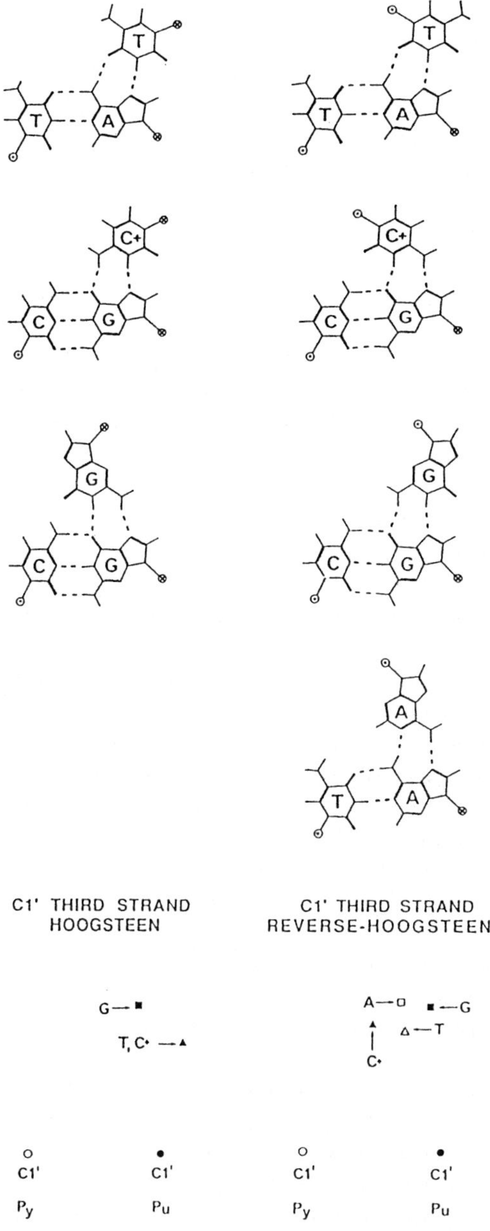

Fig. 1. *Upper part*: Hydrogen-bonding interactions in base triplets described in the text. The left and right columns correspond to *Hoogsteen* and *reverse Hoogsteen*-like configurations, respectively. The orientation of the third strand with respect to the polypurine strand is indicated by a cross (parallel orientation) or a dot (antiparallel orientation) at the position of the C1' atom of the third strand. All nucleotides adopt the *anti* glycosidic conformation. *Lower part*: Position of the C1' atoms of third strand with respect to the C1' atoms of the Watson–Crick pyrimidine(**Py**)·purine(**Pu**) base pairs (T·A or C·G). In *Hoogsteen* configuration (left), the following symbols are used: filled triangle for T and C+, filled square for G. In *reverse Hoogsteen* configuration (right): open triangle for T, filled triangle for C+, open square for A, and filled square for G. Isomorphism is only observed for the T·AxT and C·GxC+ base triplets in *Hoogsteen* configuration.

third strands is expected to shift from parallel to antiparallel orientation when the number of $^{5'}ApG^{3'}$ and/or $^{5'}GpA^{3'}$ steps increases in the target sequence; indeed, this has been experimentally observed [12]. Further energy-minimization studies suggest that the antiparallel orientation is preferred for a 10-base triplet triple helix composed of 50% C·GxG and 50% T·AxT base triplets when there are more than three $(^{5'}ApG^{3'} + {}^{5'}GpA^{3'})$ steps in the target sequence, but that the parallel orientation is preferred otherwise (J. S. Sun, results to be published elsewhere).

It now clearly appears that both the isomorphism of base triplets and the replacement of the protonated C·GxC+ base triplets are two key factors which should be taken into account in the design of base triplets with natural and/or modified nucleotides in order to form stable triple helices under physiological conditions [8,15].

Despite all these efforts, the so-called *antigene strategy*, based on the intermolecular formation of triple helices, still suffers an inherent restriction in that the DNA target sequence should be a relatively long stretch of polypurine·polypyrimidine double helix (about 20 to 30 base pairs) in order to ensure a stable triple helix formation under physiological conditions. Substantial progress has been made in extending the range of double-helical DNA recognition *via* triple helix-forming oligonucleotides. There are two main classes of DNA target sequences for forming triple helices which differ from the canonical polypurine·polypyrimidine tract: (i) the polypurine·polypyrimidine sequence is interrupted by a single or double base pair inversion; (ii) the target sequence is composed of adjacent and alternating oligopurine·oligopyrimidine tracts.

In the first case, triple helices can still be formed by using an appropriate natural or modified nucleotide at the site of a single base pair inversion which is considered as a single mismatch of base triplets. The rules for skipping a single mismatch with a limited loss of affinity are now well established [16–20], mostly in the (C,T)-motif triple helix. However, the stability of a single mismatch is highly dependent on its nearest-neighbors [18]. A double base pair inversion can also be overcome by virtue of the dimerization of two triple helix-forming oligonucleotides [21].

In the second case, several short triple helices can be formed in which the third strand oligonucleotides are bound to the corresponding mini-oligopurine·oligopyrimidine tracts, all these mini-third strand oligonucleotides can be linked together as an entire third strand oligonucleotide in order to achieve cooperative binding. In this way, such an oligonucleotide binds to the target DNA composed of adjacent and alternating oligopurine·oligopyrimidine tracts, and zigzags along the major groove, switching from one oligopurine strand to another at the $^{5'}$purine-pyrimidine$^{3'}$, $^{5'}$pyrimidine-purine$^{3'}$ step (hereafter designated to as $^{5'}RpY^{3'}$ or $^{5'}YpR^{3'}$ junction, respectively). This work is devoted to the development of this so-called *switched triple helix* formation.

Two distinct approaches have been described to form a switched triple helix: (i) the mini-triple helices are formed by motifs involving the same hydrogen-bonding interactions (*Hoogsteen* or *reverse Hoogsteen*-like configuration), usually the (C,T)-motif [22–24]; (ii) different motifs are used to form mini-triple helices involving alternating *Hoogsteen* and *reverse Hoogsteen*-like configurations [12,25–27].

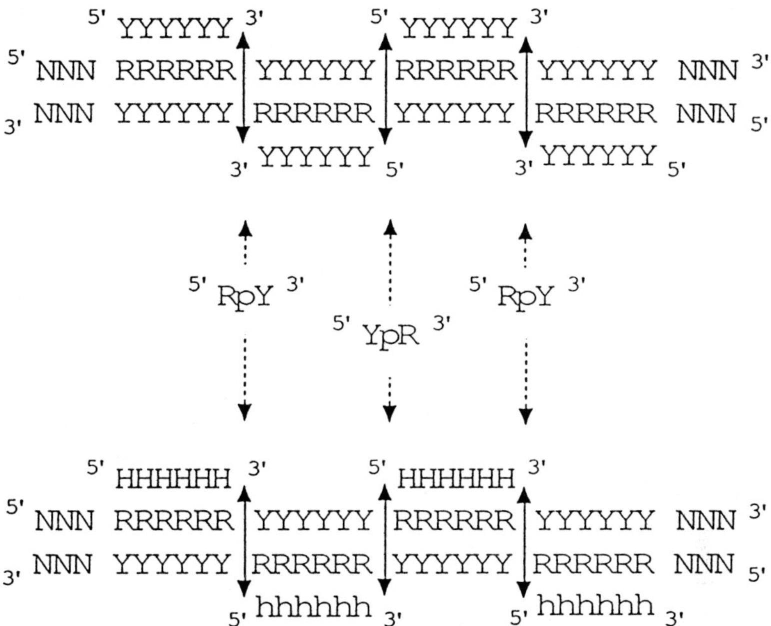

Fig. 2.   Principles of the formation of switched triple helix. The following symbols are used: R, Y and N stand for purine, pyrimidine and any nucleotide, respectively; H and h indicate that the nucleotides in the third strand are engaged in either *Hoogsteen*-like or *reverse Hoogsteen*-like hydrogen-bonding interactions, respectively. The target DNA sequence is composed of alternate and adjacent oligopurine·oligopyrimidine fragments. Two types of junctions occur: $^{5'}$RpY$^{3'}$ and $^{5'}$YpR$^{3'}$. Two approaches are possible to form a switched triple helix: (i) *Upper part*: the mini-triple helices are formed by motifs involving the same hydrogen-bonding interactions, i.e. *Hoogsteen* configuration, as in the (C,T)-motif triple helix; and (ii) *Lower part*: different motifs are used to form mini-triple helices involving alternating *Hoogsteen*-like and *reverse Hoogsteen*-like configurations.

In the first approach, a mini-oligopurine strand runs antiparallel to its neighbor mini-oligopurine strands which are located on the Watson–Crick complementary strands in the target sequence, and the orientation of the third strand is always parallel with respect to the oligopurine strands. Thus the 3′ or 5′ ends of two neighboring mini-third strand oligonucleotides meet at the $^{5'}$RpY$^{3'}$ or $^{5'}$YpR$^{3'}$ junction, respectively (Figure 2, upper part). Consequently, the use of a suitable linker is necessary to tether together these mini-triple helix-forming oligonucleotides at the junction steps.

By contrast, the second approach consists of combining alternately different triple helix motifs involving either *Hoogsteen* or *reverse Hoogsteen*-like configuration. Since the orientation of a triple helix motif in *Hoogsteen*-like configuration is always parallel with respect to its target oligopurine strand, and antiparallel for any other motif in *reverse Hoogsteen*-like configuration, the 5′ (or 3′) end of a mini-third strand oligonucleotide encounters the 3′ (or 5′) end of its neighbor third strand oligonucleotide at the junction (Figure 2, lower part). Thus, 'artificial' linkers are not required at the junction of mini-oligopurine·oligopyrimidine tracts, provided the conformational constraints are properly removed. Consequently, a standard oligonucleotide which can easily be synthesized without any further chemical modification, can bind on such an extended target sequence by formation of switched triple helix.

There are two types of junctions, i.e. $^{5'}RpY^{3'}$ and $^{5'}YpR^{3'}$. Within each type of junction, there are only three distinct junction steps. They are $^{5'}ApT^{3'}$, $^{5'}GpC^{3'}$ and $^{5'}GpT^{3'}$ (which is equivalent to $^{5'}ApC^{3'}$) for the $^{5'}RpY^{3'}$ junction, $^{5'}TpA^{3'}$, $^{5'}CpG^{3'}$ and $^{5'}TpG^{3'}$ (which is also equivalent to $^{5'}CpA^{3'}$) for the $^{5'}YpR^{3'}$ junction. A comprehensive knowledge of how to handle all these six junction steps correctly would theoretically open the door for extension of the DNA recognition by triple helix-forming oligonucleotides. This attractive approach, which merits a thorough examination, is the subject of the present modeling study which is aimed at optimizing the cooperative binding of the so-called switched triple helix-forming oligonucleotide by using appropriate nucleotide(s) at the junction between two adjacent mini-triple helices. This study has allowed us to propose a *switch code* which describes the rules for forming switched triple helices. Practical applications to extend DNA recognition by this new generation of molecularly tailor-made triple helix-forming oligonucleotide will be discussed.

Methodology

Molecular modeling by conformational energy-minimization was carried out using the *JUMNA* (version 7) program package [28]. Neither water nor positively charged counterions were explicitly included in the energy minimization. However, their effects were simulated by a sigmoidal, distance-dependent, dielectric function [29] and by assignment of a half negative charge for each phosphate group. Computations were carried out on a Silicon Graphics 4D/420GTXB dual processor workstation.

The coordinates of triple helices were derived from the previously published B-like triple helix [13,14] which is now widely supported by many NMR and vibrational spectroscopic studies. Typically, the triple helices were 10 base triplets in length with the junction step occuring at the center. The three last base triplets at both ends were restrained to a mononucleotide symmetry in order to decrease the end effects and to focus on the effect at the junction (the central four base triplets). A manual docking was sometimes required around the junction in order to establish an appropriate interaction and to avoid strong steric clashes during initial steps of minimization. The total complexation energy (E_{TOT}) was decomposed in terms of intermolecular interactions ($E_{DH\text{-}III}$) between the third strand (III) and the target double-helical DNA (DH), as well as the conformational deformation energy of the double helix (ΔE_{DH}) and the third strand (ΔE_{III}). Conformational deformation energies were evaluated as the difference between the corresponding energetic components before and after complexation of the third strand. It should be noted that this evaluation is approximative, especially for that of the third strand since the conformation of a free single-stranded DNA is generally less well defined than a free double-stranded DNA. However, these approximations were necessary to take into account the effect of base composition.

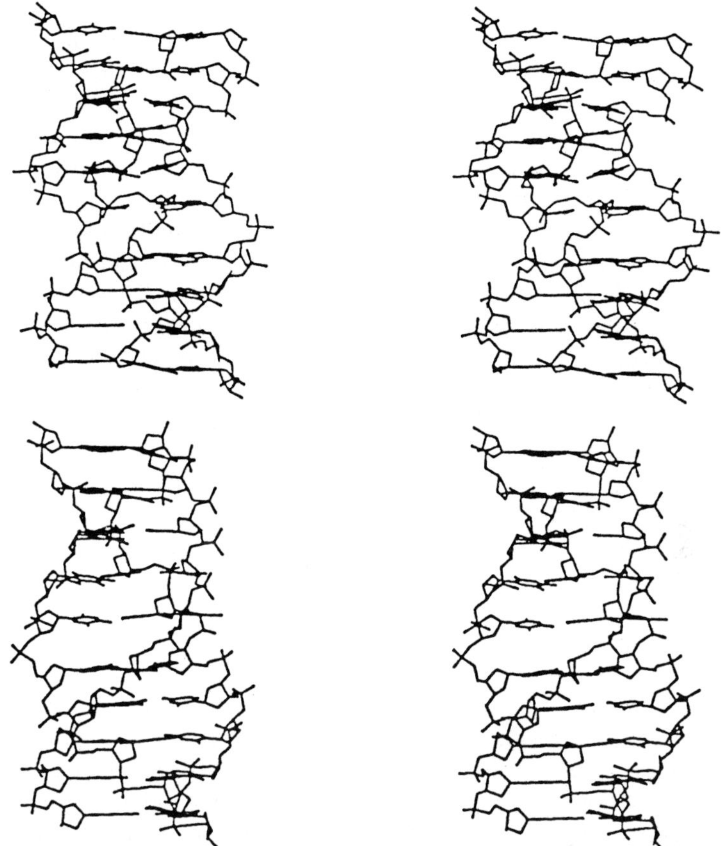

Fig. 3. Stereoview of switched triple helices at the $^{5'}\text{GpT}^{3'}/^{5'}\text{ApC}^{3'}$ (upper) and at the $^{5'}\text{TpG}^{3'}/^{5'}\text{CpA}^{3'}$ (lower) junctions. The triple helices have been partially energy-minimized around the junctions. Hydrogen atoms are not drawn for clarity. They illustrate the conformational characteristics at the $^{5'}\text{RpY}^{3'}$ and $^{5'}\text{YpR}^{3'}$ junctions (see text for details).

Results and Discussion

Molecular modeling reveals that the conformational requirement for spanning over the $^{5'}\text{RpY}^{3'}$ type of junctions is distinctly different to that of the $^{5'}\text{YpR}^{3'}$ type of junction. Figure 3 illustrates the conformational difference between these two types of junctions. Different approaches were used to cope with these conformational requirements.

In $^{5'}\text{RpY}^{3'}$ type of junction, the 5' and 3' ends of two mini-triple helices can be easily joined together by using a single phosphodiester unit. Commonly, a strong zigzag of the third strand occurs at the junction, as illustrated by the $^{5'}\text{GpT}^{3'}$ junction (Figure 3, upper part) in which the switched triple helix is formed by combination of the T·AxT base triplet in *Hoogsteen* configuration and the C·GxG base triplet in *reverse Hoogsteen*-like configuration. The extent of backbone distortion of the third strand at the junction depends on the constituent base triplets on both sides, particularly the relative postion of the C1' atoms in the third strand at the junction. Generally, the backbone distortion can be significantly reduced by removing one appropriate nucleotide at the junction (see below).

In contrast to the $^{5'}$RpY$^{3'}$ steps, the 5' and 3' ends of two mini-triple helices are almost too far away in the $^{5'}$YpR$^{3'}$ junction steps, as illustrated by the $^{5'}$TpG$^{3'}$ junction (Figure 3, lower part) in which the switched triple helix is formed by combination of the C·GxG base triplets in *Hoogsteen* configuration and the T·AxT base triplets in *reverse Hoogsteen*-like configuration. It can be seen that the backbone is under stretching distortion. Therefore, additional nucleotides are required to facilitate the switching over of the third strand at the $^{5'}$YpR$^{3'}$ junction step. The guideline of the molecular design was aimed at simultaneously bridging the gap between two adjacent mini-triple helices by using the backbone of the additional nucleotide(s) as a linker, and at seeking suitable hydrogen bonds between the base moiety of the additional nucleotide and the base triplets at the junction in order to consolidate the junction and to increase the specificity.

I. $^{5'}$RpY$^{3'}$ TYPE OF JUNCTIONS

I.1. $^{5'}$ApT$^{3'}$ Junction Step

In Table I, all triplexes result from the combination of the T·AxT base triplet in *Hoogsteen* configuration as encountered in a canonical, parallel, (C,T)-motif triple helix on the left half of the target sequence, and the T·AxA or T·AxT base triplet in *reverse Hoogsteen*-like configuration, which can be either in an antiparallel (G,A)- or (G,T)-motif triple helix on the right half of the target sequence.

For the combination of the (C,T)- and (G,A)-motifs (triplexes AT1-AT3), the total complexation energies (see Methodology section for details of energy decomposition) are improved by removing one nucleotide at either side of the junction (AT2 and AT3), as compared to that of the full length triplex (AT1). In addition, it seems that the removal of one adenosine in *reverse Hoogsteen*-like configuration is better than one thymidine in *Hoogsteen* configuration (AT2 *versus* AT3). This preference is mainly ascribed to stronger interactions between the third strand and the target double-helical DNA of the triplex AT2 than that of the triplex AT3, since a T·AxT base triplet in *Hoogsteen* configuration has a lower conformational energy (more stable) than that of a T·AxA base triplet in *reverse Hoogsteen*-like configuration (Sun, unpublished results).

It can be seen that a complete destacking of the base triplets following upon a strong backbone distortion at the junction, occurs for the full length triplex, AT1 (Figure 4, upper part), while the removal of one adenine at the junction recovered partially base stacking (Figure 4, lower part).

When a combination of the (C,T)- and (G,T)-motifs (triplexes AT4-AT6) was considered, all triplexes have quite similar complexation energies. However, there is a slight tendency in favor to the triplexes in which one thymidine was removed at either side of the junction.

TABLE I

Analysis of the complexation energies of the energy-minimized switched triple helices at the $^{5'}$ApT$^{3'}$ junction step (see text for details). The underlined letters schematize those nucleotides interacting with the purine strand in the target sequence through *reverse Hoogsteen*-like hydrogen bonds. The hyphen (-) symbolizes a standard phosphodiester unit. Energies are given in kcal mole^{-1} (1 kcal = 4.18 kJ) unit

$5'$ApT$^{3'}$ junction :	$5'$AAAAATTTTT$3'$ $\quad$ $3'$TTTTTAAAAA$_{5'}$				
Triplex	Strand III	ΔE_{DH}	$E_{DH\text{-}III}$	ΔE_{III}	E_{TOT}
AT1	$5'$ TTTTT<u>AAAAA</u>$^{3'}$	+48	-165	+52	-65
AT2	$5'$ TTTTT<u>-AAAA</u>$^{3'}$	+32	-157	+34	-91
AT3	$5'$ TTTT<u>-AAAAA</u>$^{3'}$	+28	-148	+31	-89
AT4	$5'$ TTTTT<u>TTTTT</u>$^{3'}$	+32	-161	+56	-73
AT5	$5'$ TTTTT<u>-TTTT</u>$^{3'}$	+19	-150	+57	-74
AT6	$5'$ TTTT<u>-TTTTT</u>$^{3'}$	+16	-150	+59	-75

I.2. $^{5'}$GpC$^{3'}$ Junction Step

In the Table II, the triplexes GC1-GC4 result from the combination of two C·GxG base triplets, one in *Hoogsteen*-like configuration, and another in *reverse Hoogsteen*-like configuration. Therefore, the first C·GxG base triplet is involved in a parallel (G,T)-motif triple helix on the left half of the target sequence, while the second C·GxG base triplet can either be in an antiparallel (G,T)- or (G,A)-motif triple helix on the right half of the target sequence.

When one guanosine in the third strand was removed at the $^{5'}$GpC$^{3'}$ junction step, either in *reverse Hoogsteen* or *Hoogsteen*-like part as in the triplexes GC3 and GC4, respectively, their complexation energies (E_{TOT}) are lower than that of the full-length triplex, GC1, due to a greater reduction of deformation energies in the double-helical DNA (ΔE_{DH}) and in the third strand (ΔE_{III}) than that of intermolecular interaction ($E_{DH\text{-}III}$) between the target DNA and the third strand. In addition, as previously observed in the $^{5'}$ApT$^{3'}$ junction step, the removal of one guanosine in *reverse Hoogsteen*-like (triplex GC3) is better than that in *Hoogsteen*-like configuration (triplex GC4).

However, it appears that the full-length triplex GC2 is energetically preferred, in which the guanine located at the $^{5'}$GpC$^{3'}$ junction in the motif involving *reverse Hoogsteen*-like hydrogen bonds (in the right half part of the target sequence), adopts a *syn* glycosidic conformation. Thus, this C·GxG base triplet is actually in

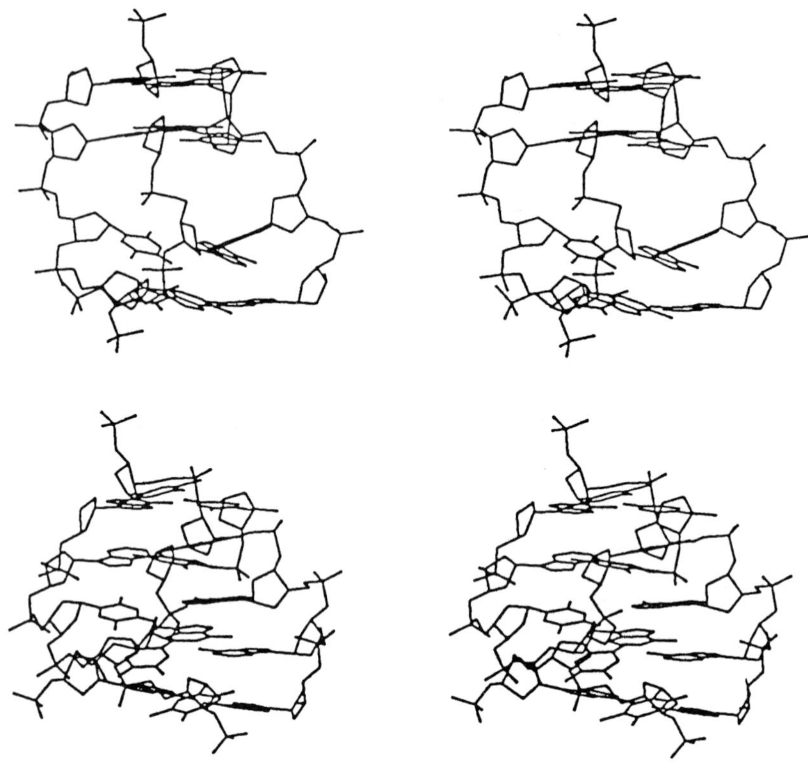

Fig. 4.   Stereoview of energy-minimized switched triple helices at the $^{5'}$ApT$^{3'}$ junction step. *Upper part*: the triplex AT1 made by the juxtaposition of the canonical T·AxT base triplets in *Hoogsteen* (above the junction) and the T·AxA base triplets in *reverse Hoogsteen*-like (below the junction) configurations. *Lower part*: the triplex AT2 in which the first adenosine in the previously *reverse Hoogsteen*-like part (see the triplex AT1) was removed. Hydrogen atoms are not drawn for clarity (see text for details).

Hoogsteen-like configuration. By doing so, a significant reduction of deformation energies with a limited loss in intermolecular interaction, as compared to that of the corresponding triplex GC1, is obtained.

Figure 5 depicts the switching over of the third strand at the $^{5'}$GpC$^{3'}$ junction step for triplexes GC1 and GC2. It can be seen that the backbone distortion is reduced by flipping over one guanine from *anti* to *syn* glycosidic conformation in the previously *reverse Hoogsteen*-like part of the junction (GC2 *versus* GC1). In doing so, the base stacking and the hydrogen-bonding are much well preserved at the junction, as revealed by a detailed energy analysis (not shown).

The triplexes GC5-GC8 are variants of the triplex GC1 but one or two C·GxC+ base triplets are introduced at the junction by replacement of the corresponding C·GxG base triplets in either *Hoogsteen* or *reverse Hoogsteen*-like configuration. It appears that the replacement of one C·GxG base triplet by a C·GxC+ base triplet at the junction, both in *reverse Hoogsteen* configuration (triplex GC5), is the most beneficial.

TABLE II

Analysis of the complexation energies of the energy-minimized switched triple helices at the $^{5'}$GpC$^{3'}$ junction step (see text for details). The underlined letters schematize those nucleotides interacting with the purine strand in the target sequence through *reverse Hoogsteen*-like hydrogen bonds. The hyphen (-) symbolizes a standard phosphodiester unit. The **bold** letter shows that nucleotide adopts a *syn* glycosidic conformation. Energies are given in kcal mole^{-1} (1 kcal = 4.18 kJ) unit

$5'$GpC$3'$ junction :	$^{5'}$GGGGGCCCCC$^{3'}$				
	$_{3'}$CCCCCGGGGG$_{5'}$				
Triplex	Strand III	ΔE_{DH}	E_{DH-III}	ΔE_{III}	E_{TOT}
GC1	$^{5'}$GGGGG<u>GGGGG</u>$^{3'}$	+58	-268	+83	-127
GC2	$^{5'}$GGGGGG<u>GGGG</u>$^{3'}$	+50	-263	+72	-141
GC3	$^{5'}$GGGGG-<u>GGGG</u>$^{3'}$	+47	-245	+66	-132
GC4	$^{5'}$GGGG-<u>GGGGG</u>$^{3'}$	+48	-246	+69	-129
GC5	$^{5'}$GGGGG<u>**C**GGGG</u>$^{3'}$	+47	-294	+80	-167
GC6	$^{5'}$GGGGG<u>**CC**GGG</u>$^{3'}$	+56	-320	+122	-142
GC7	$^{5'}$GGGG<u>**C**GGGGG</u>$^{3'}$	+65	-297	+93	-139
GC8	$^{5'}$GGGG<u>**CC**GGGG</u>$^{3'}$	+79	-331	+131	-121

Another possibility consists of combining the C·GxC+ base triplets in a (C,T)-motif triple helix in *Hoogsteen* configuration and the C·GxG base triplets in a (G,T)- or (G,A)-motif triple helix in *reverse Hoogsteen*-like configuration. However, the solution is the same as that for the $^{5'}$GpT$^{3'}$ junction (see below), since the C·GxC+ base triplet is isomorphous to the T·AxT base triplet in *Hoogsteen* configuration.

I.3. $^{5'}$GpT$^{3'}$ or $^{5'}$ApC$^{3'}$ Junction Step

A combination of the C·GxG base triplet in *Hoogsteen*-like and other T·AxA or T·AxT base triplet in *reverse Hoogsteen*-like configuration can be used. The first base triplets take part in a parallel (G,T)-motif triple helix on the left half of the target sequence, and the second base triplets can form either an antiparallel (G,A)-(G,T)-motif triple helix on the right half of the target sequence. Similar to the $^{5'}$ApT$^{3'}$ junction step, the most efficient way for switching over a $^{5'}$GpT$^{3'}$ or a $^{5'}$ApC$^{3'}$ junction step, consists of removing an adenosine or a thymidine at the

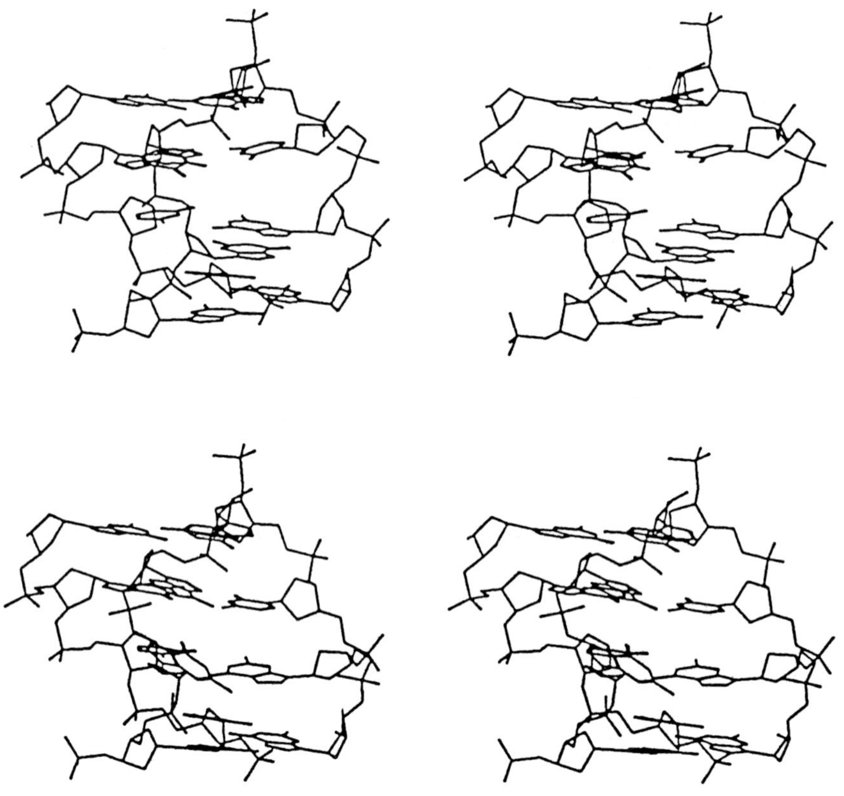

Fig. 5.   Stereoview of energy-minimized switched triple helices at the $^{5'}$GpC$^{3'}$ junction step. *Upper part*: the triplex GC1 made of the juxtaposition of the C·GxG base triplets in *Hoogsteen*-like (above the junction) or in *reverse Hoogsteen*-like (below the junction) configurations. *Lower part*: the triplex GC2 in which the first guanine in the previously *reverse Hoogsteen*-like part (see the triplex GC1) was flipped over from *anti* to *syn* glycosidic confomation. Hydrogen atoms are not drawn for clarity (see text for further detail).

reverse Hoogsteen side of the junction (triplexes GT2 and GT5 in Table III, respectively).

An alternative option is to combine the T·AxT base triplet in *Hoogsteen* configuration, and the C·GxG base triplet in *reverse Hoogsteen*-like configuration. An example would be a (C,T)-motif triple helix on the right half of the target sequence joined to an antiparallel (G,A)- or (G,T)-motif triple helix on the left half of the target sequence (see the triplexes GT7-GT10 in the Table III). The energetically preferred triplex involves a guanine adopting a *syn* glycosidic conformation (in the triplex GT8), exactly like the similar combination for coping with $^{5'}$GpC$^{3'}$ junction step (Table II, triplex GC2).

TABLE III

Analysis of the complexation energies of the energy-minimized switched triple helices at $^{5'}GpT^{3'}/^{5'}ApC^{3'}$ junction step (see text for details). The underlined letters schematize those nucleotides interacting with the purine strand in the target sequence through *reverse Hoogsteen*-like hydrogen bonds. The hyphen (-) symbolizes a standard phosphodiester unit. The bold letter shows that nucleotide adopts a *syn* glycosidic conformation. Energies are given in kcal mole^{-1} (1 kcal = 4.18 kJ) unit

$^{5'}GpT^{3'}$ or $^{5'}ApC^{3'}$ junction :	$^{5'}GGGGGTTTTT^{3'}$ $_{3'}CCCCCAAAAA_{5'}$				
Triplex	Strand III	ΔE_{DH}	E_{DH-III}	ΔE_{III}	E_{TOT}
GT1	$^{5'}$GGGGG<u>AAAAA</u>$^{3'}$	+61	-221	+58	-102
GT2	$^{5'}$GGGGG-<u>AAAA</u>$^{3'}$	+29	-212	+35	-148
GT3	$^{5'}$GGGG-<u>AAAAA</u>$^{3'}$	+35	-197	+31	-131
GT4	$^{5'}$GGGGG<u>TTTTT</u>$^{3'}$	+60	-232	+79	-92
GT5	$^{5'}$GGGGG-<u>TTTT</u>$^{3'}$	+27	-203	+59	-116
GT6	$^{5'}$GGGG-<u>TTTTT</u>$^{3'}$	+25	-189	+64	-100
GT7	$^{5'}$TTTTT<u>GGGGG</u>$^{3'}$	+43	-207	+68	-97
GT8	$^{5'}$TTTTT**G**<u>GGGG</u>$^{3'}$	+43	-207	+59	-105
GT9	$^{5'}$TTTTT-<u>GGGG</u>$^{3'}$	+57	-191	+61	-73
GT10	$^{5'}$TTTT-<u>GGGGG</u>$^{3'}$	+39	-191	+61	-92

II. $^{5'}YpR^{3'}$ TYPE OF JUNCTIONS

II.1. $^{5'}TpA^{3'}$ Junction Step

As mentioned above, in contrast to the $^{5'}RpY^{3'}$ type of junctions, additional nucleotides are required to facilitate the spanning over of the third strand at the $^{5'}YpR^{3'}$ type of junctions. For the $^{5'}TpA^{3'}$ junction step (Table IV), the solution consists of combining the T·AxA (in the triplexes TA1-TA6) or T·AxT base triplet (in the triplexes TA7-TA12) in *reverse Hoogsteen*-like configuration, in an antiparallel (G,A)- or (G,T)-motif triple helix formed on the left half of the target sequence, and the canonical T·AxT base triplet in *Hoogsteen* configuration, in a parallel (C,T)-motif triple helix on the right half of the target sequence.

In general, the addition of an extra nucleotide provides lower complexation

                                     JIAN-SHENG SUN

TABLE IV

Analysis of the complexation energies of the energy-minimized switched triple helices at the $^{5'}$TpA$^{3'}$ junction step (see text for details). The underlined letters schematize those nucleotides interacting with the purine strand in the target sequence through *reverse Hoogsteen*-like hydrogen bonds. The *italic* letters indicate that those nucleotides are primarily used as linkers to bridge the gap between *Hoogsteen* and *reverse Hoogsteen*-like parts of the third strand. Energies are given in kcal mole^{-1} (1 kcal = 4.18 kJ) unit

$^{5'}$TpA$^{3'}$ junction : 5′ TTTTTAAAAA3′

3′ AAAAATTTTT5′

Triplex	Strand III	ΔE_{DH}	E_{DH-III}	ΔE_{III}	E_{TOT}
TA1	5′ <u>AAAAA</u>TTTTT3′	+56	-173	+36	-81
TA2	5′ <u>AAAAA</u>*A*TTTTT3′	+51	-179	+31	-97
TA3	5′ <u>AAAAA</u>*G*TTTTT3′	+51	-204	+35	-118
TA4	5′ <u>AAAAA</u>*C*TTTTT3′	+43	-181	+27	-111
TA5	5′ <u>AAAAA</u>*U*TTTTT3′	+27	-169	+49	-93
TA6	5′ <u>AAAAA</u>*TTTTTT*3′	+37	-180	+45	-98
TA7	5′ <u>TTTTT</u>TTTTT3′	+59	-180	+60	-61
TA8	5′ <u>TTTTT</u>*A*TTTTT3′	+31	-192	+75	-86
TA9	5′ <u>TTTTT</u>*G*TTTTT3′	+35	-209	+77	-97
TA10	5′ <u>TTTTT</u>*C*TTTTT3′	+39	-191	+65	-87
TA11	5′ <u>TTTTT</u>*U*TTTTT3′	+43	-191	+76	-72
TA12	5′ <u>TTTTT</u>*TTTTTT*3′	+34	-187	+93	-60

energy as compared to that without any extra nucleotide. The chemical nature of the added nucleotide plays an important role. Guanosine turns to be the best candidate for each case (triplexes TA3 and TA9, respectively), while the second best is cytidine (triplexes TA4 and TA10, respectively). These improvements are mainly ascribed to a stronger intermolecular interaction between the third strand and the target DNA double helix, than that of the triplexes without any extra nucleotide (TA1 and TA7, respectively).

Figure 6 reveals the hydrogen-bonding interactions between the 4-carbonyl

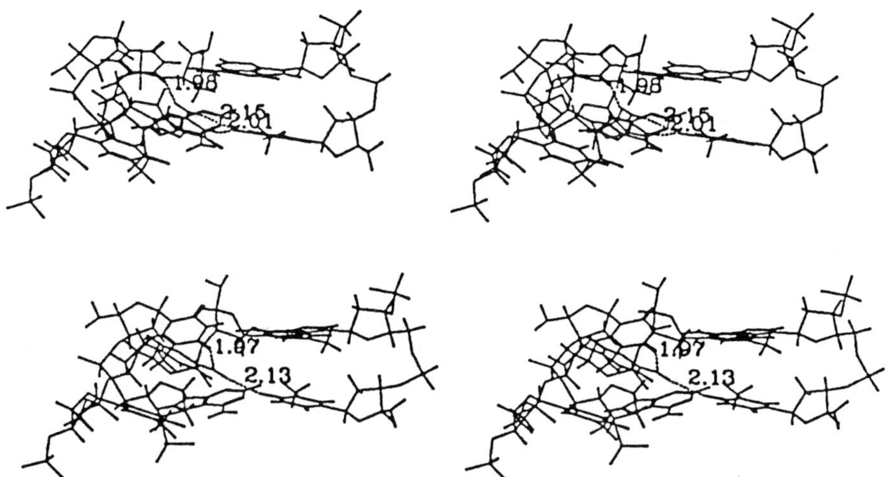

Fig. 6.   Stereoview of energy-minimized switched triple helices at the $^{5'}$TpA$^{3'}$ junction step. *Upper part*: the triplex TA3, in which the extra guanosine forms hydrogen bonds with the 4-carbonyl group of both thymidines located at both parts of the junction (see the text for details). *Lower part*: the triplex TA10, in which the extra cytidine also forms a pair of hydrogen bonds with the 4-carbonyl group of both thymidines located at both parts of the junction (see text for details).

group of the thymidines located at both sides of the junction and the 2-amino and 1-imino groups of the extra guanosine, as highlighted in the triplex TA3 (Figure 6, upper part). There are also similar hydrogen-bonding interactions between the 4-carbonyl group of the thymidines located at both sides of the junction and the 4-amino group of the extra cytidine, as illustrated in the triplex TA10 (Figure 6, lower part).

Therefore, it appears that an appropriate nucleotide (guanosine or cytidine) does play a dual role, as expected from the molecular designing. First, its backbone acts as a linker joining together *Hoogsteen* and *reverse Hoogsteen* parts of the third strand. Second, its base is engaged in a rather specific interaction with the DNA substrate at the junction level.

II.2. $^{5'}$CpG$^{3'}$ Junction Step

In Table V, a combination of two C·GxG base triplets has been considered. The first one is in *reverse Hoogsteen*-like configuration which can be involved either in an antiparallel (G,T)- or (G,A)-motif triple helix on the left half of the target sequence, while the second one is in *Hoogsteen*-like configuration, in a parallel (G,T)-motif triple helix on the right half of the target sequence.

The triplexes CG2-CG6 contain an extra nucleotide (A, G, U, T and C, respectively) at the $^{5'}$CpG$^{3'}$ junction step. All of them have lower complexation energies than that of the triplex CG1 without any extra nucleotide, mainly due to a stronger intermolecular interactions between the third strand and the DNA substrate. In the triplex CG7, two cytidines were introduced at the junction. It

TABLE V

Analysis of the complexation energies of the energy-minimized switched triple helices at the $^{5'}CpG^{3'}$ junction step (see text for details). The underlined letters schematize those nucleotides interacting with the purine strand in the target sequence through *reverse Hoogsteen*-like hydrogen bonds. The *italic* letters indicate that those nucleotides are primarily used as linkers to bridge the gap between *Hoogsteen* and *reverse Hoogsteen*-like parts of the third strand. Energies are given in kcal mole^{-1} (1 kcal = 4.18 kJ) unit

| $^{5'}CpG^{3'}$ junction : | $^{5'}CCCCCGGGGG^{3'}$ | | | | |
| | $^{3'}GGGGGCCCCC_{5'}$ | | | | |
Triplex	Strand III	ΔE_{DH}	$E_{DH\text{-}III}$	ΔE_{III}	E_{TOT}
CG1	$^{5'}$ <u>GGGGG</u>GGGGG$^{3'}$	+43	-284	+76	-165
CG2	$^{5'}$ <u>GGGGG</u>*A*GGGGG$^{3'}$	+67	-333	+69	-197
CG3	$^{5'}$ <u>GGGGG</u>GGGGGG$^{3'}$	+64	-322	+85	-173
CG4	$^{5'}$ <u>GGGGG</u>*U*GGGGG$^{3'}$	+52	-301	+51	-198
CG5	$^{5'}$ <u>GGGGG</u>*T*GGGGG$^{3'}$	+54	-299	+58	-187
CG6	$^{5'}$ <u>GGGGG</u>*C*GGGGG$^{3'}$	+55	-318	+54	-209
CG7	$^{5'}$ <u>GGGGG</u>*CC*GGGGG$^{3'}$	+59	-303	+53	-191

can be seen that both the nature and the number of the extra nucleotide(s) play important roles. As a matter of fact, the introduction of a single extra cytidine in the third strand at the junction step which acts as a linker, contributes to the most energetically preferred switched triple helix (triplex CG6) as compared to the other single nucleotides (CG2-CG5), while the presence of two extra cytidines gives a less favored triplex (CG7).

The model of the best switched triple helix, CG6, around the $^{5'}CpG^{3'}$ junction step is shown in Figure 7. It can be seen that the phosphodiester backbone of the extra cytidine in the third effectively acts as an appropriate linker bridging the gap between the *Hoogsteen* and *reverse Hoogsteen* parts of the switched triple helix (Figure 7). In addition, the cytosine of this extra nucleotide (*C*) forms hydrogen bonds with the second neighbor C·GxG base triplet in *reverse Hoogsteen*-like configuration. It is interesting to see that this extra cytidine, fulfilling the envisioned design concepts mentioned above, provides not only the necessary length for linking two parts of the switched triple helix-forming oligonucleotide, but also confers a specific recognition of the $^{5'}CpG^{3'}$ junction step.

An alternative way for targeting this $^{5'}CpG^{3'}$ junction is to combine the C·GxC+ base triplet in *Hoogsteen* configuration, as in a (C,T)-motif triple helix on the right

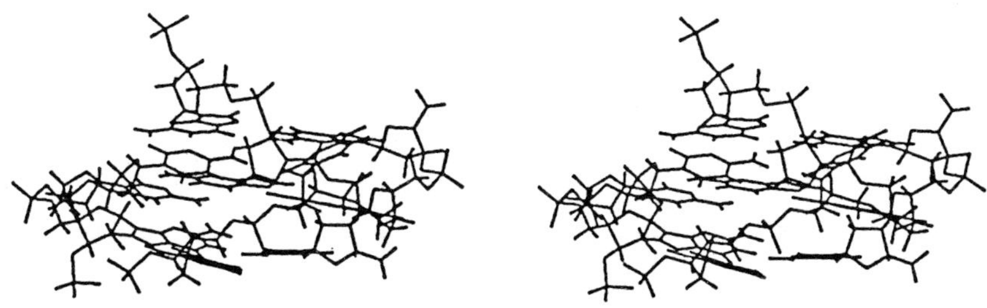

Fig. 7. Stereoview of energy-minimized switched triple helices at the $^{5'}CpG^{3'}$ junction step. The triplex CG6, in which the extra cytidine (*C*) forms hydrogen bonds with the second neighbor C·GxG base triplet in *reverse Hoogsteen*-like configuration at the top of the figure (see text for details).

half of the target sequence, and the C·GxG base triplet in *reverse Hoogsteen*-like configuration, as in an antiparallel (G,A)- or (G,T)-motif triple helix on the left half of the target sequence. However, the solution is the same as that for the $^{5'}TpG^{3'}$ junction (see below), since the C·GxC+ base triplet is isomorphous to the T·AxT base triplet in *Hoogsteen* configuration.

II.3. $^{5'}TpG^{3'}$ or $^{5'}CpA^{3'}$ Junction Step

A combination of the T·AxA or T·AxT base triplet in *reverse Hoogsteen*-like and another C·GxG base triplet in *Hoogsteen*-like configuration can be used. The first base triplet can take part either in an antiparallel (G,A)- or (G,T)-motif triple helix on the left half of the target sequence, and the second base triplet form a parallel (G,T)-motif triple helix on the right half of the target sequence. Similar to the $^{5'}TpA^{3'}$ junction step, the most efficient way for switching over a $^{5'}TpG^{3'}$ or a $^{5'}CpA^{3'}$ junction step, consists of introducing an extra nucleotide, preferentially a guanosine, at the junction (triplexes TG3 and TG9 in Table VI, respectively). However, things are little different.

It turns out that an additional purine (adenosine or guanosine) at the junction (triplexes TG2 and TG3) suffices to give the best switched triple helix-forming oligonucleotide for the first combination (triplexes TG1-TG6). This probably stems from the fact that their base moiety ensures good base stacking between the terminal bases located at both sides of the junction, as suggested by a conformational and energetical examination of the triplexes TG2 and TG3. Instead, there is only one hydrogen bond between the 4-carbonyl group of the thymidine involved in the T·AxA base triplet at the junction and the 2-amino group of the additional guanosine, or 4-amino group of the cytidine involved in the C·GxG base triplet at the junction and the N1 atom of the additional adenosine (result not shown). This probably accounts for the relative lack of specificity in this combination.

For the second combination (triplexes TG7-TG12), the additional guanosine clearly is the best choice, retaining some degree of specificity, which can be explained by the formation of a pair of hydrogen bonds between the 2-amino and 6-carbonyl groups of the extra guanosine, and the 4-carbonyl group of the thymid-

TABLE VI

Analysis of the complexation energies of the energy-minimized switched triple helices at the $^{5'}$TpG$^{3'}$/$^{5'}$CpA$^{3'}$ junction step (see the text for details). The underlined letters schematize those nucleotides interacting with the purine strand in the target sequence through *reverse Hoogsteen*-like hydrogen bonds. The *italic* letters indicate that those nucleotides are primarily used as linkers to bridge the gap between *Hoogsteen* and *reverse Hoogsteen*-like parts of the third strand. Energies are given in kcal mole^{-21} (1 kcal = 4.18 kJ) unit

5'TpG3' or 5'CpA3' junction :	5' TTTTTGGGGG3' 3' AAAAACCCCC5'				
Triplex	Strand III	ΔE_{DH}	$E_{DH\text{-}III}$	ΔE_{III}	E_{TOT}
TG1	5' <u>AAAAA</u>GGGGG3'	+54	-227	+38	-135
TG2	5' <u>AAAAA</u>*A*GGGGG3'	+46	-275	+39	-190
TG3	5' <u>AAAAA</u>*G*GGGGG3'	+49	-278	+42	-187
TG4	5' <u>AAAAA</u>*C*GGGGG3'	+49	-256	+37	-170
TG5	5' <u>AAAAA</u>*U*GGGGG3'	+40	-252	+51	-161
TG6	5' <u>AAAAA</u>*T*GGGGG3'	+40	-253	+50	-163
TG7	5' <u>TTTTT</u>GGGGG3'	+42	-232	+81	-109
TG8	5' <u>TTTTT</u>*A*GGGGG3'	+33	-248	+87	-128
TG9	5' <u>TTTTT</u>*G*GGGGG3'	+19	-246	+85	-142
TG10	5' <u>TTTTT</u>*C*GGGGG3'	+35	-238	+72	-131
TG11	5' <u>TTTTT</u>*U*GGGGG3'	+37	-243	+114	-92
TG12	5' <u>TTTTT</u>*T*GGGGG3'	+29	-241	+100	-112
TG13	5' <u>GGGGG</u>TTTTT3'	+126	-238	+55	-57
TG14	5' <u>GGGGG</u>*T*TTTTT3'	+58	-240	+64	-118
TG15	5' <u>GGGGG</u>*C*TTTTT3'	+53	-237	+55	-129
TG16	5' <u>GGGGG</u>*CC*TTTTT3'	+49	-254	+37	-168
TG17	5' <u>GGGGG</u>*T*TTTTTT3'	+56	-257	+56	-145

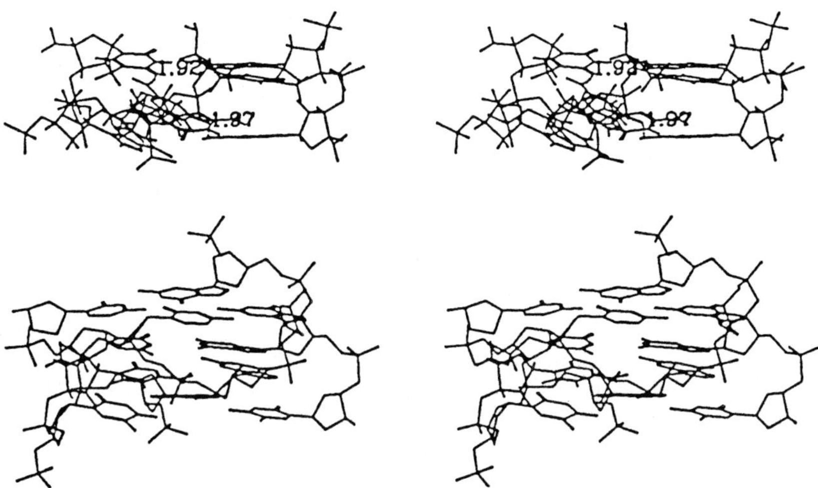

Fig. 8.   Stereoview of energy-minimized switched triple helices at the $^{5'}TpG^{3'}/^{5'}CpA^{3'}$ junction step. *Upper part*: the triplex TG9, in which the extra guanosine forms a pair of hydrogen bonds with the 4-carbonyl group of the thymidine and the 4-amino group of the cytidine, located at the either side of the junction (see text for details). *Lower part*: the triplex TG16, in which two extra cytidines form hydrogen bonds with the two C·GxG base triplets in *reverse Hoogsteen*-like configuration near the junction (at the top of the figure). Hydrogen atoms are not drawn for clarity (see text for further details).

ine and the 4-amino group of the cytidine, both located at the either side of the junction in the T·AxA and C·GxG base triplets, respectively (Figure 8, upper part).

A third combination is to choose the C·GxG base triplet in *reverse Hoogsteen*-like configuration which can be either in an antiparallel (G,A)- or (G,T)-motif triple helix on the left half of the target sequence, and the T·AxT base triplet in *Hoogsteen* configuration, as in a parallel, canonical (C,T)-motif triple helix on the right half of the target sequence. Quite similar to the $^{5'}CpG^{3'}$ junction step, the best solution is to add two cytidines at the junction (in the triplex TG16) which can form hydrogen bonds with the first two C·GxG base triplets in *reverse Hoogsteen*-like configuration near the junction (Figure 8, lower part). In the literature, the use of two additional thymidines as a linker at a conformationally equivalent junctions, has experimentally proven to be much better than the use of only one or no thymidines [25]. The current modeling study is in agreement with the experimental evidence (the triplex TG17 *versus* the triplexes TG14 and TG13).

Finally, an alternative way to target this $^{5'}TpG^{3'}$ junction is to combine the C·GxC+ base triplet in *Hoogsteen* configuration which occurs in a well charactized (C,T)-motif triple helix on the right half of the target sequence, and the T·AxA and T·AxT base triplet in *reverse Hoogsteen*-like configuration, as in an antiparallel (G,A)- or (G,T)-motif triple helix on the left half of the target sequence. As stated before, the solution is the same as that for the $^{5'}TpA^{3'}$ junction (see

Table IV, triplexes TA3 and TA9, respectively), since the C·GxC+ base triplet is isomorphous to the T·AxT base triplet in *Hoogsteen* configuration.

III. SWITCH CODE: A PRACTICAL GUIDE FOR THE FORMATION OF SWITCHED TRIPLE HELIX

To sum up, an exhaustive and systematic theoretical study has been carried out by means of molecular modeling using conformational energy-minimization, in order to optimize the sequences of switched triple helix-forming oligonucleotides for ensuring efficient binding over any of six junction steps.

The unanticipated results from the theoretical computations bears on the conformational complexity at either $^{5'}RpY^{3'}$ and $^{5'}YpR^{3'}$ junctions, depending on the combination of triple helix motifs on either side of the junction. There are many variations such as the removal of one nucleotide at the junction, the juxtaposition of two chosen motifs, as well as the addition of one or even two nucleotides at the junction. It involves different natural nucleotides. It should be noted that the second neighboring base triplets around the junction may also affect the stability of the switched triple helices, as illustrated in the cases of the triplexes CG6 and TG16. The appropriate choice of both the motifs of triple helices and the nucleotide at the junction depends on the experimental conditions.

Mechanistically, the switching over the junction may involve the intramolecular base stacking of the oligonucleotide in the third strand, and the cross-strand hydrogen bonding between the extra nucleotide(s) and the nucleotides on the underlying double-helical DNA located at the junction.

The compilation of the results presented in Tables I–VI constitutes a predictive code called *switch code*. This *switch code* is aimed at describing the rules for choosing the appropriate sequences of the switched triple helix-forming oligonucleotides, thus facilitating *antigene strategy* developments on the more extended target double-helical DNA sequences than the currently required, canonical polypurine·polypyrimidine tract. The experimental work to check the validity of these predictions are currently under way in our laboratory. It is very encouraging that all the predictions which have been so far checked were backed up by experimental evidence (results will be published elsewhere).

Theoretically, the knowledge of the present *switch code* together with that for circumventing a single and/or double base pair inversions will virtually solve the problem of how to recognize, sequence-specifically, the major groove of any DNA sequence by triple helix-forming oligonucleotides. It is wise to note that the major inherent difficulty in practice may come from the formation of intramolecular, hairpin-like, structures of the triple helix-forming oligonucleotides which now contain all four natural nucleotides. However, a compromise could be achieved by choosing a somewhat less favored solution regarding the switch code (i.e. the combination of motifs of triple helices and the nucleotides at the junction) which would prevent the formation of stable self-structures of the oligonucleotides. Thus, the overall binding of the switched triple helix-forming oligonucleotides would be improved after such a global system optimization.

Conclusion

This paper reports on the molecular design of a versatile approach using oligonucleotides to sequence-specifically target the major groove of the double-helical DNA composed of alternating, adjacent fragments of oligopurine·oligopyrimidine tracts, through formation of switched triple helix. The ultimate objective is to find the optimal solution using natural nucleotides for switching over any of six junction steps. A comprehensive set of the best sequences for each junction step has been established. This so-called *switch code* describes the rules for forming stable switched triple helices, which helps to circumvent current limitations on DNA target sequence. It will hopefully help molecular biologists to further develop the new *antigene strategy* for controlling the gene expression.

Acknowledgement

The author wishes to thank Drs. Richard Lavery and Krystyna Zakrewska in Institut de Biologie Physico-Chimique (Paris) for software assistance in molecular modeling, Profs. Thérèse Garestier and Claude Hélène for helpful discussions, and Dr. Gail Silver for careful reading and comments.

References

1. G. Felsenfeld and A. Rich: *Biochim. Biophys. Acta.* **26**, 457 (1957).
2. T. Le Doan, L. Perrouault, D. Praseuth, N. Habhoub, J. L. Decout, N. T. Thuong, J. Lhomme, and C. Hélène: *Nucleic Acids Res.* **15**, 7749 (1987).
3. H. E. Moser and P. B. Dervan: *Science* **238**, 645 (1987).
4. M. Cooney, G. Czernuszewicz, E. H. Postel, S. J. Flint, and M. E. Hogan: *Science* **241**, 456 (1988).
5. L. J. Maher III, B. Wold, and P. B. Dervan: *Science* **245**, 725 (1989).
6. F. Birg, D. Praseuth, A. Zerial, N. T. Thuong, U. Asseline, T. Le Doan, and C. Hélène: *Nucleic Acids Res.* **18**, 2901 (1990).
7. G. Duval-Valentin, N. T. Thuong, and C. Hélène: *Proc. Natl. Acad. Sci. U.S.A.* **89**, 504 (1992).
8. J. S. Sun and C. Hélène: *Curr. Opin. in Struct. Biol.* **3**, 345 (1993).
9. P. A. Beal and P. B. Dervan: *Science* **251**, 1360 (1991).
10. D. S. Pilch, C. Levenson, and R. H. Shafer: *Biochemistry* **30**, 6081 (1991).
11. R. H. Durland, D. J. Kessler, S. Gunnell, M. Duvic, B. M. Pettitt, and M. E. Hogan: *Biochemistry* **30**, 9246 (1991).
12. J. S. Sun, T. de Bizemont, G. Duval-Valentin, T. Montenay-Garestier, and C. Hélène: *C. R. Acad. Sci. Paris, Série III*, **313**, 585 (1991).
13. M. Ouali, R. Letellier, F. Adnet, J. Liquier, J. S. Sun, R. Lavery, and E. Taillandier: *Biochemistry* **32**, 2098 (1993).
14. M. Ouali, R. Letellier, J. S. Sun, A. Adhebat F. Adnet, J. Liquier, and E. Taillandier: *J. Am. Chem. Soc.* **115**, 4264 (1993).
15. J. F. Milligan, S. H. Krawczyk, S. Wadwani, and M. D. Matteucci: *Nucleic Acids Res.* **21**, 327 (1993).
16. L. C. Griffin and P. B. Dervan: *Science* **245**, 967 (1989).
17. J. L. Mergny, J. S. Sun, M. Rougée, T. Montenay-Garestier, F. Barcelo, J. Chomilier, and C. Hélène: *Biochemistry* **30**, 9791 (1991).
18. L. L. Kiessling, L. C. Griffin, and P. B. Dervan: *Biochemistry* **31**, 2829 (1992).
19. K. Yoon, C. A. Hobbs, J. Koch, M. Sardaro, R Kutny, and A. L. Weis: *Proc. Natl. Acad. Sci. U.S.A.* **89**, 3840 (1992).
20. P. A. Beal and P. B. Dervan: *Nucleic Acids Res.* **20**, 2773 (1992).

21. M. D. Distefano and P. B. Dervan: *J. Am. Chem. Soc.* **113**, 5901 (1991).
22. D. A. Horne and P. B. Dervan: *J. Am. Chem. Soc.* **112**, 2435 (1990).
23. A. Ono, C. N. Chen, and L. S. Kan: *Biochemistry* **30**, 9914 (1991).
24. B. C. Froehler, T. Terhorst, J. P. Shaw, and S. N. McCurdy: *Biochemistry* **31**, 1603 (1992).
25. P. A. Beal and P. B. Dervan: *J. Am. Chem. Soc.* **114**, 4976 (1992).
26. S. D. Jayasena and B. Johnston: *Nucleic Acids Res.* **20**, 5279 (1992).
27. S. D. Jayasena and B. Johnston: *Biochemistry* **32**, 2800 (1992).
28. R. Lavery: 'Junctions and Bends of Nucleic Acids: A New Theoretical Modeling Approach', in W. K. Olson, R. H. Sarma, M. H. Sarma and M. Sundaralingham, Eds., *Structure and Expression*. Vol. 3, *DNA Bending and Curvature*, Adenine Press, New York, p. 191 (1988).
29. R. Lavery, H. Sklenar, K. Zakrewska, and B. Pullman: *J. Biomol. Struct. Dyn.* **3**, 989 (1986).

Construction of a DNA Four-Way Junction: Design and NMR Spectroscopy

CORNELIS ALTONA and JEROEN A. PIKKEMAAT
Leiden Institute of Chemistry, Gorlaeus Laboratories, P.O. Box 9502, NL – 2300 Leiden, The Netherlands

Abstract. In 1964 Holliday postulated the formation of cruciform structures (four-way junctions) in duplex DNA as intermediate in genetic recombination. Since then, many biochemical and biophysical investigations were directed at solving questions concerning structural details of stable four-way junctions. Thus far, NMR spectroscopy played a minor part in these investigations on account of the minimum size of the molecule (expressed as the number of nucleotide residues) that was thought necessary to produce a stable cruciform structure. Indeed, the smallest four-way junction studied thus far by NMR methods was built from four separate DNA strands, each containing 16 nucleotides, a total of 64. Obviously, with such a large structure one runs into assignment problems. We considered the possibility of constructing a stable four-way junction from a single strand of DNA. The underlying idea was to make use of our detailed knowledge of the building principles of stable minihairpin loops. These loops, containing only two nucleotides to bridge the gap between antiparallel strands, are maximally stable in DNA sequences like 5′-d(-C-TT-G-), where C and G form a normal Watson–Crick base pair and the two T residues cross the minor groove to form the minihairpin loop. Three of such miniloops could in principle cap three arms of the cruciform. The fourth arm would have an open end. The problem to be solved is to find the minimum length that is required to insure stability of the three closed arms and of the fourth open arm. We were successful with a structure that has three short stems (four base pairs each) and an open-end stem consisting of eight base pairs, a total of 46 residues. NMR experiments, carried out on this molecule in the presence of Mg^{2+}, showed details of folding which have never been observed before.

Key words. DNA four-way junction, ^{1}H-NMR spectroscopy, conformational analysis, hairpin loops.

Introduction

The postulate [1] that branched DNA species are key intermediates in DNA rearrangements, such as genetic recombination, has inspired many researchers to concentrate their studies on the biochemical role of these branches. Much has been learned about the four-way junction (cruciform, HJ) in particular [2]. Cruciform structures (Figure 1) arise in inverted-repeat sequences in supercoiled DNA. It is easily appreciated that in such a structure extensive branch migration can and will occur (mobile HJ). However, it was found possible to construct HJs by chemical synthesis of appropriate DNA sequences designed to exclude branch migration as far as possible [3]. In this way so-called stable four-way junctions can be studied. In the presence of Mg^{2+} full base pairing in a four-way junction is preserved [4].

Modern techniques that are used to model the structure of the junction [2] include gel electrophoresis, fluorescence energy transfer (FRET), neutron scattering and electric birefringence. The electrophoretic mobility was found to be highly sensitive to the type of cation present and to its concentration; these observations were interpreted to mean that multivalent cations stabilize a compact structure [5]. The use of the FRET approach entails the attachment of a donor

A. Pullman et al. (eds.), Modelling of Biomolecular Structures and Mechanisms, 289–304.
© 1995 *Kluwer Academic Publishers. Printed in the Netherlands.*

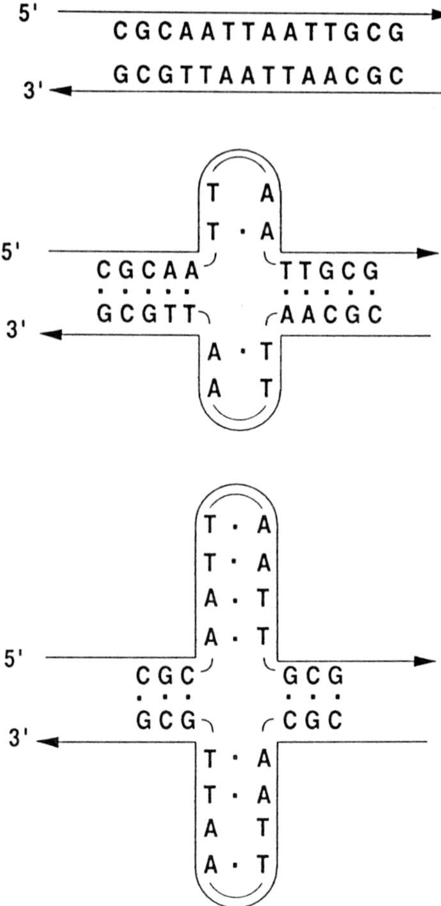

Fig. 1. Schematic representation of the formation of cruciform structures (mobile four-way junctions) at inverted-repeat sequences in duplex DNA as postulated by Holliday [1].

(fluorescein) to the 5′-end of one arm of the junction and of an acceptor (rhodamine) to another arm [6]. The efficiency of FRET between the fluorophores is inversely proportional to their mutual distance. For a given junction the procedure is repeated for each of the possible end-to-end vectors AB, AC, AD, BC, BD, and CD, Figure 2. From electrophoresis experiments it was already known [7] that the symmetry of the compact structure of the junction is lower than tetrahedral. The FRET approach [6] tells us that (in presence of Mg^{2+}) the structure folds into an *antiparallel* H-like form, Figure 2 (also called X-form [2] or U-form; there seems to be no concensus concerning the nomenclature of the various rotational isomers). Note that Figure 2 shows four different folded forms representing the four possibilities: two with a parallel folding (**1** and **2**), two with an antiparallel folding (**3** and **4**). Let us consider **1** as an example. Here strands I and III show continuous helicity and run in an antiparallel fashion with respect to one another.

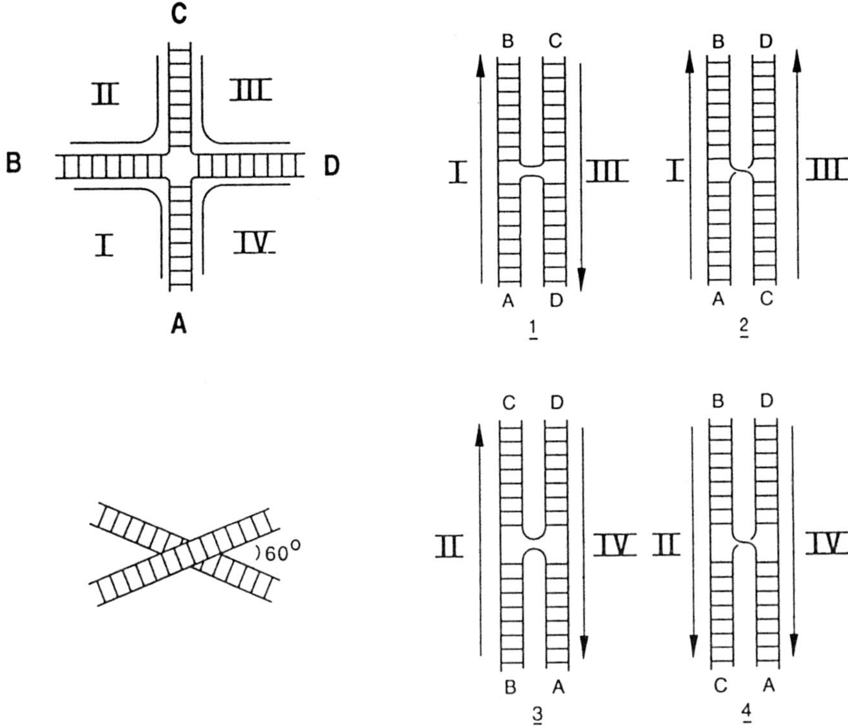

Fig. 2. Hypothetical structures of four-way junctions. The four single strands are indicated by the Roman numerals I–IV, the four base-paired arms by A–D. In the presence of a stabilizer like Mg^{2+} the 'open' X-like structure folds into one (or several) more compact H-like structure(s), see text. It is believed [2] that the through-stacked duplexes meet at an angle of approximately 60° (insert).

The consequence is that strands II and IV each make a sharp U turn. Here FRET would show a short distance between the 5′-termini of arms A and D and similarly between those of arms B and C. An alternative antiparallel form can be imagined as well (2), i.e. one in which strands II and IV show continuous helicity and I and III adopt the U turn.

In a similar vein the existence of two parallel H-like folded forms (X-crossing) can be postulated, but all biophysical evidence obtained thus far on relatively small four-way junctions is consistent only with an antiparallel structure [2]. Detailed molecular mechanics calculations on the possible structures of the four-way junction have been published [8]. It is postulated that the helices in the H structures are not strictly antiparallel, but are inclined at an angle of about 60° (Figure 2, insert). However, many questions concerning the local stereochemistry and themodynamical stability at the point of strand exchange remain open.

Few NMR studies on stable four-way junctions have been reported [4,9,10]. The main reason for this scarcity lies in the complexity of the NMR spectra displayed by the model compounds employed thus far, i.e. junctions obtained by mixing four separate 16-mer strands of DNA in a 1:1:1:1 ratio, giving a structure with eight base pairs in each arm or 64 residues in total. Now the detailed assignment of resonances poses a formidable challenge [9,10]. Even if these assignments may well be ultimately successful, it will readily be appreciated that model compounds containing 32 base pairs do not lend themselves to NMR analyses of a more routine nature with the aid of the present-day generation of 600 MHz spectrometers. However, such routine studies are precisely what is necessary in order to bring the specific powers of NMR into play, for example to put to the test various hypotheses concerning relations between base sequence, detailed structure of the branch point, and relative thermodynamic stability of the two allowed antiparallel junction conformers.

At this point it should be noted that the design of a synthetic four-way DNA structure having eight base pairs in each (open-ended) arm was dictated by considerations of the thermodynamic stability of each arm: even in the case that the junction structure does not severely destabilize base pairing at this critical point, one would not expect shorter duplex arms, consisting of A·T and G·C pairs, to display a T_m well above 40 °C.

With this information the stage is set for the description of an entirely different approach to the rational design of a much smaller DNA oligonucleotide that was expected to fold into a stable four-way junction. This design is based upon the properties of minihairpin loops.

Structure and Stability of Minihairpin Loops

We considered the possibility of constructing a stable immobilized four-way junction from a single strand of DNA. The underlying idea was to make maximum use of our accumulated knowledge of the building principles of stable minihairpin loops [11–19]. In order to clarify our initial considerations, previous experimental findings and conclusions [11–19] are briefy reiterated. Minihairpin loops are defined as loops that consist of only two nucleotides. These two residues cross the minor groove of the double-helical stem. The first minihairpin loop was discovered by accident: the sequence 5′-d(C*GC*-GT-GC*G) was designed to investigate the effect of a double G × T mismatch on the proclivity of CG sequences to revert from B to Z duplex DNA at high Na$^+$ or Mg^{2+} concentrations; C* = m^{5}C in the abovementioned compound. However, at low DNA concentration (0.4 mM DNA) and low ionic strength this sequence exclusively prefers to adopt a monomeric hairpin form, which consists of a stem of three Watson–Crick base pairs bridged by the -GT- residues across the minor groove of the stem [11,13,14]. The high thermodynamic stability of this minihairpin loop (T_m 59 °C) came as a surprise. These findings are in striking contrast with earlier beliefs which held that in oligonucleotides loops containing fewer than three residues are sterically impossible for the simple reason that two residues can span a phosphate-phosphate distance of no more than about 14 Å, whereas the distance to be bridged in a *regular* B-DNA duplex is more than 18 Å. The solution to this apparent mystery was soon obvious: in this DNA minihairpin loop (as well as in all the others that were subsequently studied, without exception) a sharp turn occurs at the 3′,5′-loop stem junction, which turn involves an unusual γ^t torsion angle in the G(6) residue of the example discussed. A rotation from the usual γ^+ to γ^t at this position swings the attached phosphate into the minor groove over a distance of 4–5 Å and the remaining gap of 13–14 Å is easily bridged by two nucleotides [11,13,14]. An analogous sequence having a potential double G × A mismatch showed the same behavior (T_m of the hairpin form 56 °C), although in this case competition with the duplex is more severe [12].

In both DNA mismatched sequences mentioned, the NMR data are consistent with a continuous stacking of the loop residues on the 3′-end of the stem, C*(3) in the above examples. In the course of time this particular type of loop structure was classified as H1-type hairpin-loop folding [15,16].

A second type (H2) of hairpin conformation was discovered soon afterwards. This type is exemplified by the stable minihairpin loops that occur (under suitable conditions of low DNA and salt concentration) in the self-complementary octamers 5′-d(CGC-TA-GCG) and 5′-d(CG^{a}C-TA-GCG), ^{a}C = arabinofuranosyl cytidine [15,16]. These H2-type loops are distinct from H1-family loops in that the first base in the loop, T(4), does not stack upon C(3) but is held in the minor groove by a hydrogen-bonding network; instead, the second loop residue (A(5) in this example) stacks upon C(3) [15,16].

Subsequent work of the Leiden group concentrated on three aspects of loop formation:

(a) The importance of the nature of the closing base pair, adjacent to the loop; note that a C·G closing pair is present in all of the above examples. Therefore, the closing pairs T·A, G·C, and A·T were investigated.

(b) What is/are the determinant(s) that cause(s) the selection of either the H1 or the H2 family loop structure? This aspect was covered by systematic variation [18] of the two loop residues in 5′-d(CGC-X1X2-GCG): X1/X2 = T/T, T/C, C/T, A/C, A/A, in addition to the previously studied combinations G/T, G/A and T/A.

(c) The relative thermodynamic stabilities of the various hairpin-loop structures [18].

The following findings are pertinent to the present design study of a 'small' four-way junction.

(a) The closing base pair that lends maximum stability to a minihairpin loop is C·G, followed at some distance by T·A. Reversal of these pyrimidine·purine (Y·R) base pairs into G·C or A·T (R·Y) results in a marked thermodynamic destabilization of the minihairpin structure, or even in (partial) disruption of the closing base pair [17,18,20] to give a four-membered loop.

(b) For minihairpin loops that contain a favorable C·G closing base pair an H1 family miniloop is preferred in all cases where the first loop residue X1 is a purine (R), regardless of the nature (R or Y) of the second loop residue X2; the formation of a H2 family loop requires a pyrimidine at the first loop position, again regardless of the nature of the second residue in the loop.

(c) In hairpin loops that share the (unmodified) octameric sequence 5′-d(CGC-X1X2-GCG) the relative thermodynamic stabilities (T_m) cover the range 47–56 °C; the nature of the residue at the second loop position now does

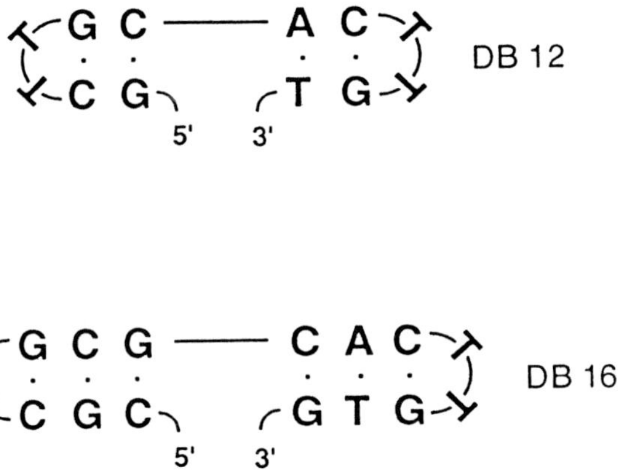

Fig. 3. The sequences and folding of the two dumbbell DNA molecules DB12 and DB16.

play a role: a pyrimidine here appears to induce maximum stability of the minihairpin structure.

The extraordinary ease of formation of minihairpin loops is exemplified by two DNA sequences: the 12-mer 5′-d(GC-TT-GCAC-TT-GT) (DB12), and the 16-mer 5′-d(CGC-TT-GCGCAC-TT-GTG) (DB16), Figure 3. At room temperature both single strands readily form a dumbbell with continuous base-base stacking across the nick; the double helix that is created is capped on each side by a -TT-minihairpin loop [19]. The thermal stabilities (T_m values) of each half of the dumbbells could be measured separately by the use of chemical shift vs temperature profiles: DB12 45 °C, DB16 66 °C in the C·G-rich halves, four to six degrees less in the 'A·T' halves. Clearly, stability considerations point to the use of minimally eight residues in each arm: three base pairs and two loop residues.

Finally, it should be mentioned here that H1 and H2 family miniloops can be recognized unambiguously by their distinct NOESY patterns [14–18]. Recognition of an H2-type hairpin is particularly facile for a miniloop that contains the -TT-sequence [17,18]. Unusual NOESY patterns involving the methyl groups of the thymine bases are unmistakable markers, see below. Moreover, the presence of two neighboring T residues serves to prevent base-pair migration. The design of the dumbbells DB12 and DB16, mentioned above, followed these principles.

Design of a 'Small' Four-Way Junction J4

Our experiments with DNA octamers that yield stable minihairpin structures (see the previous section) led us to consider the possibility to construct a stable four-way junction (J4) from a single strand of DNA. Three miniloops could cap three duplex arms of the prospective cruciform, the fourth duplex arm would necessarily have an open end. The resulting four-way junction should remain immobilized, i.e. correspond to a unique base-pairing scheme with the exclusion of alternative

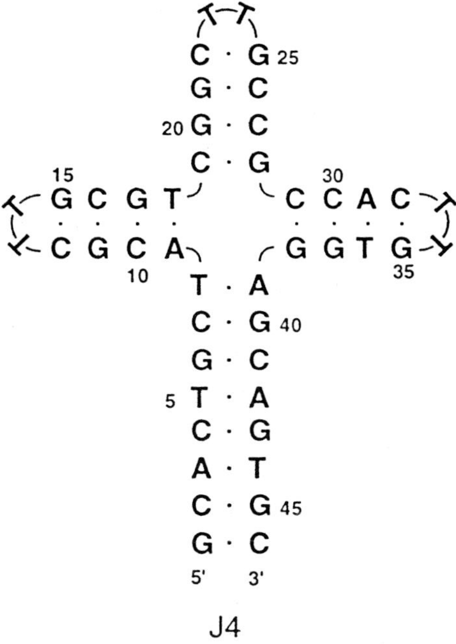

Fig. 4. The DNA sequence of synthetic J4. The molecule is designed as a continuous single strand of 46 residues. Arm A with eight base pairs represents the 'open end', arms B, C and D each consist of four potential base-pairs capped by a T-T sequence.

possibilities, and display sufficient thermodynamic stability (T_m of each arm >45 °C).

In the design described in the present paper a conservative approach was preferred. Consideration of thermodynamic stability was given precedence over our wish to construct the smallest possible J4 molecule. It is not known with certainty whether or not the folding of the junction induces a destabilization of each arm. If it does, arms consisting of three base pairs each, capped by a miniloop, might not remain intact at temperatures best suited for NMR spectroscopy. Therefore, we chose to base our first design on three capped arms, each of which contains four base pairs. The closing base pair at the loop positions had to be C·G and each miniloop was allotted two thymidines. Thus, all the three short arms of the cruciform contain at the closing end the sequence 5'-d(C-TT-G). Moreover, in the present design use is made of our previous experiences with the dumbbell DB16, which has an A·T pair next to the closing C·G pair in one of the two arms. This serves to distinguish the two arms of the dumbbell by the creation of chemical-shift differences. For the same reason an A·T pair is introduced into arm D of the perspective four-way junction. In fact, the analogy was carried further: arms B and D of J4 have sequences of eight consecutive residues in common with the corresponding arms in DB16, Figure 4.

Arm C differs from arm B in that base pair G(20)·C(27) in arm C replaces the pair C(10)·G-(17) in arm B. The long, open-ended, arm A needed eight base pairs and in the design we took care to create on the one hand as far as possible an alternate purine-pyrimidine sequence in arm A (in order to facilitate ^{1}H resonance assignments) which sequence on the other hand should not give rise to branch migration. An extreme example of branch migration should be considered in any design of a junction that is based upon the single-strand concept. This is the possibility that a competing duplex can exist. Imagine, for example, that both arms B and D are merged into a continuous helix running from arm A to arm C. The more complementary residues in arms B and D are positioned in an inverted-repeat relationship, the more likely this is to occur.

Since little is known concerning the structure at the branch point, the selection of the eight central residues was guided somewhat by educated guesswork. Since the rules that govern the choice of stacking conformer are not understood as yet [8], we wished to construct a central sequence that differs from sequences investigated previously by others [9,10,21,22]. In the final design J4 consists of a single strand of 46 residues (Figure 4), that is, 18 residues less than the 64-residue junction partly analyzed by Chazin *et al.* [9,10].

NMR Spectroscopy of the J4 Molecule

EXPERIMENTAL

The synthesis of the J4 sequence was carried out on a 5 μmol scale using standard phosphoramidite chemistry. After purification the sample was lyophilized and taken up in D_2O: strand concentration 1.4 mM, added NaCl 50 mM, $MgCl_2$ 5 mM, pD 7.0 (uncorrected). The sample was annealed by heating to 65 °C and slow cooling over several hours. One- and two-dimensional (TOCSY with 40 ms spin-lock pulse; NOESY with mixing times 75 and 350 ms) ^{1}H-NMR spectra were measured (Bruker AM 600 spectrometer). In addition, the following spectra were collected: (i) a 600 MHz NOESY spectrum in H_2O, showing the imino-proton resonances (10–15 ppm); (ii) a one-dimensional ^{31}P spectrum in D_2O at 121 MHz; (iii) a three-dimensional TOCSY-NOESY spectrum in D_2O at 600 MHz. Tetramethylammonium ion (TMA) was used as primary internal reference for the ^{1}H spectra. All ^{1}H chemical shifts reported were converted to the DSS scale by the addition of 3.18 ppm to the observed shifts.

GENERAL OBSERVATIONS

Minimum line widths and maximum spectral resolution was found in the temperature range of 27–35 °C. The spectra pointed to the occurrence of a single conformational species. An important observation was already made at this stage: five signals at low field were prominent in the ^{1}H spectrum of J4 in H_2O, Figure 5. As these resonances are highly characteristic of the imino protons of T residues involved in Watson–Crick base pairing, we concluded that in the presence of Mg^{2+} all A·T pairs that are possible in the J4 molecule are actually formed, including the two at the center of the junction, Figures 6 and 7. Since the three hairpin loops are recognizable at first glance (*vide infra*), it is already apparent at this stage of our study that J4 adopts an intact four-way junction structure. However,

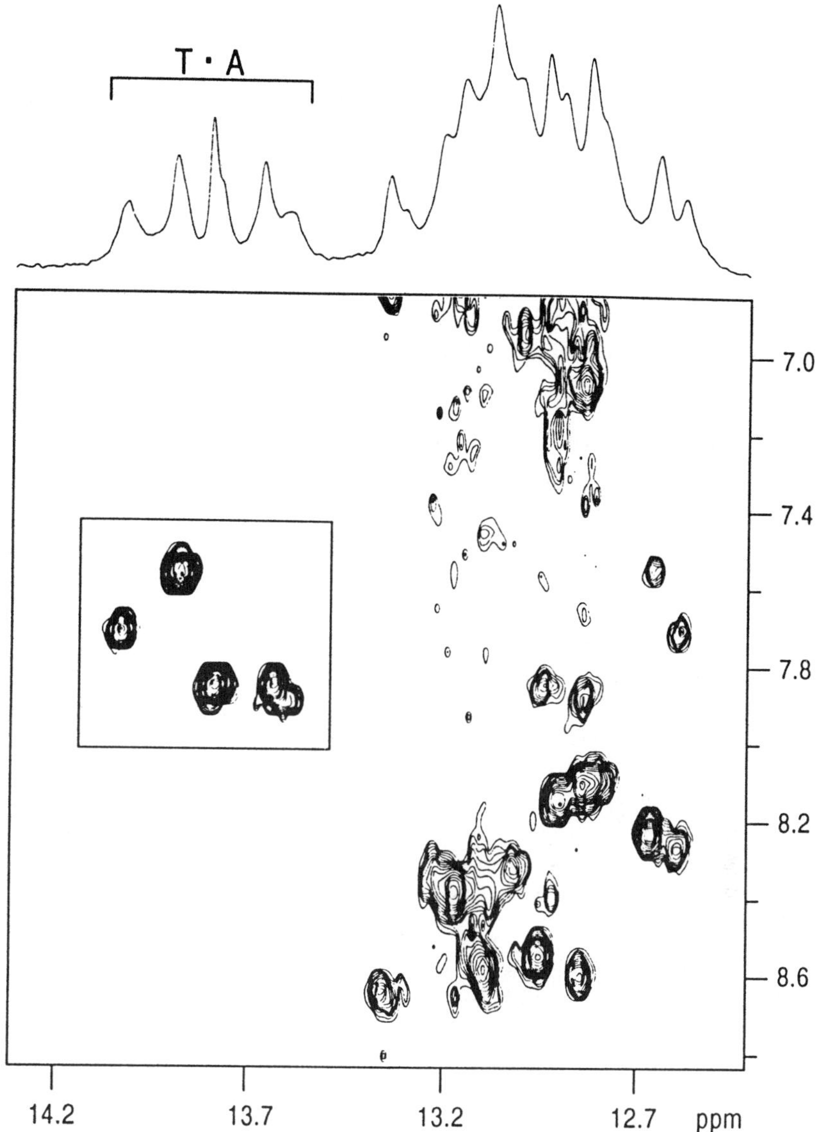

Fig. 5. Part of the NOESY spectrum of J4 in H_2O. The boxed region contains the five cross peaks that are ascribed to the five intact T-A base pairs (NH3(T)–H2(A)).

from these data alone no differentiation is yet possible between competing conformations J4-I and J4-II, Figure 7.

ASSIGNMENTS

The spectral assignment was divided into three parts: (i) the minihairpin regions; (ii) the self-complementary stem regions; (iii) the central region of the junction.

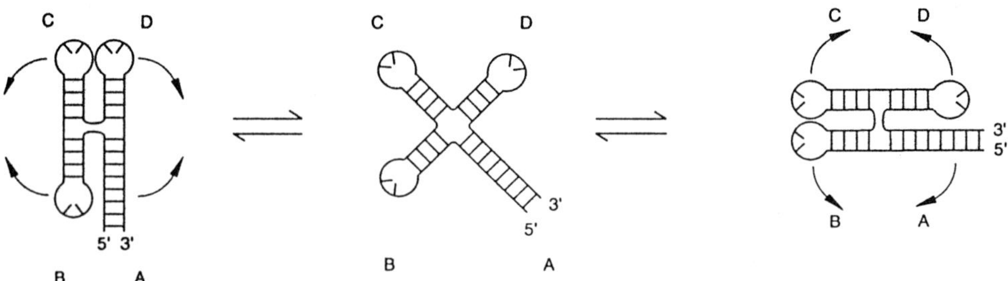

Fig. 6. Schematic representation of the two antiparallel H-like conformations that can arise in J4 in the presence of Mg^{2+}, see also Figure 7.

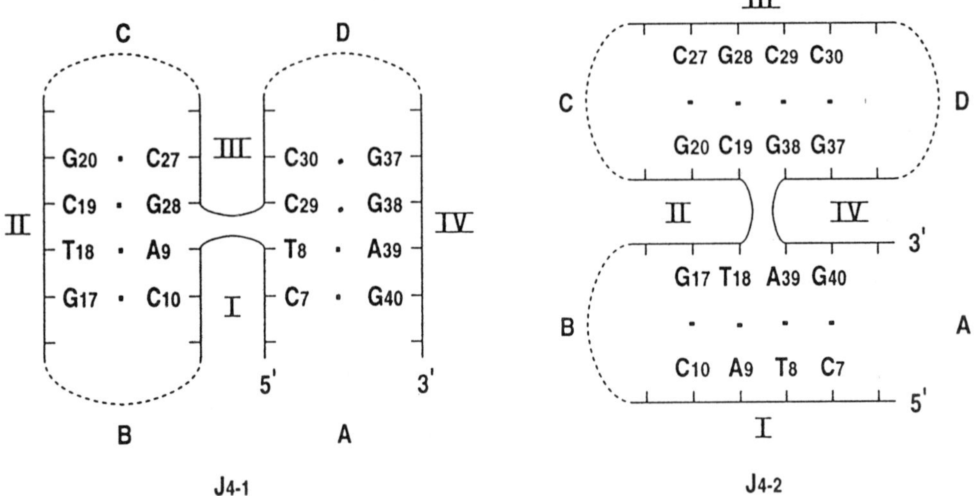

Fig. 7. The two possible through-stacking schemes, J4-I and J4-II, that must be considered for antiparallel cross-over structures. Only J4-I is actually adopted by J4, compare Figure 9.

In the final stage of this study the following signals were assigned: all base protons (H5, H6, H8, CH$_3$, except H2), H1′, H2′, H2″ and most H3′ of the sugars. In addition, various marker signals typical for the H2-type hairpin-loop structure (H4′, H5′, H5″ at specific positions) were identified. Full details of the assignment procedure and chemical-shift tables will be presented elsewhere [19]. Here it suffices to reiterate the highlights step by step.

Step 1. The easiest part was the recognition of the three minihairpin-loop regions. From our earlier NMR studies [17–19] on stem-loop-stem 5′-d(C(n)-T(n + 1)T(n + 2)-G(n + 3)) sequences the essential fingerprints are known with confidence: H2-type miniloops with this sequence invariably display short distances between the H5 methyl protons of the T residues in the loops to other protons.

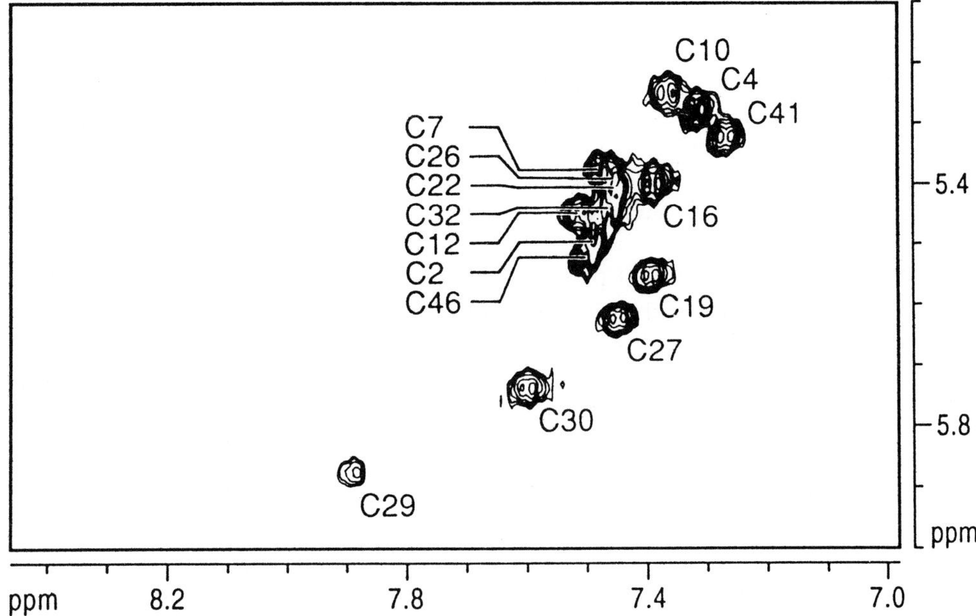

Fig. 8. Part of the TOCSY spectrum of the intact four-way J4 (Mg^{2+} added) showing the H5–H6 (J-coupled) connectivities. Note that H5 and H6 of C29 at the center of the junction display anomalous low-field chemical shifts compared to the bulk of the cytidines.

In particular, NOESY cross peaks occur between the signals of MeT(n + 2) and H1′, H2′, H2″, H5, H6 of residue C(n), also between MeT(n + 2)–H3′T(n + 1) and MeT(n + 1)–H4′C(n). These peculiar interactions occur because the T(n + 1) residue is positioned in the minor groove of the duplex stem, as a result the T(n + 2) base stacks upon the C(n) residue. All relevant NOEs that are predicted for the three 5′-d(CTT) sequences were found in the NOESY spectrum (not shown). Independent corroboration for the conclusion that the J4 structure indeed contains the three expected minihairpin loops comes from the study of the ^{31}P-NMR spectrum. All H2-family miniloops investigated thus far [15–18] display a downfield ^{31}P signal accompanied by an upfield resonance, removed by 0.8 ppm and −1.6 ppm from the main bulk of the ^{31}P signals, respectively. The ^{31}P spectrum of J4 clearly shows three separate downfield and three separate upfield resonances at the expected positions (not shown).

Step 2. The assignment of the ^{1}H signals in the stem regions A, B, C, D, Figure 7, offered quite a challenge, since 40 different stem residues are present in J4. The assignment was aided by the fact that in our design of J4 as many alternating purine-pyrimidine steps as possible were introduced. This alternation has the advantage that R-Y steps can be clearly identified from characteristic interresidue NOEs of the type H1′/H2′/H2″(n) – CH$_3$/H5/H6/H8(n + 1). The 15 H5–H6 connections, obtained from the TOCSY spectrum (Figure 8), were especially helpful in this respect. In the next step of the analysis triplet and quadruplet base sequences were identified. Finally, complete interresidue-intraresidue NOE walks through the various strand sequences were achieved (not shown). As expected, each interresidue NOE walk terminated at the point of the sharp turn at the 3'-end of each miniloop. The signals of the three dC residues at the 5'-end of the loops, C(12), C(22), and C(32), were assigned individually at this stage and thus the resonance positions of the T residues in each loop were also known. The NOE patterns are entirely consistent with B-DNA type helices.

Step 3. During the assignment procedure followed for the stem region no *a priori* assumptions were made concerning the mode of stacking at the center of the four-way junction of J4. From the current interpretation of the biophysical experiments referred to in the Introduction of this contribution [2–6], we conjectured from the outset of this study that a verdict on the choice of the J4 molecule for either the J4-I or the J4-II folding pattern (Figure 7) could be obtained from interresidue NOEs between the eight central residues, provided that the center behaved as expected. This conjecture proved to be correct.

During the assignment procedure it became clear that relatively intense NOE peaks of the type $H1'(n)$–$H6/H8(n + 1)$ occurred at the sites of the T(18)–C(19) and the G(38)–A(39) steps. This finding is only consistent with a folding as depicted for the J4-I conformer, Figure 7. A systematic survey of the NOE maps, in which the cross peaks between $H1'(n)$, $H2'/H2''(n)$, and $H6/H8(n + 1)$ are collected, revealed no less than 16 NOE connections that are consistent with continuous 5'–3' stacking of residues T(18)–C(19), G(28)–A(9), G(38)–A(39) and T(8)–C(29), Figure 9 (spectra not shown). These findings are in complete accord with a folding consistent with the J4-I structure, Figure 7.

An exhaustive search for NOEs that would be consistent with the J4-II conformer yielded as the only possible candidate a single low-intensity peak (free from overlap with other cross peaks) that could perhaps arise from a H2''G(38)–H6C(19) interaction. It appears reasonable to assume that this peak is a spurious one, however, because no less than ten other overlap-free positions occur in the NOESY spectra where one would expect cross peaks indicative of through-stacking according to model J4-II. None were found.

Conclusion

The carefully designed J4 DNA molecule which contains only 46 nucleotides, indeed folds into a stable four-way junction (immobilized Holliday junction). Complete NMR assignments are obtained for the non-exchangeable base protons, for the 1', 2', and 2'' sugar protons and for most 3' protons. A study of the signals of the exchangeable imino protons reveals that all A·T pairs expected are present, including the two located at the center of the junction.

Most resonances of residues at the center are broadened. This corroborates an earlier finding in the proton NMR spectra of two 64-residue junctions [9]. The molecular basis for this line-broadening effect remains unclear at the present time.

The sequence-specific assignments of the J4 molecule permitted the discovery of 16 interresidue NOE connections in the center of the four-way junction. These

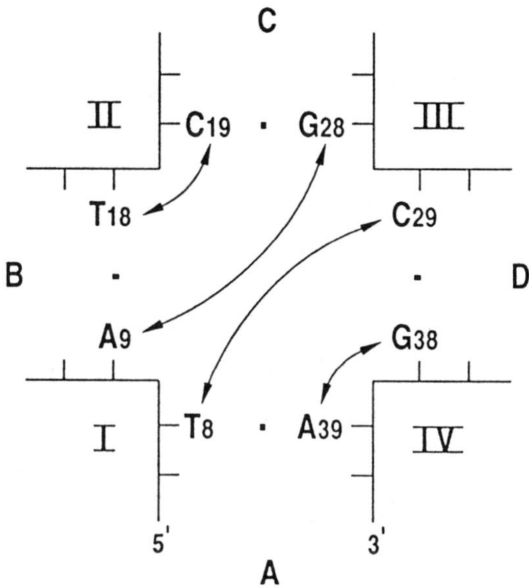

Fig. 9. Interresidue NOE contacts in the center of the intact J4 four-way junction. These contacts agree with base-base stacking of arm A on arm D and of arm B on arm C (J4-I model), compare Figure 7.

NOEs show conclusively that the structure is constituted from two through-stacked helical strands and of two strands that are bent in the center in such a way that two interstrand stacking steps occur: T(8)–C(29) and G(28)–A(9). Thus, two B-DNA double helices exist side by side (Figure 10), in accordance with models developed by Lilley *et al.* [2,5,6,8] and by others [3,4,7] on the basis of various biophysical techniques. Only one of the two theoretically possible through-stacking patterns (J4-I) is realized by J4. If conformer J4-II exists at all, it must be present in an insignificant amount. This is the first time that detailed information of this kind is available from standard NMR experiments in a relatively short time.

Perspectives

It was postulated [8] that the minor groove must open to permit strand exchange while avoiding close contact of the crossed backbones. NMR techniques are suited in principle to supply an experimental picture of widening and narrowing of grooves *via* NOE distance constraints. However, we refrained from the measurement of the necessary NOE *vs* time build-up curves because of rather strong overlap of many of the most interesting NOESY cross peaks in J4. Studies aimed at such measurements would greatly profit if smaller four-way junctions were available. For this reason we give high priority to the design and synthesis of DNA molecules that are promising candidates.

The duplex parts of J4 are connected by two interhelical phosphate-diester bridges. Our data do not allow a judgement on the relative orientation of the two

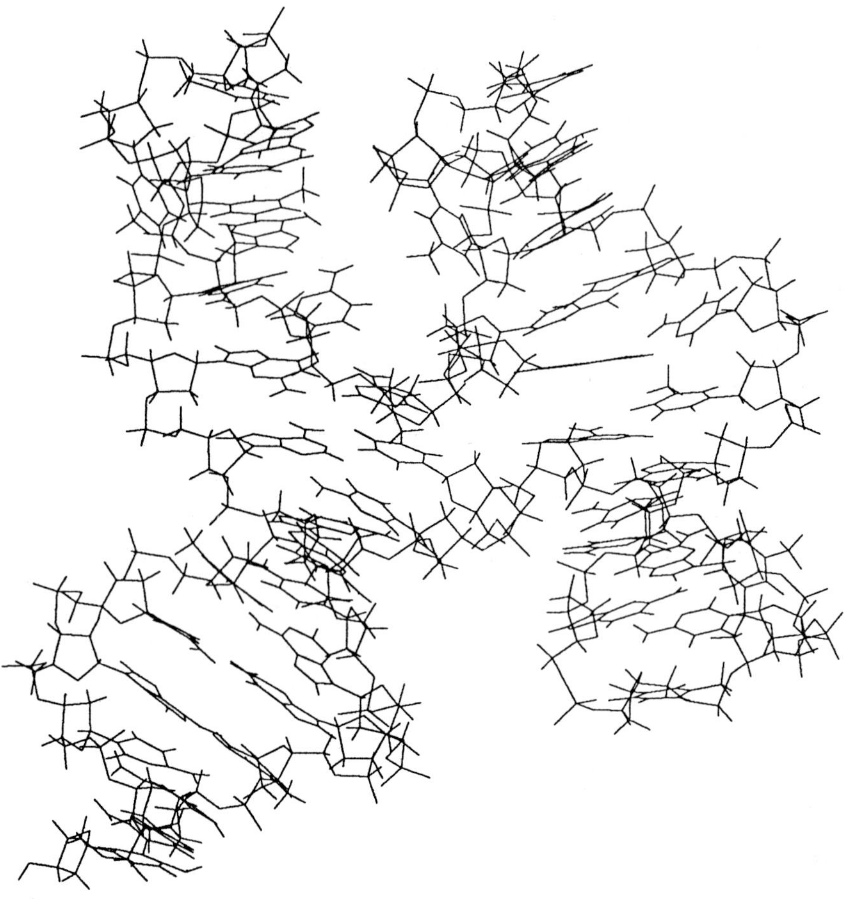

Fig. 10. Stick model of the J4-I cross-over junction conformation. The antiparallel model was generated with the aid of the QUANTA program [27] and energy-minimized with the use of the AMBER 3.0 [28] molecular-mechanics force field. Improvements on this model must be postponed until the time that NOE and coupling constant constraints are available that can be applied to molecular dynamics calculations, see text.

stacking domains: parallel or antiparallel, nor on the value of the postulated angle between the duplexes, see Figure 2, insert. Since the helix-helix connection proceeds *via* the phosphate bridges in which many NMR active nuclei participate in one way or another (C2′-C5′, H4′, H5′, H5″, ^{31}P) it is easy to see that such a judgement could benefit from NMR analysis of site-specific ^{13}C isotope-enriched deoxyribose sugars involved in these intrahelical bridges (for example: isotopic enrichment of T8, A9, G28, and C29 in J4). The ^{13}C-enriched junction would allow us to assign the all-important C2′, C3′, C4′, C5′, H4′, H5′, H5″, and central ^{31}P resonances and to measure build-up curves of the possible interbridge NOEs. Equally important, the various vicinal coupling constants along the bridges can then be determined. The latter are a principal source of direct information. This point is illustrated by the following example: A molecular mechanics calculation of a four-way junction [8] suggests that the torsion angles ϵ(C4'-C3'-O3'-P) at the 5'-sides of the bends adopt ϵ^- rotamers (ϵ 289–298°). This theoretically derived feature would be amenable to direct experimental verification through measurement of ^3J(C4'P) and ^3J(C2'P). The relevant ^3J(CP) Karplus parameters are well known [23,24]. In a similar manner ^3J(CP), ^3J(HP) and ^3J(HH) would provide information on the sugar geometry and on the preferred rotamers along torsion angles γ and β at the point of strand exchange.

A worthwhile aim of the future researches described above is to arrive at a good starting position from where ^3J/NOE-constrained molecular-dynamics trajectories [15,25,26] can be employed as a reliable source of information concerning local mobility and conformational freedom in four-way junctions. This would provide a framework for modelling how the four-way junction interacts with enzymes like Holliday resolvases and other ligands.

Acknowledgements

We thank H. van den Elst and Prof. J. H. van Boom (Leiden) for the synthesis of J4 and C. Erkelens (Leiden), J. J. M. Joordens and Dr. S. S. Wijmenga (Nijmegen) and Dr. W. Bermel (Bruker, Germany) for their aid with NMR experiments at various stages. This research was supported by The Netherlands Foundation for Chemical Research (SON) with financial aid from The Netherlands Organization for Scientific Research (NWO).

References

1. R. Holliday: *Genet. Res.*, **5**, 282 (1964).
2. D. R. Duckett, A. I. H. Murchie, A. Bhattacharyya, R. M. Clegg, S. Diekmann, E. von Kitzing, and D. M. J. Lilley: *Eur. J. Biochem.*, **207**, 285 (1992), and refs. cited therein.
3. N. R. Kallenbach, R.-I. Ma, and N. C. Seeman: *Nature* **305**, 829 (1983).
4. D. E. Wemmer, A. J. Wand, N. C. Seeman, and N. R. Kallenbach: *Biochemistry* **24**, 2745 (1985).
5. S. Diekmann and D. M. J. Lilley: *Nucleic Acids Res.* **14**, 5765 (1987).
6. R. M. Clegg, A. I. H. Murchie, A. Zechel, C. Carlberg, S. Diekmann, and D. M. J. Lilley: *Biochemistry* **31**, 4846 (1992).
7. J. R. Cooper and P. J. Hagerman: *J. Mol. Biol.* **198**, 711 (1987).
8. E. von Kitzing, D. M. J. Lilley, and S. Diekmann: *Nucleic Acids Res.* **18**, 2671 (1990).
9. S-M. Chen, F. Heffron, W. Leupin, and W. J. Chazin: *Biochemistry* **30**, 766 (1991).
10. S-M. Chen, F. Heffron, and W. J. Chazin: *Biochemistry* **32**, 319 (1993).
11. L. P. M. Orbons, G. A. van der Marel, J. H. van Boom, and C. Altona: *Nucleic Acids Res.* **14**, 4187 (1986).
12. L. P. M. Orbons, G. A. van der Marel, J. H. van Boom, and C. Altona: *Eur. J. Biochem.* **170**, 225 (1987).
13. L. P. M. Orbons, G. A. van der Marel, J. H. van Boom, and C. Altona: *J. Biomol. Struct. Dyn.*, **4**, 939 (1987).
14. L. P. M. Orbons, A. A. van Beuzekom, and C. Altona: *J. Biomol. Struct. Dyn.* **4**, 965 (1987).
15. J. M. L. Pieters, E. de Vroom, G. A. van der Marel, J. H. van Boom, T. M. G. Koning, R. Kaptein, and C. Altona: *Biochemistry* **29**, 788 (1990).
16. J. M. L. Pieters: *Thesis*, Leiden (1989).
17. J. H. Ippel, V. Lanzotti, A. Galeone, L. Mayol, J. E. van den Boogaart, J. A. Pikkemaat and C. Altona: *J. Biomol. Struct. Dyn.* **9**, 821 (1992).
18. J. H. Ippel: *Thesis*, Leiden (1994).

19. J. A. Pikkemaat and C. Altona: in preparation; J. A. Pikkemaat, H. van den Elst, J. H. van Boom, and C. Altona: *Biochemistry* **33** (1994), in press.

20. M. J. J. Blommers, J. A. L. I. Walters, C. A. G. Haasnoot, G. A. van der Marel, J. H. van Boom and C. W. Hilbers: *Biochemistry* **28**, 7491 (1989).

21. D. R. Duckett, A. I. H. Murchie, S. Diekmann, E. von Kitzing, B. Kemper and D. M. J. Lilley: *Cell*, **55**, 79 (1988).

22. P. Guo, M. Lu and N. R. Kallenbach: *Biopolymers*, **31**, 359 (1991).

23. P. P. Lankhorst, C. A. G. Haasnoot, C. Erkelens, and C. Altona: *J. Biomol. Struct. Dyn.* **1**, 1387 (1984).

24. P. P. Lankhorst, C. A. G. Haasnoot, C. Erkelens, H. P. Westerink, G. A. van der Marel, J. H. van Boom, and C. Altona: *Nucleic Acids Res.* **13**, 927 (1985).

25. D. F. Mierke and H. Kessler: *Biopolymers* **33**, 1003 (1993).

26. A. E. Torda, R. M. Brunne, T. Huber, H. Kessler, and W. F. van Gunsteren: *J. Biomol. NMR* **3**, 55 (1993).

27. QUANTA-CHARMm programs, version 3.0, Molecular Simulations, Burlington Mass., U.S.A.

28. S. J. Weiner, P. A. Kollman, D. T. Nguyen, and D. A. Case: *J. Comput. Chem.* **7**, 230 (1986).

On the Role of Single-Stranded Adenines in RNA-RNA Recognition

E. WESTHOF
Equipe de modélisation et de simulation d'acides nucléiques, UPR 9002, SMBMR, IBMC du CNRS, 15, rue R. Descartes, F-67084 Strasbourg-Cédex, France.

Abstract. Catalytic RNA molecules are metalloenzymes which require only water and magnesium ions for their biological function. The specificity and precision of their catalytic activity stem from their intricate three-dimensional structures which are attained and maintained through RNA-RNA interactions. In several instances, adenine residues of single-strands, internal loops or hairpin loops, have been implicated in tertiary contacts either unspecifically to the sugar-phosphate backbone or specifically to the shallow groove side of G-C base pairs. Examples from each of the three families of catalytic RNAs are described.

Key words: RNA catalysis, RNA tertiary structure, computer modelling.

Introduction

Catalytic RNA molecules belong to three families: the RNAseP [1], the small viral ribozymes like the hammerhead [2] or hepatitis delta virus [3,4], and the self-splicing introns [5]. Each family corresponds to a specific chemical attack on the phosphodiester linkage (see Figure 1). In the RNaseP family, the attacking species, an activated water molecule, reacts with the phosphodiester linkage to yield 3'-OH and 5'-phosphate extremities. In the small ribozymes, an activated species abstracts a proton from the O2'-H hydroxyl group of a ribose which itself will react with its 3'-phosphate group to yield a 2',3'-cyclic phosphate (which subsequently leads to 3'-phosphate and O2'-H hydroxyl groups) and a 5'-OH extremity. Self-splicing introns are divided into two groups. In group I introns, the catalytic process is started by the presence of a guanosine cofactor; while in group II introns, a ribose group within the intron reacts at the 5'-exon: intron junction to give a lariat. In the self-splicing introns, the activated species abstracts a proton from either a O2'-H hydroxyl group (group II) or a O3'-H terminal hydroxyl group (groups I and II) which then attacks another phosphodiester linkage leading to the formation of a new phosphodiester bond and a 3'-OH extremity. Structurally, the most investigated catalytic RNAs are the self-splicing introns of group I.

The Group I Self-Splicing Introns

On the basis of sequence comparisons, a detailed molecular model was proposed, first, for the catalytic core of group I introns [6] and, secondly, for entire group I introns [7]. In the model of the catalytic core, seven contacts between the substrate (helix made of the 5'-exon and a complementary sequence of the intron called the Internal Guide Sequence) and the enzyme were proposed (Figure 2). Except for one, all those contacts involved adenine residues of single-stranded regions

A. Pullman et al. (eds.), Modelling of Biomolecular Structures and Mechanisms, 305–313.

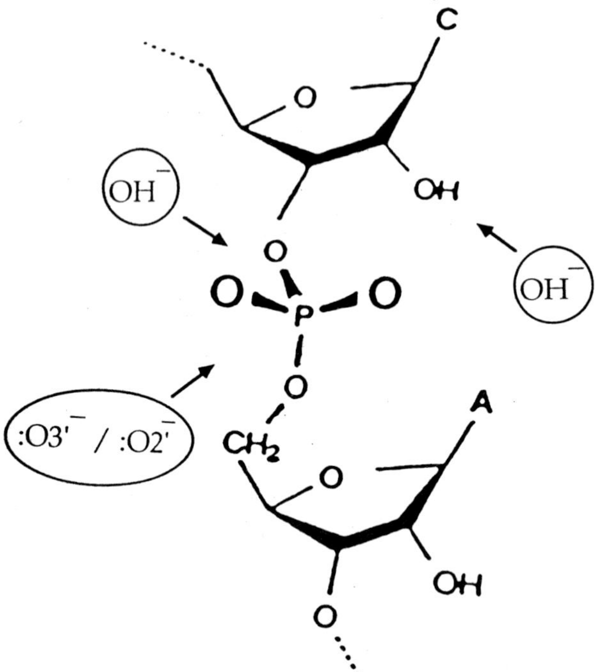

Fig. 1.   The three pathways for attacking a phosphodiester linkage. An activated water molecule (top left) can attack directly leading to a 3'-OH and a 5'-phosphate groups as occurring in RNAse P catalysis. In small ribozymes (hammerhead or HDV), an activated water molecule (top right) abstracts a proton from the 2'-OH group of a ribose which then attacks its attached phosphodiester giving a 2',3'-cyclic phosphate and a 5'-OH groups. Finally, in self-splicing introns, the 2'-OH or 3'-OH hydroxyl groups of an adjacent residue attacks the phosphodiester leading to the formation of a new phosphodiester linkage.

interacting with the sugar-phosphate backbone of the helical substrate, especially 2'-OH hydroxyl groups. Specifically, adenines A114 and A115 of an internal loop would bind O2'-H of G22 and U21 in the IGS, respectively, while adenines A302 and G303 of a single-strand would bind G25 on the IGS and position −3 on the 5'-exon. Several elegant experiments [8–12] have brought strong evidence for most of those predictions. They have also shown [8,9] that residues A301 and A302 contact G25 and −3, instead of residues A302 and G303 as originally proposed [6].

An original RNA-RNA motif was discovered in the course of the sequence analysis of group I introns [6]. It involves a tetraloop of the type -GNRA- (where N stands for any nucleotide and R for any purine). Those loops are extremely frequent on a statistical basis in structured RNAs (as are -UNCG- tetraloops) [13]. Molecular modelling based on chemical probing had shown that -GNRA- loops are closed by an unusual G-A pair between the first and the fourth residues of the loop with N2(G)----N7(A) and N3(G)----N6(A) hydrogen bonds [14]. This structure was later established by NMR methods [15]. From the analysis of sequences of group I introns, it was proposed that such loops bind in the shallow

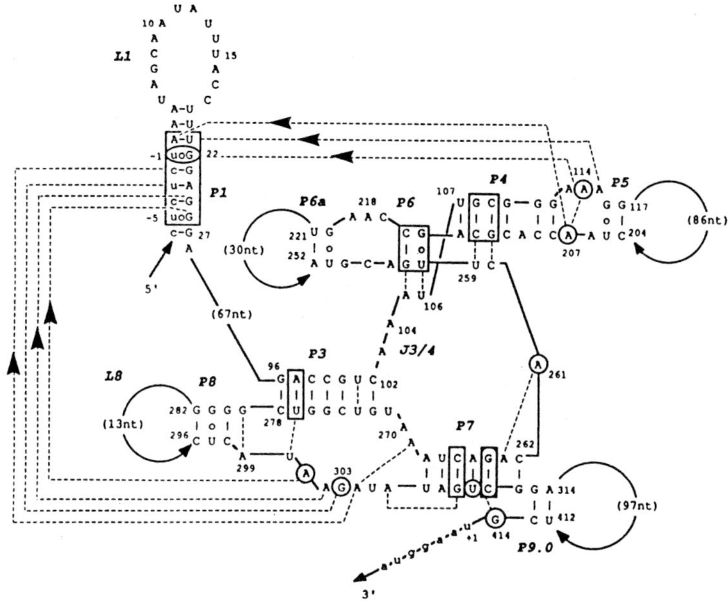

Fig. 2. Secondary structure of the catalytic core of the *Tetrahymena* group I intron (from Reference 6) with arrows indicating the adenine contacts between the core and the substrate (helix P1).

groove of RNA helices to G-C base pairs via N2(G)----N1(A) and N3(G)----N6(A) hydrogen bonds. Recently, experimental evidence supporting this RNA-RNA motif were published [7,16–17]. Figure 3 displays the 2D and Figure 4 the 3D structure of the group I intron *sunY* from bacteriophage T4 which presents such a -GNRA- shallow groove interaction between loop L9 and helix P5.

The RNAse P of *Escherichia coli*

Recently, together with S. Altman, we have modelled the three-dimensional structure of the M1 RNA of *Escherichia coli* [18]. In the secondary structure [19] of that 377 nucleotide long molecule, there are two pseudoknots, one of which is highly conserved, and four tetraloops of -GNRA- type (Figure 5). Comparative chemical probing on a wild type molecule and on a G89A mutant revealed several conformational alterations between nucleotides 90 and 239 in the RNA, as catalytically active as the wild type, but defective in the association with the protein component [20]. Closer examination indicated an increased DMS attack in a -GNRA- loop at the end of helix P6. We therefore concluded that loop L6 interacted, in a manner similar to that discovered in group I introns, with the shallow groove of G89 in helix P9b. The presence of a -GNRA- loop at that position is conserved [21] in reported sequences, except in Gram positive bacteria (*Bacillus*). This interaction added further 3D constraints to the modelling. We then wondered whether one of the conserved -GNRA- loop could not interact with the substrate, i.e. the precursor tRNA. This search was elicited by the fact

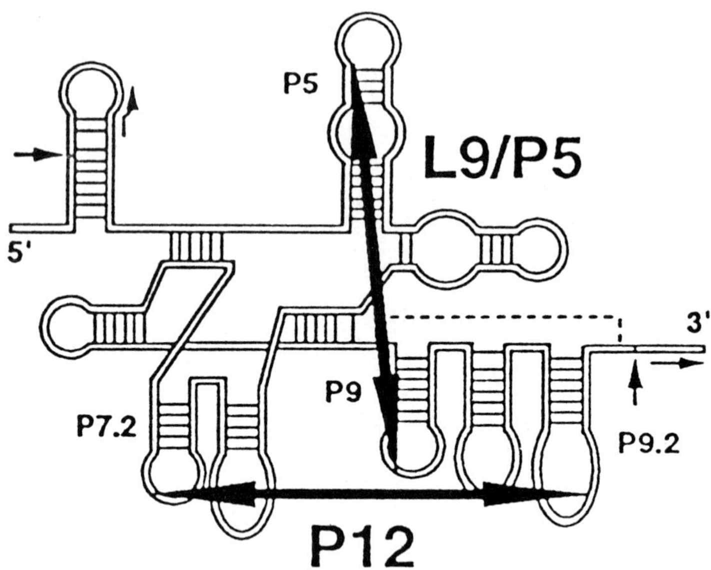

Fig. 3.   Schematic secondary structure of the *sunY* group I intron illustrating two long-range interactions: a loop-loop interactions (P12), not discussed in the text, and the interaction between the L9 - GNRA- loop and the shallow groove of helix P5.

that all elongator tRNAs possess an invariant G53-C61 base pair at the end of the thymine helix [22]. The proposed contact between M1 RNA and its tRNA substrate (Figure 5) would explain the strong conservation throughout evolution of the G53-C61 base pair (in addition to its structural role in the maintenance of the thymine loop). A highly conserved -GNRA- loop occurs indeed at the end of helix P3. Again, in Gram positive bacteria, it is not conserved but, in those sequences, another -GNRA- appears at the end of a new helix, which removes the possibility of forming a pseudoknot (P13) and which could play the same role as loop L3 in the other instances. Recently, experimental evidence have appeared showing that loop L3 interacts with the end of the T helix in the tRNA substrate (23). Unspecific interactions between adenines in the internal bulge of P7 and the sugar-phosphate backbone of the T-helix of the precursor tRNA were also suggested by the modelling.

The Hepatitis Delta Virus (HDV) Ribozyme

Although the genome of human hepatitis delta virus is a covalently closed, circular single-stranded RNA of about 1700 nucleotides, the catalytic domain, responsible for the autocatalytic cleavages occurring during replication, has barely 90 nucleotides [3,4]. The now accepted secondary structure is highly unusual and is centered on a pseudoknot between an internal bulge and the 3'-end single-stranded region [24]. Figure 6 shows that mutations in three main regions lead to decreased activity: the junctions between I and IV, between IV and II, as well as hairpin III [25]. In

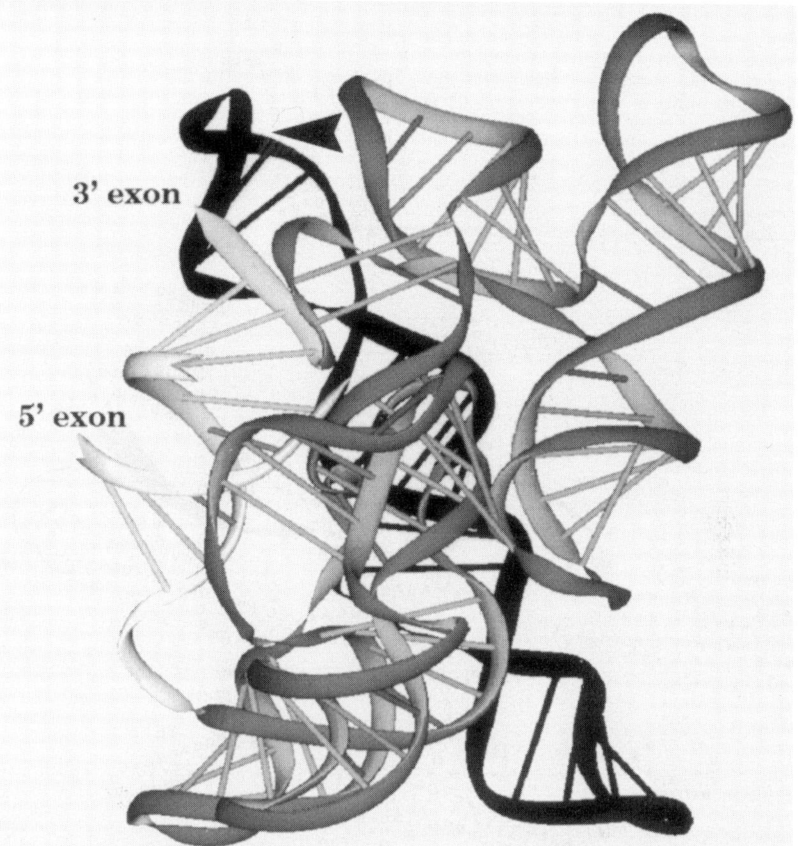

Fig. 4. Ribbon diagram of the three-dimensional model of the *sunY* intron. The modelling is described in Reference 7. The arrow indicates the L9/P5 contact. The drawing was made using DRAWNA [30].

the 3D model (Figure 7), recently proposed on the basis of such mutagenesis data [25], junction between IV and II winds its way between hairpin III and the cleavage site (which is at the base of helix I). This conformational path brings the last two adenine residues of the junction between IV and II in contact with the shallow groove of the G-C base pairs in helix III. The suggested interactions were of the same type as those identified in the -GNRA-/shallow groove motif.

Conclusions

The unravelling of the RNA-RNA interactions underlying the compact and complex tertiary folds of structured RNA molecules started with the crystal structures of transfer RNA molecules. The structures showed non-Watson–Crick pairings, intercalation, triple helix formation, as well as hydrogen bonds between O2'-H hydroxyl groups and bases or sugar-phosphate backbone atoms [26]. Since then, our understanding has not progressed much. Despite the fact that the existence of pseudoknots was theoretically envisaged a long time ago [27], their feasibility and the proofs of their existence in several structures RNAs are more recent

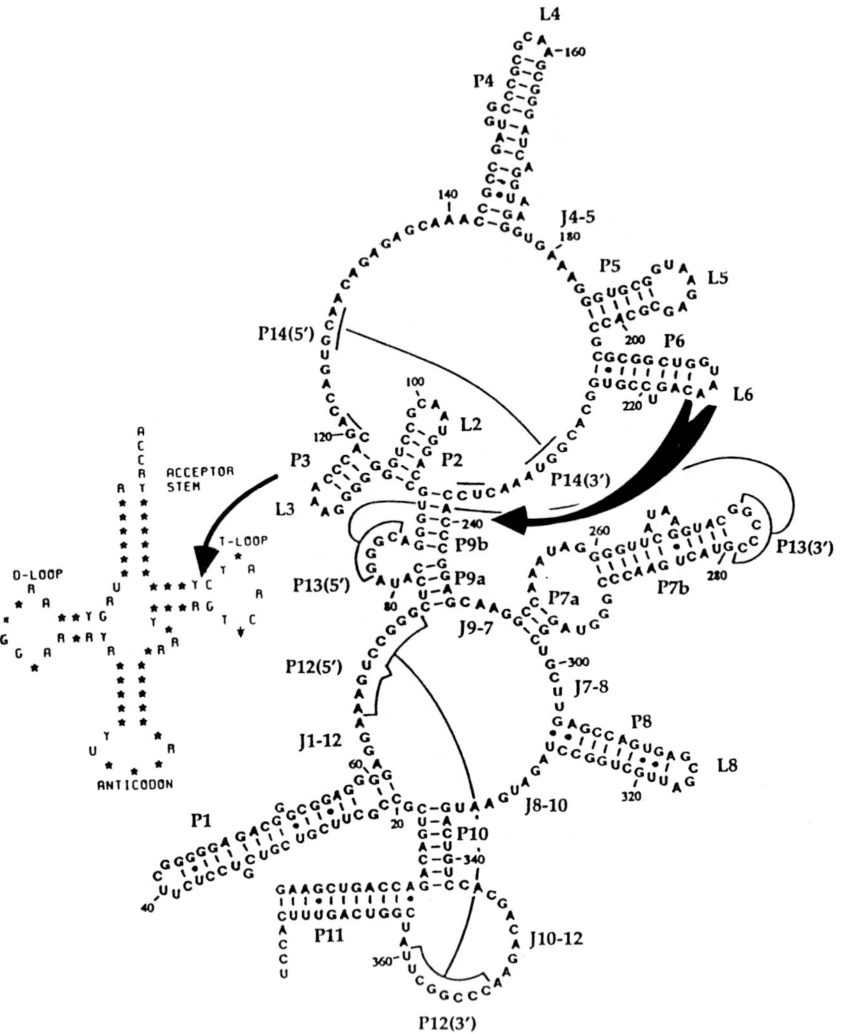

Fig. 5.   Secondary structure of M1 RNA from *Escherichia coli* illustrating the two -GNRA- contacts between L6 and P9b as well as between L3 and the invariant base pair G53-C61 of the tRNA substrate [18].

[28,29]. The first evidence for pseudoknots came from sequence comparisons and, especially, the analysis of compensatory base changes between Watson–Crick complementary residues maintaining a helix. The importance of some tetraloops, like the -GNRA- and -UNCG- loops, was also detected by sequence comparisons. The recognition potential of -GNRA- loops was also recognized by sequence comparisons [6]. Modelling showed the feasibility of the proposed tertiary contacts and mutagenesis together with chemical probing data established the existence of this new RNA-RNA tertiary motif [7,16].

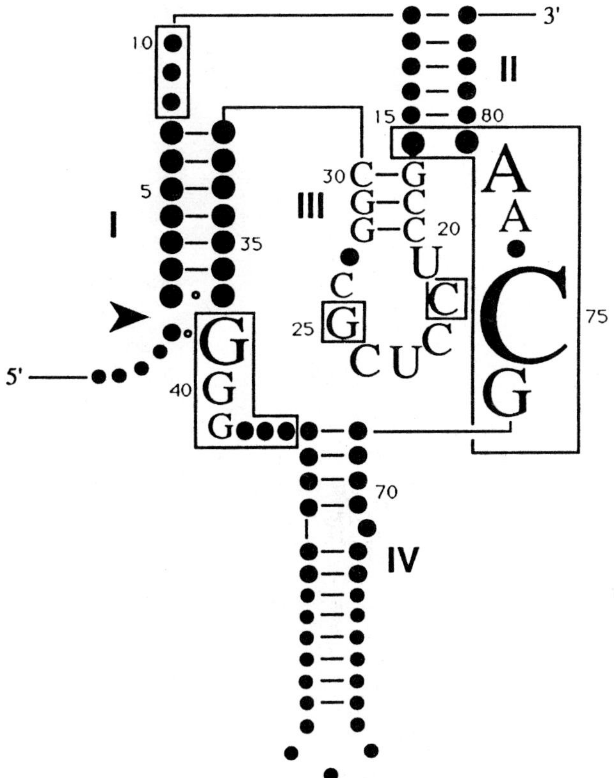

Fig. 6.   Secondary structure of HDV RNA illustrating the residues which lead to interference in autocatalysis (the size is proportional to the effect). Adapted from [25].

Acknowledgements

The work reviewed here was done in a collaborative effort with François Michel (C.G.M.-C.N.R.S.) and Luc Jaeger for the group I introns, with Sidney Altman (Yale University) for the RNAse P molecule, and with Kyle Tanner (GENSET, Paris) for the hepatitis delta virus RNA. My participation at the Symposium was made possible by a travel grant from the Comité National de Biophysique.

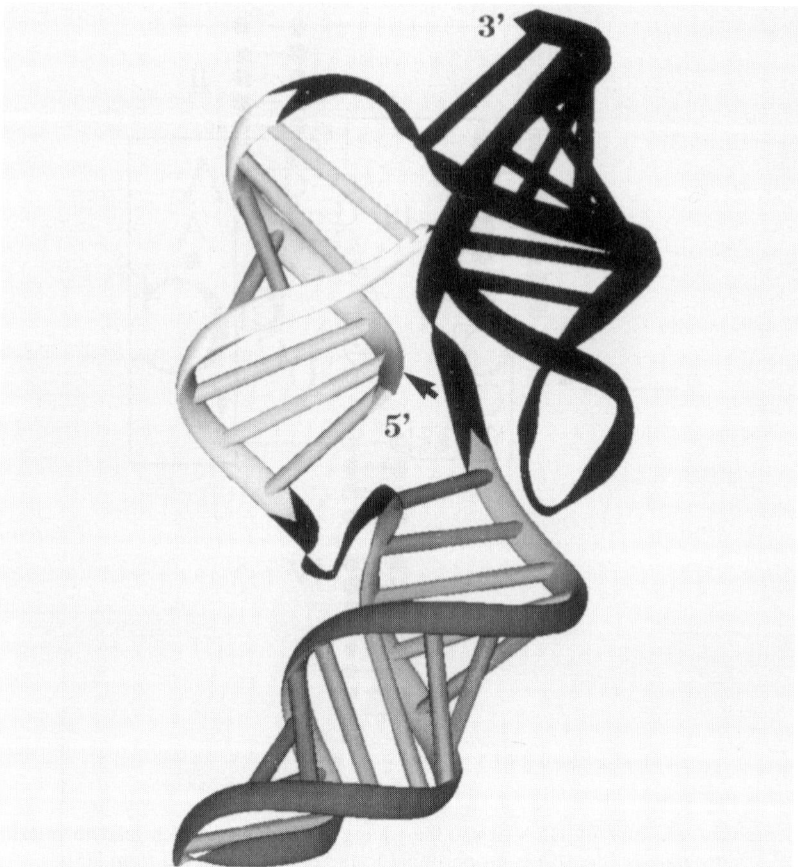

Fig. 7.   Ribbon diagram of the three-dimensional moel of HDV RNA. The 5′ and 3′ extremities are shown. The cleavage site is one base down from the 5′ extremity (indicated by the arrow). The modelling is described in Reference 25. The drawing was made using DRAWNA [30].

References

1. S. Altman, L. Kirsebom, and S. Talbot: *FASEB J.* **7**, 7 (1993).
2. R. H. Symons: *Ann. Rev. Biochem.* **61**, 641 (1992).
3. M. D. Been: *TIBS* **19**, 251 (1994).
4. N. K. Tanner: in *The Unique Hepatitis Delta Virus. The Molecular Biology Intelligence Unit*, G. Dinter-Gottlieb (Ed.), Austin, Texas, R.G. Landes, (1994, in press).
5. T. R. Cech: in *The RNA World*, R. F. Gesteland and J. F. Atkins (Eds.), Cold Spring Harbor Laboratory Press, p. 239 (1993).
6. F. Michel and E. Westhof: *J. Mol. Biol.* **216**, 585 (1990).
7. L. Jaeger, E. Westhof, and E. Michel: *J. Mol. Biol.* **234**, 331 (1993).
8. A. M. Pyle and T. R. Cech: *Nature* **350**, 628 (1991).
9. A. M. Pyle, F. L. Murphy, and T. R. Cech: *Nature* **358**, 123 (1992).
10. J. F. Wang and T. R. Cech: *Science* **6**, 526 (1992).
11. J. F. Wang, W. D. Downs, and T. R. Cech: *Science* **260**, 504 (1993).
12. S. A. Strobel and R. Cech: *Nature Struct. Biol.* **1**, 13 (1994).
13. C. R. Woese, S. Winker, and R. Gutell: *Proc. Natl. Acad. Sci. USA* **87**, 8467 (1990).

14. E. Westhof, P. Romby, P. J. Romaniuk, J. P. Ebel, C. Ehresmann, and B. Ehresmann: *J. Mol. Biol.* **207**, 417 (1989).
15. H. A. Heus and A. Pardi: *Science* **253**, 191 (1991).
16. L. Jaeger, F. Michel, and E. Westhof: *J. Mol. Biol.* **236**, 1271 (1994).
17. F. L. Murphy and T. R. Cech: *J. Mol. Biol.* **236**, 49 (1994).
18. E. Westhof and S. Altman: *Proc. Natl. Acad. Sci. USA* **91**, 5133 (1994).
19. J. W. Brown and N. R. Pace: *Nucleic Acids Res.* **20**, 1451 (1992).
20. H. Shiraishi and Y. Shimura: *EMBO J.* **7**, 3817 (1988)
21. J. W. Brown, E. S. Haas, D. G. Gilbert, and N. R. Pace: *The Ribonuclease P sequence Database* (March 1994).
22. S. Steinberg, A. Misch, and M. Sprinzl: *Nucleic Acids Res.* **21**, 3011 (1993).
23. J. M. Nolan, D. H. Burke, and N. R. Pace: *Science* **261**, 762 (1993).
24. A. T. Perrotta and M. D. Been: *Nature* **350**, 434 (1991).
25. N. K. Tanner, S. Schaff, G. Thill, F. Petit-Koskas, A.-M. Crain-Denoyelle, and E. Westhof: *Curr. Biol.* **4**, 488 (1994).
26. W. Saenger: *Principles of Nucleic Acid Structure*, Springer -Verlag, New York (1984).
27. R. G. Richards: *Eur. J. Biochem.* **10**, 36 (1969).
28. C. W. A. Pleij: *TIBS* **15**, 143 (1990).
29. E. Westhof and L. Jaeger: *Curr. Opn. Struct. Biol.* **2**, 237 (1992).
30. C. Massire, C. Gaspin, and E. Westhof: *J. Mol. Graphics* **12**, 201 (1994).

A Computer Simulation Study of the Relation between Lipid and Probe Behaviour in Bilayer Systems

HADASS EVIATAR,[1] UULKE A. VAN DER HEIDE,[2] and YEHUDI K. LEVINE[3]
*Debye Research Institute and Department of Molecular Biophysics, Buys Ballot Laboratory,
University of Utrecht, P.O. Box 80.000, 3508 TA, Utrecht, The Netherlands*

Abstract. Computer simulations are presented of the behaviour of elongated probe molecules anchored to the interface of bilayers of dipalmitoylphosphatidylcholine (DPPC) above the phase transition of the hydrocarbon chains. The simulations mimic the behaviour of the fluorescent probe 1-[4-(trimethyl-ammonio)phenyl]-6-phenyl- 1,3,5-hexatriene (TMA-DPH) and Cholestane spin label (CSL) in the DPPC bilayers. In contrast to any experimental technique the simulations follow the behaviour of both the lipid molecules and the probe within the bilayer structure. Thus the relation between the behaviour of the probe molecules and the order and dynamics of the lipid chains can be studied in detail. We find that the presence of probe molecules, at the low concentrations used experimentally, causes only a marginal perturbation in the intrinsic properties of the lipid chains. The simulations presented substantiate the conventional prescription for describing the orientational behaviour of probe molecules in lipid bilayers in terms of a local effective orienting potential. They show further that the potential arises from the confinement of the probe molecules between long segments of lipid chains in elongated free-volume cavities within the bilayer structure. Consequently the description of the rotational dynamics needs to be refined in order to take into account the combined effect of the restricted free rattling motions of the probes within the free-volume cavities and the orientations of the cavities themselves relative to the normal to the bilayer plane. The time scale of the motions of the cavities within the bilayer is determined by the rotational motions of long segments of the lipid chains. These observations justify the use of rigid probe molecules such as TMA-DPH and Cholestane spin labels for monitoring the orientational order and dynamics in lipid bilayer systems.

Key words. Lipiol bilayer, computer simulation, Monte Carlo method, membrane dynamics.

Introduction

Probe techniques such as ESR spectroscopy and fluorescence depolarization are widely used for the study of the orientational order and rotational dynamics in membrane systems. The experiments monitor the behaviour of extraneous probe molecules embedded in the membrane structure at the smallest possible concentration compatible with an acceptable signal/noise ratio. A commonly used ESR probe is Cholestane Spin Label (CSL), in which the nitroxide group is rigidly bound to the nucleus of the cholestane molecule so that the A-tensor anisotropy is large along an axis perpendicular to the long axis of the molecule. The ESR lineshape of CSL thus reflects both the tumbling motion of the long axis of the molecule relative to the external magnetic field and the rotational motion about

[1] *Present Address*: Institute for Biodiagnostics, National Research Council of Canada, Winnipeg, Canada.
[2] *Present Address*: Department of Medical Physiology, Utrecht University, Utrecht, The Netherlands.
[3] Author for correspondence.

A. Pullman et al. (eds.), *Modelling of Biomolecular Structures and Mechanisms*, 315–331.

this axis. Fluorescence depolarization experiments often employ the rigid elongated probe molecules (1-[4-(trimethylammonio)-phenyl]-6-phenyl-1,3,5-hexatriene, TMA-DPH), incorporated in macroscopically oriented bilayer systems. However, TMA-DPH only monitors the tumbling motion since the absorption transition dipole moment lies along the long molecular axis. A common feature of both these probe molecules is that they are anchored at the aqueous interface of the membrane.

The tacit assumptions underlying the models used for the analysis of the experiments are that the long axis of each probe molecule tumbles in a mean field orienting potential [1–4], and that the rotational motion about the long axis of the molecule is independent of the tumbling of the long axis. The orienting potential for a rigid, cylindrical molecule in a lipid bilayer, where the bilayer normal is now the z-axis of the system, may be defined in terms of Legendre polynomials in the polar angle β:

$$\frac{U(\beta)}{kT} = \sum_{L \geq 0} \lambda_L P_L(\cos \beta). \tag{1}$$

As the interactions involved in both ESR and fluorescence depolarization are second rank symmetrical tensors, only the even rank order parameters such as $\langle P_2 \rangle$ and $\langle P_4 \rangle$ can be determined experimentally [1,4]. It is therefore customary to neglect the odd values of L and to truncate the series after one or two terms. Most software for the simulation of ESR spectra or the fluorescence anisotropy decay curves of cylindrical probe molecules in lipid bilayers [1,4,5] makes use of the horizontal plane of symmetry of the bilayer, so that the orientating potential is symmetric about $\pi/2$. The orientational distribution function is the normalized Boltzmann distribution corresponding to the orienting potential, Equation (1), using only the terms with $L = 2$ and 4. The typical values found for λ_2 and λ_4 in fits of experimental spectra are such that there is a barrier of high energy at $\beta = \pi/2$, resulting in a very low probability that the molecules will actually flip from one half of the bilayer to the other. It is therefore to be expected that a bilayer in which the distribution function for the probe molecules is very sharp and symmetrical about $\beta = \pi/2$ will be equivalent to two monolayers stacked back to back, each with its own distribution function which has a maximum at $\beta = 0$ and decreases monotonically to 0 at $\beta = \pi/2$.

The mean field prescription for describing the behaviour of probe molecules in lipid bilayers has been successful in describing not only ESR spectral lineshapes, but also fluorescence depolarization experiments on probes incorporated in macroscopically oriented bilayer systems. Both probe techniques yield the same information about molecular order and rotational dynamics for probes anchored at one end in the aqueous interfaces. However, with vesicles, great difficulties have been encountered in the interpretation of fluorescence anisotropy decays [6,7]. In particular, the experiments could only be described using orientational distribution functions which contained a sizable probe population at $\beta = \pi/2$ as well as a maximum at $\beta = 0$. This result runs counter to the amphiphilic properties of the TMA-DPH molecule. In marked contrast, the ESR spectra of CSL in vesicles could be simulated excellently using a purely unimodal distribution [5], which decreases monotonically from a maximum at $\beta = 0$ to a minimum at $\beta = \pi/2$. As we expect the same physical processes to determine the motions of both fluorescent molecules and ESR spin probes, we need to address the question as to why the ESR spectra of probe molecules in lipid systems can be simulated well using a unimodal distribution. We now need to bear in mind that fluorescence depolarization techniques monitor the time-correlation functions of the rotational motions of the transition dipole moments of the probe molecule. Thus the information about the dynamics is obtained from direct fits of the experimental data to model predictions. On the other hand, molecular motions modulate the ESR spectral lineshapes of the nitroxide probes, so that the dynamics are extracted by a somewhat cumbersome procedure involving simulations of the lineshapes within the framework of the Stochastic Liouville Equation [2–5].

A recent re-evaluation of time resolved fluorescence depolarization measurements in vesicles and in macroscopically aligned multibilayer systems has led to the formulation of an alternative description of the motion of a probe molecule in a lipid bilayer [8,9]. The compound motion (CM) model involves the motion of the molecule within a cage which itself is slowly tumbling in the bilayer structure. This description is based on the Slowly Relaxing Local Structure model advanced by Freed and coworkers for the simulation of ESR lineshapes of probe molecules in nematic liquid crystals [10]. The cage may be envisaged as a free volume cavity between the lipid molecules, so that its position and orientation change with time. The total motion of the molecule reflects both its own rapid reorientation within the cage and the slow tumbling of the cage. Consequently all the time-correlation functions corresponding to the compound motion contain two widely separated timescales. The conventional BRD model does not allow for such timescale combination, but accounts for this effect by introducing two probe populations. In this way fast probe motions within each local potential minimum are combined with the slow transitions between them. The interesting question now is why this mode of motion has not been found in simulations of ESR spectra.

A better insight into the information provided by probe techniques about membrane systems, can be gained using computer simulations techniques. In contrast to any experimental technique the simulations follow the behaviour of both the lipid molecules and the probe within the bilayer structure. Thus the relation between the behaviour of the probe molecules and the order and dynamics of the lipid chains can be studied in detail. Moreover, the simulations afford a direct comparison between the local information revealed about the order of C—D bonds along the chains by ^{2}H-NMR experiments and the orientational behaviour of long chain segments in the bilayer. In this way the simulations can be used to relate experimental data obtained from selective experiments such as NMR and from probe molecule techniques. An added advantage of this approach is that it visualizes the effects of the incorporation of probe molecules on the order and dynamics of the lipid molecules. The trajectories of the time evolution of the molecular orientations $\Omega(t)$ obtained from the simulations can be used both for the evaluation of the time-correlation functions monitored in fluorescence depolarization experiments as well as for the simulation of ESR spectra using the time-domain algorithm recently developed by us [11].

We shall here use lattice Monte Carlo dynamics (MCD) methods for following simultaneously the conformational fluctuations of the lipid chains in bilayer systems and the rotational motions of the probe molecules between them. For the sake of computational efficiency, only the intramolecular conformational energy of the lipid chains and excluded volume interactions are taken into account. We find that the presence of probe molecules causes a marginal perturbation in the intrinsic properties of the lipid chains. The time-correlation functions obtained from the simulations substantiate the compound motion prescription. They indicate furthermore, that the orientational order and rotational rates of the probe molecules are determined to a large extent by the fluctuations in the free-volume cavities of the bilayer structure. Importantly, the orientational order and the timescale of the tumbling motion of the long molecular axis is found to reflect that for long chain segments in the bilayer structure. Interestingly, we find that the BRD model based on the orienting potential of Equation (1) yields good fits to the CSL spectra calculated from the trajectories. This is surprising as the model is based on a much simpler type of motion than that revealed by the time-correlation functions. We believe that this insensitivity of the spectral lineshapes to the details of underlying physical processes arises from the similarity of the effects produced by the rotational motion about the long molecular axis and the fast rattling motion of the long axis in the free-volume cavity.

Method of Simulation

a. MONTE CARLO SIMULATIONS OF BILAYER SYSTEMS

The Monte Carlo (MCD) technique for simulating the dynamic behaviour of hydrocarbon chains of lipids in monolayer and bilayer structures has been discussed and validated in detail previously [12–15] and only the salient points will be summarized here.

The method makes use of a representation of hydrocarbon chains on a cubic lattice. The bilayer was constructed of two monolayers each containing 64 model lipid chains. Two of the molecules in each monolayer were subsequently replaced by rigid probes. A unified atom representation was used, with the first headgroup atom of each chain being dressed with a larger excluded volume envelope than the atoms depicting the methylene groups. The bilayer was contained in a Monte Carlo box having a square cross-section in the XY-plane of the lattice and extending along the positive Z-axis. The impenetrable interfaces of the monolayers forming the bilayer system were taken to lie parallel to the XY-plane of the lattice, at heights $Z = 0$ and $Z = T_{bil}$, with T_{bil} the bilayer thickness. The molecules were attached to the interfaces with the first atom being allowed to move vertically by two lattice units. Periodic boundary conditions were imposed on the positions of all the beads in the XY-plane only.

The conformational dynamics of the lipid chain is considered to arise from a superposition of local structural rearrangements constrained by the bond lengths. The elementary move involves the transfer of a pair of adjacent atoms chosen at random to different lattice sites. The first and last bonds are allowed to undertake random orientations. The lack of restrictions on the positions of the headgroup atoms in the XY-plane resulted in a lateral translational motion of the chains across that plane.

Each random elementary move attempted by the atoms is subjected to three acceptance tests:

1. A move is only accepted if the final lattice positions of the dressed atoms are unoccupied.
2. The new configuration was accepted with a probability P given by the symmetric scheme:

$$P = \exp(-E_{\text{new}}/kT)/\{\exp(-E_{\text{old}}/kT) + \exp(-E_{\text{new}}/kT)\}$$

3. Chain configurations with g^+g^- or g^-g^+ conformations about adjacent bonds are rejected.

The MCD algorithm for a bilayer containing M hydrocarbon chains each consisting of N atoms was executed in the following way. Given a particular configuration of the bilayer, Ω_j, $M \times N$ local conformational moves are attempted at random with each atom in the system having an equal chance of being picked. These $M \times N$ moves generate a new configuration of the bilayer, Ω_{j+1}. The cycle is now repeated using Ω_{j+1} as the starting configuration. The fundamental time-step of the algorithm is defined as the time required for the bilayer to undergo a transition from configuration Ω_j to Ω_{j+1}. It is important to realize that this time step of the MCD algorithm cannot be related to an absolute scale.

b. SIMULATION OF PROBE MOTION

The probe molecules were represented as rigid, solid objects 21 (or less) lattice units long. Their cross-section was taken to be a square with a side of five lattice units. The molecules are anchored to the bilayer interface at one end and allowed to undergo random rotational and translational motions. The moves were subjected to a single acceptance test, namely that the final lattice positions were unoccupied.

The vertical translational motions were restricted to three lattice units relative to the bilayer interfaces, but no restrictions were placed on their lateral motions in the XY-plane. The axis of the cylindrically symmetric probe molecule is represented by the vector $\mathbf{R}$, Figure 1. The tip of the vector $\mathbf{R}$ is allowed to jump over all the lattice points enclosed in a spherical shell with an internal radius of 35 lattice points and an external radius of $(36^2 - 1)^{1/2}$ units. This procedure is efficient as it allows the use of a lookup table for selecting the random rotational motions of the vector, by jumping to a randomly chosen destination from any starting position in the spherical shell. Small angular jumps are realized by restricting the linear displacement of the tip to less than 5 lattice units. The internal radius of the shell was chosen as a useful compromise between the size of the lookup table and the smoothness of the distribution of vector orientations. Given a particular orientation $\Omega(t_1)$ of the vector $\mathbf{R}$, a new orientation $\Omega(t_2)$ is chosen at random from the space of allowed orientations. The algorithm of the simulations of bilayer containing probe molecules is similar to the one described above for the lipid

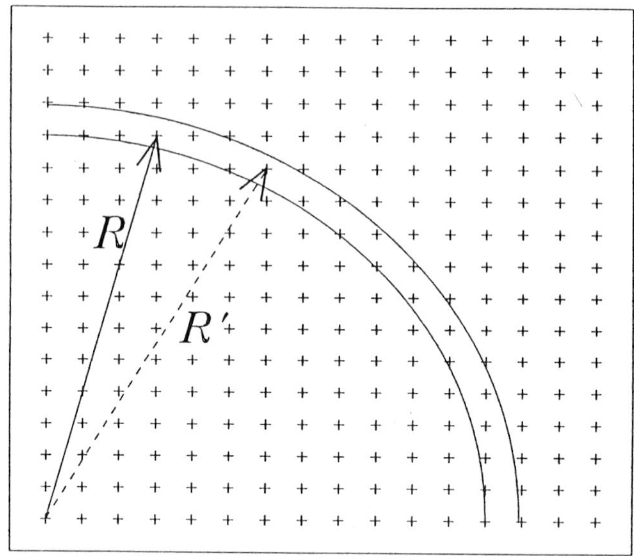

Fig. 1.   MCD scheme for the rotational motion of the vector **R** on a cubic lattice. The vector changes to orientation **R'** by allowing its tip to jump to lattice points within the spherical shell. Small-angle diffusion is attained by limiting the size of the jumps.

chains. Now a move of each probe molecule is attempted once every simulation cycle leading to a new bilayer configuration.

c. EVALUATION OF TIME-CORRELATION FUNCTIONS FROM THE MCD TRAJECTORIES

We have typically generated a trajectory consisting of 2^{17} bilayer configurations each separated by 32 elementary time-steps at a temperature of 330 K for the calculation of ESR spectra. A significantly shorter trajectory containing 2^{14} configurations sufficed for the calculations of the time-correlation functions. The calculation of the longest trajectory for a bilayer of 124 chains each consisting of 18 atoms and 4 probe molecules required around 90 hours on a Silicon Graphics Iris Indigo workstation. With this choice, the statistical fluctuations in the calculated order parameters for the probe molecules and chains obtained from runs starting with different initial bilayer configurations varied by less than 2%. Furthermore, the decay of the time-correlation functions for every atom of the chain and for the long axis of the probe molecules was found to be independent of the sampling frequency of the trajectory generated. For the sake of simplicity we shall henceforth take the time t to be measured in steps along the trajectory of orientations rather than in elementary time steps of the MCD algorithm.

In order to characterize the rotational motions of the probe molecules we have used the trajectories to evaluate three orientational time-correlation functions $G_{kl}(t)$, $k, l = 0, 1, 2$

$$G_{kl}(t) = \langle\langle D^2_{kl}(\Omega_{BM0})D^{2*}_{kl}(\Omega_{BMt})\rangle\rangle.\tag{2}$$

Here D^L_{mn} are Wigner rotation matrix elements [16] and Ω_{BM0} and Ω_{BMt} denote the orientation of the Z-axis of the probe molecule at time $t = 0$ and t, resectively, in the bilayer frame. The double bracket indicates an average over the probe or lipid molecules. The amplitudes of the orientational time-correlation functions at time $t = 0$ can be expressed as linear combinations of the order parameters $\langle P_2\rangle$ and $\langle P_4\rangle$ in a model-independent way. It is important to note now that the decay of the correlation functions $G_{k0}(t)$, $k = 0$, 1, 2, is described by the diffusion coefficient $D_\perp$ of the tumbling of the long molecular axis only. On the other hand, the decay of the correlation functions $G_{k2}(t)$, $k = 0$, 1, 2, is described by both $D_\perp$ and $D_\parallel$, the diffusion coefficient for rotation about the long axis.

The time-correlation functions were fitted to the predictions of the BRD and CM models using a nonlinear Marquardt optimization procedure. It will also prove convenient when dealing with the motion of long chain segments to consider the total time-correlation function $G_0(t)$:

$$G_0(t) = G_{00}(t) + 2G_{10}(t) + 2G_{20}(t).\tag{3}$$

We note that this time-correlation function is the fluorescence anisotropy decay measured in vesicle systems [1].

d. SIMULATIONS OF ESR LINESHAPES FROM THE MCD TRAJECTORIES

The simulation of ESR absorption spectra from the trajectories generated by the MCD calculations was carried out as described previously [11]. The reorientational motion of the probe molecule relative to the static magnetic field results in stochastic behaviour of the spin Hamiltonian $H(t)$. The algorithm essentially consists of the numerical solution of the time-dependent Schrödinger equation, i.e., the evaluation of the time evolution over the trajectory of the spin wave function $\psi(t)$:

$$\psi(t) = \exp\left[i\int_0^t H(t')\mathrm{d}t'\right]\psi(0).\tag{4}$$

This wave function is used to calculate the expectation value of the complex transverse magnetic moment of the electron. The autocorrelation function of the expectation value is the free induction decay (FID) of the magnetization. The Fourier transform of this FID is the required spectrum.

The time dependence of the Hamiltonian is obtained simply from the Euler angles comprising the trajectory. As the MCD simulation has no absolute time scale, we are free to choose the value of the time increment δt between trajectory steps when calculating the spectra. A choice in the picosecond region results in motionally narrowed spectra, whereas a longer δt allows the calculation of spectra in the slow motion regime and in the rigid limit. However, if the loss of correlation in the trajectory is too slow, the spectra quickly approach the rigid limit.

For each motional regime, X-band ($B_0 = 3175$ G) spectra were calculated for multibilayer systems with the normal oriented at 0, $\pi/6$, $\pi/3$, $\pi/2$ to the external magnetic field. Experimental spectra generally have a resolution of dB $-$ 0.1 G; matching this resolution would require the FID in the simulation to be calculated for a very long time, according to the relationship $dB \propto 1/\Delta t$, where Δt is the duration of the FID. A resolution ≈ 0.2 G, requiring a total duration of each FID $1.78\,\mu$sec, was found to be a useful compromise between resolution and computer time. The spectra were then interpolated to achieve a resolution of 0.1 G. Each spectrum consumed about 25 minutes of CPU time on a Silicon Graphics Iris Indigo workstation. To mimic experimental spectra, which show residual effects not caused by motion, Gaussian and Lorentzian broadening were added. This had the added advantage of a significant smoothing of the spectra, allowing good spectra to be calculated with fewer points. The magnetic parameters for CSL were: $\mathbf{g}$ = diag (2.0081, 2.0024, 2.0061), $\mathbf{A}$ = diag (5.6 G, 33.7 G, 5.3 G).

Results and Discussion

a. EQUILIBRIUM PROPERTIES

MCD simulations were carried out in order to model the behaviour of bilayers of dipalmitoylphosphatidylcholine (DPPC), at 330 K, above the phase transition temperature of the hydrocarbon chains. Only the intramolecular torsional potentials and excluded volume interactions were implemented. The simulations were performed on varying the side of the Monte Carlo box in the XY-plane from 54 to 72 lattice units at a constant temperature. In this way the effective area of each of the 128 lipid chains was changed over a wide range. The order parameters S, $S = \frac{1}{2}\langle 3 \cos^2 \beta - 1 \rangle$ of the C—D bonds of the lipid chains were found to decrease on increasing the side of the Monte Carlo box, in agreement with experimental data [17]. Here β denotes the angle between the C—D vector and the normal to the bilayer interfaces, the lattice Z-axis. It now needs to be recognized that the C—D vector in a saturated hydrocarbon chain is oriented tetrahedrally relative to the plane defined by the sp^3—sp^3 bonds attached to the atom [13,14].

We have found that the order parameter profile of the C—D bonds in DPPC in a box of side 54 lattice units in the XY-plane, Figure 2, tracks well the experimental profile observed using ^{2}H-NMR. The orientational order of the C—D vectors is also reflected in the preferential alignment of the end-to-end vectors of the chains in a direction parallel to the bilayer normal, $S \approx 0.5$.

Two probe molecules were now introduced into each monolayer containing 31 lipid molecules. This effective molecular ratio is significantly higher than that used experimentally (about 1:250). This high ratio was used with the sole object of keeping the simulation time within reasonable bounds. Simulations for a DPPC bilayer were now carried out as a function of the length of the model probe molecules. Probe molecules less than 11 lattice units long showed no significant orientational order with a low order parameter S for their Z-axes. It is important to note here that such a distribution simply implies an equal distribution of the long molecular axes over the surface a sphere whose pole axis is the normal to the bilayer surface.

The probe molecules start to exhibit substantial orientational order only on increasing their length to 15 lattice units and above. Indeed, the experimental values for the order parameters, $S \approx 0.6$, found for TMA-DPH and Cholestane spin labels are obtained for probes 21 lattice units long in agreement with experi-

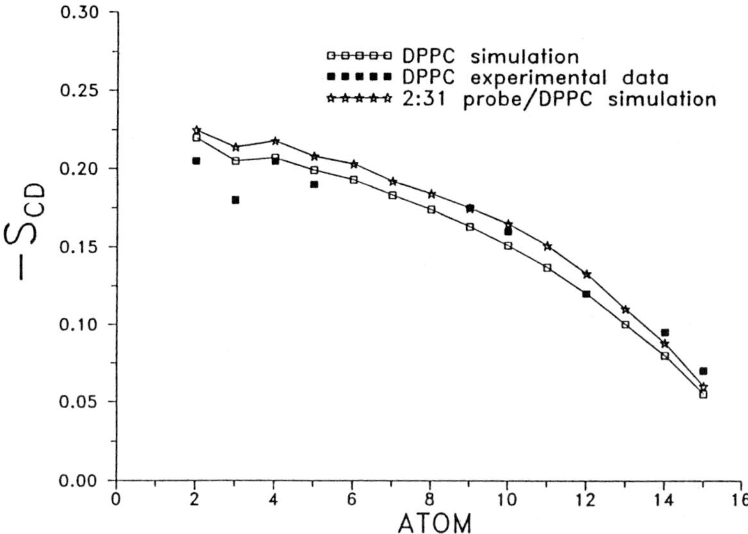

Fig. 2. A comparison of the order parameter profile of C—D bonds along the hydrocarbon chains in bilayers of DPPC measured in ^{2}H-NMR experiments and simulated using MCD. Also shown is the effect of incorporating probe molecules (2:31 molar ratio) anchored at the aqueous interfaces.

ments on oriented bilayer systems [18]. We shall henceforth only consider probes of this length.

The number density of probes relative to the bilayer normal for DPPC bilayers is shown in Figure 3. The number density simply gives the fractional number of times the Z-axis of the probe is found at a given angle β with respect to the normal to the monolayer surface during the simulation. The normal was taken to be pointing outwards towards the water phase. The number density for the bilayer as a whole can be obtained by reflecting that shown in Figure 3 about $\beta = \pi/2$. Conventionally, the orientational distribution function $f(\beta)$ is presented rather than the number density $n(\beta)$, $n(\beta) = f(\beta) \sin \beta$. The general form of the number density reflects the fact that the probe axes are actually distributed over the surface of a sphere and in no way implies a collective molecular tilt relative to the bilayer normal.

The number density distribution can be seen to broaden and moreover to shift towards higher angles on increasing the side of the Monte Carlo box from 54 to 72 lattice units, Figure 3. This change corresponds to a reduction in the order parameter of the probe from 0.60 to 0.42. The order parameters of both the C—D bonds over the length of the hydrocarbon chains and the end-to-end chain vectors also decrease significantly on increasing the effective area per molecule.

The finding that only long probe molecules are oriented by the lipid matrix can be rationalizing within the framework of free-volume cavities between the conformationally disordered lipid chains. The short probe molecules can be accommodated easily within free-volume cavities in the bilayer structure. However, these cavities are either too small or do not have a simple geometrical form for containing

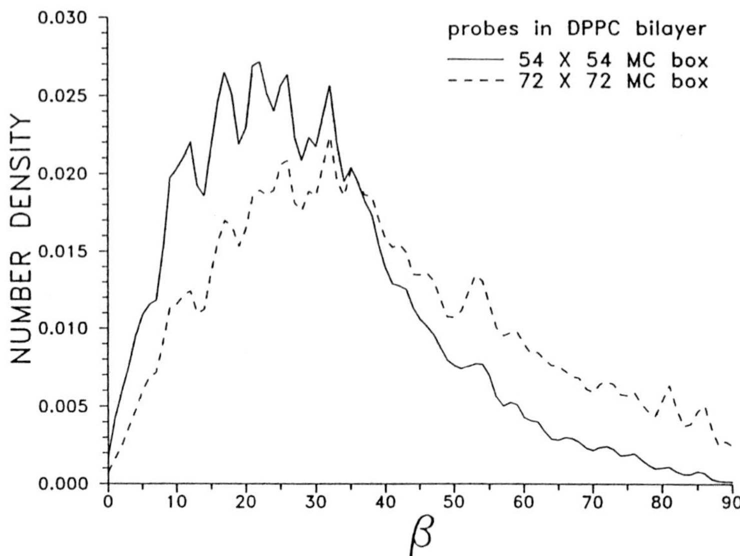

Fig. 3.   The fractional number of times (number density) the long axis of an anchored probe molecule is encountered at an angle β relative to the bilayer normal along the trajectory obtained from MCD simulations. The number densities for the two monolayers have been combined by reflection about the symmetry plane at $\beta = \pi/2$. The ripples are caused by the underlying lattice used in implementing the rotational jumps.

the long probe molecules. Consequently, the rotational motions of the long probe molecules are determined by the availability of large free-volume cavities created by the conformational motions of the lipid chains. Such large cavities are more likely to exist in a direction parallel to long segments of hydrocarbon chains than across them, in a direction parallel to the bilayer surface. As the long chain segments lie on average perpendicular to the bilayer surface, we expect the long probe molecules to exhibit a preferential orientation along the normal to the bilayer. Indeed, the order parameters for the end-to-end vectors of the chains in DPPC bilayers are some 10–15% higher than the order parameters of the probe molecules, in agreement with our working hypothesis.

b. PERTURBATION OF BILAYER AND PROBE MOLECULES

The order parameter profile for the C—D bonds of the lipid chains in the presence of two probes per 31 lipid molecules, is only slightly higher than that found for the pure bilayer, Figure 2. This effect is akin to the increase in the order parameters of the C—D bonds induced by cholesterol molecules in lipid bilayers. Indeed, we have previously reproduced the effects of cholesterol on the order parameters by introducing these probe molecules into lipid bilayers up to a molar ratio of 1:1 [12].

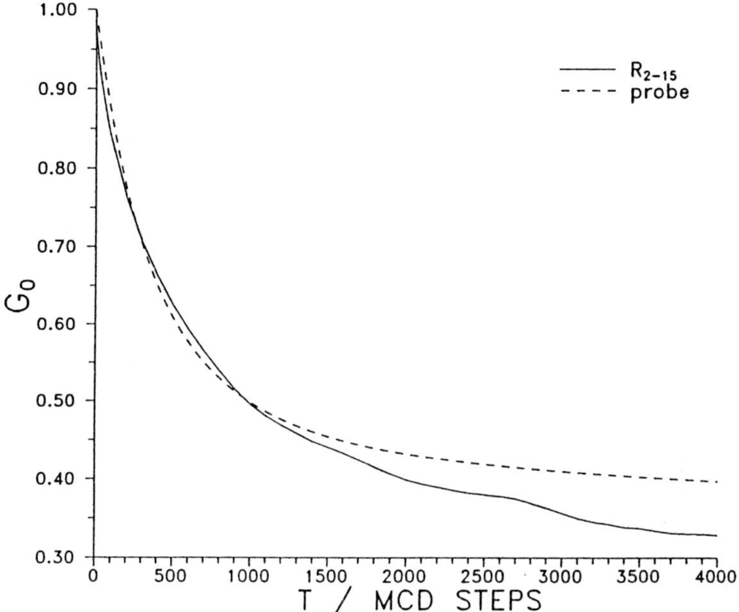

Fig. 4. The decay of the orientational time-correlation function G_0 (see text) of the vector connecting atoms 2 and 15 of the hydrocarbon chains in bilayers of DPPC and for the long axes of the probe molecules. Also shown is the corresponding time-correlation function extracted from the CM model for the slow motions of the long axes of the probe molecules in the bilayer.

c. PURE LIPID BILAYER – DYNAMIC PROPERTIES

In order to gain insight into the dynamic properties of the lipid chains which are monitored by the probe molecules, we shall now consider the rotational motions of large chain segments. The orientational time-correlation function $G_0(t)$ for the vectors $\mathbf{R}_{2-15}$ joining atoms 2 and 15 in chains of DPPC is shown in Figure 4. The time-correlation function decays to a constant plateau at long time whose level is given by S^2, with S the corresponding order parameter of the vector. The time-correlation functions decay faster and the level of the long time plateau drops on increasing the size of the Monte Carlo box. The behaviour is indicative of lower orientational order and higher rotational rates on increasing the effective area per molecule of the lipid chains.

d. PERTURBATION OF BILAYER DYNAMICS BY PROBE MOLECULES

The decay of the orientational time-correlation functions for the chain segments shown are marginally faster in the presence of probe molecules at a molar concentration of 2:31 lipid molecules within the level of the statistical uncertainties. The generalized correlation time obtained from the area under the curves decreases by less than 10%. The simulations thus indicate again that the incorporation of non-interacting probe molecules in the bilayer structure causes minimal perturbations to the lipid chain dynamics.

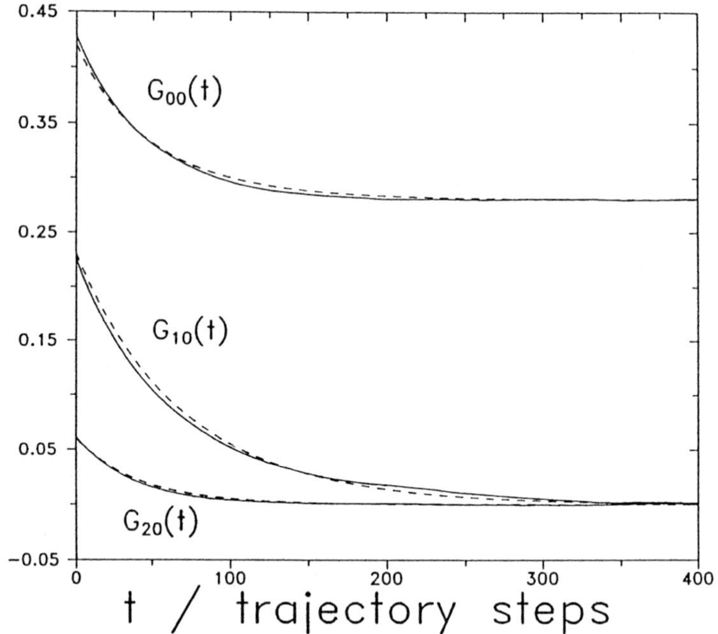

Fig. 5.   BRD model fits to the correlation functions G_{k0} calculated from the MCD trajectories for a free probe tumbling in an orienting potential. The solid lines are the MCD simulations and the dashed lines give the BRD fits.

e. PROBES IN LIPID BILAYERS – FLUORESCENCE DEPOLARIZATION

We shall here analyze the time-correlation functions $G_{k0}(t)$ calculated from the trajectories in the conventional way by fitting them to the predictions of the BRD and CM models. These time-correlation functions are monitored in fluorescence depolarization experiments using the probe TMA-DPH and it is important to note that only the tumbling motion of the long molecular axis contributes to their decay.

The solutions provided by the models can now be validated by comparing the extracted order parameters with the ones obtained directly from averaging along the trajectories. Order parameters, such as $\langle P_2 \rangle$ and $\langle P_4 \rangle$, can be calculated in a straightforward way from the number density obtained from the simulations, Figure 2.

The algorithm used by us for implementing the rotational motions of the probe molecules was first of all tested by simulating the time-correlation functions $G_{k0}(t)$, $k = 0, 1, 2$, for a free probe moving subject to the orienting potential of Equation (1). We have found excellent agreement between the simulated decay curves and those obtained from a numerical solution of the BRD for a variety of potentials, Figure 5.

The time-correlation functions, $G_{k0}(t)$, $k = 0, 1, 2$, calculated along the MCD trajectory of the probe in the bilayer could not be fitted satisfactorily to the predictions of the BRD model using the orienting potential of Equation (1), Figure

TABLE I

Model parameters extracted from a nonlinear least squares fit to the orientational time correlation functions G_{k0}

Model	$\langle P_1 \rangle$	$\langle P_2 \rangle$	$\langle P_3 \rangle$	$\langle P_4 \rangle$	$D_c = \tau^{-1}$ (MCD steps)$^{-1}$	$D_\perp$ (MCD steps)$^{-1}$
Simulation	0.733	0.479	0.273	0.175		
BRD	0.0	0.506	0.0	0.159	–	1.3×10^{-3}
CM inside cavity	0.960	0.883	0.775	0.644	9.7×10^{-2}	–
CM cavity in bilayer	0.816	0.568	0.368	0.243	–	8.4×10^{-4}
CM overall order	0.783	0.501	0.285	0.157		

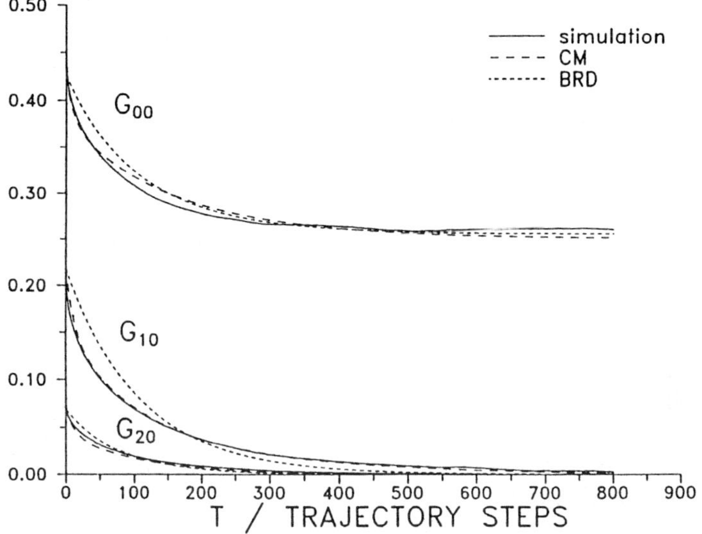

Fig. 6. The decay of the orientational correlation functions G_{k0} (see text) of the long axes of the probe molecules relative to the normal to the surfaces of a DPPC bilayer. Shown are the decays calculated from the MCD simulations and the fits obtained form the CM and BRD models. The model parameters extracted from the fits are given in Table I.

6. While the extracted order parameters $\langle P_2 \rangle$ and $\langle P_4 \rangle$ are in excellent agreement with those obtained from the trajectories, Table I, the BRD model was unable to reproduce the initial fast decay of the time-correlation functions. Note, however, that the BRD model is capable of reproducing the even rank order parameters only by virtue of the symmetry of the orienting potential about $\beta = \pi/2$.

The recently proposed CM model [8,9] provides a good fit of both the decay of the time-correlation functions, Figure 6, and all the order parameters up to rank 4, Table I. As expected the order parameters are found to decrease on increasing the effective area per molecule with a concomitant increase in the rotational diffusion coefficients.

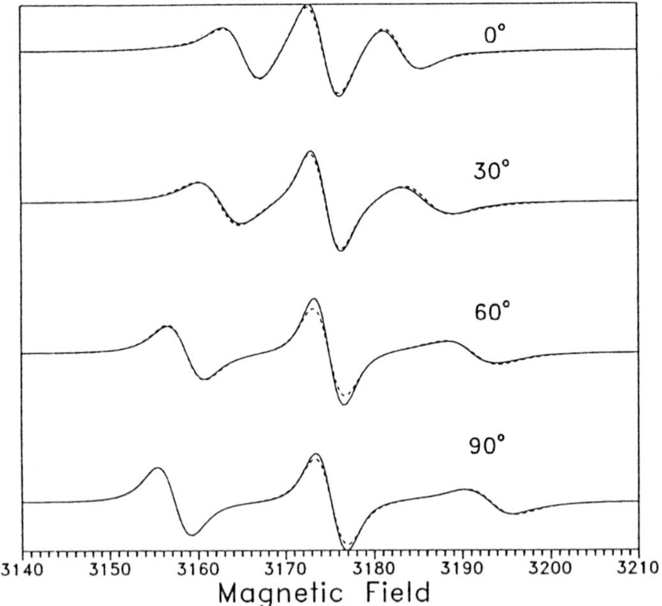

Fig.7.   BRD model fits to the ESR spectral lineshapes calculated from the MCD trajectories. The spectra were calculated in the slow-motion regime for oriented bilayers as a function of the orientation of the normal to their plane relative to the applied magnetic field.

f. PROBES IN LIPID BILAYERS – ESR SPECTRAL LINESHAPES

The ESR spectra of CSL in oriented bilayer systems simulated from the trajectories, Figure 6 were fitted using the BRD in the fast (time step along the trajectory $\delta t = 0.027$ ns) and slow ($\delta t = 0.108$ ns) motional regimes using conventional software [4,5]. The method involves the solution of the Stochastic Liouville equation (SLE) using an expansion of eigenfunctions of the rotational diffusion operator. This approach breaks down for slower motions due to the clustering of the eigenvalues. The spectral fits obtained for both motional regimes were of a quality generally considered to be acceptable, Figure 7. The small discrepancies between the BRD spectra and those generated from the MCD trajectories are most noticeable at angles intermediate between 0 and $\pi/2$ and are primarily manifested in the positions of the ESR spectral lines rather than their widths. Their magnitude is similar to that usually found between lineshape fits and actual experimental spectra. An improvement in the fits can be achieved in an arbitrary way by a small change (<0.3 G) in the largest component (perpendicular to the molecular axis) of the hyperfine tensor.

These findings are in marked contrast to the problems encountered above with the time-correlation functions $G_{k0}(t)$.

The key to resolving this discrepancy must be sought in the fact that the ESR spectra of CSL reflect both the tumbling motion of the long molecular axis and the fast rotational motion about it, while the fluorescence depolarization data only monitors the tumbling motion. For this reason we have also analyzed the time-correlation functions $G_{k2}(t)$ with the BRD model since their decay is determined by both rotational diffusion coefficients $D_\perp$ and $D_\parallel$. Surprisingly, we find that the BRD model provides an excellent fit to the MCD data. This arises simply because the fast rotation about the long molecular axis has the same effect on the decay of the time-correlation functions as the fast rattling motion of the long molecular axis in the free-volume cavity. Consequently, the BRD model cannot differentiate between the two modes of motion. This insensitivity to the details of underlying motional process is manifested in the ESR spectra and for the same reason. Interestingly, the values of the even order parameters and diffusion coefficients recovered from the spectral fits in the slow motion regime were virtually identical to those extracted from the BRD fits of the time-correlation functions $G_{k2}(t)$:

	$\langle P_2 \rangle$	$\langle P_4 \rangle$	$D_\parallel \times 10^8$ (rad^2/sec)	$D_\perp \times 10^8$ (rad^2/sec)
$G_{k2}(t)$	0.50	0.17	2.3	0.13
ESR simulations	0.49	0.18	1.7	0.14

Note that we have here used the value of δt to scale the diffusion coefficients extracted from the time-correlation functions. Consequently, the diffusion coefficients for the fast motion regime are a factor 4 higher than those given above. The MCD simulations thus strongly suggest that the experimental ESR spectra of CSL spin labels in lipid bilayer systems are not sensitive to the details of the underlying physical dynamic processes as manifested in the time-correlation functions. The reason for this is the similarity in the effects of the fast rotational diffusion about the long molecular axis and the fast rattling motions between the lipid chains on the CSL spectral lineshapes. In order to test the sensitivity of the ESR spectral lineshapes to the compound motions we used the same trajectories to simulate spectra from a fictitious probe whose magnetic hyperfine tensor is oriented with its highest anisotropy parallel to the long molecular axis. This is achieved simply by interchanging the y- and z-components of the magnetic tensors of the nitroxide molecule. Now the ESR spectrum is most sensitive to the tumbling of long molecular axis, a situation similar to that encountered with the fluorescent probe TMA-DPH. The quality of the spectral fits of these spectra using the BRD model with the parameters above is considerably worse than for CSL in the fast time regime ($\delta t = 0.027$ ns). Interestingly, good fits are obtained in the slow motion regime ($\delta t = 0.108$ ns). It would thus appear that the ESR spectral lineshapes are affected in a complex way by the details of the underlying motional processes.

What Do Probe Molecules Monitor in Lipid Bilayers?

The simulations presented above support the conventional prescription for describing the orientational behaviour of probe molecules in lipid bilayers in terms of a local effective orienting potential. They indicate, however, that the potential arises from the confinement of the probe molecules between long segments of lipid chains in elongated free-volume cavities within the bilayer structure. In this sense the orienting potential concept needs to be refined in order to take into account the combined effect of the restricted free rattling motions of the probes within the free-volume cavities and the orientations of the cavities themselves relative to the normal to the bilayer plane. The time scale of the motions of the cavities within the bilayer is determined by the rotational motions of long segments of the lipid chains.

This separation of motional modes forms the key to answering the question as to what physical processes the probe molecules monitor within the lipid bilayer structure. It was shown above that the widely used BRD model which assumes that the probe molecules move within a simple orienting potential does not account for the decay of the time correlation functions yielded by the simulations. On the other hand, the CM model, which explicitly includes the fast motions within the cavities as well as the slow reorientations of the cavities in the bilayer structure correctly describes the correlation function decays.

In order to check this picture, we evaluated the time-correlation function $G_r(t)$ but taking into account only the slow motion of the probe molecules as extracted from the CM model. The time-correlation function, Figure 4, can be seen to decay on the same timescale as that of the time-correlation function corresponding to the motion of the interatomic vector $\mathbf{R}_{2-15}$ for the DPPC chains. Moreover, the two functions decay to the same long time plateau, thus indicating a similar orientational order relative to the normal to the bilayer plane.

Our findings justify the use of rigid probe molecules such as TMA-DPH and Cholestane spin labels for monitoring the orientational order and dynamics in lipid bilayer systems. However, it must be borne in mind that CSL should only be used in investigations of general trends such as the variation of the order parameters and rotational rates on changing the temperature of chemical composition of the lipid bilayer.

References

1. C. Zannoni, A. Arcioni, and P. Cavatorta: *Chem. Phys. Lipids* **32**, 179 (1983).
2. P. L. Nordio and U. Segre: 'Rotational Dynamics', in *The Molecular Physics of Liquid Crystals*, G. R. Luckhurst and G. W. Gray (Eds.), Academic Press, New York, p. 411 (1979).
3. M. Evans, G. J. Evans, W. T. Coffey, and P. Grigolini: *Molecular Dynamics*, Wiley, New York (1982).
4. J. H. Freed: 'Theory of Slow Tumbling ESR Spectra for Nitroxides', *Spin labelling; Theory and Applications*, L. Berliner (Ed.), Academic Press, New York (1976).
5. L. J. Korstanje, E. E. van Faassen, and Y. K. Levine: *Biochim. Biophys. Acta* **982**, 196 (1989).
6. H. van Langen, G. van Ginkel, and Y. K. Levine: *Liquid Cryst.* **3**, 1301 (1988).
7. H. van Langen, Y. K. Levine, M. Ameloot, and H. Pottel: *Chem. Phys. Lett.* **140**, 394 (1987).
8. D. A. van der Sijs, E. E. van Faassen, and Y. K. Levine: *Chem. Phys. Lett.* **216**, 559 (1993).
9. U. A. van der Heide, M. A. M. J. van Zandvoort, E. E. van Faassen, G. van Ginkel, G., and Y. K. Levine: *J. Fluorescence* **4**, 269 (1993).
10. C. Polnaszek and J. H. Freed: *J. Phys. Chem.* **79**, 2283 (1975).
11. H. Eviatar, E. E. van Faassen, Y. K. Levine, and D. I. Hoult: *Chem. Phys.* **181**, 369 (1994).
12. D. A. van der Sijs and Y. K. Levine: *J. Chem. Phys.* **100**, 6783 (1994).
13. Y. K. Levine, A. Kolinski, and J. Skolnick: *J. Chem. Phys.* **98**, 7581 (1993).
14. Y. K. Levine: *Molec. Phys.* **78**, 619 (1993).

15. A. Rey, A. Kolinski, J. Skolnick, and Y. K. Levine: *J. Chem. Phys.* **79**, 1240 (1992).
16. M. E. Rose: *Elementary Theory of Angular Momentum*, Wiley, New York (1957).
17. A. Seelig and J. Seelig: *Biochemistry* **13**, 4839 (1974).
18. L. J. Korstanje, E. E. van Faassen, and Y. K. Levine: *Biochim. Biophys. Acta* **980**, 225 (1989).

Molecular Modeling Studies on the Ribosome

STEPHEN C. HARVEY, ARUN MALHOTRA and ROBERT K.-Z. TAN
*Department of Biochemistry and Molecular Genetics, University of Alabama at Birmingham,
Birmingham, Alabama 35294 U.S.A.*

Abstract. In the absence of a high resolution crystal structure for the ribosome, numerous research groups are carrying out low resolution structural studies using neutron diffraction, electron microscopy, fluorescence energy transfer, chemical crosslinking, chemical footprinting studies, and other methods. We have developed a computer-based refinement method for incorporating these data into low resolution three-dimensional models. The method is based on a molecular mechanics approach, with proteins represented by spherical particles of suitable diameter and the ribosomal RNA represented by a string of spherical pseudoatoms, one for each nucleotide. Experimental data are used to derive constraints that are introduced through a special force field (potential function). Models are refined by simulated annealing. Since every term in the force field is quadratic, any model that satisfies all of the input data has an energy of zero; higher energies indicate residual unsatisfied constraints. The residual energy provides a quantitative statement of model quality and can be used to identify conflicts in the experimental data. The method has been applied to the refinement of a low resolution model for the 30S subunit (the small subunit) of the *E. coli* ribosome. Since this is a very underdetermined system, the range of acceptable models has also been explored. This provides an estimate of the resolution of the structure, which is about 15 Å overall, with the uncertainty in position of individual nucleotides ranging from about 5 Å to 50 Å.

Key words. Ribosome, molecular mechanics, structure refinement.

Introduction

Molecular modeling methods, including distance geometry and molecular mechanics (energy minimization, molecular dynamics, Monte Carlo, and simulated annealing) have been used for many years to assist in refinement of high resolution models of molecular structures as determined by X-ray crystallography and NMR [1–3]. They are also used as predictive tools in studies on structure-function relationships on proteins and nucleic acids [4]. At the highest level of resolution, these methods deal with all-atom models.

In order to reduce computational costs, 'reduced representations' are often used. The simplest of these is the common united atom approximation, in which hydrogen atoms are incorporated into pseudoatoms representing methyl, methylene, amino, imino and other groups. More commonly, united atoms are used to represent all hydrogen-containing groups except hydrogen bond donors, where the hydrogens are explicitly included. Both of these approximations substantially reduces the number of atoms that must be considered in modeling proteins and nucleic acids, with the corresponding increase in computational efficiency. The latter preserves the ability to treat hydrogen bonding more accurately than would be possible if all hydrogens were eliminated from the description.

For lower resolution structures, pseudoatoms can represent larger functional groups. This approach was used a number of years ago for investigating a simple model for protein folding, with each pseudoatom representing an entire amino

A. Pullman et al. (eds.), Modelling of Biomolecular Structures and Mechanisms, 333–338.
© 1995 *Kluwer Academic Publishers. Printed in the Netherlands.*

acid [5] More recently, a reduced representation has been introduced with three pseudoatoms per DNA basepair, for simulations on DNA supercoiling [6]. Three pseudoatoms were chosen because three points define a plane, and this description allows the monitoring of basepair roll, tilt, and twist, permitting the inclusion of sequence-dependent variations in the structure of the double helix, particularly the curvature that is caused by tracts of A_n ($n \approx 4$–6) repeated in phase with the period of the double helix.

Here we describe a reduced representation for macromolecular assemblies containing a mixture of RNA and protein molecules (ribonucleoprotein particles, or RNPs) and its application to the refinement of the structure of the small subunit of the *E. coli* ribosome. This article focusses on a concise overview of the methodology and a brief discussion of the results.

A reduced representation of RNPs is necessary for studying systems as large as ribosomes and their subunits, because of their size. The small subunit of the *E. coli* ribosome contains 21 proteins and a large RNA molecule, the 16S RNA, which has 1542 nucleotides; this represents about 10^5 atoms and a total molecular mass of about 10^6 D. This is considerably larger than the systems traditionally studied with molecular mechanics approaches. Furthermore, an all-atom representation would be inappropriate. Although some high resolution crystallographic data are available for this system, suitable heavy atom derivatives that would permit correct phases to be determined have not yet been obtained [7]. The only reliable and interpretable data that are available come from a wide variety of low resolution methods, including neutron diffraction, electron microscopy (EM), immuno-EM, phylogenetic studies, DNA hybridization, and chemical crosslinking and footprinting. These data dictate that the final structural models cannot accurately specity the positions of individual atoms. Thus, a reduced representation that shows the positions of larger structural elements is more appropriate than one that shows individual atoms explicitly.

Methods and Results

The positions of the 21 proteins have been determined by triangulation from neutron diffraction experiments that measure distances between different pairs of labeled proteins, in a series of pairwise studies [8]. The question addressed by our studies is, then, how is the 16S RNA strung through the protein scaffold? We use spherical pseudoatoms to represent the individual proteins, with the diameters of these determined by the known molecular weights and a standard hydrated protein density. In the RNA regions of our model, each nucleotide is also represented by a single pseudoatom. These are located at the positions of the phosphate groups, so graphic images that connect successive pseudoatoms give recognizable double helices; large pseudoatoms are also placed down the center of double helices to provide suitable exclusion volumes for the stems. As will be shown later, the available data are only sufficient to define the position of individual nucleotides with an average resolution of about 15 Å, and the resolution of some regions approaches 5 Å, so this is a suitable level of detail.

The secondary structure of the RNA is known from phylogenetic studies [9,10]. This defines a set of internucleotide distances, since each stem of the secondary structure can be assumed to have the geometry of a standard right-handed A-RNA double helix. Other constraints can also be derived from this assumption, since the double helix defines pseudo-bond angles between triplets of phosphates and pseudo-torsions among quartets of phosphates. To guarantee that the double helices have the proper chirality, the torsion angles are treated as improper torsions [4]. Each double helix containing N atoms is defined by exactly 3N internal constraints, so it moves as a nearly rigid object during structure refinement. The final model will therefore specify the position and orientation of each double helical stem, along with the positions of individual nucleotides from the single stranded regions connecting the stems.. RNA-RNA tertiary contacts have been determined from phylogenetic studies [10] and photocrosslinking [11]. These provide additional internucleotide distance constraints. Some of these data indicate intramolecular contacts indirectly, particularly those cases where two nucleotides in the 16S RNA both crosslink to the mRNA/tRNA complex, while there is no evidence of direct crosslinking capability between them. These indirect indications of proximity can also be converted into suitable distance constraints for model building.

Chemical crosslinking reagents are popular probes of ribosome structure [12,13]. They provide information on RNA-RNA contacts and on RNA-protein contacts. RNA-protein contacts can also be inferred by footprinting studies, in which the reactivity of the intact subunit to a chemical probe is compared to that of a reconstituted subunit from which a particular protein has been omitted [14]. The distance information from crosslinking and footprinting experiments is treated in our models as pseudobonds.

The shape of the 30S ribosomal subunit has been visualized by image reconstruction of data from electron microscopy [15]. Since conventional molecular mechanics force fields [4] do not include terms for shape information, we have developed a special term that expresses shape in terms of spherical harmonics and incorporated it into our force field [16]. In essence, this term assigns an energy of zero to a nucleotide if it lies anywhere inside the defined surface, and a positive energy if it lies outside. The energy penalty increases with increasing distance from the molecular surface. The gradient of the energy is calculated analytically, so the force associated with this shape term is available for energy minimization, molecular dynamics, and so on.

The parameters of the force field are easily derived. All interparticle distance constraints are modeled as pseudobonds. The ideal bond length is equal to the measured distance, and the associated force constant is inversely proportional to the uncertainty in the distance (see below). Here we present a few examples. The crystal structure of tRNA [17] was used to determine the ideal interphosphate distances and the associated uncertainty for double helical regions (e.g. 5.67 Å ± 0.3 Å for successive phosphates on the same strand). Tertiary interactions established by phylogenetics [10], crosslinks [11–13], and footprinting [14] are represented by a mean distance that includes the diameters of the interacting pseudoatoms and the length of the probe, along with an appropriate uncertainty. A well established tertiary contact might have an uncertainty as small as ±2 Å, while a less reliable contact could have an uncertainty as large as ±8 Å. Similar considerations apply to means and variances for angles and improper torsions for double helical regions, which were also derived from the crystal structure of tRNA [17]. For volume exclusion effects, the uncertainty in protein radii is ± 5 Å for all proteins except S1, which is known to be elongated rather than spherical. The radius of S1 was therefore given an uncertainty of ± 15 Å. Since the surface of the ribosome is defined only approximately in the electron microscopy studies [15], and since the spherical harmonics approximation introduces an additional approximation, the constraints associated with the shape are given an uncertainty of ± 10 Å. Uncertainties in the positions of individual proteins and in interprotein distances have been given in detail by Capel *et al.* [8], who made the original measurements, so these were incorporated directly into our models. Parameter derivation is discussed more fully elsewhere [18]. A full list of all distances and uncertainties is given in the paper that presents and analyzes our final models in detail [19].

Of course, the molecular mechanics protocol requires the values of force constants for the pseudobonds and other terms in the force field. These are derived from the statistical uncertainties by noting that, for any degree of freedom with a harmonic constraint, a full thermodyramic ensemble would have values for that coordinate that are normally distributed about the specified mean, with a variance s^2 given by

$$s^2 = RT/a$$

where R is the universal gas constant, T is the absolute temperature, and a is the harmonic force constant. Thus, the specification of the temperature (300 K) and the uncertainty in the measurement allows the calculation of the force constant.

As stated before, there are far too few data to specify the structure at high resolution. It is therefore necessary to generate a collection of models, each of which is compatible with the data, to establish a consensus model, and to determine the uncertainty (resolution) of the final model. We do this by beginning refinement with a wide range of starting RNA conformations. Each conformation is a random walk chain whose center of mass is superposed on the center of mass of the protein scaffold; the constraint information is then added, and the model is refined by simulated annealing, which uses a combination of Monte Carlo with varying temperature and energy minimization. All calculations were carried out with *yammp*, a modular molecular mechanics package [20]. Refinement of a typical model requires about 36 CPU hours on a Silicon Graphics 220/420 hybrid workstation.

The collection of final models was analyzed by examining the various components of the residual energy (to identity consistently unsatisfied constraints that suggest conflicts in the experimental data), by cluster analysis [21,22], and by visual inspection of graphical models generated with the *ribbons* program [23]. We also examined alternative structural assumptions, including putative stem stacking patterns in the RNA and the argument [24,25] that protein S20 is misplaced in the protein map [8] and should be placed much lower in the body of the subunit, near S17. Although these alternative assumptions produced small local changes in the model, they did not substantially alter the overall structure or the conclusions drawn from the modeling studies.

The models have the same general features as previously constructed manual models [4,26–29], with the 5′ domain of the RNA in the body of the subunit, the central domain divided between the body and the platform, and the 3′ major domain occupying the head of the subunit. But there are significant differences in local details, particularly in the active site region. These differences are due to additional experimental data that were not available when the hand-built models were first proposed. Of particular significance are the positions of helices 44 and 45, which we find in the platform, facing into the cleft between the platform and the head. Since the Shine–Dalgarno sequence and residue C1400 (which is known to be in contact with the anticodon loop of the P site tRNA) are immediately adjacent to these helices, the positions predicted by our model are appropriate for the required interactions with the mRNA and the two tRNAs. Other features of our models, such as the radius of gyration and the distance between the center of mass of the RNA and that of the proteins, are in reasonably good agreement with experimental values. A detailed examination of the models is given in a recent publication [19].

Superposition of the final collection of acceptable models shows that 75% of the RNA helices have a root mean square uncertainty in position of less than 15 Å. This is reasonable resolution, considering the limited quantity of data and the fact that they come from a variety of low resolution experimental methods. This level of resolution also provides a *post hoc* justification for the simplified reduced representation that we chose, with one pseudoatom for each protein and one pseudoatom per nucleotide.

In summary, a new refinement method has been developed for low resolution modeling of RNAs and RNPs. It uses a reduced representation and is based on a special potential function (force field) derived directly from low resolution experimental methods. The protocol is used to develop a range of models consistent with the experimental data and to determine a consensus model. It also provides quantitative statements about the resolution of the model and identifies conflicts in the experimental data. Originally developed for application to the ribosome [19], the method has recently been used to determine the structure of the RNA component of a catalytically active RNA, RNase P [M. E. Harris, J. M. Nolan, A. Malhotra, J. W. Brown; S. C. Harvey and N. R. Pace, *submitted*]. We expect that the method will have broad application as a refinement tool to assist in low resolution structural studies on many different RNAs and RNPs.

Acknowledgments

This research was supported by a grant from the National Science Foundation (DMB-90-05767).

References

1. A. Jack and M. Levitt: *Acta Cryst.* **A34**, 931 (1978).
2. J. H. Konnert and W. A. Hendrickson: *Acta Cryst.* **A36**, 344 (1980).
3. A. T. Brünger, J. Kuriyan, and M. Karplus: *Science* **235**, 458 (1987).
4. J. A. McCammon and S. C. Harvey: *Dynamics of Proteins and Nucleic Acids,* Cambridge University Press, London (1987).

5. M. Levitt: *J. Mol. Biol.* **104**, 59 (1976).

6. R. K.-Z. Tan and S. C. Harvey: *J. Mol. Biol.* **205**, 573 (1989).

7. A. Yonath, W. Bennett, S. Weinstein, and H. G. Wittman: 'Crystallography and Image Reconstruction of Ribosomes', in W. E. Hill, A. Dahlberg, R. A. Garrett, P. B. Moore, D. Schlessinger and J. R. Warner (Eds.), *Ribosomes, Structure, Function and Genetics*, University Park Publishers, Baltimore, p. 134 (1990).

8. M. S. Capel, M. Kjeldgaard, D. M. Engelman, and P. B. Moore: *J. Mol. Biol.* **200**, 65 (1988).

9. C. R. Woese, R. R. Gutell, R. Gupta, and H. F. Noller: *Microbiol. Rev.* **47**, 621 (1983).

10. R. R. Gutell and C. R. Woese: *Proc. Natl. Acad. Sci. USA* **87**, 663 (1990).

11. J. B. Prince, B. H. Taylor, D. L. Thurlow, J. Ofengand, and R. A. Zimmerman: *Proc. Natl. Acad. Sci. USA* **79**, 59 (1982).

12. R. Brimacombe: *Biochemistry* **27**, 4207 (1988).

13. R. Brimacombe, P. Mitchell, M. Osswald, K. Stade, and D. Bochkariov: *FASEB, J.* **7**, 161 (1993).

14. S. Stern, B. Weiser, and H. F. Noller: *J. Mol. Biol.* **204**, 447 (1988).

15. A. Verschoor, J. Frank, M. Radermacher, T. Wagenknecht, and M. Boublik: *J. Mol. Biol.* **178**, 677 (1984).

16. A. Malhotra, R. K.-Z. Tan, and S. C. Harvey: *J. Comp. Chem.* **15**, 191 (1994).

17. B. Hingerty, R. S. Brown, and A. Jack: *J. Mol. Biol.* **124**, 523 (1978).

18. A. Malhotra, R. K.-Z. Tan, and S. C. Harvey: *Biophys. J.* **66**, 1777 (1994).

19. A. Malhotra and S. C. Harvey: *J. Mol. Biol.* **240**, 308 (1994).

20. R. K.-Z. Tan and S. C. Harvey: *J. Comp. Chem.* **14**, 455 (1993).

21. R. A. Becker, J. M. Chambers, and A. R. Wilks: *The New S Language. A Programming Environment for Data Analysis and Graphics*, Wadsworth and Brooks/Cole, Pacific Grove CA (1988).

22. D. Gautheret, F. Major, and R. Cedergren: *J. Mol. Biol.* **229**, 1049 (1993).

23. M. Carson: *J. Mol. Graph.* **5**, 103 (1987).

24. R. Brimacombe: *Biochimie* **73**, 927 (1991).

25. G. Schwedler, R. Albrecht-Ehrlich, and K.-H. Rak: *Eur J. Biochem.* **217**, 361 (1993).

26. A. Expert-Bezançon and P. L. Wollenzien: *J. Mol. Biol.* **184**, 53 (1985).

27. R. Brimacombe, J. Atmadja, W. Stiege, and D. Schüler: *J. Mol. Biol.* **199**, 115 (1988).

28. K. Nagano, M. Harel, and M. Takezawa: *J. Theor. Biol.* **134**, 199 (1988).

29. M. Oakes, L. Kahan, and J. A. Lake: *J. Mol. Biol.* **211**, 907 (1990).

The Molecular Mechanics Program DUPLEX: Computing Structures of Carcinogen Modified DNA by Surveying the Potential Energy Surface

BRIAN E. HINGERTY
Health Sciences Research Division, Oak Ridge National Laboratory, Oak Ridge TN 37831, U.S.A.

SUSE BROYDE
Biology Department, New York University, New York NY 10003, U.S.A.

Abstract. The nucleic acids molecular mechanics program DUPLEX has been designed with useful features for surveying the potential energy surface of polynucleotides, especially ones that are modified by polycyclic aromatic carcinogens. The program features helpful strategies for addressing the multiple minimum problem: (1) the reduced variable domain of torsion angle space; (2) search strategies that emphasize large scale searches for smaller subunits, followed by building to larger units by a variety of strategies; (3) the use of penalty functions to aid the minimizer in locating selected structural types in first stage minimizations; penalty functions are released in terminal minimizations to yield final unrestrained minimum energy conformations. Predictive capability is illustrated by DNA modified by activated benzo[*a*]pyrenes.

Key words. Molecular mechanics, DNA, carcinogens, structures, potential energy surface.

Carcinogenesis by Polycyclic Aromatic Chemicals

Polycyclic aromatic hydrocarbons and amines are chemicals present in the environment. Included among them are substances that are known to cause cancer. Chemical carcinogenesis by these substances is now understood to involve a series of initiating events which include the following steps [1–4]: (a) enzymatic conversion of the proximate carcinogen to a reactive species, the ultimate carcinogen; (b) the ultimate carcinogen may be quenched by reaction with solvent and/or it may be enzymatically detoxified, in which case no harm results; (c) the ultimate carcinogen may react with DNA to form one or more covalently linked adducts. These adducts can affect the shape of the DNA, and we believe that the conformation of the modified DNA plays a key role in the outcome of the next steps; (d) the adduct may be repaired or by-passed by mechanisms that are error-free or error-prone; (e) if error-free repair fails to occur, the modified DNA may fail to act as a faithful template during replication. A mutation, a change in base sequence can result; in combination with other such events this eventually can produce cell transformation and tumors. Of course, many biological processes occur between the initial steps and the appearance of tumors.

Benzo[*a*]pyrene (Figure 1), a substance present in automobile exhaust, is a carcinogenic polycyclic aromatic hydrocarbon that is of great interest. Among its metabolic activation products are a pair of mirror image molecules known as (+) and (−) *anti*-benzo[*a*]pyrene-diol-epoxide (BPDE) (Figure 1). The (+) enantiomer is highly tumorigenic in mice while the (−) member of the pair is not; the (+)

A. Pullman et al. (eds.), *Modelling of Biomolecular Structures and Mechanisms*, 339–347.
© 1995 *Kluwer Academic Publishers. Printed in the Netherlands.*

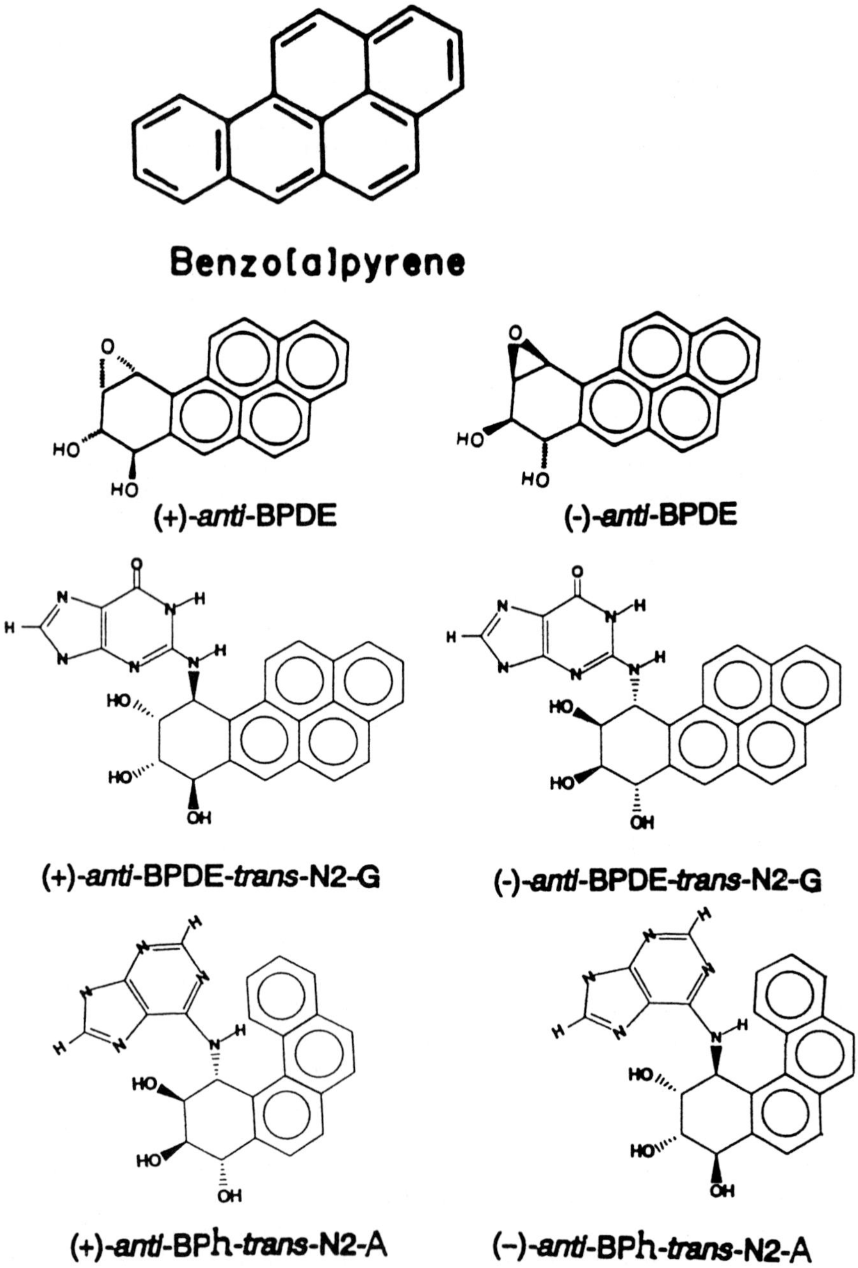

Benzo[a]pyrene

(+)-*anti*-BPDE

(-)-*anti*-BPDE

(+)-*anti*-BPDE-*trans*-N2-G

(-)-*anti*-BPDE-*trans*-N2-G

(+)-*anti*-BPh-*trans*-N2-A

(–)-*anti*-BPh-*trans*-N2-A

Fig. 1.   Structures of polycyclic aromatic molecules and DNA adducts.

enantiomer is also considerably more mutagenic in mammalian systems. Both (+) and (−)-*anti*-BPDE react with DNA to form covalent adducts. The major adduct for the (+)enantiomer is to the amino group of guanine, by trans epoxide opening; the (−) enantiomer also forms such an adduct (Figure 1). Since (+)-*anti*-BPDE is tumorigenic, its major adduct may well be an important contributor to the carcinogenicity of this molecule; however, the corresponding adduct of the non-tumorigenic (−)-*anti*-BPDE is not harmful. Because such chemically similar DNA adducts can lead to such different biological outcomes, it is considered likely that adduct structure plays a key role in governing its mutagenicity and ultimate tumorigenicity. The extensive early literature on the benzo[*a*]pyrenes is reviewed in references [3–5].

We will use our conformational studies on the fascinating (+) and (−)-*trans-anti*-BPDE adduct pair to illustrate our approach, via the molecular mechanics program DUPLEX, to survey the potential energy surface of carcinogen modified DNAs.

The Molecular Mechanics Program DUPLEX

The task in predicting nucleic acid structures *de novo*, with no experimental input, by molecular mechanics is so formidable that it is usually not attempted by most workers in the nucleic acids arena. To make predictions one must find a way to survey the multi-dimensional surface of the potential energy to locate the structure that corresponds to the lowest energy, and other structures accessible at ambient temperature. This multiple-minimum problem limits the application of both molecular mechanics and molecular dynamics to macromolecules.

DUPLEX deals with this problem for minimization in a number of ways: (1) It employs internal rather than Cartesian coordinates as the minimization variables. The internal coordinates are the flexible torsion angles that describe the large movements that govern DNA shape. Limiting the allowed motions to these variables (while fixing naturally very small movements like bond lengths, bond angles and those dihedral angles that vary little in nature, such as those in an aromatic moiety) reduces the number of variables that must be simultaneously optimized from $3N - 6$ ($N =$ No. of atoms) in Cartesian space to 7 for a nucleic acid residue; it also avoids the prediction of chemically unrealistic geometries. Furthermore, this simplification exacts little cost for our application: determining conformations of DNA modified covalently by polycyclic aromatic chemicals. Of course, for other types of investigations, such as studies of reaction pathways, all degrees of freedom must be allowed. (2) We have developed strategies for searching the potential energy surface by minimization. These can employ very large numbers of trials, using arbitrary or selected combinations of the torsion angle variables as starting parameters for small nucleic acid subunits. The resulting minima are energy ranked and then built to larger energy minimized structures by various building strategies [6–12]. These include combinatorial build-up, embedding small units in canonical helices, and adding favored carcinogen orientations found in small units to larger standard helices, all with energy minimization. This procedure again reduces the number of variables in the searches. It is especially useful for helical DNA, whose shape is governed by short range forces. Some of our building strategies are philosophically derived from the build-up technique for polypeptides pioneered in Scheraga's laboratory [13,14]. (3) Penalty functions are employed as searching tools to aid the minimization algorithm in locating structures on the potential energy surface satisfying specified distance bounds between atom pairs. These functions are useful for finding structures with interproton distances that are within bounds of measured NMR data, or for locating structures with designated hydrogen bonding patterns either within a given DNA strand or between strands. In contrast to many other approaches, the penalty functions are released in the final minimization stage, and unrestrained structures that fail to retain the target distances are rejected. Thus, accepted structures are unrestrained low energy minima.

The specific penalty functions we employ are the following: for restraints on interproton distances

$$F_N = W_N \sum_1^n (d - d_N)^2 \tag{1}$$

$$F_{NN} = W_{NN} \sum_1^n (d - d_{NN})^2 \tag{2}$$

W_N and W_{NN} are adjustable weights (usually in the range of 10–30 kcal/mol Å^2), d is the current value of the interproton distance, d_N is a target upper bound, and d_{NN} is a target lower bound. Equation (1) is implemented when d is greater than d_N and Equation (2) is implemented when d is less than d_{NN}. The summation is over the nNMR derived distance bounds. These functions can also be used as overall goodness-of-fit indexes for a given structure in relation to the upper or lower bounds, with the d values now being achieved distances. For restraints that guide the minimizer in locating selected hydrogen bonding patterns, the penalty function, unique to DUPLEX, is

$$F_{HB} = W \sum_1^n [(d - d_0)^2 + (1 + \cos(\tau))^2 + p] \tag{3}$$

If not implemented at a given site, this function can also be used to locate denatured structures, which can be important in chemically modified DNAs. Here, d is the current value of the hydrogen bond length, d_0 is its ideal value, τ is the current value of the angle about the hydrogen participating in the hydrogen bond, and $p = |c_1 - c_2|^2$, where c_1 and c_2 are unit vectors perpendicular to the planes of the hydrogen bonded bases. The function is summed over all n hydrogen bonds. It can also be used to evaluate the overall quality of hydrogen bonds achieved in a model, with α and τ now being the model values.

The force field employed by DUPLEX is one that has been developed for nucleic acids in the laboratory of Olson [15–17], with potential functions and parametrization now widely accepted in the nucleic acids community. Details are given in Reference 6. In DUPLEX, counterion condensation can be mimicked by reduced partial charges on pendant phosphate oxygens [6], or by explicit metal ions [18]; solvent is modeled by distance dependent dielectric functions developed by Srinivason and Olson [17] and by Hingerty and coworkers [18].

Studies of (+) and (−)-*trans-anti*-Benzo[*a*]pyrene-diol-epoxide Adducts

Our original studies for the two BPDE adducts of Figure 1 employed no experimental information other than the force field parameters [5,7]. Many thousands of energy minimization trials were first performed for a modified deoxydinucleoside monophosphate d(CpG); additional hundreds of trials were carried out for single stranded trimers, and further trials were carried out with modified duplex trimers that explored the orientation space of the carcinogen-base linkage in that context. Finally, important conformers located from these searches were embedded in duplex 12-mers with energy minimization. The hydrogen bond penalty function was employed in all duplex trials, to search for structures with Watson–Crick, Hoogsteen, or no base pairing at the modification site, and Watson–Crick base pairs at all other loci (except at the site next to the lesion where the possibility of no pairing was also explored). The penalty function was released in terminal minimizations and final structures were energy ranked.

Three types of structures resulted from these trials: structures which placed the pyrenyl moiety in: (1) the B-DNA minor groove; (2) the B-DNA major groove; or (3) a base-displaced position, with the modified guanine removed from its normal, base-stacked position and the pyrenyl moiety located in its place. A striking observation in all three structural types was that the pyrenyl group was always oriented oppositely in the enantiomer pair adducts: whether in the minor or major groove, the (+) enantiomer adduct pointed in the 5'-direction of the modified strand, while it was oriented in the 3'-direction for the (−) enantiomer adduct. For the base-displaced conformers, the long axis of the pyrene ring in the (+)-adduct pointed toward the major groove, while in the (−) adduct conformer of this type it pointed toward the minor groove. Energetically, the structure in which the pyrene was in the B-DNA minor groove was lowest in energy for the (+) adduct; for the (−) adduct the minor and major groove type structures were about equal in energy. Base displaced structures were higher energy. Figures 2 and 3 show these structures.

These pairs of adducts were subsequently synthesized in a duplex 11-mer by Monique Cosman in the laboratory of N. Geacintov, and analyzed by high resolution NMR studies in the laboratory of D. J. Patel [19,20]. Our computed minor groove structures served as starting conformations for minimizations with the NMR derived distance restraints. The NMR distances were achieved in a single round of minimization and were retained once all restraints were released. The root mean square deviations between the predicted starting structures and the final NMR structures, following sequence and length adjustment of the predicted structures, were only about 1 Å.

The prediction that adducts of enantiomer pair polycyclic aromatic hydrocarbon diol-epoxides are oppositely oriented with respect to the DNA helix may be a more general phenomenon. Benzo[c]phenanthrene (BPh) is another carcinogenic polycyclic aromatic hydrocarbon of current interest. It is also metabolically activated to *anti*-diol-epoxides, like BPDE. The (+) and (−)-anti-benzo[c]phenanthrene-diol-epoxide can form (+) and (−) *trans-anti* adducts to the amino group of adenine (Figure 1). Reference 11 summarizes the literature on the benzo[c]phenanthrenes. Synthesis and high resolution NMR studies in the laboratories of N. Geacintov and D. J. Patel, respectively, together with our computational searches which incorporated the NMR data, have shown that the (+) adduct is intercalated between intact base pairs on the 5' side of the modified strand [11], while the (−) adduct is intercalated on the 3' side [21].

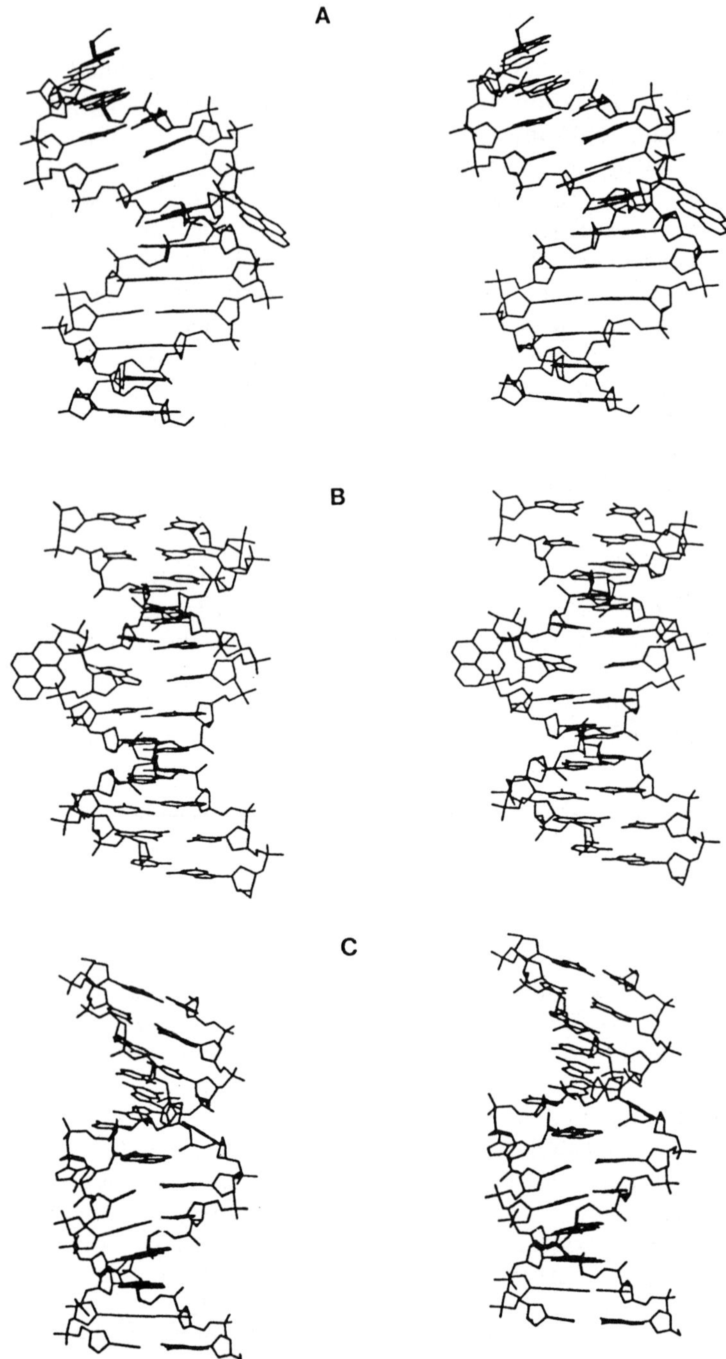

Fig. 2.   Structural types computed for (+)-*anti*-BPDE-*trans*-N2-dG adducts: (A) pyrenyl moiety in B-DNA minor groove; (B) pyrenyl moiety in B-DNA major groove; (C) pyrenyl moiety in base-displaced position. Reproduced from Figure 2 of *Cancer Research* **51**, 3482–3492, with permission.

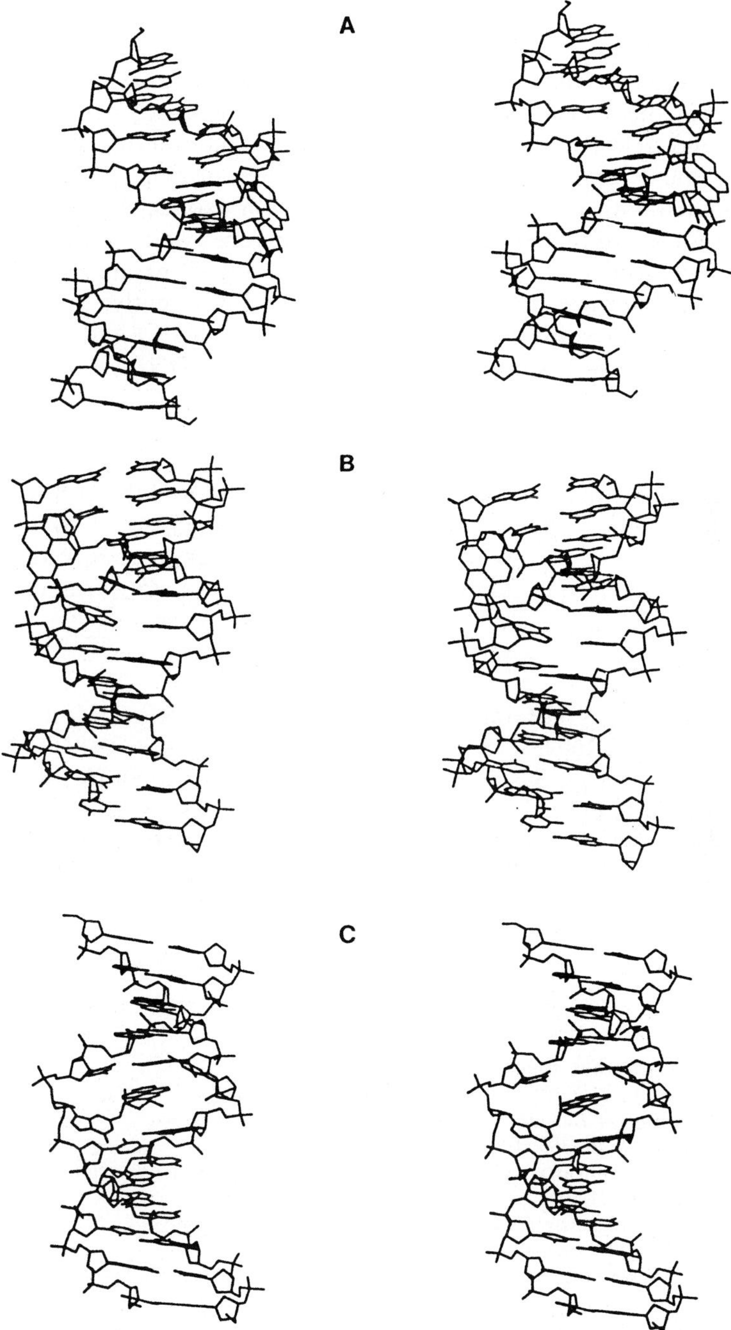

Fig. 3.   Structural types computed for (−)-*anti*-BPDE-*trans*-N2-dG adducts: (A) pyrenyl moiety in B-DNA minor groove; (B) pyrenyl moiety in B-DNA major groove; (C) pyrenyl moiety in base-displaced position. Reproduced from Figure 3 of Cancer Research **51**, 3482–3492, with permission.

Biological Implications

Differing orientations of adducts of mirror-image molecules could well result in different treatments of the lesions by replication or repair enzymes. Indeed, it has been shown in the Geacintov laboratory that exonucleases which digest DNA from the 3'-end treat BPDE modified single strands differently than do those that digest from the 5'-end, depending on the isomer present. Specifically, single strands modified with the (+) adduct slow down a exonuclease more, as would be expected for a 5'-directed orientation of the pyrene, while those modified with the (−) adduct, which is 3'-directed, slow down a 3'-5' exonuclease more [22]; this shows that enzymes do treat the adduct pair differently, at least *in vitro*. Perhaps the strikingly different biological properties of (+) and (−)-*anti*-BPDE may result, in some part, from different orientations in DNA.

Note

The present manuscript has been authored by a contractor of the U.S. Government under contract number DE-AC05-84OR21400. Accordingly, the U.S. Government retains a non-exclusive royalty-free license to publish or reproduce the published from of the contribution, or allow others to do so for U.S. Government purposes.

Acknowledgement

Supported in part by NIH Grants CA 28038 (S. B.) and RR 06458 (S. B.), DOE Grant DE FG0290ER60931 (S. B.), DOE contract DE-AC05-84OR21400 with Martin Marietta Energy Systems (B. E. H.), and DOE, Office of Health and Environmental Research, Field Work Proposal ERKP931 (B. E. H.). Calculations were carried out on Cray Supercomputers at the DOE's National Energy Research Supercomputer Center and the NSF's San Diego Supercomputer Center. We thank Professor Nicholas Geacintov and Professor Robert Shapiro, Chemistry Department, New York University, for on-going helpful dicussions and advice.

References

1. C. C. Harris: *Cancer Res.* (*Suppl.*) **51**, 50235 (1991).
2. S. M. Cohen and L. B. Ellwein: *Cancer Res.* **51**, 6493 (1991).
3. A. H. Conney: *Cancer Res.* **42**, 4875 (1982).
4. R. Harvey: *Am. Sci.* **70**, 386 (1982).
5. S. B. Singh, B. E. Hingerty, U. C. Singh, J. P. Greenberg, N. E. Geacintov, and S. Broyde: *Cancer Res.* **51**, 3482 (1991).
6. B. E. Hingerty, S. Figueroa, T. L. Hayden, and S. Broyde, *Biopolymers* **28**, 1195 (1989).
7. B. E. Hingerty and S. Broyde: *Biopolymers* **24**, 2279 (1985).
8. D. Norman, P. Abuaf, B. E. Hingerty, D. Live, S. Broyde, and D. J. Patel: *Biochemistry* **28**, 7462 (1989).
9. S. F. O'Handley, D. G. Sanford, R. Xu, C. C. Lester, B. E. Hingerty, S. Broyde, and T. R. Krugh: *Biochemistry* **32**, 2481 (1993).
10. A. M. Carothers, W. Yuan, B. E. Hingerty, S. Broyde, D. Grunberger, and E. G. Snyderwine: *Chem. Res. Toxicol.* **1**, 209 (1994).
11. M. Cosman, R. Fiala, B. E. Hingerty, A. Laryea, H. Lee, R. Harvey, N. E. Geacintov, S. Broyde, and D.J. Patel: *Biochemistry* **32**, 12488 (1993).

12. R. Shapiro, B. E. Hingerty, and S. Broyde: *J. Biomolec. Struct. Dynam.* **7**, 493 (1989).
13. M. Pincus, R. Klausner, and H. Scheraga: *Proc. Natl. Acad. Sci. USA* **79**, 5107 (1982).
14. K.D. Gibson and H.A. Scheraga, 'The multiple minimum problem in protein folding', In *Structure and Expression Volume I: From Proteins to Robisomes*, R. H. Sarma and M. H. Sarma (Eds.), Adenine Press, p. 67 (1988).
15. W. K. Olson and A. R. Srinivasan: 'Classical energy calculations of DNA, RNA and their constituents', in Landolt–Bornstein, *Numerical Data and Functional Relationships in Science and Technology, Group VII.* Volume 1D, W. Saenger (Ed.), Springer-Verlag, Berlin, p. 415 (1990).
16. E. R. Taylor and W. K. Olson: *Biopolymers* **22**, 2667 (1983).
17. A. R. Srinivasan and W. K. Olson: *Fed. Proc.* **39**, 2199 (1980).
18. B. E. Hingerty, R. H. Ritchie, T. L. Ferrell, and J. E. Turner: *Biopolymers* **24**, 427 (1985).
19. M. Cosman, C. de los Santos, R. Fiala, B. E. Hingerty, S. B. Singh, V. Ibanez, L. Margulis, D. Live, N. E. Geacintov, S. Broyde, and D. J. Patel: *Proc. Natl. Acad. Sci. USA* **89**, 1914 (1992).
20. C. de los Santos, M. Cosman, B. E. Hingerty, V. Ibanez, L. Magulis, N. E. Geacintov, S. Broyde, and D. J. Patel: *Biochemistry* **31**, 5245 (1992).
21. M. Cosman, R. Fiala, B. E. Hingerty, A. Laryea, H. Lee, R. G. Harvey, S. Amin, N. E. Geacintov, S. Broyde, and D. J. Patel: in preparation.
22. B. Mao, B. Li, S. Amin, M. Cosman, and N. E. Geacintov: *Biochemistry* **32**, 11785 (1993).

Simulations of Molecular Mechanisms in Radiation Damage to DNA

ROMAN OSMAN,* CHUNG F. WONG and KAROL MIASKIEWICZ[1]
Department of Physiology and Biophysics, Mount Sinai School of Medicine of the City University of New York, One Gustave L. Levy Place, New York, NY 10029, U.S.A.

Abstract. The article reviews recent attempts to integrate the description of the molecular events in radiation damage to DNA. The stages of indirect damage include diffusion of reactive radicals produced by energy deposition, their collision and interaction with DNA, and induction of conformational changes in the damaged DNA. The collision events, studied with Brownian dynamics simulations, depend on the electrostatic field and the accessibility of various sites in DNA. Quantum chemical calculations of H-abstraction from sugar illustrate that this process depends on the nature of the abstracted hydrogen. Molecular dynamics simulation of damaged DNA provide the molecular details of changes in its dynamic and conformational properties. The integration of these methodologies will become the basis for a unified approach to simulations of radiation damage to DNA.

Key words. DNA radiation damage, indirect damage, Brownian dynamics, H-abstraction, molecular dynamics.

An attempt to develop a molecular understanding, derived from first principles, of the stages that bridge between the deposition of high energy and the formation of permanently damaged DNA requires an integrated approach which takes into account the specific physicochemical properties of the individual stages. The process of DNA damage can be somewhat arbitrarily divided into three stages that encompass: (i) formation of reactive species in the environments of DNA; (ii) their diffusion towards DNA and interaction with it; and (iii) the conversion of an initial damage to a permanent one. From a chemical point of view the diffusion and the interaction, termed *indirect damage*, are one of the most interesting aspects of the nature of radiation damage to DNA because indirect damage is controlled by rate constants of the reactions between reactive species and DNA. These rate constants depend on the collision of reactive species with DNA, on the rates of subsequent chemical reactions, and on scavenging rates of these radicals. Also, since the interaction between the reactive species and DNA is on the same time scale as the conformational changes in DNA, the dynamic properties of DNA have an important influence on the rates and overall yield of indirect damage.

To approach this problem from first principles one needs to apply a combination of theoretical methods especially designed to address the specific problems in each of the stages. For example, the characterization of the collision between DNA and the reactive particles produced from radiolysis of water can be studied with Brownian Dynamics (BD) simulations. These yield rate constants and probability

* Author for correspondence.
[1]*Present Address*: Pacific Northwest Laboratory, Richland WA 99352.

A. Pullman et al. (eds.), Modelling of Biomolecular Structures and Mechanisms, 349–363.
© 1995 *Kluwer Academic Publishers. Printed in the Netherlands.*

distribution of diffusion controlled collisions influenced by long range forces between DNA and the reactive particles. The collisions often lead to reactions in which chemical bonds are broken or formed. For example, OH· radicals produced by radiolysis of water can add to the double bonds of nucleic acid bases and can abstract hydrogens from sugars. The consequence of such reactions is the formation of initial lesions in DNA, which is equivalent to transferring the reactivity of the radicals to the DNA molecule. The rate constants for these reactions are composed from the diffusion controlled collisions and from the efficiency of the chemical events, which are determined by energy barriers for the reactions. Since new bonds are formed and previously existing bonds are being disrupted in the course of these reactions, the energetic barriers must be calculated with quantum chemical methods. Finally, to construct an integrated representation of the interaction of reactive species with DNA we must take into account the dynamic fluctuations of DNA, which produce a distribution of structures. These fluctuations are on a similar time scale as the diffusion of the reactive species towards the DNA and the chemical reactions that follow the collision. Thus, the results from BD simulations and quantum chemical studies must be supplemented by Molecular Dynamics (MD) simulations to provide an integrated description of collisions and interactions between reactive particles and dynamically active DNA. Such an integrated approach, in addition to providing rates and probabilities of distribution of initial damage in DNA, can become the theoretical basis for the evaluation of the importance of oxygen in damage 'fixing' and of scavengers in providing protection against indirect damage to DNA.

We describe here the application of the three methodologies to specific examples related to radiation damage to DNA. These examples can serve as the initial basis in the process of integrating the methodologies to provide a unified approach to simulations of radiation damage to DNA

1. Brownian Dynamics Simulations

The Brownian dynamics simulation method is being increasingly used for studying the relative diffusional encounters of biomolecules [1–15]. Several important developments have made it easier to carry out BD simulations and to improve the validity of models for such simulations. These developments include the Ermak–McCammon algorithm, which allows simulations of Brownian trajectories rather than time-dependent probability distributions of diffusing particles [16], the method of Northrup *et al.*, which provides estimates of diffusion rate constants by simulating Brownian trajectories in a finite rather than an infinite region [14], and the use of the Poisson–Boltzmann equation for deriving the interaction forces between diffusing molecules [17–26]. These methods have been described in details in literature, and here we summarize only key ideas relevant to our studies of the relative diffusion between DNA and DNA-damaging agents.

Knowledge of the electrostatic potential is the basis for calculating the forces that act between a molecule and the diffusing particle. The electrostatic field around the target molecule is calculated by solving the Poisson–Boltzmann equation (PBE). The solution is carried out by dividing the Cartesian space into two regions. An inner region of low dielectric constant and an outer region with a high dielectric constant, usually that of water. The boundary between the inner and the outer regions is defined by the shape of the target molecule outlined by the van der Waals radii of its atoms.

The PBE is given by

$$-\nabla(\epsilon(r)\nabla\Phi(r)) = \rho(r) - en_\infty \sinh\left(\frac{e\Phi(r)}{k_bT}\right) \tag{1}$$

where $\epsilon(r)$ is the dielectric coefficient, $\Phi(r)$ the electrostatic potential, $\rho(r)$ the charge density, e the electronic charge, n_∞ the density of ions, k_b the Boltzmann constant, and T the absolute temperature.

To simplify the solution of Equation (1), the PBE is usually linearized to give

$$-\nabla(\epsilon(r)\nabla\Phi(r)) + \epsilon(r)\kappa^2\Phi(r) = \rho(r) \tag{2}$$

where κ is the inverse Debye–Hückel screening length, given by

$$\kappa^2 = \frac{Ie^2}{k_bT\epsilon} \tag{3}$$

and I is the ionic strength.

Although analytical solutions of the PBE are possible for molecules with regular shape, numerical solutions of the PBE are required for most molecules of biological interest, which have a complex surface. The linearized PBE can be solved very efficiently by a numerical finite-difference method. The partial differential equation expressed in Equation (2) is thus transformed into a set of linear equations, the solution of which gives the electrostatic potential at a set of points on a cubic grid.

Once the potential $\Phi(r)$ of the target molecule is obtained, the force $F(r)$ acting on the moving particle is calculated by the test charge method, in this method, the effects of solvent exclusion volume of the diffusing molecule and reaction field effects are ignored. This is a reasonable approximation in our applications in which the diffusing particles (e_{aq}^-, H or OH radicals) are much smaller than the target DNA.

The trajectory of a diffusing particle is simulated by the Ermak–McCammon algorithm [16], which the new position is calculated as:

$$r = r_0 + \frac{1}{k_bT}DF(r_0)\Delta t + R \tag{4}$$

where r is the new position vector obtained after a time step of duration Δt, D is a diffusion tensor, usually approximated by the relative diffusion constant D obtained as the sum of the diffusion constants of the two molecules, $F(r_0)$ is the force acting on the diffusing particle at r_0 (calculated from the electrostatic potential at r_0 using the test charge method), and R is a Gaussian random displacement with zero mean and a variance of

$$\langle R^2 \rangle = 2D\Delta t \tag{5}$$

R results from the stochastic force exerted on the diffusing particle by the random collisions with solvent molecules.

Based on the method of Northrup *et al.* [14], one can obtain a bimolecular diffusion rate constant by simulating Brownian trajectories in a finite region. In a BDS, the smaller molecule is placed at random on the surface of a sphere (*b* surface) which divides the space into inner and outer regions. The value of *b* is chosen such that the electrostatic potential is centrosymmetric in the outer region ($r > b$). The trajectory of the diffusing particle is simulated by the Ermak–McCammon algorithm [16] and the trajectory is terminated when one of the following conditions is satisfied:

1. The trajectory moves far away from the target and crosses another surface defined by a sphere with radius *q* (*q* surface) where $q > b$.
2. The trajectory comes close to one of the reactive centers of the target molecule.

From a large number of trajectories, one can obtain a probability β that a diffusing molecule from the *b* surface reacts with the target molecule rather than diffuse to the *q* surface. The probability β is used to calculate another *P*, which corresponds to the probability that a particle starting at the *b* surface will react with the target molecule rather than diffuse to infinity. The bimolecular diffusion rate constant *k* can then be obtained from

$$k = k_D(b)P \tag{6}$$

where $k_D(b)$ is the rate constant for the two molecules to come to a separating distance *b*. Since *b* is chosen such that the electrostatic potential $\Phi(r)$ is centrosymmetric for $r \geqslant b$, $k_D(b)$ can be calculated from [27]

$$k_D(b) = \frac{4\pi}{\displaystyle\int_b^\infty \frac{\exp[\Phi(r)/k_b T]}{r^2 D}\, dr}. \tag{7}$$

The probability *P* is related to β by

$$P = \frac{\beta}{1 - (1 - \beta)\Omega} \tag{8}$$

where Ω is

$$\Omega = \frac{k_D(b)}{k_D(q)}. \tag{9}$$

Based on extensive tests, we have developed optimized parameters for simulating the diffusional encounters between DNA or their constituents and DNA-damaging agents, such as e_{aq}^-, OH radicals, or H radicals. Based on experimental findings, we chose all carbon atoms of the bases as potential reaction centers for e_{aq}^-. For modeling the hydroxyl radical reactivity, the carbon atoms of double bonds in the bases and $H_{4'}$ of the sugar were selected as sites of attack. The reaction distance criteria, i.e., the distance between the diffusion particle and the reaction center within which a reaction is assumed to occur, were as follows: the distance between e_{aq}^- and base carbons was 1.9 Å; the distance from OH radical

TABLE I

Comparison of simulated and experimental values of k

Molecule	e_{aq}^-			OH	
	Simulated without ES force	Simulated with ES force	exp	Simulated	exp
Bases					
Ade	8.4 ± 0.5	12.6 ± 0.6	9	8.8	4.4
Cyt	8.1 ± 0.4	10.1 ± 0.4	13	8.0	4.7
Gua	10.5 ± 0.5	11.0 ± 0.4	–	8.6	–
Thy	10.1 ± 0.3	11.2 ± 0.5	18	7.9	5.6
Ura	10.1 ± 0.6	11.4 ± 0.3	15	8.3	5.2
d-Nucleotides					
5'-dAMP	9.2 ± 0.5	5.9 ± 0.2	3.8^a	7.7	3.5
5'-dCMP	8.6 ± 0.4	6.9 ± 0.4	6.8^a	6.8	4.4
5'-dGMP	9.0 ± 0.4	6.0 ± 0.3	–	8.0	5.0
5'-dTMP	8.2 ± 0.5	6.0 ± 0.4	1.5^a	6.6	5.3^a
DNA[b]					
$(dA)_{12} \cdot (dT)_{12}$	1.72 ± 0.02	0.26 ± 0.04	0.06–0.14	1.2	0.3 0.8
Dickerson					
dodecamer[c]	1.66	0.271	0.06–0.14	1.2	0.3–0.8

k in $10^9\,\mathrm{M}^{-1}\,\mathrm{s}^{-1}$ per nucleotide base at $25\,^\circ\mathrm{C}$, $I = 100\,\mathrm{mM}$; SEM values represent 95% confidence limits.
[a]Experimentally determined k from ribonucleotides.
[b]Rate constants given per nucleotide.
[c]d(CGCGAATTCGCG)$_2$.

to base carbon was 3.5 Å and to $H_{4'}$ 2.7 Å. All the diffusing particles were assumed to be spherical and the reaction distances were measured from the centers of the spheres to the centers of the reaction sites For e_{aq}^- and the hydroxyl radical, the experimental values of the diffusion coefficients were used: $4.9 \times 10^{-5}\,\mathrm{cm}^2/\mathrm{s}$ and $2.9 \times 10^{-5}\,\mathrm{cm}^2/\mathrm{s}$, respectively [28]. The hydrodynamic radii of the target biomolecules were estimated from the Stokes–Einstein equation according to the method of Bowen *et al.* [29] and of Garcia de la Torre applied to short DNA fragments [30]. Using the optimized parameters, rate constants have been calculated and a comparison with experimental results is shown in Table I. They offer some insights into the quality of the results obtained from the Brownian dynamics simulations.

As can be seen from Table I, the calculated rate constants are of the correct order of magnitude compared to experimental values. They also correctly predict the trend of reduction in diffusion-controlled rate constants upon going from neutral bases to the highly charged DNA. Nevertheless, several shortcomings of the current simulations can be detected. For example, the simulations give an incorrect trend for collision rate constants with different bases. This can be partly attributed to the estimates of hydrodynamic radii using the Stokes–Einstein equation, to the neglect of hydrodynamic interactions, to the inaccurate solutions of the PBE due to large grid spacings, and to the assumption that DNA can be treated as a rigid molecule in the course of BD simulations. Interestingly, the collision rate constants between hydroxyl radicals and DNA or its constituents appear to be consistently larger than the experimental values. This may suggest that these reactions are diffusion influenced rather than diffusion controlled, and therefore it would be necessary to account for the fact that not every collision of a hydroxyl radical can lead to a reaction.

1.a. BROWNIAN DYNAMICS SIMULATIONS OF e_{aq}^{-}

One of the major effects of the diffusion of e_{aq}^{-} in the neighborhood of DNA is the electrostatic field produced by the polyelectrolyte nature of DNA. We have therefore conducted simulations in the presence and the absence of forces produced by the electrostatic field (ES) generated by DNA. The data in Table I show that there is no significant difference between the simulation without and with ES forces for the neutral nucleic acid bases. These expected results derive from the fact that e_{aq}^{-} only feels small electrostatic forces when it reaches a short distance from the bases, but at a larger distance their behavior is a completely random Brownian motion unaffected by the electrostatic potential generated by the bases. In clear distinction, the rate constants for collision with DNA are reduced when electrostatic interactions are turned on, indicating the effect of the strong repulsive electrostatic forces between e_{aq}^{-} and the phosphate groups. The decrease in rate constants has been previously attributed to a reduced collision frequency of the particle with a given base in a poly-deoxynucleotide or DNA [31], and has been termed the 'DNA shielding effect' [32].

The analysis of the BD simulation of the collision of e_{aq}^{-} with a dodecamer of a typical B-DNA, $dA12\cdot dT_{12}$, shows the relative probabilities of attack at different sites of the DNA by a solvated electron. The results are shown in Figure 1. Both the control simulation (without ES force; Figure 1A) and the regular simulation (with ES force; Figure 1B) show that the electrons prefer to attack the ends of the double helix, reflecting the larger exposure of these reactive sites. However, in the presence of electrostatic forces, the relative reactivities at the neutral ends of the helix are nearly doubled compared to the simulation in the absence of these forces. The relative reactivities in the center of the helix, on the other hand, are substantially decreased, except the third and fourth bases which present slightly higher relative reactivities than the control. An examination of the electrostatic potential on the molecular surface of the DNA molecule indicates that near the center of the helix it has much more negative values than at the helix ends, providing an explanation for the low relative reactivities at the center of the helix. A more detailed examination of the electrostatic potential around the third, the fourth, and the ninth base shows that they reside underneath the neutralized end phosphates and thus the repulsive forces in this region are much smaller.

An analysis of specific sites on thymine shows that the relative reactivities are comparable for the C_2, C_4, C_5, and C_6 atoms in the absence of ES forces. However, electrostatic interactions substantially increase the reactivities of the C_6 sites on thymine. In the absence of ES the most likely site for attack on adenine is C_8 showing approximately a three-fold higher probability than other sites. In the presence of ES forces the probability of attack on C_8 sites of the adenine is reduced by a factor of 2, whereas C_6 is increased by 50%. The reactivities of the C_2, C_4,

A)

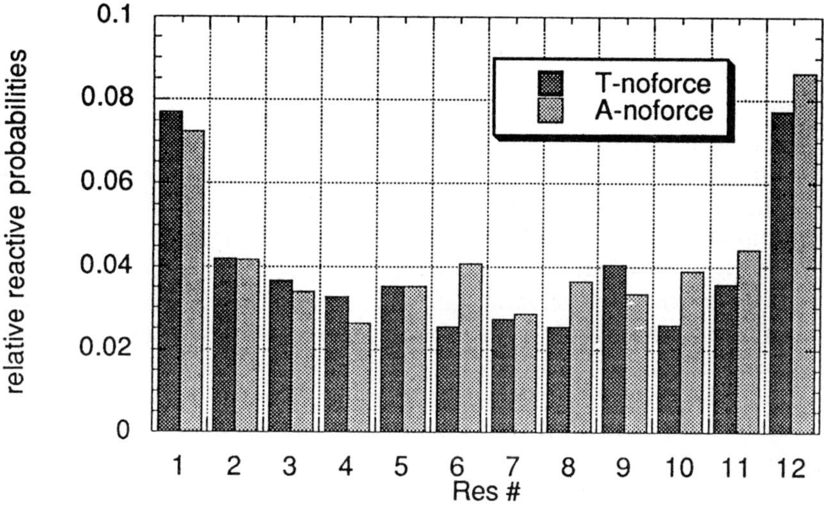

B)

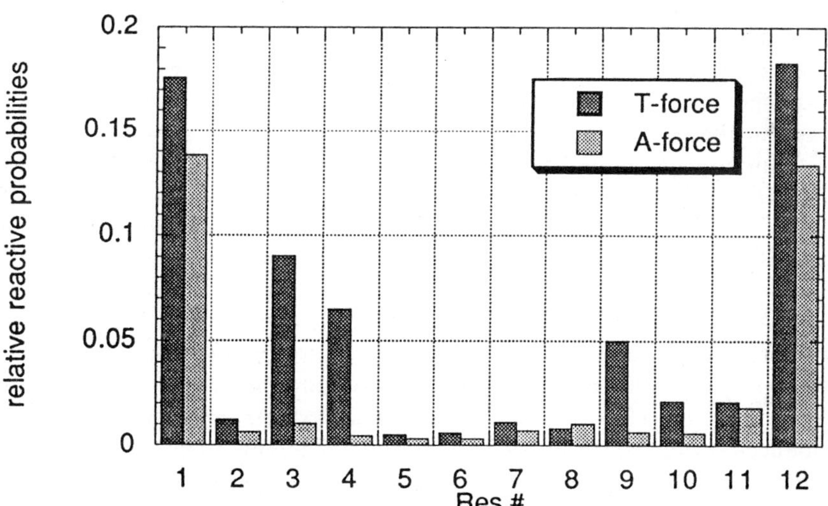

Fig. 1. Relative reactive probabilities in $dA_{12} \cdot dT_{12}$ as a function of residue number (Res#). (A) results from a Brownian dynamics simulation without the inclusion of electrostatic forces; (B) results from a simulation including electrostatic forces.

and C_5 sites are nearly independent of the ES forces. Clearly, ES forces have an important effect on the relative reactivities of DNA sites to attack by solvated electrons.

1.b. BROWNIAN DYNAMICS SIMULATION OF OH RADICAL

In simulations of collisions with the hydroxyl radical, the calculated rate constants consistently overestimate the experimental data (Table I). This may suggest that the activation barriers for the addition reactions to the bases or for hydrogen abstractions from the sugars are clearly non-negligible. This finding is in accordance with quantum chemical calculations which show that the activation energy for $H_{4'}$ hydrogen abstraction by hydroxyl radical is of the order of 4.0 kcal/mole (see below). Examination of the trajectories from the simulations shows that the hydroxyl radical may collide with other hydrogen atoms of the sugar. Thus, in spite of the fact that the results presented in Table I include only $H_{4'}$ as a reactive site on sugar, the overestimated rate constants support the conclusion that reaction barriers have a significant effect on these reactions. Little difference is observed among the simulated rate constants of hydroxyl radicals reacting with the bases. This is in accordance with experimental findings [33].

Our simulations show that OH reactivity decreases as the molecules increase in size. This trend is similar to that observed in the simulations for e_{aq}^-, which is consistent with decrease in exposure of the reaction sites in DNA. The extent of this effect is smaller for the OH than for the e_{aq}^- attack. This may be due to the presence of reactive sites on the sugar moieties for OH but not for e_{aq}^-. Furthermore, since there are no strong electrostatic repulsions between DNA and hydroxyl radicals, the decrease in the rate constants in the highly charged DNA compared to the neutral bases is much smaller than in the e_{aq}^- case.

The simulation presented here for the e_{aq}^- constitute an important test of the ability of BD simulations to represent the relative probabilities of collision with various sites on DNA. Clearly, under the strict condition where the incorporation of electrostatic forces is important, the simulations show that these forces are affecting the results. Furthermore, the results from the simulation also demonstrate that the proper sensitivity of different bases and different sites on the bases are strongly affected by electrostatic forces. The results from the simulations with OH radical give overestimated rate constants, probably due to the neglect of reaction activation barriers. However, it is important to remember that the biological consequences from OH damage are farther reaching than from e_{aq}^-. Thus an integration of the rate constants calculated from collision events (BD) with those obtained from chemical reactions (see below) is an essential part in a unified representation of radical damage to DNA.

2. Quantum Chemical Calculation of Rate Constants

2.a. RATE CONSTANTS OF H-ABSTRACTION IN GAS PHASE

We have previously conducted *ab initio* quantum chemical studies of the process of hydrogen abstraction from methanol and ethanol in the gas phase [34]. The computations of the structures at selected points on the potential energy surface (reactants, products and transition states) were performed at a high level of theory (MP4/6–311G*) an the rate constants were calculated with a zero curvature approximation of the transition state theory with quantum corrections for tunneling [35]. These studies demonstrated that the calculated rate constants are in very good agreement with experiment over a wide range of temperatures. Based on the good agreement, we used these calculations to predict the rate constant for the abstraction of the hydrogen from C_β of ethanol, which is not known from experimental measurements.

2.b. H-ABSTRACTION FROM DEOXYRIBOSE

We have conducted quantum chemical studies to characterize the structures and properties of carbon centered radicals of deoxyribose produced by H-abstraction [36]. For each radical, two puckered conformations were observed, which were significantly different from those of deoxyribose. In radicals centered on C_1 and C_2 the pseudorotation angle of the five membered ring was shifted toward higher values compared to that of deoxyribose. In C_3 and C_4 radicals the shift in the pseudorotation angle was in the opposite direction. Also the puckering amplitude was significantly decreased in the radicals compared to deoxyribose, indicating a flattening of the five-membered ring. The C—H bond enthalpies in deoxyribose at MP2/6–31G* level showed that H-abstraction from C_1, C_3, and C_4 require similar energies, whereas the H-abstraction from C_2 is less favorable by 3–4 kcal/mole. Furthermore, it was shown that the contributions from C—H bond strengths have to be combined with accessibility in nucleotides and DNA to yield reasonable estimation of relative rate constant for H-abstraction from different carbon atoms of deoxyribose. When both effects are taken into account the abstraction of $H_{4'}$ in B-DNA is predicted to be roughly 20 times faster than that of $H_{2'}$.

To examine this question further we have calculated the absolute rate constant for H-abstraction from two positions of the model of deoxyribose that was used in our previous studies [36]. We chose to study the abstraction of $H_{4'}$ and $H_{2'}$ because these reactions lead to two carbon centered radicals of deoxyribose, which may act as precursors for β-cleavage of the bond between $C_{3'}$ and $O_{3'}$ and the formation of strand breaks. Strand breaks are one of the most deleterious chemical changes in the structure of DNA because the continuity of the backbone is disrupted. Double strand breaks that occur in a short distance from each other have been shown to be one of the most lethal damages to DNA [37].

We have optimized the structures of the reactants and the transition states using GAUSSIAN92 with the 6-31G basis set. Single point calculations of the optimized structures were then performed at the MP2/6-31G* level. Zero point energies were calculated from the frequencies of the optimized structures. The rate constants were evaluated at the zero-curvature approximation of the transition state theory and were corrected for tunneling.

The interaction of an OH radical with the sugar produces a stable complex whose structure is shown in Figure 2. From this initial structure the OH radical can turn toward $H_{4'}$ or toward $H_{2'}$ and engage in a hydrogen abstraction process.

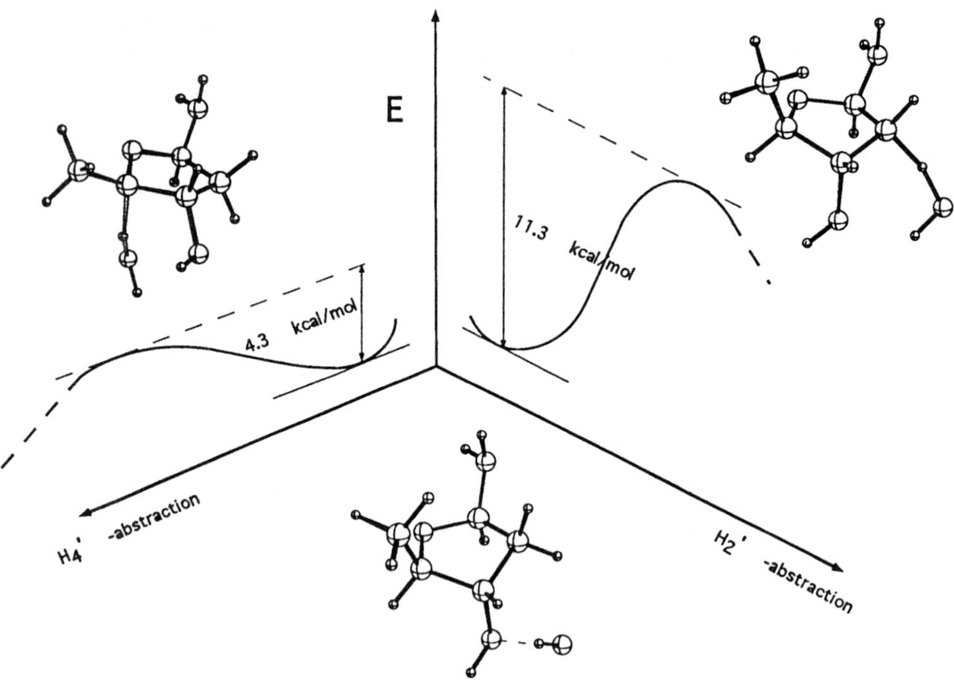

Fig. 2. Diagramatic representation of the two possible H-abstraction in a model of deoxyribose. The structures shown are: the ground state complex between deoxyribose and OH radical (center); transition state for $H_{4'}$-abstraction (left); transition state for $H_{2'}$-abstraction.

TABLE II

Barriers and rate constants for H-abstraction from deoxyribose

Abstracted Atom	Barrier (Kcal/mole)	Classical k (sec^{-1})	k tunnel (sec^{-1})
$H_{4'}$	4.25	2.91×10^9	2.27×10^{10}
$H_{2'}$	11.28	4.91×10^3	1.31×10^8

The two possible transition structures were optimized at the same level as described above and are shown in Figure 2. The energies and the frequencies of the optimized structures were then used to calculate the barriers, the classical rate constants and the rate constants corrected for tunnelling. The results are shown in Table II.

The calculated barriers demonstrate the difference between $H_{4'}$ and $H_{2'}$. It is much more difficult to abstract $H_{2'}$ than $H_{4'}$; the difference in the barriers is about 7 kcal/mole. This is consistent with the calculated C—H bond strengths, although the estimated difference was only 3–4 kcal/mole [36]. The data in Table II illustrate another important point in calculating rate constant for H-abstraction. The classical ks reflect the large difference in the barriers, which translates to a ratio of nearly six orders of magnitude in the rate constants. However, because the moving H-atom during the abstraction is much lighter than the rest of the molecule, tunneling becomes very important, especially when the barrier increases. The rate constants corrected for tunneling show a ratio of 173. Clearly, even at this approximate level of calculation, in which solvent effects on structure and energetics were not included, the calculations indicate that the abstraction of $H_{4'}$ should be much faster than of $H_{2'}$. The abstraction of $H_{4'}$ acquires a larger importance when its higher accessibility than that of $H_{2'}$ is taken into account [36]. Furthermore, the enhancement of β-cleavage initiated from the $C_{4'}$ radical due to stabilization of the resulting carbocation compared to the process initiated from the $C_{2'}$ radical may also contribute to the lower probability of strand break formation produced by $C_{2'}$ sugar radicals [37].

The quantum chemical calculation of potential energy surfaces for the interaction of reactive species with models of DNA components is essential for computing rate constants for reactions that damage DNA. They are also important in providing a rigorous formulation of the connection between collision events and chemical reactions. Since part of the chemical reactivity is determined by accessibility, which in turn depends on dynamic fluctuation of DNA, molecular dynamics simulations are necessary to evaluate the importance of the dynamics of DNA on the mechanisms for DNA damage by reactive species.

3. Molecular Dynamics Simulation of Damaged DNA Aqueous Solution

We have conducted molecular dynamics simulations of the DNA dodecamer d(CGCGAATTCGCG)$_2$ with various forms of pyrimidine lesions to study their effect on DNA structure. For comparison, a parallel simulation of the native form of DNA was also conducted. All the lesions were introduced to replace the thymine in position 7 (T_7) and they represent various stages and forms of damaged pyrimidines. For example, 5-hydroxy-6-thyminyl radical represents an initial lesion formed by the addition of OH to thymine [38]. We have also studied the structural changes in DNA effected by stable lesions produced by the addition of OH and/or H radicals to thymine. Among these were 5-hydroxy-5,6-dihydrothymine, 6-hydroxy-5,6-dihydrothymine (thymine photohydrate), 5,6-dihydrothymine, and 5,6-dihydroxy-5,6-dihydrothymine (thymine glycol) [39].

All the simulations were conducted in aqueous environment, which consisted of approximately 1500 water molecules. To neutralize the charge of the DNA, 22 Na^+ ions were placed around the DNA at the bifurcating positions of the OPO angle at a distance of 5 Å from the phosphorus atom. With periodic boundary conditions, the resulting box was of approximate dimensions $53 \times 35 \times 35$ Å^3. The energy of the system was minimized and then the system was heated to 300 °K over a period of 2 ps followed by an equilibration stage in which the velocities were repeatedly assigned from a Maxwell–Boltzmann distribution. Finally, a 120 ps simulation was carried out at constant volume and the temperature was maintained around 300 °K by coupling the system to a bath with a coupling constant τ of 0.1. A time step of 1 fs was used and the non-bonded interactions were updated every 0.02 ps. A cutoff of 10 Å was used in the evaluation of non-bonded interactions.

The structural changes in DNA produced by introducing a 5-hydroxy-6-thyminyl radical in its sequence can be characterized by local and global distortions. Local

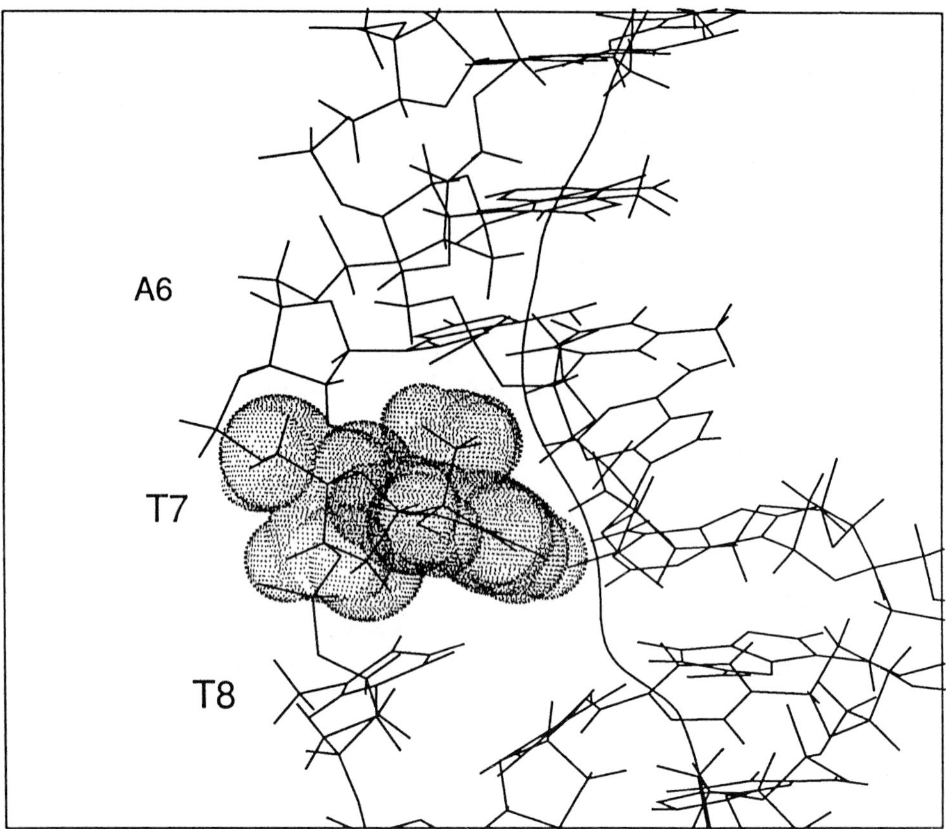

Fig. 3. Closeup of the area near the lesion represented by 5-hydroxy-6-thyminyl radical (T$_7$) shown in space filling representation to emphasize the bulky methyl group pointing in the direction of the neighboring 5′-adenine (A$_6$). The 3′-thymine (T$_8$) is shown displaced from the normal position. The kink and the bending of the DNA can be seen by the solid line representing the helical axis. The structure shown in a frame at the end of the simulation (120 ps).

distortions are produced by the stereochemical changes on C$_5$ of thymine to which the OH radical was added. Because the stability of an equatorial OH group is more favorable than the axial orientation [40], the methyl group is forced into an axial position. Such an orientation presents a bulky methyl group in the 5′-direction, which impinges on the adjacent adenine (A$_6$) and forces its displacement from its original position. At the same time the position of T$_7$ is displaced toward the 3′-direction, virtually leaving no room for T$_8$ to occupy its normal place and displacing it out of the normal sequence. As a consequence, Watson–Crick hydrogen bonds in base pairs A$_6$–T$_{19}$, T$_7$–A$_{18}$, and T$_8$–A$_{17}$ are severely disrupted. The local changes around the point of damage are shown in Figure 3. Furthermore, the steric disturbance around T$_7$ is translated into a global change in DNA structure. A kink is produced around the point of damage (see Figure 3) and the shape of the

DNA is bent, as reflected in an overall shortening of the end-to-end distance of the dodecamer by 14% compared to the native DNA.

The simulations of the same sequence with the stable lesions result in structures that generally belong to the B-DNA family. Base-pairing and stacking are reasonably well preserved and sugar puckering are usually in the N and S minima without ever moving into the energetically unfavorable W region. It is possible that the difference in the resulting structures is due to the periodic boundary conditions that were applied in these simulations.

In spite of the fact that generally the structures seem to belong to the B-DNA family, there are significant perturbations in the DNA structure in the vicinity of the lesions. The extent of the perturbations, however, depends on the specific lesion. The distortions are negligible for 5,6-dihydrothymine and the sequence containing this lesion looks very much like the native sequence. Small perturbations can be seen with the 6-hydroxy-5,6-dihydrothymine, which are larger than those with dihydrothymine. However, the full extent of the distortion is observed for 5-hydroxy-5,6-dihydrothymine and for thymine glycol. The major structural difference between the various lesions is the degree of the out-of-plane distortion of the thymine ring and the bulk in the axial direction. In 5,6-dihydrothymine both axial positions at the C_5 and C_6 are occupied by hydrogens. In 6-hydroxy-5,6-dihydrothymine the axial position on C_5 is a hydrogen, but on C_6 the axial substituent is an hydroxy group [40]. In DNA containing the two lesions that produce the largest structural distortions, i.e., 5-hydroxy-5,6-dihydrothymine and thymine glycol, the methyl group at C_5 occupies an axial conformation. This axial methyl seems to be responsible for the large distortions in the structure, which is expressed by an increased distance between the lesioned base and its 5′-neighbor, the *rise* parameter, which becomes 7.5 Å in the lesioned DNA compared to 3.5 Å in native DNA. In a complementary way the *rise* parameter between the lesioned base and the 3′-neighbor is reduced to ~2.0 Å. Another structural characteristic of the changes produced by the lesions that contain a 5-hydroxy group and an axial methyl is a negative inclination (~ −60°) of the base with respect to the helical axis. This value in native DNA is close to 0°. As a consequence of these local distortions certain hydrogen bonds in the base pair A_6–T_{19} (5′ to the lesion) are significantly disturbed almost to the point of complete dissociation. These structural perturbations, while quantitatively different, share a great deal of qualitative similarities with the local changes produced by 5-hydroxy-6-thyminyl radical, the initial lesion formed by the addition of OH radical to thymine. Clearly, the bulky methyl group pointing in the 5′-direction is responsible for most of the structural perturbations.

The pattern of structural perturbations to DNA induced by the lesions correlates well with observed biological consequences of damaged DNA. The negligible changes induced with 5,6-dihydrothymine agree well with the fact that this lesion does not affect DNA replication [41] and is not recognized by the UvrABC complex [42], which recognizes bulky lesions to DNA. The small extent of polmerase I inhibition induced by the presence of 6-hydroxy-5,6-dihydrothyminein DNA [43] agrees with the small perturbations in the structure of DNA produced by this lesion. The extensive distortions induced in DNA by thymine glycol are in excellent agreement with the efficient blockade of DNA synthesis produced by this lesion

[44–46]. Furthermore, the inability of thymine glycol 5′-triphosphate to act as a substrate for DNA synthesis [46] is also in agreement with extensive non-planar nature of the thymine glycol. Finally, the perturbations in base stacking in the vicinity of the thymine glycol lesion are in full agreement with the proposed recognition properties of the UvrABC repair enzyme complex [47].

4. Conclusion

We have illustrated the application of three methodologies to the investigation of molecular mechanisms of radiation damage to DNA. Brownian dynamics simulations provide an estimate of the diffusion controlled rate constants of collision between reactive particles, such as e_{aq}^- and OH radical, and DNA. They also provide a relative probability of reactivity of different sites on the DNA derived from the distribution of trajectories that terminate on various DNA sites. The consistent overestimation of diffusional rate constants for the collision of OH with DNA and its components suggests that chemical barriers control the reactions of this particle with DNA. Clearly, a derivation of a theory that incorporates diffusional with chemical rate constants will be needed to provide a more complete representation of these processes.

Derivation of chemical rate constants has been demonstrated on small molecules as well as on the sugar constituents of DNA. Quantum chemical calculations provide parts of potential energy surfaces that control the rates of chemical reactions. These calculations are therefore an essential component in generating the information about the classical and the quantum mechanical rate equations in which tunneling through the barriers is very important.

Both collisions between reactive species and DNA and the chemical reactions that ensue depend on the dynamical properties of DNA. These have beeen studied with molecular dynamics simulations, which showed the molecular basis for the biological effects produced by lesions in DNA. Such structural perturbations can be responsible for inhibition or blockage of DNA synthesis. They can also become the molecular signals that are recognized by various repair enzymes, which maintain the functionality of DNA by restoring it to its original form.

Acknowledgement

This work was supported by the U.S. Department of Energy Grant DEFG02-88ER60675. We acknowledge the contributions of our collaborators to various parts of the work described here.

References

1. S. A. Allison, G. Ganti, and J. A. McCammon: *Biopolymers* **24**, 1323 (1985).
2. S. A. Allison and J. A. McCammon: *J. Phys. Chem.* **89**, 1072 (1985).
3. M. E. Davis, J. D. Madura, J. Sines, B. A. Luty, S. A. Allison, and J. A. McCammon: *Meth. Enzymol.* **202**, 473 (1991).
4. E. Dickinson: *Chem. Soc. Rev.* **14**, 421 (1985).
5. B. A. Luty, R. C. Wade, J. D. Madura, M. E. Davis, J. M. Briggs, and J. A. McCammon: *J. Phys. Chem.* **97**, (1993).

6. S. H. Northrup, J. O. Boles, and J. C. L. Reynolds: *Science* **241**, 67 (1988).
7. K. Sharp, R. Fine, and B. Honig: *Science* **236**, 1460 (1987).
8. K. Sharp, R. Fine, K. Schulten, and B. Honig: *J. Phys. Chem.* **91**, 3624 (1987).
9. E. D. Getzoff, D. E. Cabelli, C. L. Fisher, H. E. Parge, M. S. Viezzoli, L. Banci, and R. A. Halliwell: *Nature* **358**, 347 (1992).
10. T. Head-Gordon and C. L. Brooks: *J. Phys. Chem.* **91**, 3342 (1987).
11. J. D. Madura and J. A. McCammon: *J. Phys. Chem.* **93**, 7285 (1989).
12. J. A. McCammon and S. C. Harvey: *Dynamics of Proteins and Nucleic Acids*, Cambridge University Press, New York (1987).
13. J. A. McCammon: *Science* **238**, 486 (1987).
14. S. H. Northrup, S. A. Allison, and J. A. McCammon: *J. Chem. Phys.* **80**, 1517 (1984).
15. R. Tan, T. N. Truong, J. A. McCammon, and J. L. Sussman: *Biochemistry* **32**, 401 (1993).
16. D. L. Ermak and J. A. McCammon: *J. Chem. Phys.* **69**, 1352 (1978).
17. M. E. Davis and J. A. McCammon: *J. Comp. Chem.* **10**, 386 (1989).
18. M. K. Gilson, K. A. Sharp, and B. H. Honig: *J. Comp. Chem.* **9**, 327 (1987).
19. M. K. Gilson and B. H. Honig: *Nature* **330**, 84 (1987).
20. B. Jayaram, K. A. Sharp, and B. Honig: *Biopolymers* **28**, 975 (1989).
21. I. Klapper, R. Hagstrom, R. Fine, K. Sharp, and B. Honig: *Proteins*, **1**, 47 (1986).
22. A. Nicholls and B. Honig: *J. Comp. Chem.* **12**, 435 (1991).
23. K. Sharp and B. Honig: *Ann. Rev. Biophys. Biophys. Chem.* **19**, 301 (1990).
24. K. Sharp, J. Jean-Charles, and B. Honig: *J. Phys. Chem.* **96**, 3822 (1992).
25. J. Warwicker and H. C. Watson: *J. Mol. Biol.* **157**, 671 (1982).
26. M. K. Gilson, M. E. Davis, B. A. Luty, and J. A. McCammon: *J. Phys. Chem.* **97**, 3591 (1993).
27. D. F. Calef and J. M. Deutch: *Ann. Rev. Phys. Chem.* **34**, 493 (1983).
28. E. J. Hart and M. Anbar: *The Hydrated Electron*, Wiley, New York (1970).
29. W. J. Bowen and H. L. Martin: *Archives Biochem. Biophys.* **107**, 30 (1964).
30. M. M. Tirado, C. L. Martinez, and J. Garcia de la Torre: *J. Chem. Phys.* **81**, 2047 (1984).
31. P. C. Shragge, H. B. Michaels, and J. W. Hunt: *Radiation Res.* **47**, 598 (1971).
32. J. F. Ward: in *Advances in Radiation Biology* Vol. 5, p. 181 (1975).
33. A. J. Bertinchamps, J. Hutterman, W. Kohnlein, and R. Teoule, Eds., *Effects of Ionizing Radiation on DNA*, Springer-Verlag, Berlin (1978).
34. L. Pardo, J. Banfelder, and R. Osman: *J. Am. Chem. Soc.* **114**, 2382 (1992).
35. D. G. Truhlar, A. D. Isaacson, and B. C. Garret: in *Theory of Chemical Reaction Dynamics*, M. Baer (Ed.), CRC Press, Boca Raton, FL, p. 66 (1984).
36. K. Miaskiewicz and R. Osman: *J. Am. Chem. Soc.* **116**, 232 (1994).
37. C. von Sonntag: *The Chemical Basis of Radiation Biology*, Taylor & Francis, Lonodn (1987).
38. K. Miaskiewicz and R. Osman: *Rad. Prot. Dosim.* **52**, 149 (1994).
39. K. Miaskiewicz, J. Miller, R. Ornstein, and R. Osman: *Biopolymers* (1995, in press).
40. K. Miaskiewicz, J. Miller, and R. Osman: *Int. J. Radiat. Biol.* **63**, 677 (1993).
41. H. Ide, L. A. Petrullo, Z. Hatahet, and S. S. Wallace: *J. Biol. Chem.* **266**, 1469 (1991).
42. Y. W. Kow, S. S. Wallace, and B. Van Houten: *Mutation Res.* **235**, 147 (1990).
43. T. Ganguly and N. J. Duker: *Mutation Res.* **293**, 71 (1992).
44. J. M. Clark and G. P. Beardsley: *Nucleic Acids Res.* **14**, 737 (1986).
45. J. M. Clark and G. P. Beardsley: *Biochemistry* **26**, 5398 (1987).
46. H. Ide, Y. W. Kow, and S. S. Wallace: *Nucleic Acids Res.* **13**, 8035 (1985).
47. B. van Houten and A. Snowden: *BioEssays* **15**, 1 (1993).

Molecular Similarity and Dissimilarity

W. GRAHAM RICHARDS
Physical Chemistry Laboratory, Oxford University, Oxford OX 1 3QZ, U.K.

Abstract. Molecular similarity provides a quantitative measure of the resemblance of molecules. It is of use in the design of mimetics and in finding relationships between biological activity and molecular properties which incorporate three-dimensional features. Dissimilarity between enantiomeric forms can rationalize the relative potencies of optical isomers.

Key words. Molecular similarity, mimetics, neural networks, eudismic ratios.

Introduction

Molecular similarity aims to give a quantitative answer to the question of how closely one molecule resembles another. Among its possible applications are:

1. Finding bioisosteres and designing mimetics
2. Optimal superimposition of molecules
3. Quantitative structure activity relationships

Mimetic design is a major activity in the pharmaceutical industry, both by attempting to make non-peptide mimics of natural interactions (e.g. parts of a cytokine which bind to a receptor) and in the design of stable transition state analogues. This latter approach to drug design was anticipated in nature by antibiotics such as penicillin which is a mimic of a transition state recognised by enzymes involved in building bacterial cell walls.

Superposition of molecular structures is a key step in the development of a notion of receptor structure from a knowledge of small molecules which will bind to it and also a crucial first step in three dimensional structure-activity studies.

It is in the area of structure-activity studies and quantitative relationships where molecular similarity has had most impact.

Similarity Indices

A wide variety of similarity indices can be defined [1–3], but the two most commonly employed are due to Carbo [1] and Hodgkin [2]. They measure similarity R, or R' between molecules A and B in terms of some property p of each molecule measured at the same point in the space surrounding the superimposed molecules, integrating over all space and normalizing the index to give a range of 0 to 1 (or in some instances -1 to $+1$).

$$R_{\mathrm{AB}} = \frac{\int p_A p_B \, dv}{(\int p_A^2 \, dv)^{1/2}(\int p_B^2 \, dv)^{1/2}}$$

and

A. Pullman et al. (eds.), Modelling of Biomolecular Structures and Mechanisms, 365–369.
© 1995 *Kluwer Academic Publishers. Printed in the Netherlands.*

$$R'_{AB} = \frac{2 \int p_A p_B \, dv}{\int p_A^2 \, dv + \int p_B^2 \, dv}$$

There are attractions in the case of the first formula if p is an electron density in that it is based on quantum mechanics, but if other properties are used (such as electrostatic potential or field) then the second formulation has marginal advantages.

Electrostatic potential, being more discriminating than charge density, is perhaps the most widely employed molecular property. It has the disadvantage of producing singularities at nuclei (since the $1/r$ term goes to infinity). Formerly this was countered by computing the p values only outside the van der Waals surface of the molecules. A second problem was that the integrals were evaluated numerically by placing the superimposed molecules A and B in a gridded box. This has difficulties due to the crudity of the grid and its limited extent. Finer grids and larger boxes very much slow down computation. Both problems can be avoided by the realisation [4] that $1/r$ can be fitted by Gaussian functions and the integrals performed analytically. This gives an infinite grid with zero spacing and the Gaussians do not go to infinity when r is zero. This technique is faster and more reliable when finding best overlap by adjusting relative positions since derivatives may be used to give an analytical solution to the problem.

The electrostatic potential may be envisaged as some sort of indicator of the attractive part of the interaction of a small molecule with a biological macromolecule. The repulsive interaction is dominated by shape and is much more discriminating. Meyer [5] introduced quantitative measures of shape similarity using a gridded box and numerical solutions based on counting points lying inside or outside molecular surfaces. The shape can, however, also be expressed in an analytical form using gaussians to fit atomic charge densities obtained from *ab initio* wave functions [6].

Similarity in QSAR

The beauty of the similarity index is that it compares three dimensional properties of pairs of molecules but gives a single number. A whole series of n molecules may thus be compared with each other to give an $n \times n$ similarity matrix. If both shape and electrostatic potential are compared for the same series there will be an $n \times 2n$ similarity matrix. The columns of these matrices may be used as input to a symmetric neural network of the type illustrated in Figure 1.

In this unsupervised net the net is trained so that the output matches the input with the data being squeezed through two central nodes, labelled x and y. This is a means of data reduction. If the values are presented as xy plots, then striking structure activity patterns emerge. For example, taking the well-known series of steroids [7], frequently used as a test case in this area, the xy plots clearly distinguish compounds of high binding affinity from medium and low binding members of the series [8], as illustrated in Figure 2(a) and (b).

The similarity matrices may also be subjected to partial least squares analysis to provide not merely a correlation with structure but a full three-dimensional quantitative structure-activity relationship. The results are striking giving some

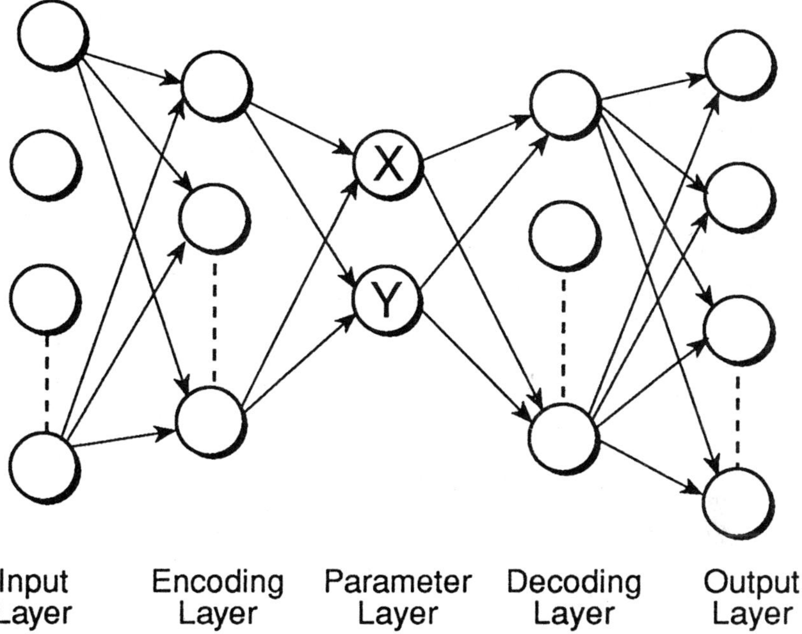

Fig. 1.   Schematic representation of the neural network.

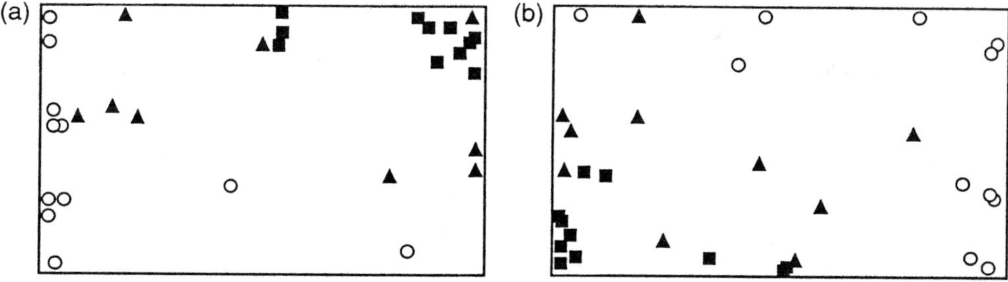

Fig. 2.   Neuron plots from the neural network: (a) from electrostatic potential similarity, (b) from electrostatic potential and shape similarity. (High affinity ■; intermediate affinity ▲; low affinity □.)

advantages over the widely applied method of comparative molecular field anlaysis [9] while not matching all the qualities of the latter. The use of similarity matrices is much faster since the matrices are sized on the basis of the number of compounds in the series rather than on a large grid of widely spaced points. Statistically the correlations may be superior [10]. On the other hand the CoMFA technique can highlight regions of the space surrounding a molecule which play a crucial role in binding where similarity matrices cannot, and interpretation of the statistical model is more difficult.

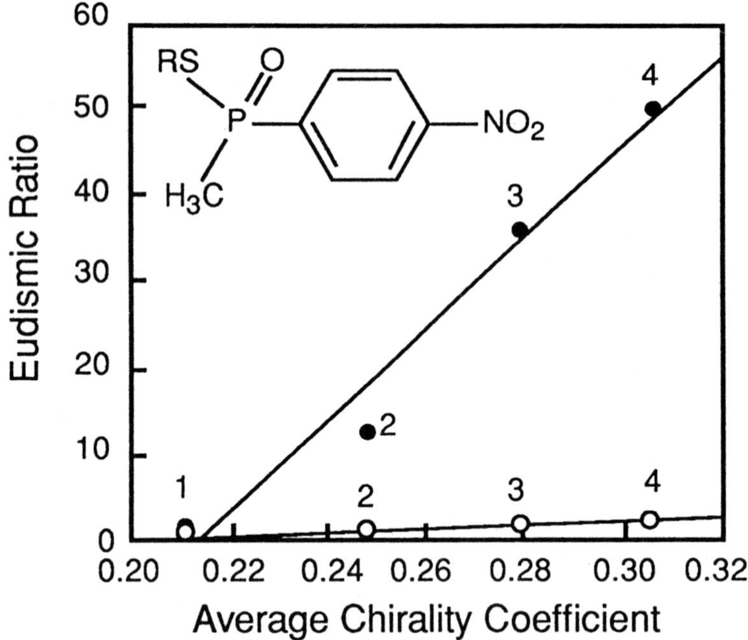

Fig. 3.   A linear correlation between the ERs and the averaged chirality coefficients of *S*-alkyl-*p*-nitrophenyl methylphosphonothiolates (R = (1) methyl, (2) ethyl, (3) *n*-propyl, (4) *n*-butyl). ● AcChE: $y = -115 + 540x$, $r = 0.987$. ○ BuChE: $y = -4.8 + 24.5x$, $r = 0.952$.

Molecular Dissimilarity

Although chirality is generally thought of as an all or none property, it is possible to take the value of molecular dissimilarity expressed as (1-similarity) between an enantiomeric pair as a quantitative measure of relative chirality [11] or chirality coefficient. The value of such a measure is that it can produce impressive correlations between this chirality coefficient and the relative biological potency or binding of an enantiomeric pair, the so-called eudismic ratio.

Interest in the relative potency of optical isomers derives from the pressure on the pharmaceutical industry to develop single optical isomers as products since in general only one form is active while the other is at best useless and at worst damaging. Ever since the work of Pfeiffer [12] it has been realized that there is a rough correlation between the logarithm of the eudismic ratio (or eudismic index) and the logarithm of average human dose of a drug. The rationale for this is simple: tightly binding compounds require low dose, but the site can distinguish three-dimensional shape differences accurately; weak or loose binders can fit less precisely.

Using the chirality coefficient within a single series it is possible to produce a direct (rather than log–log) correlation. Two examples are given in Figures 3 and 4 in which averaged chirality coefficient of shape and electrostatic potential are shown to correlate with the eudismic ratio.

The importance of these correlations is the hope that they offer in the prediction

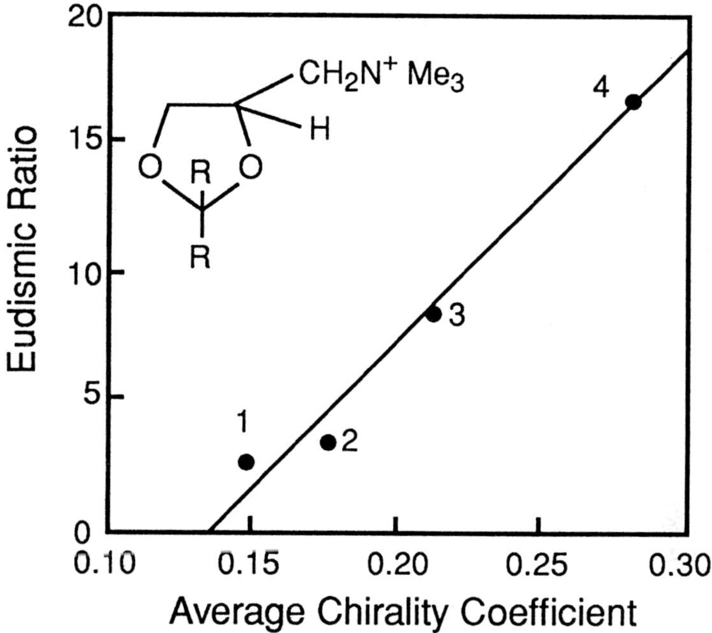

Fig. 4. A linear correlation between the ERs and the averaged chirality coefficients of muscarinic 1,3-dioxolanes (R = (1) isopropyl, (2) ethyl, (3) methyl, (4) hydrogen). $y = 16 + 114x$, $r = 0.986$.

of the relative potency of chiral forms and the means by which the eudismic ratio might be increased or decreased by design.

Acknowledgements

The similarity studies here used the program ASP of Oxford Molecular Ltd and the work has been conducted in part pursuant to a contract with the National Foundation for Cancer Research.

References

1. R. Carbo, L. Leyda, and M. Arnau: *Int. J. Quantum. Chem.* **17**, 1185(1980).
2. E. E. Hodgkin and W. G. Richards: *Int. J. Quantum Chem. Quantum Biol. Symp.* **14**, 105 (1987).
3. A. C. Good: *J. Molec. Grnph.* **10**, 144 (1992).
4. A. C. Good, E. E. Hodgkin, and W. G. Richards: *J. Chem. Inf. Comput. Sci.* **32**, 188 (1992).
5. A. Y. Meyer and W. G. Richards: *J. Comput. Aided Mol. Design* **5**, 426 (1991).
6. A. C. Good, E. E. Hodgkin, and W. G. Richards: *J. Chem. Inf. Comput. Sci.* **33**, 112 (1993).
7. A. J. Hopfinger: *J. Am. Chem. Soc.* **102**, 7196 (1980).
8. A. C. Good, S.-S. So, and W. G. Richards: *J. Med. Chem.* **36**, 433 (1993).
9. R. D. Cramer, D. E. Patterson, and J. D. Bunce: *J. Am. Chem. Soc.* **110**, 5959 (1988).
10. A. C. Good, S. J. Peterson, and W. G. Richards: *J. Med. Chem.* **36**, 2929 (1993).
11. A. Seri-Levy and W. G. Richards: *Tetrahedron Asym.* **4**, 1917 (1993).
12. C. C. Pfeiffer: *Science* **124**, 29 (1956).

Biomolecules at Phase Boundaries

PETER AHLSTRÖM,[1,2] JUKKA LAUSMAA,[1] PATRIK LÖFGREN[1] and HERMAN
J. C. BERENDSEN[2]
[1]*Department of Applied Physics, Chalmers University of Technology and University of Göteborg,
S-41296 Göteborg, Sweden*
[2]*Bioson Research Institute and Laboratory of Biophysical Chemistry, University of Groningen,
Nijenborgh 4, NL-9747 AG Groningen, the Netherlands*

Abstract. Experimental and theoretical studies of biomolecules at water surfaces and metal surfaces
are presented. We studied lecithin molecules (monolayers) and phospholipase A_2 at a water surface
with molecular dynamics (MD) simulations. Results were compared with those obtained for a pure
water surface. We also studied amino acids at a TiO_2 surface with thermal desorption spectroscopy in
the presence and absence of water.

Key words. Di(decanoyl)phosphatidyl choline, lecithin, surfactant monolayers, phospholipase A_2,
thermal desorption spectroscopy, titanium dioxide.

1. Introduction

The interaction between biomolecules and phase boundaries is of great interest
for several reasons. Many biomolecules, like phospholipids and bile salts, are
amphiphilic and act as surfactants between a aqueous and a fatty phase. Others,
like several lipases and phospholipases require an interface in order to function
or to achieve their maximum activity [1]. The interactions between biomolecules
and solid surfaces (e.g., metals and polymeric materials) play an decisive role in
the field of biocompatibility, i.e. the interaction between man-made materials and
living tissue.

2. Simulations of Water–Air Interfaces

We have performed molecular dynamics (MD) simulations of a series of systems
having an air–water interface as a common denominator. The systems were in
increasing complexity, a pure water slab, a water slab with a monolayer of phospholipids at each side and the same system with a phospholipase A_2 molecule
inserted at one of the monolayers. Results from these simulations were compared
to (the scarce) experimental results for this type of systems, e.g., surface tensions.
The results from the phospholipase simulation were also compared to results from
a simulation of phospholipase with a monomeric substrate [2]. As far as possible
the same simulation parameters were used for the different systems, cf. Table I.

2.1. PURE WATER SLABS

As a reference to the simulations of phospholipid systems we performed two
simulations of a 4 nm thick water slabs with a periodic repeat parallel to the slab
(box size 4×4 nm). Two different water models, the SPC [3] and the SPC-E [4]

A. Pullman et al. (eds.), Modelling of Biomolecular Structures and Mechanisms, 371–379.
© 1995 *Kluwer Academic Publishers. Printed in the Netherlands.*

TABLE I

Simulation parameters and thermodynamic averages. The solute includes NaCl, di(decanoyl)lecithin and phospholipase (where present). RMS deviations are given within parenthesis where appropriate. In all simulations a twin range cut-off of the interactions was used. Interactions between atoms within 0.8 nm from each other were evaluated every time step. For atoms with a distance between each other between 0.8 and 1.2 nm only electrostatic interactions were calculated. These were updated at every neighbour list update (i.e. every tenth time step). In all these simulations the time step was 2 fs. Bonds were kept rigid with the SHAKE algorithm [5] and the temperature was coupled to an external bath [6] with a coupling time $\tau_T = 0.1$ ps. For a more detailed description of the common simulation parameters, see [7]

Simulation	SPC	SPC-E	ML	ML2	PLA2
Number of lecithins	0	0	2×42	2×42	2×42
Number of Na^+Cl^-	0	0	8	8	7
Number of H_2O	2111	2111	4412	4412	3790
Simulation length (ps)	200	200	155	155	155
Analyzed period (ps)	100	100	75	75	50
Solvent temperature (K)	307(3)	309(3)	308(2)	308(2)	308(2)
Solute temperature (K)	–	–	298(4)	298(4)	298(3)
Surface tension (mN/rn)	59(7)	72(6)	58(8)	62(17)	62(8)

were used. They both have the same geometry and the same Lennard–Jones parameters (centered on the oxygen) but differ slightly in charges, the hydrogen charge in the SPC-E model is +0.4238 vs. +0.4100 in the SPC model. This minor change in dipole moment leads to a notable change in the surface tension as calculated from the pressure in different directions (see Table I); the surface tension rises from 59 mN/rn with the SPC model to 72 mN/m (close to the experimental value) with the SPC-E model. As a comparison it can be mentioned that the generalized van der Waals theory [12] predicts that the surface tension depends linearly on the dipole moment raised to the fourth power under certain simplifying conditions [11]. The simulation results suffer from large uncertainties and, since the forces, and thus the pressures, are calculated with a truncation of the interaction at a cut-off distance, these surface tensions would need a long-range correction. Such a correction has been calculated for Lennard–Jones liquids (cf., e.g. [8,9]) and at present we are developing such a correction for dipolar systems [10] based on the generalized van der Waals theory. In lack of such a correction (which probably would increase the surface tension) and with the large uncertainties of the simulated surface tensions in mind we chose to use the SPC model in the continued simulations since it was used in simulation of phospholipase with a monomeric substrate [2].

2.2. LECITHIN MONOLAYERS

Monolayers of surfactants on airwater interfaces show several two-dimensional phases depending on the density, or long chain molecules at least the following phases are observed (from low to high density): gas, liquid expanded (LE), liquid compressed (LC) and possibly solid. Short-chain lecithins do not appear to show any LE-LC transition but show a continuous decrease in surface tension as the surface tension as the density is increased in the liquid region. We chose to simulate di(decanoyl)lecithin(DDPC) at the surface density that corresponds to the maximum rate of phospholipase A_2 ($0.78\,nm^2$/molecule). The experimental surface pressure [1], i.e. the decrease in surface tension compared to water, at this coverage of di(decanoyl)lecithin is about 12 mN/rn giving an experimental surface tension of about 60 mN/m.

Three simulations of two monolayers of each 42 di(decanoyl)lecithins at each side of an approximately 4 nm thick water slab with different force fields were performed. We found that a force-field with reduced charges on the lecithin head-groups and with a Ryckaert–Bellemans potential for the tails [13,14] best reproduced experimental data. This simulation is called ML in the following. As a reference we present some results from a simulation with standard Gromos potential [15] ('ML2').

In the simulation we did not note any significant difference in the surface tension between water and a lecithin-covered surface for any of the force fields (cf. Table I). This could be due to a defiance in the force field for the lecithins, namely too small repulsion between different molecules which in its turn could be due to the united atom representation used for the tail atoms. One could note here that the halved charges in the ML simulation as compared to the ML2 simulation in this respect is compensated for by the decreased van der Waals radii in the tails. However, a simulation with halved charges and the larger van der Waals radii in the tails gave a similar (slightly higher) surface tension than both the ML and the ML2 simulations. These results seem to indicate that the effect of details in the lecithin–lecithin interaction on the surface tension is minor. On the other hand, recent results from studies of decane–water interfaces [16] show large effects on surface tension of small changes of the van der Waals parameters for the interaction between decane and water. A conclusion could be that the most important factor for the surface tension is the interaction between the phases and not within each phase.

Both the ML and the ML2 simulations show very disordered monolayers with a large spread in the penetration depth of the lecithins into water and a disordered tail structure. The main difference between the simulations is a higher degree of ordering in the of the tails in the ML2 simulation and an increased hydration of the the phosphate group in the ML2 simulation compared to the ML simulation. This comes down to a more gel-like structure in the ML2 simulation than in the ML simulation, similar to what is observed for lipid bilayers [17]. This is also reflected in the fraction gauche dihedral interactions in the tails, about 17% in the ML2 and 27% in the ML simulation.

The reorientation of the phospholipids is slow with correlation times for the head group dipole vectors in the order of 0.2 ns in the ML and 0.3 ns in the ML2 simulation. Also the diffusion of the lecithins in the surface plane is slow with diffusion coefficients in the order of $0.5 \times 10^{-9}\,m^2\,s^{-1}$ (ML) or $0.2 \times 10\,m^2\,s^{-1}$ (ML2) in the surface plane. These values are, however, very approximate since the mean square displacement does not fully approach a straight line during the short analyzable period. Measurements on tracer diffusion in di(palmitoyl)lecithin(DPPC) monolayers [18] yields a ten times lower diffusion coefficient at low surface pressures. In part his fact represent the difference between DDPC and

DPPC but also the uncertain determination of the diffusion coefficient in the simulation is important. It has also been noted in several MD simulations that diffusion coefficients tend to come out higher than the experimental ones. This could be due to inaccuracies in the force fields, notably the lack of polarisability, cf. [19].

2.3. PHOSPHOLIPASE A_2 AT LECITHIN MONOLAYERS

Phospholipase A_2 ('PLA2') is an enzyme that degrades phospholipids at the so-called A_2 position. Phospholipase A_2 is highly stereospecific and degrades only 3-sn-phosphoglycerides. The mechanism reminds of the mechanism of serine proteases but is assumed to involve a water molecule as the nucleophile. Close to the catalytic site, a calcium ion is bound which is essential for the activity of the enzyme. The rate of degradation of the phospholipids increase by orders of magnitude for an aggregated substrate if its surface density is not too high, i.e. PLA2 will degrade micelles and monolayers with a moderate density but not bilayers.

The last simulation in this series was to a simulation in which we inserted a phospholipase molecule with its so-called interfacial recognition site ('IRS') directed towards one of the monolayers from the ML simulation, cf. Figure 1. The phospholipase was placed such that one protruding lecithin molecule was halfway into the active site of the phospholipase. During the course of the simulation the attraction between the PLA2 and the phospholipids due to van der Waals interactions steadily increased whereas the electrostatic attraction was more fluctuating but on average slowly increasing.

The properties of the lecithin monolayer during this simulation were analyzed similarly to the ML simulation above. The lateral diffusion coefficient of the lecithins might be somewhat lower than in the ML simulation but with the large uncertainties the difference is barely significant. Also the inclination angles of the tails and and head groups (compared to the monolayer normal) remained grossly unchanged.

The protein mainly retains its secondary structure and the R-factor (the r.m.s deviation of the structure as compared to the X-ray structure after optimal superposition) of the backbone is about 0.19 nm. Three regions show larger fluctuations around the average structure than the other regions, namely:

(1) the β-sheet around residue 80 that was poorly defined in the crystal structure of the porcine enzyme;
(2) the surface loop around residue 65 which is known experimentally to be very flexible and not crucial to the enzyme stability;
(3) The region around the calcium-binding loop.

The large fluctuations in the last region are combined with the calcium ion losing most of its carbonyl ligands. This is obviously due to problems with the force field and has been noted in other simulations of phospholipase as well. We believe that the reason is that the carbonyl ligands to the calcium ion are highly polarized in the real protein but not in the simulation. The charges used in this simulation were adopted for protein simulations in which the electric fields on the carbonyl

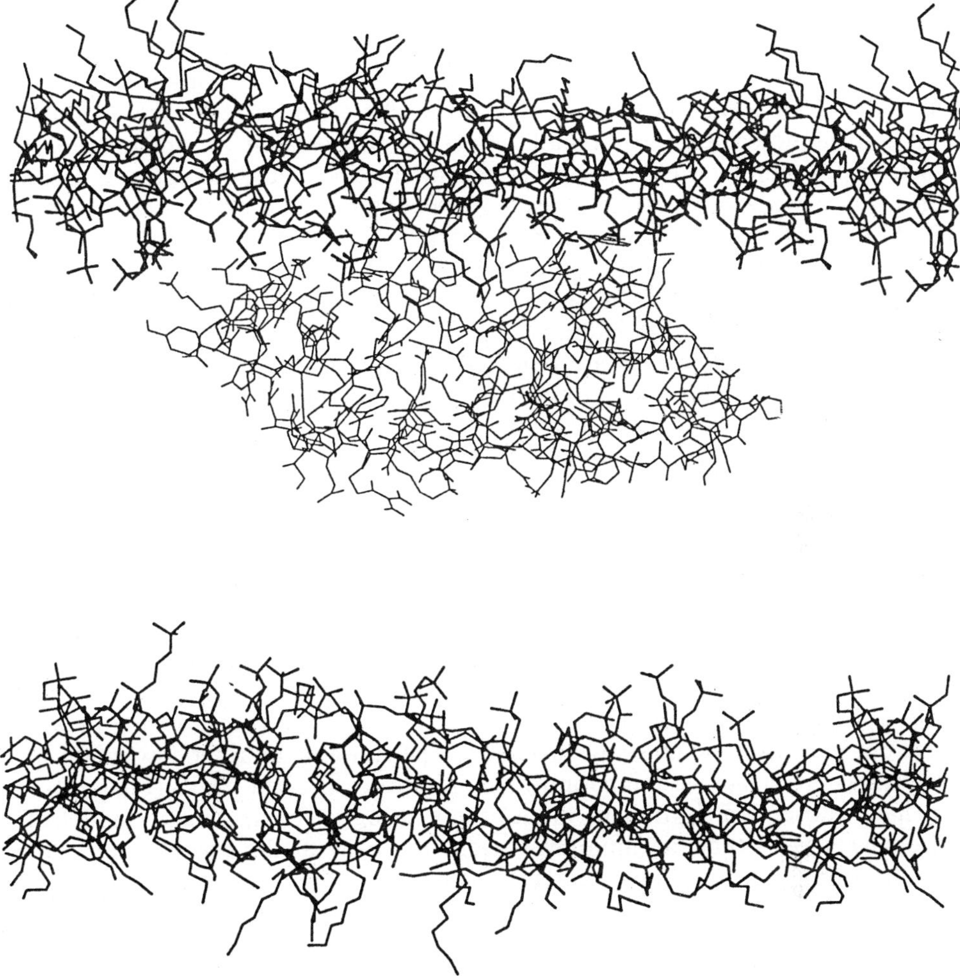

Fig. 1.   One configuration of the phospholipase and the monolayers in the PLA2 simulation (thick
lines: lecithins; thin lines: phospolipase; for clarity water molecules are not drawn).

groups are by far not as high as when they are ligands to a calcium ion. We tried
to change the van der Waals parameters in the line with earlier successful simulations of calcium binding proteins [20] but this did not show to be sufficient to keep
the calcium ligands. It has been proposed to introduce a much higher charge on
the carbonyl oxygens ($-0.58e$ instead of $-0.38e$) and correspondingly on the
carbonyl carbons in order to correctly describe ion binding [21]. We believe that
the best (most physical) solution is to include polarizabilities for all atoms in the
system. This requires, however, a complete reparametrisation of the force field
and work is on its way to find the optimal solution to this problem.

3. Solid Surfaces

In order to have a simple and clear-cut system to model we have turned our attention to studies of amino acids and water adsorbed on polycrystalline TiO_2 surfaces. This system constitutes a simple model system for, e.g., the interactions between biomolecules and biomaterials. The coadsorption of amino acids with water will hopefully yield insights in the interaction between water and amino acids. Here we are reporting a few experimental results and describe how we intend to use them in further modeling work.

The experimental method used in these studies is 'thermal desorption spectroscopy' (TDS) in which a certain amount of a substance is adsorbed on the substrate under ultra-high vacuum conditions. Subsequently, the sample is heated slowly and the desorbed molecules are detected by a mass spectrometer. This method yields information about the activation energy for desorption for the different molecules and could indirectly give information on the interactions between different molecules. For a more complete description of the technique see, e.g. [22].

3.1. WATER ADSORPTION ON TiO_2

Different amounts, ranging from fractions of a monolayer to several multi-layers, of water were adsorbed at 120 K on a titanium foil that had been oxidized to form an approximately 0.1 m thick polycrystalline TiO_2 film at its surface (experimental details will be given elsewhere [23]). Then the sample was heated at a rate of 2 K/s and particles with a mass over charge ratio of 18 (corresponding to water) were detected with the mass spectrometer. The resulting thermal desorption spectrum (mass 18 signal vs. temperature) shows three different peaks (Figure 2) and we have subsequently tried to model these using a simple kinetic model. The modeling implies that the multilayer peak (the peak at the lowest temperature) and the monolayer peak have nearly the same desorption energy, about 46 kJ/mol. This value is close to the sublimation energy for ice (47 kJ/mol). The multilayer peak is apparently zeroth order whereas the monolayer peak either is second order or, more probable, first order with an activation energy decreasing with water coverage. This part is then about 5 kJ/mol at full coverage. However, all these measurements are based on measurements on a polycrystalline surface which contains many types of sites and defects. In order to obtain a system that is easier to model we intend to repeat these experiments on a single crystal surface.

3.2. GLYCINE ADSORPTION ON TiO_2

A more biochemically oriented study regards the adsorption of glycine and the coadsorption of glycine and water at at TiO_2 surfaces. One aim of this study is to understand the interaction between glycine and water. The TDS spectrum was measured at $m/z = 28$, a mass that has been shown to be formed in the mass spectrometer from intact glycine [24]. The TDS spectrum of pure glycine on TiO_2 shows a peak at 320 K from and one peak at about 600 K. Through comparison with peaks for other masses (e.g., $m/z = 75$, corresponding to intact glycine mol-

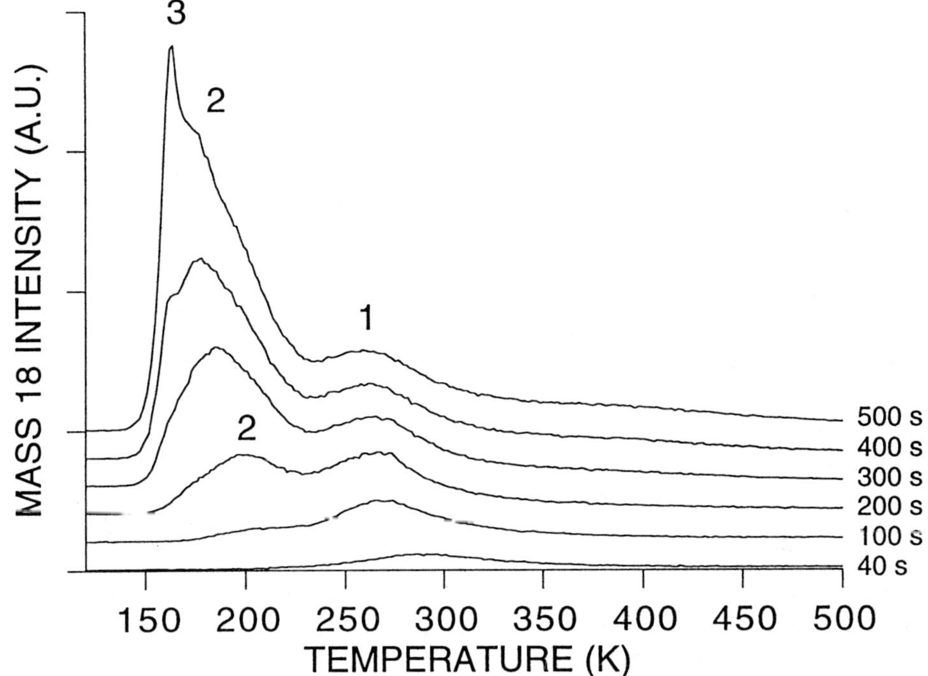

Fig 2. Thermal desorption spectra of water after adsorption of different amounts on TiO_2. The numbers to the right give the adsorption time at 120 K. This number is roughly proportional to the adsorbed amount. Note that three different peaks appear; the small one at about 250–300 K, probably corresponds to adsorption at defect sites, the peak at 180 K corresponds to monolayer adsorption and the peak at 160 K corresponds to multilayer desorption.

ecules) it can be concluded that the peak at 320 K is due to glycine molecules that desorb intact from the multilayer (with zeroth order kinetics) and are subsequently fragmented in the mass spectrometer. The peak close to 600 K is in contrast resulting from glycine molecules that are decomposed at the surface and desorb as fragments.

In a further series of experiments we wanted to study the interaction between water and glycine. In order to do so glycine and water were both adsorbed at a TiO_2 surface The order in which they are adsorbed does not give any significant difference in the results. Subsequently the TDS trace was measured (see Figure 3 in which the TDS spectrum for $m/z = 18$ is shown). We see that with increasing glycine exposure (and thus coverage) that the water peak at about 260 K is suppressed in favour of the peak at 200 K. This can be interpreted as water being displaced from the most favourable binding sites (those with the highest desorption temperature) by glycine. The glycine TDS spectrum is unaffected by the coadsorption.

We will now try to model the interactions of water and TiO_2 and later glycine, water and TiO_2 in order to be able to reproduce the binding data.

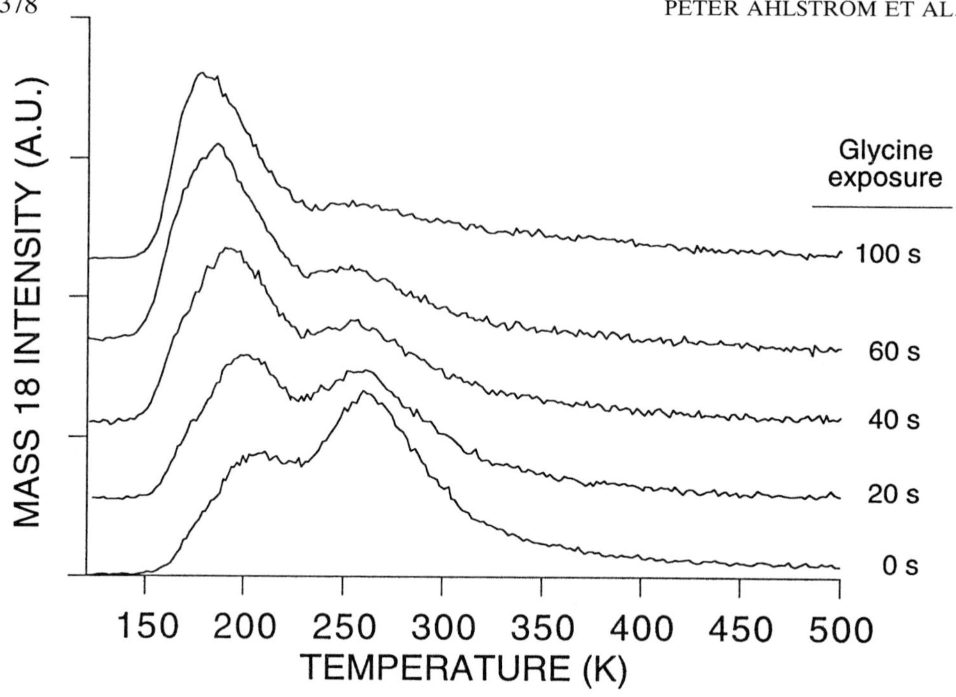

Fig. 3.   Thermal desorption spectra of mass 18 (water) after adsorption of both water and glycine to an TiO$_2$ surface. The numbers to the right give the adsorption time for glycine.

4. Conclusions

Our simulations have shown that it is possible to model large system with an atomic model. However, there is still room for improvement of the models in order to be able to reliably modeling large systems. One such improvement is the inclusion of polarisability which we believe is essential to model charged systems. Also the effects of truncating the potentials have to be taken care of. This will probably involve the use of much larger truncation radii than usual in present-day simulations combined with a mean-field treatment of particles outside the truncation radius.

Very large systems call on the use of simplified potential functions for all those parts of the system where the atomic detail as such is of no interest but merely a mean of description. Such potentials could be found from averaging the atomic detail simulations but also from experimental data. Therefore we aim at constructing an effective potential for amino acids (and later proteins) by modeling, e.g. adsorption and coadsorption data. These potentials could then be used for the study of protein adsorption in water etc.

Acknowledgements

We thank S. Jones and Dr. R. Pickersgill in Reading for fruitful cooperation regarding simulations of phospholipase A$_2$. Financial support from the European

Community (Biotechnology Action Programme), The Dutch Foundation for Chemical Research, The Swedish Research Council for Engineering Sciences and the Swedish National Board for Industrial and Technical Development is gratefully acknowledged.

References

1. C. Zografi, R. Verger, G. H. de Haas: *Chem. Phys. Liquids* **7**, 185 (1971).
2. S. T. Jones, P. Ahlström, H. J. C. Berendsen, and R. W. Pickersgill: *Biochem. Biophys. Acta* **1162**, 135 (1993).
3. H. J. C. Berendsen, J. P. M. Postma, W. F. van Gunsteren, and J. Hermans: in *Intermolecular Forces*, B. Pullman, Ed., D. Reidel, Dordrecht, p. 331 (1981).
4. H. J. C. Berendsen, J. R. Grigera, and T. P. Straatsma: *J. Phys. Chem.* **91**, 6269 (1987).
5. J. P. Ryckaert, C. Ciccotti, and H. J. C. Berendseii: *J. Comp. Phys.* **25**, 327 (1977).
6. H. J. C. Berendsen, J. P. M. Postma, W. F. van Gunsteren, A. DiNola, and J. R. Haak: *J. Chem. Phys.* **81**, 3684 (1984).
7. P. Ahlström and H. J. C. Berendsen: *J. Phys. Chem.* **97**, 13691 (1993).
8. J. Harris: *J. Phys. Chem.* **90**, 5077 (1992).
9. C. D. Holcomb, P. Clancy, and J. A. Zollweg: *Mol. Phys.* **78**, 437 (1993).
10. S. Abbas, P. Ahlström, and S. Nordholm: work in progress.
11. S. Abbas and S. Nordholm: submitted.
12. S. Nordholm amd A. D. Haymet: *Aust. J. Chem.* **33**, 2013 (1980).
13. J. P. Ryckaert and A. Bellemans: *Chem. Phys. Lett.* **30**, 123 (1975).
14. J. P. Ryckaert and A. Bellemans: *Faraday Discuss. Chem. Soc.* **66**, 95 (1978).
15. W. F. van Gunsteren and H. J. C. Berendsen: *Gromos Manual*, University of Groningen, Groningen, the Netherlands (1987).
16. A. R. van Buuren, S.-J. Marrink, and H. J. C. Berendsen: *J. Phys. Chem.* **97**, 9206 (1993).
17. E. Egberts, S.-J. Marrink, and H. J. C. Berendsen: *Eur. Biophys. J.* **22**, 423 (1994).
18. F. Caruso, F. Grieser, A. Murphy, P. Thistlethwaite, R. Urquhart, M. Almgren, and E. Wistus: *J. Am. Chem. Soc.* **113**, 4838 (1991).
19. P. Ahlström, A. Wallqvist, S. Engström, and B. Jönsson: *Mol. Phys.* **68**, 563 (1989).
20. P. Ahlström, O. Teleman, B. Jönsson, and S. Forsén: *J. Am. Chem. Soc.* **109**, 1541 (1987).
21. J. Åqvist, *J. Phys. Chem.* **96**, 10019 (1992).
22. D. A. King: *Surface Sci.* **47**, 384 (1975).
23. J. Lausmaa, P. Löfgren and B. Kasemo: in preparation.
24. S. Okude, F. Matsushima, H. Kuze T Shimizu: *Jpn. J. Appl. Phys.* **26**, 627 (1987).

Continuum-Model Studies of Redox Reactions, Complex Formation, and Electron Transfer: The Paradigm of Cytochrome c and Cytochrome c Peroxidase

HUAN-XIANG ZHOU
Laboratory of Chemical Physics, National Institute of Diabetes and Digestive and Kidney Diseases, National Institutes of Health, Bethesda, Maryland 20892, U.S.A.

Abstract. Redox reactions, complex formation, and electron transfer involving cytochrome c (cc) and cytochrome c peroxidase (CCP) are studied by using a dielectric continuum model. In this model a protein or protein complex is represented by a low dielectric region embedded in the high dielectric solvent. Variations in the character (charge, polarity, and size) and the location (interior or surface) and orientation of an amino acid are found to produce a diverse spectrum of effects on the reduction potentials of cc and CCP. The interface of the cc:CCP complex is made mostly of van der Waals rather than electrostatic contacts, yet the electrostatic interactions between the two proteins are found to give rise to experimentally observed effects of ionic strength and mutations of charged residues on the binding constant. Complex formation is found to have rather minor effects on the reduction potentials of cc and CCP, thus preserving the driving force of electron transfer. The protein matrices are found to play an important role of reducing the reorganization energy from what would have been if the redox centers were embedded directly in the solvent and thus speeding up the electron transfer.

Key words. Oxidation-reduction, complex formation, electron transfer, cytochrome c, cytochrome c peroxidase, electrostatic interactions, dielectric continuum model.

1. Introduction

Electron transfer holds a key place in respiration and photosynthesis. The past decade has witnessed a rapid progress toward understanding biological electron transfer on both the experimental [1,2] and the theoretical [3–5] fronts. However, many questions remain unanswered. The rate of electron transfer in general is given by the Marcus expression [6]

$$k_{et} = k_0 K \exp[-(-ef\Delta E^{0\prime} + \lambda)^2/4\lambda k_B T], \tag{1}$$

where K is the equilibrium constant for the donor and acceptor to form a complex, $\Delta E^{0\prime}$ is the difference in reduction potential between the acceptor and the donor in the complex (f is inserted for the conversion from eV to kcal/mol: $1\,eV = 23.1\,kcal/mol$), λ is the reorganization energy, and $k_B T$ is the product of Boltzmarn's constant and the temperature. The coefficient k_0 is related to the electronic coupling of the reactant and product states. How do amino acids of an electron-transfer protein control its reduction potential? The reduction potentials are usually measured for the donor and the acceptor in isolation, how does complex formation influence the reduction potentials of the proteins? What is the nature of reorganization in biological electron transfer?

To address these questions, I have chosen the well-characterized system involving yeast iso-1-cytochrome c (cc) and yeast cytochrome c peroxidase (CCP). The X-ray structures of cc and CCP in both the reduced (Fe^{2+}) and the oxidized (Fe^{3+})

A. Pullman et al. (eds.), Modelling of Biomolecular Structures and Mechanisms, 381–398.
© 1995 *Kluwer Academic Publishers. Printed in the Netherlands.*

states have been solved [7–10]. The X-ray structure of the complex between cc and CCP has also been solved [11]. In addition, structures for a number of CCP and cc mutants are now available [9,10,12,13]. Accompanying these structures is a wealth of experimental data on the reduction potentials of CCP, cc, and their mutants [10,13–16], the binding constant of cc with wild-type CCP under different ionic strengths [17–20] and with several CCP mutants (each bearing a Asp → Lys replacement) [18], and the rates of electron transfer from ferro-CCP to ferri-cc and from the anion radical of H_2 porphyrin cc to ferri-CCP [21]. It is hoped that by relating these data to the structures of the involved proteins, one may be able to find some answers to the questions posed earlier.

Of central importance in redox reactions, complex formation, and electron transfer are electrostatic interactions. The fact that their effects can be easily tuned through a change of buffer conditions (e.g., ionic strength) or a point mutation expands enormously our ability to study these processes. Presently two approaches exist for dealing with electrostatic interactions in biological systems. One is to use molecular dynamics or quantum simulations [3–5], the other is to use dielectric continuum models [22,23]. I have taken the latter approach as it allows for efficient treatment of the polarizability of proteins and the presence of the solvent. In a dielectric continuum model, a protein or protein complex is represented by a low dielectric region embedded in the high dielectric solvent. The polarization obtained in this model is one that is in equilibrium with the charge distribution of the protein or protein complex, consequently the free energy of the system is easily found. In contrast, to find the free energy in moleculer dynamics simulations would often require inhibitively long trajectories.

2. Dielectric Continuum Model

Through the dielectric constant, a continuum model in many ways emulates the microscopic picture of electrostatic interactions. Let us look at the simplest case, that of the interaction between a sodium ion (charge e) and a chloride ion (charge $-e$) in water. A microscopic description of the interaction is like this. When the ions are far apart, a shell of water molecules would solvate each ion so that their dipoles point away from the positive charge and point toward the negative charge. If the two ions are brought closer, the mutual influence of the solvated ions will result in some water molecules being released to the bulk and acquiring higher mobility. Consequently one anticipates a contribution from the entropy gain to the decrease in free energy upon bringing the ions together.

In a continuum-model representation of water, the change in free energy upon bringing the ions together is given by Coulomb's law

$$G_{\mathrm{el}} = -e^2/\epsilon r, \qquad\qquad (2)$$

where r is the final distance between the ions and ϵ is the dielectric constant of water. One can separate G_{el} into entropic and enthalpic components (S_{el} and H_{el}) using the usual thermodynamic relations. Noting that ϵ is the only temperature-dependent quantity, one obtains

$$TS_{el} = -(\partial \ln \epsilon / \partial \ln T)e^2/\epsilon r, \tag{3a}$$

$$H_{el} = -[1 + (\partial \ln \epsilon / \partial \ln T)]e^2/\epsilon r. \tag{3b}$$

At $T = 298$ K, $\epsilon = 78.5$ and $\partial \ln \epsilon / \partial \ln T = -1.37$. For concreteness let us take $r = 4$ Å. Equation (3a) predicts $TS_{el} = 1.45$ kcal/mol, an entropy gain, in agreement with the expectation from the microscopic picture of the ion–ion interaction. Furthermore, the continuum model predicts $H_{el} = 0.39$ kcal/mol, an enthalpy gain. This at first sight may seem counterintuitive, but is actually consistent with the microscopic picture considering that favorable charge–dipole interactions are lost upon the release of water molecules from the solvation shell to the bulk.

A second example further illustrates the ability of a continuum model to emulate a microscopic description. Consider moving a sodium ion from the bulk of water across an interface with air. In a microscopic description one would say that it is much more energetically favorable for the ion to be solvated in a shell of water molecules (with their dipoles pointing away from the ion). Thus in moving the ion toward the interface one is met with resistance. This is equivalent to say that the ion experiences a force pointing from the air side to the water side. A continuum-model representation of water and air (with dielectric constants ϵ and ϵ', respectively) predicts just such a force. In this representation, the sodium ion in the water side has an image charge $e(\epsilon - \epsilon')/(\epsilon + \epsilon')$ in the air side. As $\epsilon \gg \epsilon'$, this is a positive charge. The sodium ion thus experiences a repulsive force $e^2(\epsilon - \epsilon')/4(\epsilon + \epsilon')r^2$ from the image charge, where r is the distance from the sodium ion to the interface.

The present study of redox reactions, complex formation, and electron transfer uses a dielectric continuum model in which each protein (or protein complex) is represented by a low dielectric region (with dielectric constant ϵ_i) embedded in the high dielectric solvent (with dielectric constant ϵ) (see Figure 1). The dielectric boundary between the protein and the solvent is given by the X-ray structure of the protein. Inside the protein the electrostatic potential satisfies Poisson's equation

$$\nabla^2 V(\mathbf{r}) = -4\pi \sum_l q_l \delta(\mathbf{r} - \mathbf{r}_l)/\epsilon_i, \tag{4}$$

where q_l are the protein charges and $\mathbf{r}_l$ are their positions. In the solvent, salt ions may be present. They are taken into consideration by the Poisso–Boltzmann equation

$$\nabla^2 V(\mathbf{r}) - \kappa^2 V(\mathbf{r}) = 0 \tag{5}$$

for the electrostatic potential in the solvent. Here the Debye screening constant κ is given by $\kappa^2 = 8\pi I e^2/\epsilon k_B T$, in which I is the ionic strength.

After solving the Poisson and Poisson–Boltzmann equations for the electrostatic potentials V_l at the charge sites, one finds the electrostatic free energy of the protein by

$$G_{el} = \sum_l q_l V_l/2. \tag{6}$$

To look at the contribution of the dielectric boundary, one can separate the

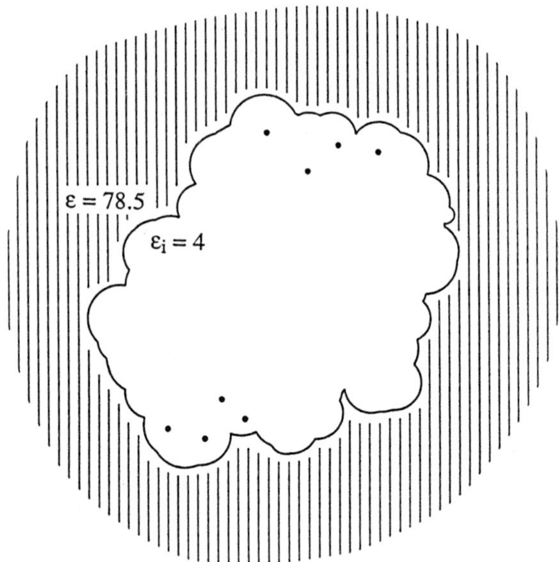

Fig. 1. The continuum model for calculating the electrostatic free energy of a protein or protein complex. Closed circles represent charges on the atoms of the protein.

electrostatic potential at each charge site into a direct-action component V_l^d and a reaction-field component V_l^r. The direct-action component is

$$V_l^d = -\sum_{l'} q_{l'}/\epsilon_i |\mathbf{r}_l - \mathbf{r}_{l'}|. \tag{7}$$

In the summation of Equation (7), the infinite self potential at each charge site should be excluded. This is justified as only the *change* in free energy (ΔG_{el}) from one state to another has significance. The self potentials are preserved and thus do not contribute to ΔG_{el}. By summing the direct-action components and the reaction-field components respectively over all the charge sites, one obtains the direct-action contribution G_d and the reaction-field contribution G_r to G_{el}.

The Poisson and Poisson–Boltzmann equations have mostly been solved by the finite-difference method [22,23], in which the infinite three-dimensional space is represented by a finite cubic lattice. I [24–26] have developed an alternative method based on the boundary-element technique, in which the Poisson and Poisson–Boltzmann equations are transformed into integral equations on the protein surface. The boundary-element method allows one to have an accurate description of the protein surface and to have charges located exactly. These two features are important in reducing numerical errors, which are of particular concern in studying the effects of single amino acids on reduction potential as the involved changes in free energy are rather small. In addition, the boundary-element method allows one to easily treat the large system presented by the complex formed between cc and CCP. Because of these considerations, this method is chosen for the present study.

Let us now express the relevant quantities in redox reactions, complex formation, and electron transfer in terms of particular free-energy changes. The reduction potential of a redox protein is related to the change in free energy from the reduced to the oxidized state (the oxidized species has an extra unit of charge distributed on the redox center). The electrostatic component of this change is

$$\Delta G_{el}^{OR} \equiv G_{el}^{O} - G_{el}^{R} = \sum_l (q_l^O V_l^O - q_l^R V_l^R)/2. \tag{8}$$

An additional component ΔG_{cen}^{OR} characterizes bonding interactions at the redox center. One particular interest of the present study is to find out how individual amino acids control the reduction potential of a protein by calculating the effects of point mutations. A point mutation affects ΔG_{el}^{OR} but not ΔG_{cen}^{OR}, thus the resulting change in reduction potential ($\Delta E^{0'}$) is directly related to the change in ΔG_{el}^{OR} ($\Delta\Delta G_{el}^{OR}$) through

$$ef\Delta E^{0'} = \Delta\Delta G_{el}^{OR}. \tag{9}$$

The same equation also holds when the effect of complex formation is considered, except that now the changes in $E^{0'}$ and ΔG_{el}^{OR} are due to complex formation with a second protein.

The equilibrium constant K for two proteins 1 and 2 to form a complex is determined by the change in free energy from the proteins in isolation to the proteins in complex. The electrostatic component of this change is

$$\Delta G_{el}^{CF} = G_{el}(1:2) - G_{el}(1) - G_{el}(2). \tag{10}$$

A change in ionic strength or a point mutation changes the binding constant (from K to K') primarily through a change in ΔG_{el}^{CF} ($\Delta\Delta G_{el}^{CF}$), thus [26]

$$-k_B T \ln(K'/K) = \Delta\Delta G_{el}^{CF}. \tag{11}$$

There actually is a relationship between the change in reduction potential upon complex formation and the binding constant. For concreteness, suppose we are interested in the change in reduction potential of protein 1 upon complexation with protein 2. Specialization of Equation (9) to the present case leads to

$$\begin{aligned}
ef\Delta E^{0'} &= \Delta G_{el}(1^{OR}:2) - \Delta G_{el}(1^{OR}) \\
&= [G_{el}(1^O:2) - G_{el}(1^R:2)] - [G_{el}(1^O) - G_{el}(1^R)] \\
&= [G_{el}(1^O:2) - G_{el}(1^O) - G_{el}(2)] - [G_{el}(1^R:2) \\
&\quad - G_{el}(1^R) - G_{el}(2)] = -k_B T \ln(K^O/K^R),
\end{aligned} \tag{12}$$

where K^O (K^R) is the binding constant of oxidized (reduced) protein 1 with protein 2.

The last quantity to be dealt with is the reorganization energy of electron transfer. Marcus [6] separated the reorganization energy into an inner part (λ_i) involving bond rearrangement of the donor and acceptor redox centers and an outer part (λ_o) involving polarization reorientation of the surrounding environment. The Marcus theory has been very successful for dealing with electron transfer in small molecules, where the outer reorganization occurs primarily in the

Fig. 2.   The continuum model for calculating the outer reorganization energy of electron transfer in a protein complex. The redox centers, represented by spheres, have dielectric constant ϵ_{op}. They are embedded in the protein matrices, which have dielectric constant ϵ_i (the dashed curve represents the interface between the donor and the acceptor). The protein complex in turn is embedded in the solvent, which has dielectric constant ϵ.

solvent. For electron transfer in proteins, the intervening protein matrices also have to reorganize. It is still not clear what is the proper way to account for the reorganization of protein matrices within a continuum model. I [27] have suggested using a continuum model that involves three dielectric media (see Figure 2). This is obtained by embedding two redox centers in the general model for a protein or protein complex (shown in Figure 1). Both redox centers have the optical dielectric constant ϵ_{op} (due to the electronic polarization and given by the square of the refractive index). For the sake of simplicity, each redox center is represented by a sphere. The outer reorganization energy is further partitioned into a protein contribution λ_o^p and a solvent contribution λ_o^s. The protein contribution is given by the outer reorganization energy when the protein matrices are extended to infinity. This approximately is

$$\lambda_o^p = e^2(1/\epsilon_{op} - 1/\epsilon_i)(1/a - 1/R), \tag{13}$$

where a is the radius of the sphere representing a redox center, and R is the separation between the redox centers. The solvent contribution λ_o^s is roughly given by the free energy of the protein complex with charge e at one sphere center and charge $-e$ at the other sphere center and with both spheres taking the dielectric constant of the protein matrices. This is $G_{el}(1:2)$ encountered earlier, except that now only two charges are present (one at each redox center).

As for the inner reorganization energy λ_i, I [27] have suggested it is given by

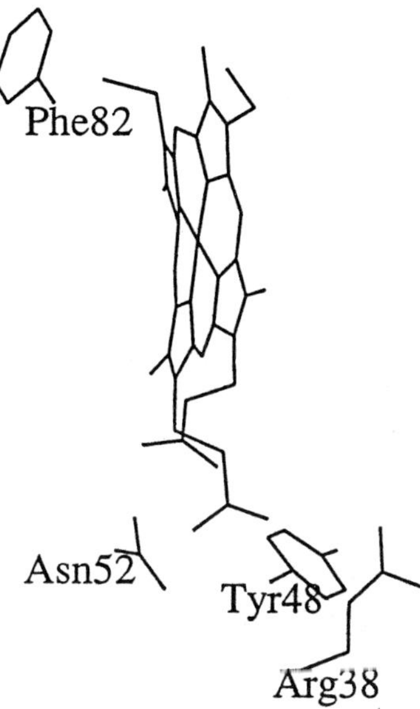

Fig. 3.   The locations of the four mutated residues relative to the heme in yeast iso-1-cytochrome *c*.

the sum of the ΔG_{cen}^{OR} components characterizing bonding interactions at the redox centers. The ΔG_{cen}^{OR} component for each redox center in turn can be obtained from the measured reduction potential and the calculated ΔG_{el}^{OR} component. Thus

$$\lambda_i = ef[E^{0\prime}(1) + E^{0\prime}(2)] - [\Delta G_{el}^{OR}(1) + \Delta G_{el}^{OR}(2)] + C, \qquad (14)$$

where C is a quantity that is invariant from one electron-transfer reaction to another.

The details of the calculations will be published elsewhere. In what follows I will present results concerning the effects of point mutations and complex formation on the reduction potentials of cc and CCP, the effects of ionic strength and point mutations on the binding constant of cc and CCP, and the reorganization energy for the electron transfer between cc and CCP. The protein dielectric constant used is 4 and the solvent dielectric constant is taken as the value for water at the appropriate temperature, which is 78.5 at room temperature ($T = 298$ K).

3. Effects of Point Mutations on Reduction Potential

3.1. POINT MUTATIONS IN cc

The effects of the following seven mutations at four positions of cc have been studied: Arg38 → Lys, His, Asn, and Ala, Tyr48 → Phe, Asn52 → Ile, and Phe82 → Ser. The locations of the four mutated residues relative to the heme are shown in Figure 3. Residues Arg38 and Tyr48 are near the pyrrole-ring A propionate of the heme, residue Asn52 is between the pyrrole-ring A and D propionates, and residue Phe82 is near the CBB methyl of the heme. The chosen mutations

TABLE I

Changes of experimental (exp) and calculated (cal) reduction potentials (in mV) and free energy contributions (in kcal/mol) due to mutations in yeast iso-1-cytochrome c^a

Mutation	$\Delta E^{0\prime}$ (exp)	$\Delta E^{0\prime}$ (cal)	$\Delta\Delta G_d^{OR}$	$\Delta\Delta G_r^{OR}$
		Arg38		
Lys	−23	−7	−0.18	0.01
His	−27	−28	−1.08	0.42
Asn	−34	−47	−6.56	5.47
Ala	−47	−58	−6.99	5.64
		Tyr48		
Phe	−22	−17	−0.39	0.01
		Asn52		
Ile	−55	−56	−1.18	−0.11
		Phe82		
Ser	−35	−32	−0.35	−0.38

[a] Experimental results for mutations at positions 38, 48, 52, and 82 are from [14], [15], [13], and [16], respectively.

occur both in the interior (positions 48 and 52) and on the surface (positions 38 and 82) of the protein. Two of them (Arg38 → Lys and His) preserve the total charge of the mutated residue but change the charge distribution, two of them (Arg38 → Asn and Ala) change the charge of the mutated residue, two of them (Tyr48 → Phe and Asn52 → Ile) change the polarity of the mutated residue, and one (Phe82 → Ser) changes both the polarity and the size of the mutated residue.

The experimental and calculated changes in reduction potential due to these seven mutations are listed in Table I. It can be seen that the two sets of results are in reasonable agreement, so the calculations can be used to find the specific roles of the residues in controlling the redox potential of cc. First notice from Table I that all the mutations lower the reduction potential, so it looks as though the protein matrix in cc is constructed to maintain the reduction potential at a high level. The reasons that the wild-type residues result in a higher reduction potential are not the same for the four positions considered.

At position 38, the side chain of the wild-type Arg residue interacts with the heme pyrrole-ring A propionate through the mediation of two water molecules. This arrangement results in the consequence that the charge of the Arg38 side chain is localized as close as possible to the heme. The strong charge–charge interaction with the heme then raises the reduction potential of cc. When Arg38 is mutated into Lys, the new residue cannot interact favorably with either of the water molecules and the charge of the Lys38 side chain is not as close to the heme as in the case of Arg38. The reduction potential is thus lowered. The same is true when Arg38 is mutated into His, which is known to be protonated at this position [14]. The difference between Lys38 and His38 is in their charge distribution. The charge of Lys38 is mainly located around the NZ atom. In protonated His38 the charge is shared between two centers, one around ND1, which has a distance roughly the same as that of the Lys38 NZ atom from the heme, and the other around NE2, which is farther away from the heme. This two-centered distribution weakens the charge–charge interactions with the heme and results in a decrease in reduction potential stronger than that from the Lys38 mutation. The difference between His38 and Asn38 lies in the extra charge of His38. For a unprotonated His38, the reduction potential is found to be the same as the value for Asn38. Finally the difference between Asn38 and Ala38 is due to the polarity of Asn38.

What is perhaps surprising is that despite the loss of a charge upon the Arg38 → Ala mutation, the resulting decrease in reduction potential is only 47 mV, corresponding to a decrease of 1.1 kcal/mol in free energy. This results from the presence of the dielectric boundary between the protein and the solvent. From Table I one can see that the decrease in the direct-action component of the free-energy change (6.99 kcal/mol) is largely offset by the increase in the reaction-field component (5.64 kcal/mol). Cutler *et al.* [14] used the protein dipole and Langevin dipole (PDLD) model to study the Arg38 → Leu mutation and arrived at a similar attribution of the small decrease in reduction potential to the dielectric boundary between the protein and the solvent.

At positions 48 and 52, the wild-type residues (Tyr and Asn, respectively) form hydrogen bonds with the heme pyrrole-ring A propionate. The formation of the hydrogen bonds results in the consequence that the dipoles of the side chains are oriented toward the heme. These dipoles are lost upon the mutations of Tyr48 → Phe and Asn52 → Ile. The resulting loss of the repulsive charge-dipole interactions between the heme and the Tyr48 and Asn52 residues then gives rise to the respective decreases of 22 and 55 mV in reduction potential. The larger decrease in the case of the Asn52 → Ile mutation is a result of the larger dipole moment from the amide group of the Asn residue.

The reason that the wild-type residue at position 82 results in a higher reduction potential is again different. Phe82 limits the exposure of the heme to the solvent. Mutation of Phe82 to a smaller Ser residue leads to the formation of a solvent channel which substantially increases the solvent exposure of the heme. This increased heme exposure accounts for more than half of the 35 mV decrease in reduction potential. The rest comes from an attractive charge–dipole interaction between the heme and the side chain of Ser82 (the dipole points away from the heme).

3.2. POINT MUTATIONS IN CCP

In wild-type CCP, the carboxylate of Asp235 forms a hydrogen bond with the ND1 atom of His175, which in turn is liganded to the heme through the NE2 atom (see Figure 4). Goodin and McRee [10] have recently measured the reduction potentials of wild-type CCP and three mutants that have Asp235 → Glu, Asn, and Ala replacements and found $E^{0\prime} = -183, -113, -79$ and -78 mV, respectively. These results create several puzzles. Both Asp235 and Glu235 have a carboxylate, yet the reduction potential of the mutant is 70 mV higher than that of the wild type. On going from Glu235 to Ala235, a buried carboxylate is lost, thus one may expect a large increase in reduction potential, yet the observed increase is only 35 mV. Similarly, on going from Asn235 to Ala235, an amide dipole is lost, yet the two mutants are found to have almost the same reduction potential.

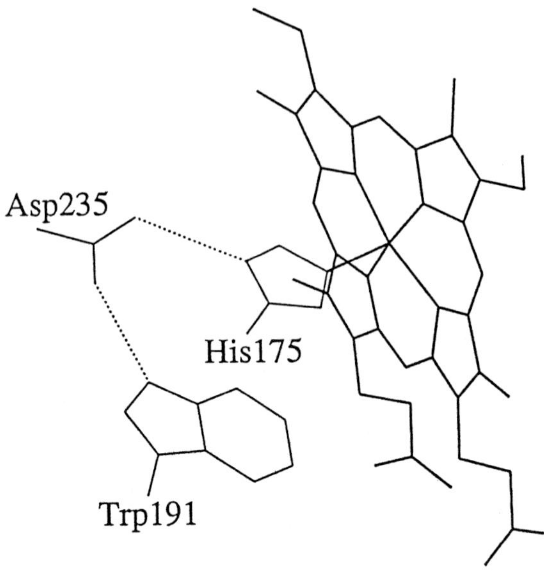

Fig. 4.   The hydrogen-bond interactions of the Asp235 carboxylate with the ND1 atom of His 175
and the NE1 atom of Trp191 in yeast cytochrome *c* peroxidase.

A clue for the 70 mV difference in reduction potential between the wild type
and the Glu235 mutant comes from resonance Raman spectroscopy [28] and
NMR [29] studies. It turns out that the strong interaction with Asp235 serves to
deprotonate His175 into an imidazolate. On the other hand, the mutant has a
resonance Raman spectrum that is characteristic of a normal imidazole in His175.
When the imidazolate character of His175 is modeled into the wild type, the
reduction potentials of the wild type and the (Glu235 mutant are found to be
separated by 77 mV, close to the experimental value of 70 mV. Thus it is the
change in the His175 character from imidazolate to imidazole that accounts for
the observed increase in reduction potential.

For the small difference of 35 mV in reduction potential between the Glu235
and the Ala235 mutants, resonance Raman spectroscopy [10,28] again suggests a
reason. Resonance Raman spectra reveal that the ferric heme of the Glu235
mutant has a six-coordinate high spin form, whereas the ferric heme of the Ala235
mutant has a six-coordinate low spin form. This means that the interaction of the
iron with its sixth ligand is much weaker in the Glu235 mutant than in the Ala235
mutant. The sixth ligand has been suggested to be a water molecule or a hydroxyl
ion. The X-ray structures of wild-type CCP (with Asp235) and the Asn235 mutant
in the ferric form [9] show that the distance from the heme iron to the nearest
water molecule decreases from 2.7 Å in the wild type to 1.9–2.0 Å in the mutant.
If this decrease in iron-water distance is used to model the strengthening of the
iron–sixth ligand interaction, the difference in reduction potential between the
Glu235 and the Ala235 mutants is found to be 30 mV, nearly the same as the
experimental result of 35 mV. If the difference in the strength of the iron–sixth

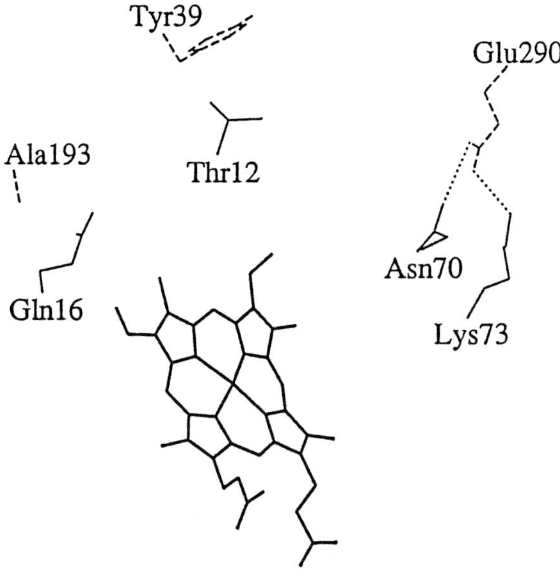

Fig. 5.   Residues in the interface between yeast iso-1-cytochrome *c* and cytochrome *c* peroxidase in the complex. The residues of cc are drawn with solid lines while those of CCP are drawn with dashed lines. Hydrogen bonds across the interface are drawn with dotted lines. The heme of cc is also shown.

ligand interaction between the two mutants were not taken into consideration, the difference in reduction potential would have been 290 mV.

The small difference in reduction potential between the Asn235 and Ala235 mutants is a result of the fact that the amide dipole of Asn52 is pointed in a direction that is almost perpendicular to the vector to the heme iron. As a result the charge–dipole interaction between the heme and Asn235 is rather weak and the loss of this dipole upon mutation does not result in a large change in reduction potential.

The study on the reduction potentials of cc and CCP shows that variations in the character (charge, polarity, and size) and the location (interior or surface) and orientation of an amino acid can produce a diverse spectrum of effects.

4. Complex Formation between cc and CCP

In the X-ray structure of the cc:CCP complex, the interface between the proteins is found to be made mostly of van der Waals, instead of electrostatic, contacts [11] (see Figure 5). This is in contrast to an earlier model built for the cc:CCP complex [30], in which three Asp residues (at positions 37, 79, and 217) of CCP were paired with three positively charged residues of cc. The lack of electrostatic contacts in the X-ray structure of the complex thus raises concern about whether the structure can be reconciled with observed effects of ionic strength and mutations of charged residues on the binding constant [17–20]. The study of the complex formation between cc and CCP is directed primarily toward this concern.

4.1. EFFECTS OF IONIC STRENGTH ON COMPLEX FORMATION

Several groups have measured the binding constant of ferri-cc and ferri-CCP. In a pioneering experiment, Leonard and Yonetani [17] found a binding constant of $K = 7.56 \times 10^4\,M^{-1}$ for the proteins in 100 mM potassium phosphate at pH = 7. More recently Corin et al. [18] (and McLendon et al. [20] also) obtained a binding constant of $K = 2.0 \times 10^7\,M^{-1}$ in 10 mM potassium phosphate at pH = 7. The results of Stemp and Hoffman [19] suggest that this maybe a lower limit for the binding constant at the low phosphate concentration. The last authors also measured the binding constant of cc and CCP in 50 mM potassium phosphate at pH = 7 and found $K = 7.14 \times 10^5\,M^{-1}$. An increase in binding constant is thus clearly seen upon a decrease in ionic strength. According to Equation (11), when the buffer concentration is lowered from 100 mM to 50 and 10 mM, the electrostatic component ΔG_{el}^{CF} of the binding energy should lower by 1.4 and at least 3.3 kcal/mol, respectively. The ionic strengths of these buffers can be estimated to be $I = 200$, 100, and 25 mM.

Ferric-cc and ferri-CCP have total charges of $7e$ and $-12e$, respectively. The opposite charges result in a large direct-action contribution of $\Delta G_d^{CF} = -473.1$ kcal/mol that is favorable for complex formation. One expects that the screening of the solvent would result in an almost complete cancellation of this contribution by the reaction-field contribution. This is indeed the case. The reaction-field contribution at $I = 200$ mM is found to be $\Delta G_r^{CF} = 464.0$ kcal/mol and when I is lowered to 100 and 25 mM, becomes 462.0 and 457.7 kcal/mol, respectively. It is comforting to note the decrease in the reaction-field contribution because that means an increase in the binding constant, and thus consistency with the experimental observations. The magnitudes of the decrease in the electrostatic component ΔG_{el}^{CF} of the binding energy are 2.0 and 6.3 kcal/mol, roughly in agreement with the corresponding experimental values.

4.2. EFFECTS OF CHARGE MUTATIONS ON COMPLEX FORMATION

As mentioned earlier, three Asp residues of CCP were paired with three positively charged residues of cc in a cc:CCP model prior to the solution of the X-ray structure. This led Corin et al. [18] to investigate the role of these residues in the complex formation of CCP with cc by mutating them into Lys. If indeed these residues are in the interface with cc, changing the sign of their charge would severely impair the interactions with their partners in cc and thus significantly reduce the binding constant. This is not what was observed by Corin et al. (in 10 mM phosphate at pH = 5.7). Of the three mutations, only Asp37 → Lys was found to reduce the binding constant by a factor of about 10. Asp79 → Lys and Asp217 → Lys were found to affect the binding constant only marginally.

Are these results consistent with the X-ray structure of the cc:CCP complex? An affirmative answer seems to be borne out directly from the locations of these three Asp residues relative to the interface of CCP with cc in the X-ray structure (see Figure 6). Asp37 is close to the interface while Asp79 and Asp217 are away from the interface. Calculations on the electrostatic component ΔG_{el}^{CF} of the binding energy for wild-type CCP and the three mutants further confirm the

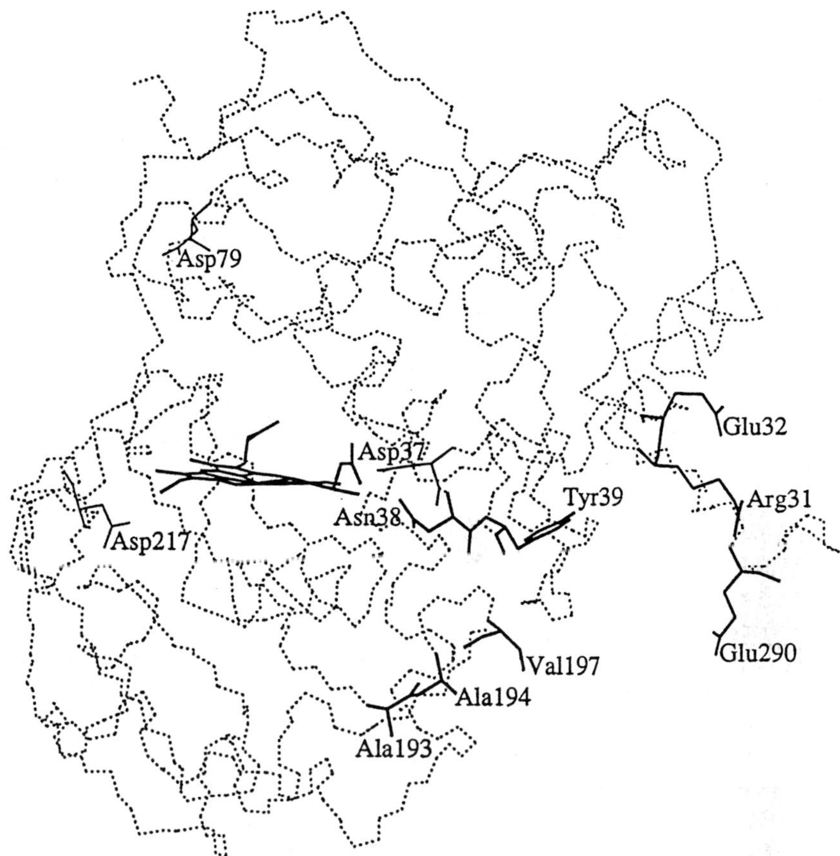

Fig. 6.   The locations of the mutated Asp residues (Asp37, Asp79, and Asp217; thin solid lines) of yeast cytochrome c peroxidase relative to the residues (thick solid lines) in the interface with yeast iso-1-cytochrome c. For the interface, several CCP residues close to those shown in Figure 5 are also shown in order to give a better sense of the interface. The backbone (dotted lines) and heme (thick solid lines in the middle of the figure) of CCP are also shown.

consistency of the X-ray structure of the cc:CCP complex with the results of Corin *et al.* For wild-type CCP, $\Delta G_{\mathrm{el}}^{\mathrm{CF}}$ at $I = 25$ mM can be found from the $\Delta G_{\mathrm{ed}}^{\mathrm{CF}}$ and $\Delta G_{\mathrm{r}}^{\mathrm{CF}}$ contributions given earlier and is -15.4 kcal/mol. Upon the mutations of Asp37 $\rightarrow$ Lys, Asp79 $\rightarrow$ Lys, and Asp217 $\rightarrow$ Lys, changes to -14.0, -15.3, and -15.1 kcal/mol, respectively. Using Equation (11), the mutation at position 37 is found to decrease the binding constant by a factor of 10.6 whereas the mutation at either position 79 or position 217 is found to decrease the binding constant by a factor of less than 2. This is entirely consistent with the results of Corin *et al.*

4.3. EFFECTS OF COMPLEX FORMATION ON REDUCTION POTENTIAL

Having reconciled the X-ray structure of the cc:CP complex with the observed effects of ionic strength and mutations of charged residues on the binding constant of the proteins, let us now examine how complex formation influences the reduction potentials of the proteins. This question is obviously important by itself, but an answer to it actually has practical value because reduction potentials for the proteins in isolation are much more easily measured than for the proteins in complex. The Marcus expression, given in Equation (1), of course requires the reduction potentials for the proteins in complex.

Rather minor effects of complex formation on the reduction potentials of cc and CCP are found. The electrostatic component of the free-energy change upon oxidation for cc alone is $\Delta G_{el}(cc^{OR}) = 1.4$ kcal/mol. After complex formation with ferri-CCP, it becomes $\Delta G_{el}(cc^{OR}:CCP) = 0.5$ kcal/mol. This results in a decrease of 0.9 kcal/mol in ΔG_{el}^{OR}, corresponding to a decrease of only 40 mV in reduction potential. Comparison of the ΔG_d^{OR} and G_r^{OR} contributions for cc in isolation and in complex with CCP reveals that the small difference in G_{el}^{OR} comes about because the difference in ΔG_d^{OR} is largely offset by the difference in ΔG_r^{OR}. For CCP alone G_{el}^{OR} is -14.79 kcal/mol. Upon complexation with ferric-cc, G_{el}^{OR} becomes -14.84 kcal/mol, virtually identical with the value for CCP alone.

The finding that complex formation has rather minor effects on the reduction potentials of proteins has direct experimental support. In an early study of cc, Vanderkooi and Erecinska [31] found that complex formation with either CCP or cytochrome b_5 did not cause any measurable change in the reduction potential of cc. More recently Burrows _et al._ [32] studied the complex formation between cc and cytochrome b_5 and observed that upon complexation the reduction potential of either protein changed by less than 20 mV.

Further support is provided by the experimental observation of Leonard and Yonetani [17] and McLendon _et al._ [20] that the binding constants of ferro-cc and ferri-cc with ferri-CCP are very similar. According to Equation (12), the change in reduction potential upon complex formation is proportional to the logarithm of the ratio between the binding constants of the oxidized and reduced forms. Thus the similar binding constants of ferro-cc and ferri-cc mean that the change in the reduction potential of cc upon complex formation is rather small.

5. Electron Transfer between CCP and cc

Cheung _et al._ [21] have measured the rates of electron transfer from ferro-CCP to ferri-cc and from the anion radical of H_2 porphyrin cc to ferri-CCP and found k_{et} to be about 1 and 150 s^{-1}, respectively. These rates, combined with the driving forces of the respective reactions, allow the reorganization energy λ to be obtained from the Marcus expression (the rates are for electron transfer in preformed complexes so the equilibrium constant K should not enter; the reorganization energy is assumed to be the same for the two complexes). The driving force for an electron-transfer reaction is given by the difference of the reduction potentials for the donor and acceptor in the complex. As the effects of complex formation on reduction potential have just been shown to be rather minor, it should be close to the difference $\Delta E^{0'}$ of the reduction potentials for the donor and acceptor in isolation. The reduction potentials of cc and CCP in isolation are about 270 and -190 mV, respectively, while the potential for the reduction of H_2 porphyrin cc to the anion radical has been measured to be -1100 mV [33]. Thus $\Delta E^{0'}$ is 460 mV for the electron transfer from CCP to cc and 910 mV from the electron transfer from H_2 porphyrin cc to CCP. Using these values of $\Delta E^{0\prime}$ and the values of k_{et} given earlier, one finds $\lambda = 37$ kcal/mol.

What constitute this reorganization energy? Earlier the reorganization energy has been separated into a protein part λ_o^p, a solvent part λ_o^s, and an inner part λ_i. The protein part λ_o^p characterizes the polarization reorientation within the protein matrices and is given by Equation (13). The iron–iron separation in the cc:CCP complex is $R = 26.5$ Å. If the radius a of a redox center is taken to be 5 Å (the size of a heme) and ϵ_{op} is assigned a value of 1.8, the protein reorganization energy is found to be $\lambda_o^p = 16.5$ kcal/mol. The solvent part λ_o^s characterizes the polarization reorientation within the solvent and can be found only by solving the Poisson and Poisson–Boltzmann equations. For the cc:CCP complex it is $\lambda_o^s = 3.5$ kcal/mol. The total outer reorganization energy is thus $\lambda_o = \lambda_o^p + \lambda_o^s = 20.0$ kcal/mol. If the redox centers are directly embedded in the solvent, the total outer reorganization energy is given by Equation (13) but with the protein dielectric constant ϵ_i replaced by the solvent dielectric constant ϵ. The larger solvent dielectric constant would result in an outer reorganization energy of 29.2 kcal/mol. Other things being equal, the 9.2 kcal/mol increase in reorganization energy would lead to a 40-fold smaller rate for the electron transfer from ferro-CCP to ferri-cc. By virtue of the small dielectric constant, the protein matrices play an important role of reducing the reorganization energy from what would have been if the redox centers were embedded directly in the solvent and speeding up the electron transfer.

From the experimental value of $\lambda = 37$ kcal/mol and the calculated value of $\lambda_o = 20.0$ kcal/mol, one finds $\lambda_i = 17.0$ kcal/mol. The inner part characterizes bonding rearrangement at the redox centers and has been related through Equation (14) to the reduction potentials $[E^{0\prime}(1)$ and $E^{0\prime}(2)]$ and the electrostatic components $[\Delta G_{el}^{OR}(1)$ and $\Delta G_{el}^{OR}(2)]$ of the oxidation-induced free-energy changes of the donor and acceptor. The value of λ_i for the electron transfer between cc and CCP allows the constant C in Equation (14) to be determined. Using $E^{0\prime}(cc) = 270$ mV and $E^{0\prime}(CCP) = -190$ mV, $\Delta G_{el}^{OR}(cc) = 1.4$ kcal/mol and $\Delta G_{el}^{OR}(CCP) = -14.8$ kcal/mol, one finds $C = 1.8$ kcal/mol.

To see whether this is a sensible approach, let us examine another electron transfer reaction: the self-exchange in cc. The iron–iron separation in a cc dimer is about 18 Å, Equation (13) thus gives $\lambda_o^p = 14.7$ kcal/mol. The solvent reorganization energy should likewise be smaller than in the cc:CCP complex and thus should be between 0 and 3.5 kcal/mol. As such the exact value of λ_o^s is not very important. For concreteness let us assume that it scales as the iron–iron distance and obtain a value 2.4 kcal/mol. The total outer reorganization energy is thus $\lambda_o = 17.1$ kcal/mol. Substituting the reduction potential and the ΔG_{el}^{OR} component of cc and the value 1.8 kcal/mol just obtained for the constant C into Equation (14), one finds $\lambda_i = 11.5$ kcal/mol. Thus the total reorganization energy is found to be $\lambda = 28.6$ kcal/mol. This is in surprising agreement with the experimental result of 28 kcal/mol found in two independent studies [34,35]. Churg *et al.* [36] have made calculations for the self-exchange in cc using a microscopic model. They found a solvent reorganization energy of 3 kcal/mol, in agreement with our estimate of 2.4 kcal/mol. However, they found much smaller values for the protein reorganization energy (2.4 kcal/mol) and the inner reorganization energy (1 kcal/mol). The resulting total reorganization energy of 6.4 kcal/mol is much smaller than the experimental result.

Using the present estimates, one can make some analysis of the difference in reorganization between the electron transfer from CCP to cc and the self-exchange in cc. Of the 9 kcal/mol difference in reorganization energy, about 6 kcal/mol is due to higher inner reorganization in the redox center of CCP and the remaining portion is due to higher outer reorganization around the redox centers in the cc:CCP complex. The higher inner reorganization for CCP may be justified considering the fact that the heme in CCP is anchored from only one side to the protein matrix (through the iron—His175 NE2 bond) whereas the heme in cc is anchored from both sides to the protein matrix (through iron—His18 NE2 and iron—Met80 SG bonds) (the heme vinyl groups in cc also form two thioether bonds with two Cys residues). The higher outer reorganization for the cc:CCP complex arises from the fact that the hemes in this complex are more isolated from each other (so the protein matrices and the solvent have to reorganize around two independent redox centers). A clearer picture on the relative importance of the various contributions (λ_i, λ_o^P, and λ_o^s) in a particular electron-transfer reaction may appear after additional results of reorganization energies are analyzed using the present approach. Experimental results are now available on the reorganization energies for the electron transfer from cytochrome b_5 to cc [33] and between the α and β subunits of hemoglobin [37], and for the self-exchange in myoglobin [38].

6. Conclusion

In this work I have used the dielectric continuum model to study redox reactions, complex formation, and electron transfer within the paradigm of yeast iso-1-cytochrome c and cytochrome c peroxidase. The role of individual amino acids in controlling the reduction potential of a protein is examined by calculating the effects of point mutations. It is shown that variations in the character (charge, polarity, and size) and the location (interior or surface) and orientation of a residue produce a diverse spectrum of effects on the reduction potentials of cc and CCP.

The interface of the cc:CCP complex is made mostly of van der Waals rather than electrostatic contacts, yet the electrostatic interactions between the two proteins are shown to give rise to experimentally observed effects of ionic strength and mutations of charged residues on the binding constant. Complex formation is found to have rather minor effects on the reduction potentials of cc and CCP, thus preserving the driving force of electron transfer. A consequence of this result is that not much error is introduced when the reduction potentials of the donor and acceptor in isolation are used to calculate the driving force for an electron-transfer reaction.

The protein matrices are found to play an important role of reducing the outer reorganization energy from what would have been if the redox centers were embedded directly in the solvent and thus speeding up the electron transfer. By relating electron transfer to redox reactions, a method for obtaining the inner reorganization energy has been proposed. A check against experimental results on the self-change in cc suggests that it is a promising attempt. Applications to other reactions will further test this method and help gain a better understanding on the nature of reorganization in biological electron transfer.

Acknowledgements

I thank A. Szabo, W. A. Eaton, and A. G. Mauk for helpful discussions. G. D. Brayer sent me the X-ray coordinates of the Ile52 and Ser82 mutants of yeast iso-1-cytochrome *c* and J. Kraut sent me the X-ray coordinates of the complex between yeast iso-1-cytochiome *c* and cytochrome *c* peroxidase. Their kindness is gratefully acknowledged.

References

1. G. McLendon: *Acc. Chem. Res.* **21**, 160 (1988).
2. Many recent experimental studies of biological electron transfer are summarized in two recent volumes of the Advances in Chemistry Series of the American Chemical Society: M. K. Johnson, R. B. King, D. M. Kurtz, Jr., C. Kutal, M. L. Norton, and R. A. Scott, Eds., *Electron Transfer in Biology and the Solid State*, Adv. Chem. Ser. Vol. 226 (1990); J. R. Bolton, N. Mataga, and O. McLendon, Eds., *Electron Transfer in Inorganic, Organic, and Biological Systems*, Adv. Chem. Ser. Vol. 228 (1991).
3. A. Warshel and W. W. Parson: *Ann. Rev. Phys. Chem.* **42**, 279 (1991).
4. K. Schulten and M. Tesch: *Chem. Phys.* **158**, 421 (1991).
5. C. Zheng, J. A. McCammon, and P. Wolynes: *Chem. Phys.* **158**, 261 (1991).
6. R. A. Marcus and N. Sutin: *Biochim. Biophys. Acta* **811**, 265 (1985).
7. G. V. Louie and G. D. Brayer: *J. Mol. Biol.* **214**, 527 (1990).
8. A. M. Berghuis and G. D. Brayer: *J. Mol. Biol.* **223**, 959 (1992).
9. J. Wang, J. M. Mauro, S. L. Edwards, S. J. Oatley, L. A. Fishel, V. A. Ashford, Ng.-h. Xuong, and J. Kraut: *Biochemistry* **29**, 7160 (1990).
10. D. B. Goodin and D. E. McRee: *Biochemistry* **32**, 3313 (1993).
11. H. Pelletier and J. Kraut: *Science* **258**, 1748 (1992).
12. G. V. Louie, G. J. Pielak, M. Smith, and G. D. Brayer: *Biochemistry* **27**, 7870 (1988).
13. R. Langen, G. D. Brayer, A. M. Berghuis, G. McLendon, F. Sherman, and A. Warshel: *J. Mol. Biol.* **224**, 589 (1992).
14. R. L. Cutler, A. M. Davies, S. Creighton, A. Warshel, G. R. Moore, M. Smith, and A. G. Mauk: *Biochemistry* **28**, 3188 (1989).
15. S. P. Rafferty, L. L. Pearce, P. D. Barker, J. G. Guillemette, C. M. Kay, M. Smith, and A. G. Mauk: *Biochemistry* **29**, 9365 (1990).
16. A. M. Davies, J. G. Guillemette, M. Smith, C. Greenwood, A. G. P. Thurgood, A. G. Mauk, and G. R. Moore: *Biochemistry* **32**, 5431 (1993).
17. J. J. Leonard and T. Yonetani: *Biochemistry* **13**, 1465 (1974).
18. A. F. Corin, G. McLendon, Q. Zhang, R. A. Hake, J. Falvo, K. S. Lu, R. B. Ciccarelli, and D. Holzschu: *Biochemistry* **30**, 11585 (1991).
19. E. D. A. Stemp and B. M. Hoffman: *Biochemistry* **32**, 10848 (1993).
20. G. McLendon, Q. Zhang, S. A. Wallin, R. M. Miller, V. Billstone, K. G. Spears, and B. M. Hoffman: *J. Am. Chem. Soc.* **115**, 3665 (1993).
21. E. Cheung, K. Taylor, J. A. Kornblatt, A. M. English, G. McLendon, and J. R. Miller: *Proc. Natl. Acad. Sci. USA* **83**, 1330 (1986).
22. K. A. Sharp and B. Honig: *Ann. Rev. Biophys. Biophys. Chem.* **19**, 301 (1990).
23. M. E. Davis and J. A. McCammon: *Chem. Rev.* **90**, 509 (1990).
24. H.-X. Zhou: *Biophys. J.* **65**, 955 (1993).
25. H.-X. Zhou: *J. Chem. Phys.* **100**, 3152 (1994).
26. H.-X. Zhou: to be published.

27. H.-X. Zhou: *J. Am. Chem. Soc.* **116**, 10362 (1994).
28. G. Smulevich, J. M. Mauro, L. A. Fishel, A. M. English, J. Kraut, and T. G. Spiro: *Biochemistry* **27**, 5477 (1988).
29. J. D. Satterlee, J. E. Erman, J. M. Mauro, and J. Kraut: *Biochemistry* **29**, 8797 (1990).
30. T. L. Poulos and J. Kraut: *J. Biol. Chem.* **255**, 10322 (1980).
31. J. Vanderkooi and M. Erecinska: *Arch. Biochem. Biophys.* **162**, 385 (1974).
32. A. L. Burrows, L. H. Guo, H. A. O. Hill, G. McLendon, and F. Sherman: *Eur. J. Biochem.* **202**, 543 (1991).
33. G. McLendon and M. R. Miller: *J. Am. Chem. Soc.* **107**, 7811 (1985).
34. R. Gupta: *Biochim. Biophys. Acta* **292**, 291 (1973).
35. D. G. Nocera, J. R. Winkler, K. M. Yocom, E. Bordignon, and H. B. Gray: *J. Am. Chem. Soc.* **106**, 5145 (1984).
36. A. K. Churg, R. M. Weiss, A. Warshel, and T. Takano: *J. Phys. Chem.* **87**, 1683 (1983).
37. S. E. Peterson-Kennedy, J. L. McGourty, J. A. Kalweit, and B. M. Hoffman: *J. Am. Chem. Soc.* **108**, 1739 (1986).
38. R. J. Crutchley, W. R. Ellis, and H. B. Gray: *J. Am. Chem. Soc.* **107**, 5002 (1985).

The Relationship between Physical Property and Function of Highly Activated Mutants of Thermolysin

SHUN-ICHI KIDOKORO, YOICHIRO MIKI, KIMIKO ENDO, and AKIYOSHI WADA
Sagami Chemical Research Center, Nishiohnuma 4-4-1, Sagamihara, Kanagawa 229, Japan

Abstract. Thermolysin mutants having a variety of amino acid at the 119th position are designed by considering electrostatic field effect upon the active area. The most activated mutant has five times higher hydrolytic activity than the wild type. Negative correlation between the activity and the thermal stability is observed. A combined effect of the flexibility of the substrate binding site and the negative electrostatic field around the site is suggested as a key to enhance the activity.

Key words. Protein engineering, thermolysin, activity, stability.

I. Introduction

One of the present targets of protein science is to reveal nature's design principle for protein molecules that governs their highly purpose oriented three-dimensional structure and physical properties. If we establish a principle which guides the construction of a protein molecule that has the desired physical property and function, its influence will be very large on both pure and applied research. This is the ultimate goal of protein engineering. In the current stage, however, our knowledge of the design principle is still poor, and rational design for higher activity or higher stability has not been successful in many cases.

It is necessary to establish a novel strategy for disclosing nature's design principle and producing useful proteins. For this purpose, we study the difference between human and nature's way of designing a machine in the following.

In design work in the mechanics and electronics fields, the property and function of the products can be precisely predicted on the basis of well-known physical principles. We are almost confident that the product will work according to our design. In this case, the process of designing is straightforward:

$$[\text{Design}] \rightarrow [\text{Production}] \rightarrow [\text{Evaluation}].$$

Nature has been designing protein molecules in a unique way, called 'natural selection'. Many kinds of slightly or greatly modified molecules have been produced in the course of evolution, and more functional and/or more stable ones have been selected to remain. A similar strategy has been used for traditional drug design, which is called 'screening'. Many newly synthesized materials are evaluated, and a few samples that seem effective for the target diseases are selected as key samples. Then they are slightly (or highly) modified and evaluated once again. In this case, the designing continues in a circuit with a feedback path:

$$[\text{Synthesis}] \rightarrow [\text{Evaluation}].$$

A. Pullman et al. (eds.), Modelling of Biomolecular Structures and Mechanisms, 399–407.
© 1995 *Kluwer Academic Publishers. Printed in the Netherlands.*

Judging from the fruits of evolution, this strategy seems to work very well. The process, however, is expensive in time and materials. While it is possible to produce useful proteins without understanding the basic design principles, the disadvantage of the screening strategy is that we accumulate no information on the design principle of protein molecules.

For the production of useful proteins and in order to understand the design principle, our strategy must take advantage of both the feedback path and the principle of rational design. We have proposed a new strategy using a 'protein engineering cycle':

$$[\text{Design}] \rightarrow [\text{Synthesis}] \rightarrow [\text{Evaluation}].$$

The most efficient way to do protein engineering is to go around this cycle rapidly, many times. Travelling around the cycle will not only give us several novel useful proteins as products but also useful knowledge about the mechanism determining the relationship between the proteins' structure, properties and functions.

Because protein synthesis has now become a routine process, we concentrate our effort on establishing the other two elements: molecular design and precise (and rapid) evaluation. In practice, several thermolysin mutants have been produced according to our design, and we have succeeded in evaluating them precisely. We have confirmed that some mutants show remarkably higher activity than the wild type, and we have found an interesting relationship between the activity and the thermal stability, as presented below.

Thermolysin is a metalloendoprotease from thermophilic bacteria, *Bacillus thermoproteolyticus*. It is one of the well-known and well-used enzymes in laboratories and industry, because of its narrow specificity for hydrophobic residues, and its high thermal stability. The amino acid sequence [1,2] and the three-dimensional structure with several inhibitors [3,4] have been determined. It has recently been pointed out, however, that its amino acid sequence deduced from the cDNA sequence of the bacteria differs from the known sequence by the two residues: the 37th residue is not aspartic acid but asparagine and the 119th is not glutamic acid but glutamine [5]. This has been confirmed by direct amino acid sequencing [6]. The corrected sequence is identical with that of the neutral protease from *B. stearothermophilus* [7]. It is interesting that the replacement of amino acid at one of the misread sites is found to cause a remarkable change of activity and stability, as presented below.

II. Molecular Design for Higher Enzymatic Activity

Enzymes work as a catalyst for chemical reactions. Although they do not change the thermodynamic equilibrium between the starting materials and the products, they accelerate the reaction rates by stabilizing the transition state of the reaction. The stabilization is realized enthalpically and/or entropically, and the Gibbs free energy of the transition state is decreased.

In order to design an enzyme for higher activity, we pay attention to the electrostatic environment around the active site. The electrostatic field is known to play an important role in terms of function and thermal stability. Several methods are utilized for calculating the field in and around the protein molecule where the electrostatic effects of water molecules are taken into account microscopically or macroscopically [8].

Figure 1 shows the profile of electrostatic potential on the surface of thermolysin. We used a macroscopic calculation method described by Nakamura and Nishida [8]. In this method, the space around the protein molecule was divided into three parts using the known three-dimensional structure of the protein: inner, outer and boundary regions of the molecule. We assigned dielectric constants for each region: 4 for the inner region, 78 for the outer, and 10 for the boundary. The Poisson equation for the inner region, Poisson–Boltzman equation for the outer, and Laplace equation for the boundary were written down and solved numerically on a Titan 750 workstation (Kubota computer, Japan). The electrostatic shielding effect by ions in the solution was taken into consideration with the screening constant of the Debye–Hückel theory. We assumed the ionic strength to be 0.01. In order to display the results of the calculations, we used both our newly developed program and a commercial software package, Quanta (Polygen, USA) on the Titan and on an IRIS 4D/70GT (Silicon Graphics, USA), respectively.

As seen in the figure, the negative electrostatic environment is prepared in the substrate binding site. This enzyme binds four calcium ions to stabilize the structure, and these binding sites are also prepared as a negative field. In the other part of the protein surface, little special pattern of the electrostatic field can be recognized. This indicates that the electrostatic field is designed uniquely around the substrate binding site or active site.

The local electrostatic field around the active site affects the enzymatic activity through three mechanisms. First, the field affects the binding affinity of the substrate to the enzyme. The positive (or negative) field is usually prepared around the active site for the negatively (or positively) charged substrate. Second, the field affects the pK values of the amino acids at the site. The protonation/deprotonation of the side chains of amino acids is sensitively affected by the electrostatic field, and pH profile of the enzyme activity is considered to be determined by the pKs of the amino acids at the active site. Because only the amino acid in a protonated (or deprotonated) state can contribute to the reaction, the pH profile may be deformed by changing the local electrostatic field. Third, the thermodynamic stability of the transition state of the reaction may be affected by the electrostatic field because the electric charges are generally changed on reaction. As the definite structure of the transition state and its electron distribution cannot be predicted, however, the qualitative evaluation of the third effect is difficult to do generally.

While the first effect and the second one concern the equilibrium of the substrate binding and the equilibrium of conversion between active and inactive enzyme, respectively, the third effect is a kinetic one. Precise evaluation of the activity of the designed enzyme can clarify these three mechanism separately.

One effective design for higher activity is to modify the electrostatic field around the active site by replacing the amino acid at the 'electrostatically effective site', and the carefuly study the property and acivity change induced by the replacement. The effective site is defined as the position where the electric change with the

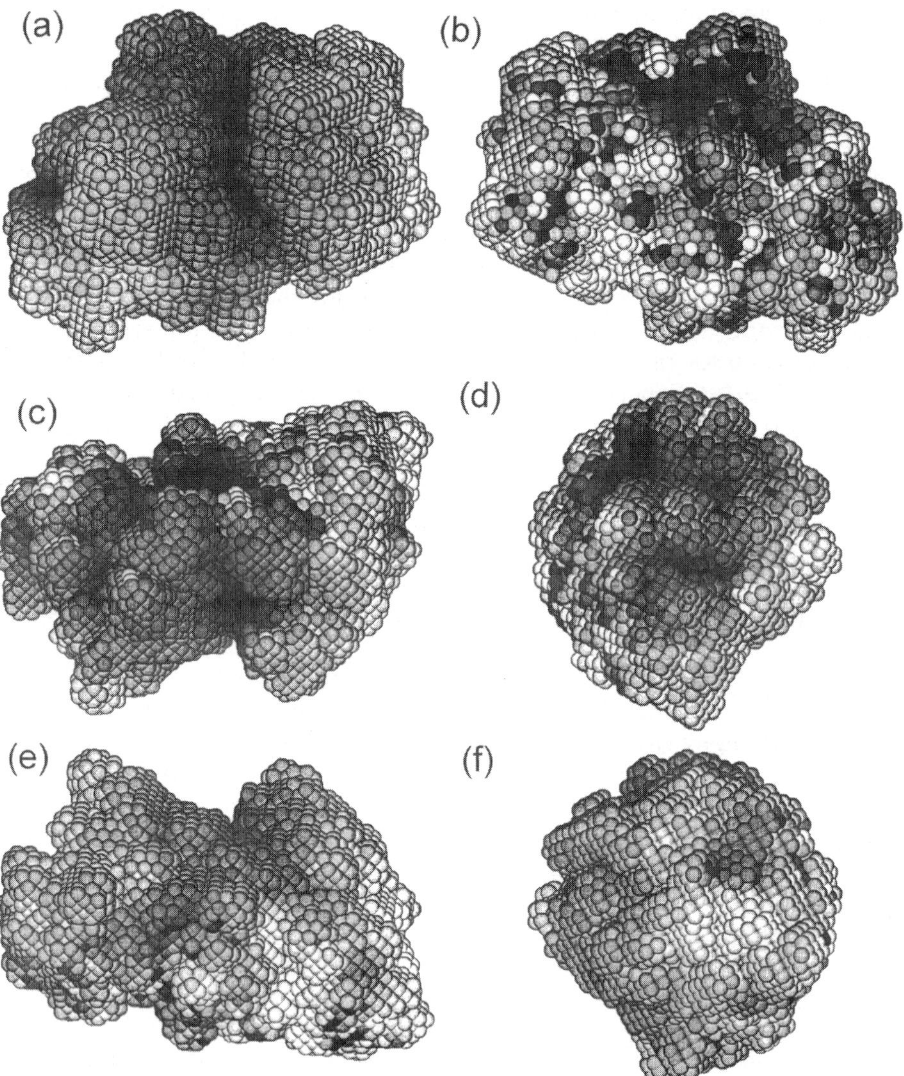

Fig. 1.   Electrostatic potential on the surface of thermolysin molecule in solution. The electric field is represented by brightness of the spheres placed on each cubic lattice point whose center is located in the boundary region of the electrostatic calculation: the darker color represents more negative potential. (a) Front view of the molecule. The substrate binds to the enzyme along the central cleft (dark region) where the potential is negative. (b) Back view. The binding sites for three calcium ions are prepared as the negative potential. (c) Top view. The substrate binding site in the front view and the calcium ion binding site in the back view are recognized. (d) Left view. The binding site for one calcium ion is located the lower part of the center of this view where the negative potential is also prepared. (e) Bottom view. No unique pattern can be recognized. (f) Right view. The positive potential is clustered in the central part of this view. The basic amino acids, lysines and arginines make a cluster: the functional and the structural meaning of this cluster is not yet known.

amino acid replacement makes a large difference to the local electrostatic field on the active site.

Using the electrostatic calculation method mentioned above, we have evaluated the contribution of every residue and selected the electrostatically effective sites against one of the essential residues for the proteolysis: the 143rd glutamic acid. The 119th is one of the selected sites. The energy difference between the deprotonation and protonation state of the 143rd glutamic acid that is contributed by negative charge of the 119th glutamic acid is evaluated as -2.2 kcal/mol. The energy corresponds to a 1.6 unit pK shift of the 143rd glutamic acid.

The 119th glutamine of the neutral protease from *B. stearothermophilus* was considered to make a hydrogen bond to the hydroxyl group of the 103rd serine [7]. Judging from the three-dimensional structure, this hydrogen bond plays an important role in fixing the peptide strand from the 111th to 117th that makes a wall of the substrate binding site. If we replace the 119th glutamine by other amino acids, such as glutamic acid, the flexibility of the binding site of the enzyme will be increased. This is expected to change the binding affinity to the substrate.

From the above consideration, we regard the 119th position as an interesting one to mutate because of the large effect of both its electrostatic and the mechanical property on the active site. We have made various mutants at this position and evaluated the activity and physical property as described in the next section.

III. Evaluation of Physical Property and Function

Site-directed mutagenesis at the 119th position of thermolysin was performed by the overlap extension method using the polymerase chain reaction by Miyake *et al.* (Tosoh Co. Ltd.) as described elsewhere in detail [9]. All the mutants but cysteine mutant were synthesized, and purified by hydrophobic and gel permeation HPLC. The purity of the tryptophan mutant was less than 90% and it could not be obtained in adequate purity for measurement. The remaining 17 mutants and wild type were obtained with good purity (above 95%) judging from gel electrophoresis. Near and far UV CD spectra of all 17 mutants coincided with those of the wild type (data not shown). This suggests that the secondary and tertiary structure is not changed by the replacement.

The electrostatic property of the molecules was evaluated by measuring the pI on IEF gel electrophoresis. The pI of the wild type was 5.0. The mutants of two basic amino acids, lysine and arginine, showed a remarkable increase of pI, as expected. The mutant of methionine showed a similar increase. As the methionine residue has no true charge, the pK of some other residues must be affected by the replacement. The other 14 mutants has almost the same pI values as the wild type. It is noteworthy that the two acidic mutants, aspartic acid and glutamic acid, showed no change in pI. This implies that the carboxyl groups of these amino acids have high pK values.

The hydrolytic activity of wild type and the mutant enzymes was evaluated by measuring the hydrolysis of N-(3-[2-furyl]acryloyl)-glycyl-L-leucine amide (FAGLA, from Sigma, USA) with a UB-35 spectrophotometer (Jasco, Japan) equipped with a personal computer PC-9801 (NEC, Japan) and thermostated circulator EL-15 (Taitec, Japan). All the activity was measured in a solution of

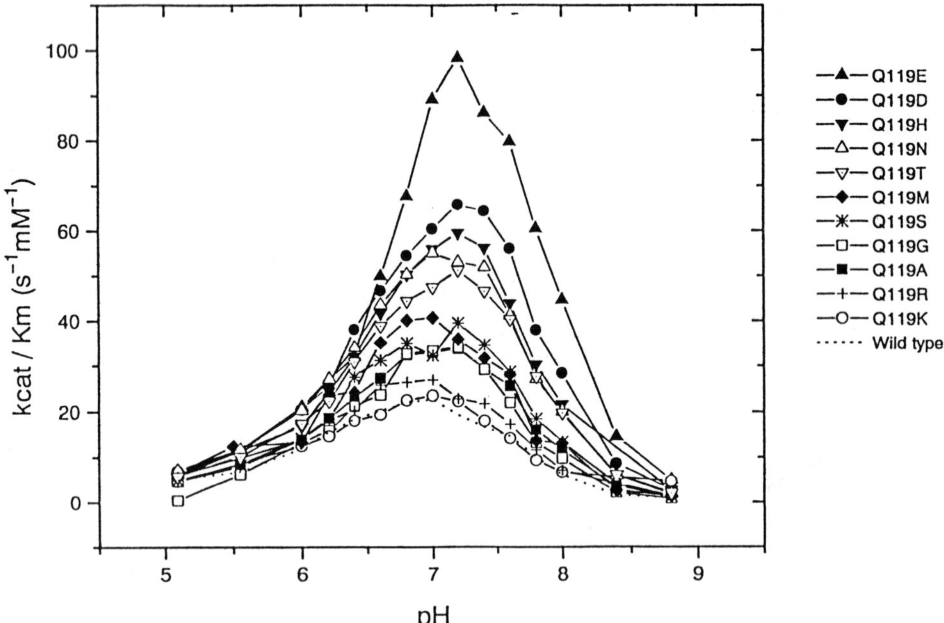

Fig. 2. The pH profile of hydrolytic activity of the 119th mutants of thermolysin against N-(3-[2-furyl]acryloyl)-glycyl-L-leucine amide (FAGLA). The solution condition is 1.8 mM FAGLA in 200 mM Tris–maleate buffer, containing 10 mM calcium chloride and 18 μM zinc sulfate at 37 °C; concentration of enzyme ranges from 0.2 to 1.0 μM. The symbols for the mutants are indicated in the figure. The dotted line indicates that of wild type.

200 mM Tris–maleate buffer with 10 mM calcium chloride and 18 μM zinc sulfate at 37 °C from pH 5 to pH 9. All the experimental data were fitted by the theoretical curves from an integral Michaelis–Menten equation within experimental error (10^{-3} optical density). The detail is described elsewhere [9].

The pH profile of the hydrolytic activity against FAGLA is shown in Figure 2. The substrate FAGLA has rather low affinity (high K_M value) for thermolysin, then k_{cat} and K_M values cannot be determined separately in solution, and only k_{cat}/K_M is obtained. Although the optimum pH of the mutant, where the activity reaches a maximum, slightly increases as its maximum activity increases, the pH dependence of every mutant is similar. The absolute value of the activity, however, is very much affected by the mutations. It is remarkable that almost all the mutants are found to be enhanced in their activity. This suggests that the breakage of the hydrogen bond of the 119th glutamine and the 103rd serine is effective in enhancing the activity, judging from the implication of the structure that only the glutamine can make the bond. The most activated mutant, Q119E, has a five-fold activity enhancement.

The thermal stability of all the mutants was evaluated with a highly sensitive differential scanning calorimeter, MCS (MicroCal, USA) at pH 8.28, 50 mM HEPES buffer containing 10 mM calcium chloide and 0.1 mM zinc sulfate. Protein concentration is 0.2–0.4 mg/ml and the scanning rate is 1 °C/min. Thermal transition of thermolysin is completely irreversible, and the transition profile depends on the scanning rate. For such a non-equilibrium system, rigorous analysis with an assumption of equilibrium, such as the deconvolution method, cannot be applied [10]. In this study, only the peak temperature, T_p, of the heat absorption curve accompanied with the transition was used as the measure of thermal stability.

It is well known that self-proteolysis occurs during the denaturation of proteases. If the irreversibility is caused by the self-proteolysis and the apparent unfolding process of the protein molecule is rate-determined by the self-proteolysis of the protease, the proteolytic activity must affect the transition profile. In this case, T_p must shift to a higher temperature with decreasing protease activity, and some corrections must be made to find the true unfolding rate of the protein.

We have checked the dependence of T_p of the wild type enzyme on the concentration of inhibitor, o-phenanthroline, known as a chelating agent of zinc, and found that the T_p showed no shift with the existence of 1 mM o-phenanthroline, although the hydrolytic activity against FAGLA decreases dramatically (data not shown). In the case of the most activated mutant, Q119E, no dependence of T_p on the inhibitor concentration is observed.

These facts indicate that we can evaluate the thermal stability with T_p without correction according to the activity change. We conclude that the rate-determining step of the thermal transition is not the self-hydrolysis process but the irreversible unfolding process of the protein.

Figure 3 shows the relationship between activity and thermal stability of the mutants. As the straight line indicates, we see a negative correlation there, without two open circles (K and R). The open circles represent the fact that the pI of these mutants deviates greatly from that of the wild type: they have more positive charges than wild type. As described in the previous section the 119th site is an effective site to make a strong electrostatic field around the active site: the 143rd glutamic acid. This might be the reason for the deviation of K and R. The other open circle is the methionine mutant. In this case, the charge of the 119th site is considered to be the same as the wild type. No deviation from the line may be reasonable from this point of view. The most activated mutant, Q119E, also deviates from the line. As the glutamic acid and the aspartic acid may be deprotonated, and charged negatively at neutral pH, the negative electrostatic field caused by the amino acid may enhance the activity in this pH region, while the positive field decreases it, as mentioned above. What causes the difference of activity between glutamic acid and the aspartic acid remains obscure, however. Furthermore, a qualitative theoretical explanation seems to be problematic for the stability and activity of each mutant.

We have designed a new substrate for thermolysin that has higher affinity for the enzyme than FAGLA for the evaluation of k_{cat} and K_M separately. Preliminary results with this new substrate suggest that the k_{cat} values of the mutants are almost the same as that of wild type and their K_M values are lowered [9]. This indicates that the activity, plotted in Figure 2, is enhanced by increasing the affinity for the substrate.

Without considering the electrostatic effect, the binding affinity may be correlated with the flexibility of binding site of the mutants, and the flexibility may be

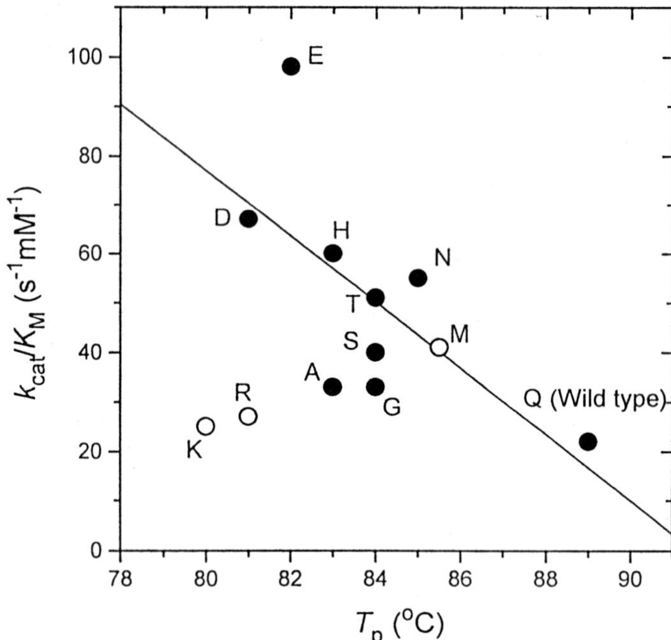

Fig. 3.   Correlation between activity and thermal stability of the 119th mutants. The maximum values of kinetic parameters k_{cat}/K_M are plotted against the peak temperature of apparent heat capacity change accompanied with thermal transition, T_p. The 119th amino acids are indicated in the figure by a one-letter representation. The three mutants, Q119K, Q119R and Q119M are plotted with open circles, which show the large pI shifts. The straight line is determined by least-squares fitting (correlation coefficient is -0.66) using all data points but K, R, and M.

negatively correlated with the thermal stability. These two correlations will result in a negative correlation, as seen in Figure 3. It may be necessary to observe the flexibility difference directly for the verification of the proposed mechanism using X-ray crystallography and NMR.

IV. Conclusion

We have succeeded in designing several highly activated mutants by paying attention to the electrostatic field around the active site. A negative correlation is found between activity and stability of the mutants. The electrostatic effect and the flexibility of the binding site may be the key property that connects these two aspects.

We have constructed our protein engineering strategy using a 'protein engineering cycle' composed of three fundamental elements: molecular design, synthesis, and evaluation. We have produced many mutants whose activity is greatly enhanced and accumulated information on the mechanism that determines the property and function of protein by travelling around our 'protein engineering cycle'.

Acknowledgement

The authors thank Drs. T. Miyake, N. Nagao, T. Yoneya, and A. Aoyama (Tosoh Co. Ltd.) for the synthesis of the mutants, and Dr. H. Nakamura (Protein Engineering Research Institute) for giving us his programs for electrostatic calculation. This work is partly supported by the Special Coordination Funds of the Science and Technology Agency of the Japanese Government.

References

1. K. Titani, M. A. Hermodson, L. H. Ericsson, K. A. Walsh, and H. Neurath: *Biochemistry* **11**, 2427 (1972).
2. K. Titani, M. A. Hermodson, L. H. Ericsson, K. A. Walsh, and H. Neurath: *Nature New Biol.* **238**, 35 (1972).
3. P. M. Colman, J. N. Jansonius, and B. W. Matthews: *J. Mol. Biol.* **70**, 701 (1972).
4. M. A. Holmes and B. W. Matthews: *J. Mol. Biol.* **160**, 623 (1982).
5. R. Marquardt, R. Hilgenfeld, and R. Keller: European Patent Application EP0418625A1.
6. Y. Miki: *J. Ferm. Bioengineer.* **77**, 457 (1994).
7. M. Kubo and T. Imanaka: *J. Gen. Microbiol.* **134**, 1883 (1988).
8. H. Nakamura and S. Nishida: *J. Phys. Soc. Jpn.* **56**, 1609 (1987).
9. S. Kidokoro, *et al.* (in preparation).
10. S. Kidokoro, H. Uedaira, and A. Wada: *Biopolymers* **27**, 271 (1988).

Molecular Dynamics Simulations of DNA and a Protein-DNA Complex Including Solvent

D. L. BEVERIDGE, K. J. McCONNELL, M. A. YOUNG, S. VIJAYAKUMAR and G. RAVISHANKER
Department of Chemistry and Program in Molecular Biophysics, Wesleyan University, Middletown, CT 06459, U.S.A.

Abstract. The results of a recent nanosecond (ns) molecular dynamics (MD) simulation of the d(CGCGAATTCGCG) double helix in water and a 100 ps MD study of the λ repressor-operator complex are described. The DNA simulations are analyzed in terms of the structural dynamics, fluctuations in the groove width and bending of the helical axis. The results indicate that the ns dynamical trajectory progresses through a series of three substates of B form DNA, with lifetimes of the order of hundreds of picoseconds (ps). An incipient dynamical equilibrium is evident. A comparison of the calculated axis bending with that observed in corresponding crystal structure data is presented. Simulation of the DNA in complex with the protein and that of the free DNA in solution, starting from the crystal conformation, reveal the dynamical changes that occur on complex formation.

Key words. MD, DNA, protein-DNA, simulation, solvent, lambda repressor, operator, GROMOS, reduced charge model.

Introduction

Molecular dynamics (MD) methodology has become an indispensable tool for exploring the motions of biological macromolecules in atomic level detail [1,2]. An historical perspective on the development and progress of the MD methodology applied to DNA and a detailed review of the hydration, structure and motion of DNA oligomers has recently been provided [3,4]. We have recently reported progress on the development of an accurate theoretical model for the simulation of DNA and Protein-DNA complexes [5]. We describe in this article further analysis of a 1 ns simulation of the oligonucleotide duplex of sequence d(CGCGAATTCGCG), which contains the target sequence for the restriction enzyme EcoRI endonuclease, and a simulation of the λ repressor-operator protein-DNA complex in water. The results are compared with experimental data from X-ray crystallography to validate the theoretical model.

MD on the d(CGCGAATTCGCG) Duplex

We recently reported a nanosecond simulation on the d(CGCGAATTCGCG) duplex in water [6] starting from the conformation of the DNA in complex with EcoRI endonuclease [7]. The simulation is based on the GROMOS86 force field [8] and the SPC model of water [9]. The GROMOS force field is well tested and has been widely employed for simulation of proteins and DNA [10–13], including several studies from this laboratory [14–16]. Our most recent MD on DNA employs a hydrogen bond potential in the form proposed by Tung *et al.* [17]. The

A. Pullman et al. (eds.), *Modelling of Biomolecular Structures and Mechanisms*, 409–423.
© 1995 *Kluwer Academic Publishers. Printed in the Netherlands.*

resulting base pair interaction energies compared closely with corresponding experimental data and *ab initio* quantum mechanical calculations [18].

The simulations described herein are based on the GROMOS force field with the above modification. The current studies utilize a longer range switching function than previously employed, and feathers the truncation of potentials over the length scale from 7.5 to 11.5 Å, thereby eliminating the tendency of charged groups to cluster at the cutoff boundary [19]. Further, the effect of counterions are treated implicitly in this set of studies, using a reduced charge of -0.24 eu on the phosphate groups, consistent with Manning's theory of counterion condensation [20]. Solvation is carried out by placing the DNA in a hexagonal prism cell and filling the void space with water to obtain an overall density of 1 gm/cc. The shape of the box chosen requires the least number of waters for solvation of a given thickness. The box is then divided into uniform grids and a water molecule is placed in each grid that has no solute atoms. The initial placement is equilibrated by a Metropolis Monte Carlo (MC) simulation, wherein the solute is held rigid and waters are allowed to relax around the solute. The MC simulations are carried out until the total and solute-solvent interaction energies converge. The simulation on the dodecamer duplex DNA is carried out using free MD on the DNA surrounded by ~3500 water molecules. The hexagonal prism elementary cell of constant volume is treated using periodic boundary conditions to model dilute aqueous solution. Velocities are rescaled when necessary (owing to conformational transitions) to produce an average kinetic energy corresponding to 300 K. The simulation was performed using the program WESDYN [21] and analyzed by means of various utilities available in MD Toolchest [22].

The simulation was monitored by one-dimensional (1D) and two-dimensional (2D) root mean square deviation (RMS) maps to determine the extent of the structural deviations from the canonical form as well as transition between conformational substates. The latter is especially informative (Figure 1), since the extent of similarity among all structures in the trajectory can be depicted graphically. Essentially, the 2-D RMS provides a measure of the deviation of every structure in the trajectory with respect to all others. Thus, a square matrix of RMS values are generated wherein the upper and bottom halves are symmetrical and, therefore, only one half of the matrix is presented. The diagonal elements represent deviation of a structure with respect to itself and consequently have an RMS value of zero. The off-diagonal elements provide a measure of the similarity of any two structures from different time points in the trajectory and their grey scaling reflects the extent to which they are similar. The results indicate that after an initial equilibration period, the MD structure resides for ~300 ps in a form ~4.5 Å RMS from canonical B form DNA. The dynamical structure then makes a distinct transition to a new form, still in the B family but somewhat more distant (~7.5 Å RMS) from the canonical form, where it remains for 180 ps. A rapid (~1.5 ps) reversible base pair opening event occurs in this structure at the T7 step, concomitant with displacements in helicoidal roll and twist. Then the dynamical structure transits to a third form, where it resides at the termination of the run. The third form, as evidenced by a cross peak in the 2-D RMS map, bears a strong resemblance to the first, indicating that the MD results appear to describe an incipient dynamical equilibrium among putative dynamical substates of the B family. The

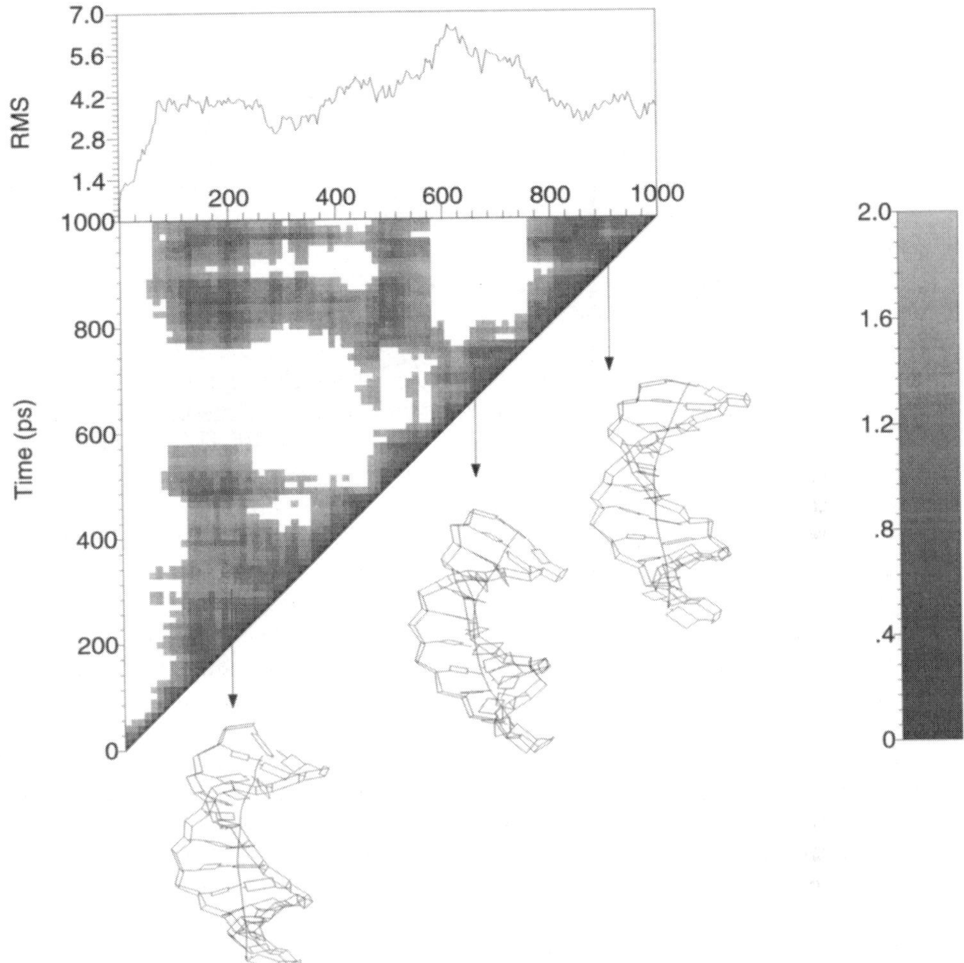

Fig. 1.   1D RMS map (top) and 2D RMS map (center) for bases in the 1 ns MD simulation of the d(CGCGAATTCGCG) duplex. Shaded areas in the 2D RMS map indicate regions of RMS deviation <2 Å. The diagonal elements represent the RMS deviation of a given structure with respect to itself (at each time point) and the off-diagonal elements represent similarity of structures from different time points in the trajectory, the gray scale reflecting the extent of the RMS deviation between them. Average MD structures for each of three putative substates are shown at bottom.

lifetime for a given structural form of DNA suggests that a minimal trajectory length of at least 200–300 ps would be necessary in order to fully relax the overall structure of the DNA in solution, given our methods and protocol. The Drew crystal structure [23] of the uncomplexed DNA containing the same sequence shows axis bending at or near the interfaces of the CG and AT tracts, and the *Eco* RI form shows, in addition to deformations at nearly the same positions as in the uncomplexed form, an extreme kink in the middle region of the structure at the A6 and T7 steps. The extreme kink in the *Eco* RI dodecamer was found

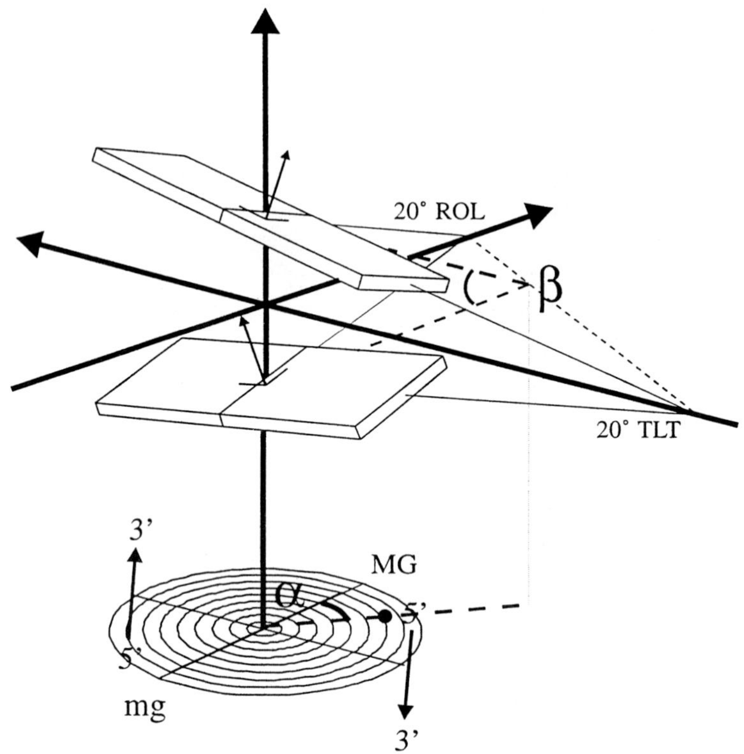

Fig. 2.   Schematic definition of the axis bending direction α and magnitude β in terms of the helicoidal parameters roll (ρ) and tilt (τ), referencing the major groove (MG), minor groove (mg) and sugar-phosphate backbone of the DNA double helix.

to relax within the first 20 ps of MD, indicating the protein-bound form to be a strained conformation rather than a metastable intermediate. This results support those of a current study by Rosenberg and coworkers [24] based on the AMBER force field.

Validation of the MD results is pursued by a comparison of calculated and observed morphological characteristics of the DNA: axis bending and groove widths. To analyze axis bending in a given structure, the magnitude β and angular direction of bending α are computed from deviations in the helicoidal parameters roll and tilt. The values of β and α for a given step can be projected onto a polar plot or 'bending dial', seen in perspective in Figure 2. Detailed analysis of multiple sequences can be carried out by superposing results from individual structures on a single bending dial for each base pair step in a DNA sequence. The use of bending dials to analyze DNA crystal structures and MD simulations has been described previously by Young et $al.$ [25].

The bending dials analysis of twenty six sequentially identical, structural homol-ogues of the Drew sequence in the Nucleic Acids Data Bank [26] are compared with the MD results in Figure 3a and 3b, respectively. The crystal structure

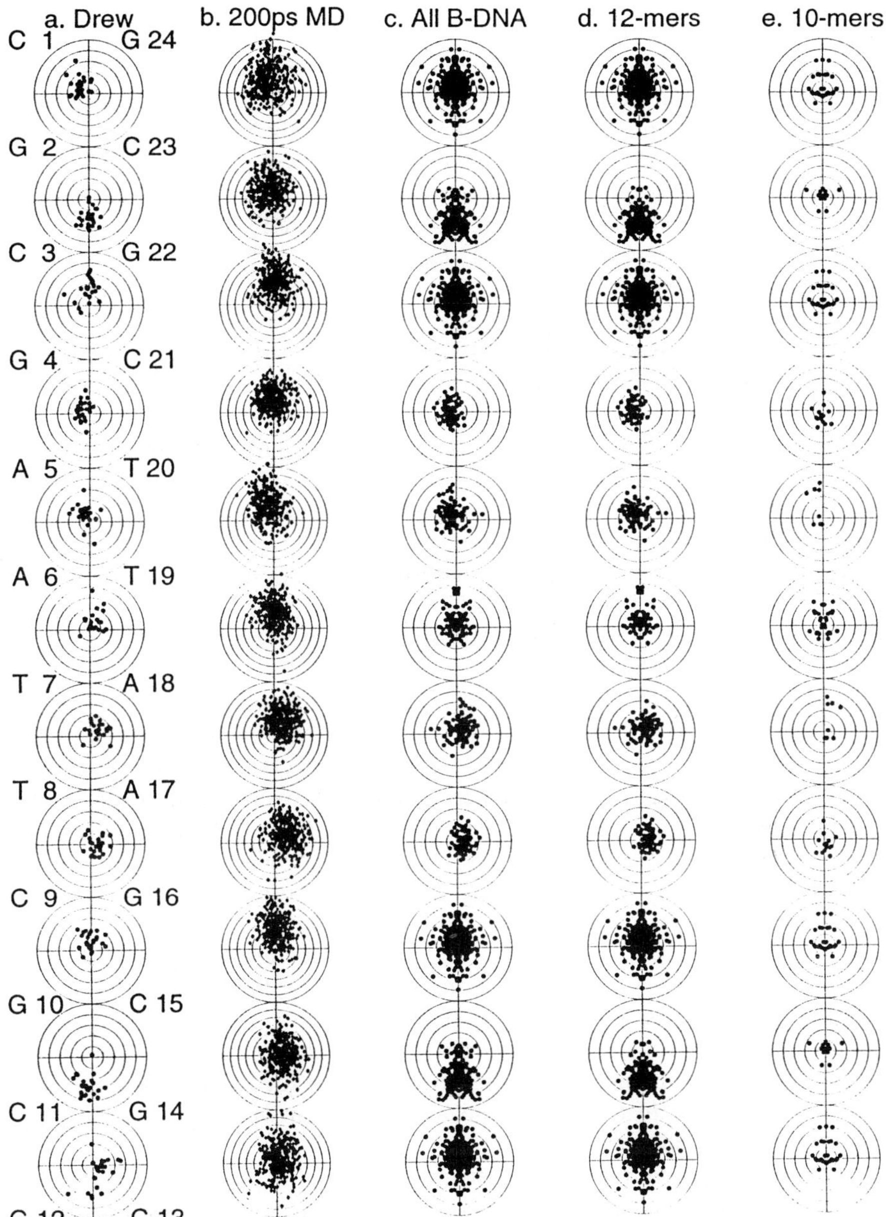

Fig. 3. Bending dials analysis [25] of (a) the 26 crystal structures of the Drew sequence d(CGCGAATTCGCG) duplex, resident in NDB [26] and (b) the last 200 ps of a 1 ns MD simulation [6]. Sequence dependent bending predictions drawn from each constituent step from (c) all B-DNAs, (d) dodecamers and (e) decamers, all resident in NDB. Each ring represents a 5° increment in axis deflection. To make trends in the analysis more discernible, crystal structure dials extend to 25°, while MD dials extend to 35°.

analyses show that bending occurs towards the major groove at the C3-G4 step, accompanied by a bending towards the minor groove in the adjoining G2-C3 step. The origin of the bending in the helicoidal parameters is inter-base pair roll. A similar effect is seen in the other flanking sequence at C9-G10 and G10-C11 steps. Analysis of axis bending from the MD trajectory includes the final 200 ps of the 1 ns simulation. The MD results indicate the major roll points to occur at C3-G4 steps and the C9-G10 steps. The deformation in each case is a bending towards the major groove. The adjoining step does not distinctly show the corresponding roll into the minor groove as observed crystallographically, a discrepancy between the calculated and observed results. The effect of crystal packing, which can be significant in DNA oligonucleotides [27–29], is not considered in the calculations.

A comparison of the results from MD with all B-DNA's in general and, specifically with dodecamers and decamer crystal structures is also informative. The results from analysis of the data in the NDB for the above three classes are summarized in Figures 3c, 3d and 3e respectively. Figure 3c contains results for the steps appearing in the d(CGCGAATTCGCG) duplex, based on instances of these steps found anywhere along the sequence in all B-DNAs. Figure 3d contains stepwise results from dodecamer sequences crystallized in the $P2_12_12_1$ space group. These columns thus approximate the generic behavior of a DNA of the above sequence based on the behavior of CG, GC, GA, AA, AT and TC steps. The results are weighted by the presence of multiple occurrences of crystal structures of any given sequence and, more specifically, by the presence of d(CGCGAATTCGCG) duplex structures. The results summarized in Figures 3c and 3d fully support the bending pattern observed among the Drew structures in Figure 3a. The decamer structures are known from crystallography to pack end to end, whereas the dodecamer structures pack with overlapping of the sequences flanking the recognition site accompanied by strong helix-helix interactions. Bending dials for the various steps appearing in decamer structures are shown in Figure 3e. Comparison of Figures 3a, 3c and 3d with 3e, the trend in CG step is supported but the trend in GC step is not, indicating that the GC behavior in Figures 3a, 3c and 3d may be due to packing effects. The MD results describing aqueous solution behavior agree better with the trends observed in the decamer crystal structures, which is presumably less influenced by packing.

Plots of the major and minor groove widths for the d(CGCGAATTCGCG) as a function of time, compared with the groove width for the crystal and canonical structures, are shown in Figure 4. The minor groove width (a) is defined as the distance between a phosphorus atom on one strand and a phosphorus atom four base steps up on the complimentary strand. The major groove width (b) is taken as a measure of the distance between a phosphorus on one strand and a phosphorus on the complimentary strand, three base steps down. The large initial changes (first 200 ps) in the minor groove widths in the center of the DNA (P7- to P10-) and changes in the major groove width at P3- and P7- probably occur from a relaxation of the deformation induced by the protein in the complex. During this relaxation process, interestingly, the minor groove width is observed to move towards a more canonical value, while the major groove widths approach a value of 22–23 Å. This width appears to be the starting value for the other major groove widths. Between 600 and 800 ps in the trajectory, a large distortion in the minor groove widths occur in the central region of the DNA. This distortion is correlated with the overall structural transition observed in the 2D-RMS map and is consistent with the appearance of a new substate in this time block. Around 800 ps, the minor groove widths return to a more canonical value as the DNA appears to revisit the initial conformation around 200 ps, evident from the cross peaks between the structures from these two time blocks. The groove width, more specifically the minor groove, thus appears to be very sensitive to conformational adjustments in the DNA. Changes in minor groove width could affect the spine of hydration [30–32], thereby altering the overall thermodynamics.

MD models of DNA are increasing in their accuracy and utility in investigating the fundamental nature of DNA structure and motions. Detailed analysis of the progress of the simulations reveals that the trajectories progress through a series of substates with lifetimes of several hundred ps, and provides evidence for the presence of an incipient dynamic equilibrium among substates on the ns time scale. The stability of the MD calculations over such a long duration suggest our model of DNA is reasonable and is encouraging to note the feasibility of studying dynamical processes occurring in the ns regime. It is likely that further detail will emerge as MD simulations are extended to successive decades of time and new motional regimes are encountered.

MD Simulation Studies of the λ Repressor-Operator Complex

The preceding series of MD simulations on DNA as well as several previously published MD simulations on proteins [10,15] demonstrate that MD models based on the GROMOS force field are reasonably accurate when applied to these systems independently. Protein-DNA complexes present an additional level of complexity to MD simulation methods, since the intermolecular interactions are likely to be governed by a highly subtle balance of forces. We seek to evaluate the accuracy with which calculations on interacting macromolecules can be carried out, using the λ repressor-operator complex as a prototype case.

A number of characteristic structural elements have now been recognized in proteins interacting with DNA, including the helix-turn-helix (HTH) motif, zinc fingers and leucine zippers [33]. The *modus operandi* of binding motifs involving recognition and regulation of genetic biochemical processes have been found to vary from one complex to another [34]. Further, regions of protein structure that are contiguous to the above motifs, some dynamically labile, have been found to contribute significantly to the binding. There is no clear indication of the structural changes, particularly in the DNA, that occur on complex formation. Collective efforts in the fields of molecular biology, biochemistry and molecular biophysics are now directed towards determining the relationship between the structural details of protein–DNA complexes and the origin of the remarkable specificity of their interactions.

Understanding the specificity of a regulatory protein to its various cognate operator sequences involves interpretive knowledge of the relative contributions of all significant factors to the free energy of binding relative to each other and to random sequence DNA. Genetic and chemical protection experiments at the molecular level and crystallography at or near the atomic level provide a knowl-

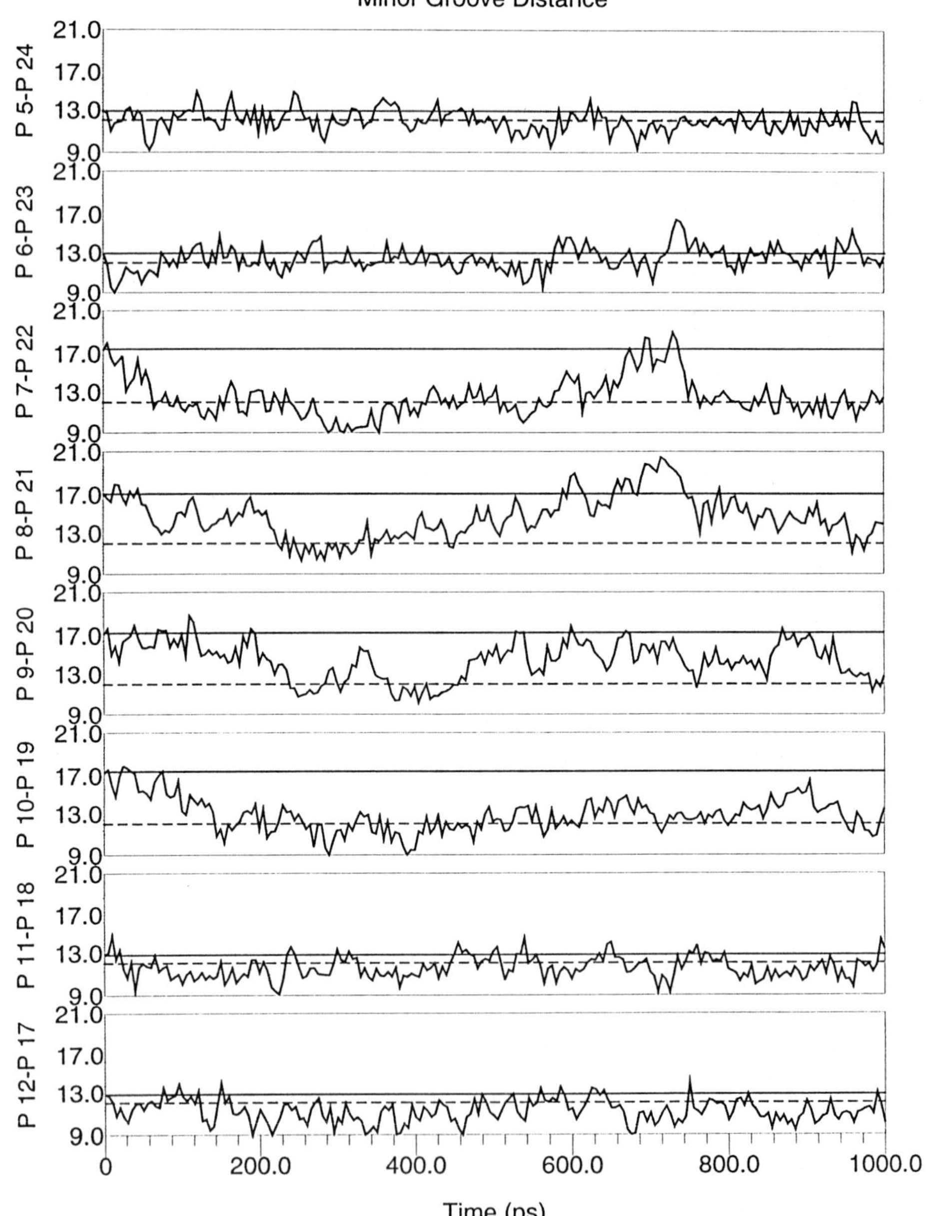

Fig. 4.   Time evolution of the groove widths from the 1 ns simulation for the (a) minor and (b) major
grooves, respectively. See text for additional details.

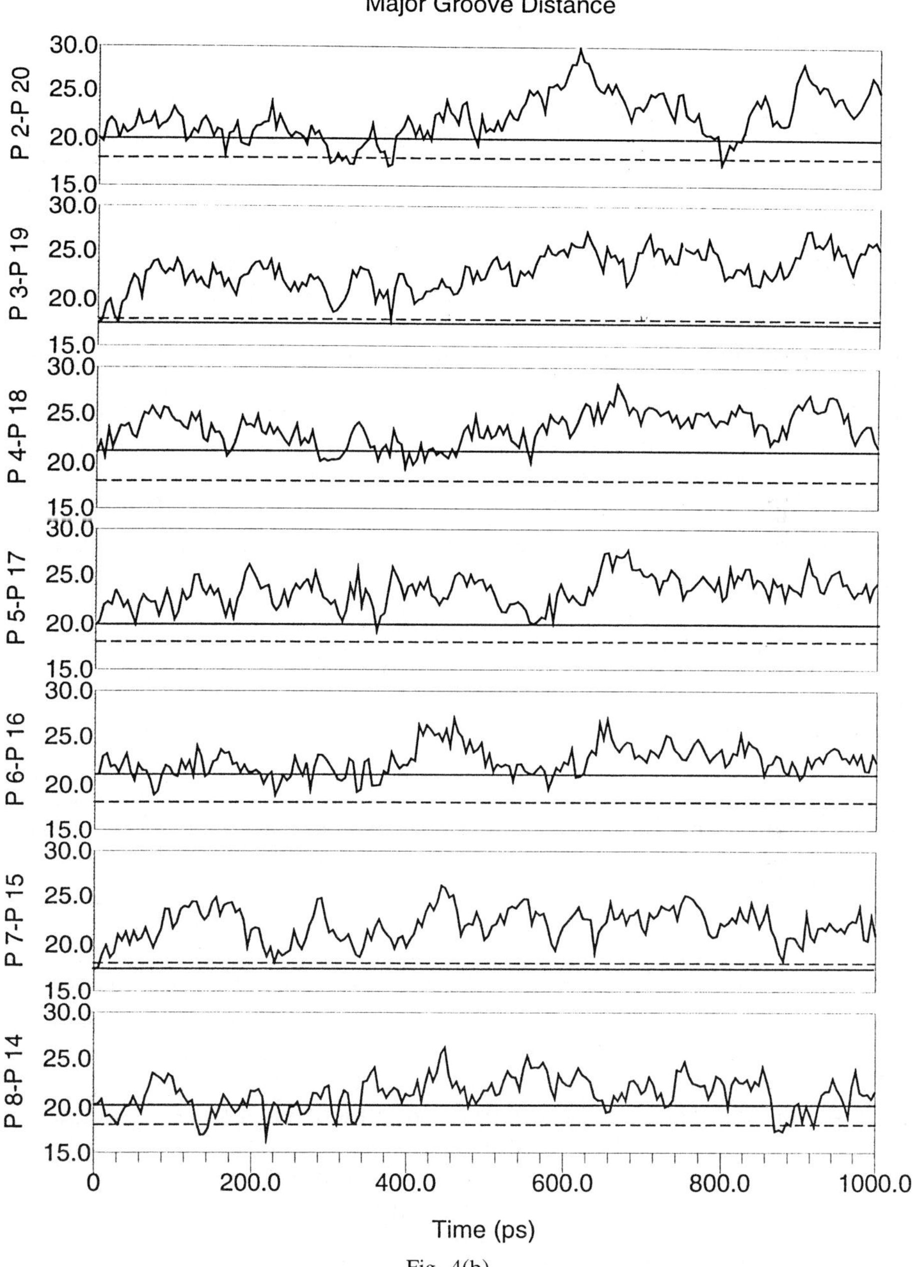

Fig. 4(b).

edge of the direct interactions involved, but not the energetics (beyond the idea that if a structure occurs, it must have a favorable free energy). Direct measurement of free energy via equilibrium binding studies provides information on the energetics, but unequivocally linking the energetics to structure is difficult, especially when subtle non-local changes are involved. Theoretical methods can, in principle, provide the link between structure and energetics. In addition, the dynamical aspects of protein–DNA complexes, not accessible to experiment, can be explored via MD studies, forming a basis for interpretation of experimental results. However, MD simulations are still limited by assumptions inherent in the underlying molecular force field, the sensitivity of results to simulation protocols and truncation of potentials, and the limited time frame that can be reasonably simulated on systems of this size. Detailed study of a prototype system is necessary to determine the capabilities and limitations of MD applied to protein–DNA interactions.

We describe herein results from separate MD studies of the repressor protein from the bacteriophage λ, its cognate duplex DNA OL1, of sequence d(TAT-CACCGCCAGTGGTA), and the repressor-operator complex. This system is of historic interest in the elucidation of the 'genetic switch' from lysogenic to lytic phases of the phage [35] and has been studied extensively from diverse points of view. The availability of crystal structures of the complex [36,37] and that of the unbound protein [38] makes this system ideally suited for inquiring into the overall accuracy and utility of MD simulation applied to a regulatory protein–DNA complex. All three simulations were carried out with water included explicitly. All protocols were similar to that of the preceding simulation on the dodecamer duplex DNA. Significant differences include the use of scaled charges on the backbone phosphates; sequestered phosphates are set to -1.0, and partially exposed phosphates have their charges scaled proportional to their solvent accessibility. The results are analyzed with respect to the intrinsic stability of the model system, groove width and axis bending of the DNA, and adaptive changes in structure, on complex formation.

The initial structure for the protein, DNA, and the complex are taken from crystal structure data [36,37]. An α carbon overlay of the protein in the uncomplexed and complexed forms is shown in Figure 5, which indicate the dynamic range of motion spanned during MD. Although the overall structure of the protein in the two simulations are similar, the extent of deformation varies considerably, being more in the case of the uncomplexed protein. The MD structures of the uncomplexed protein also indicate that the C-terminal helices, located at the protein dimer interface in the complex, are substantially destabilized. This result is consistent with the experimental observation that the free repressor does not form dimers in solution [38]. Despite the unwinding of the C-terminal helices, the calculated % helicity for the uncomplexed protein is in good agreement with experimental measurements [39]. Also, both the N-terminal arms of the dimer show significant changes with respect to the crystal conformation and appear to flop back on the protein. In the simulation of the complex, the dimer interface is more stable than in the uncomplexed protein, and the structural rearrangement of the arms remain localized within the major groove area.

A comparison of structures from the trajectory of the uncomplexed DNA and that of the DNA in complex with the protein, is presented in Figure 6. The isolated DNA remains within the B family of structures, although there appears to be some slight fraying at the ends. In general, a slight expansion of the major groove and a compression of the minor groove at all base steps is evident, except for a three base pair region in the middle segment of the DNA. Notable features of the DNA in complex include a compression of the major groove, beginning at

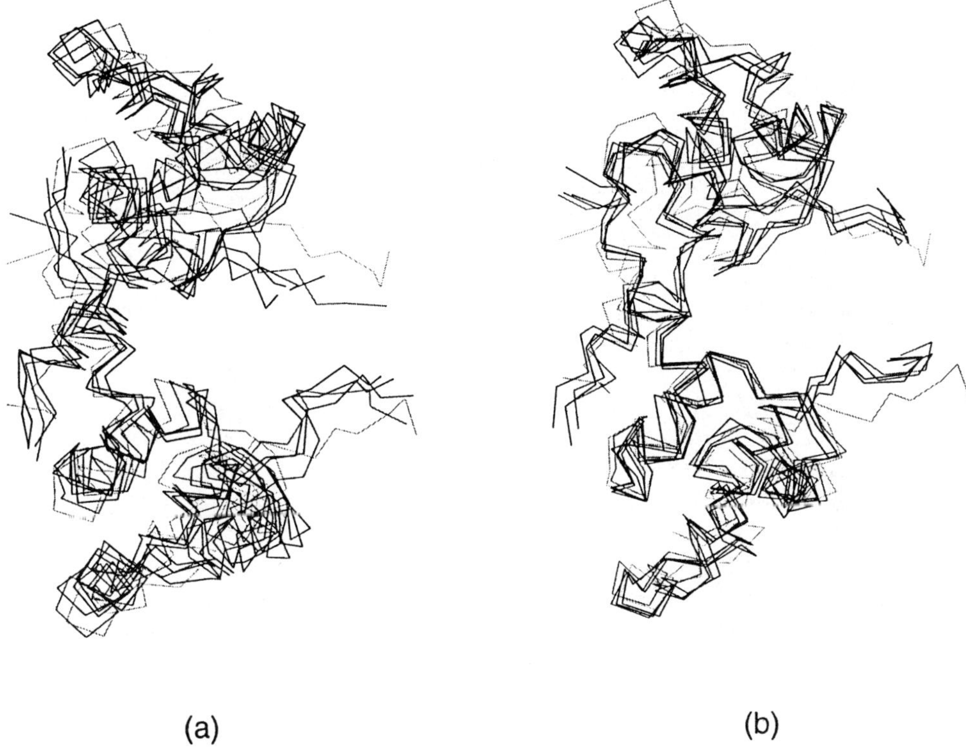

(a)   (b)

Fig. 5. Comparison of MD structures from the simulation of the free protein (a) and protein in complex (b). For each simulation, the starting structure and four snapshots from the trajectory, at equally spaced intervals, are overlaid to indicate the dynamical flexibility. The gray scaling varies from light to dark as the simulation progresses.

both ends of the DNA, and an expansion in the base steps between the two DNA binding pockets. In the minor groove, there is an expansion in the protein binding regions and at the ends. Currently we are extending each of these simulations as far as is necessary to characterize the dynamical stability, and then analyze the dynamical structures and the molecular motions in detail. The overall dynamics of the complex is shown in Figure 7. The calculations indicate the complex to be stable up to this point. Correlated motions between the recognition helices and DNA contact elements are being investigated, which will pave the way for a thorough understanding of the adaptive changes involved upon complex formation. Details of these results will be published later.

Summary and Conclusion

We have reviewed herein MD simulations on DNA oligonucleotides and a protein–DNA complex recently performed in this laboratory. MD simulation is a potentially powerful approach to the study of problems in this area, since full details of

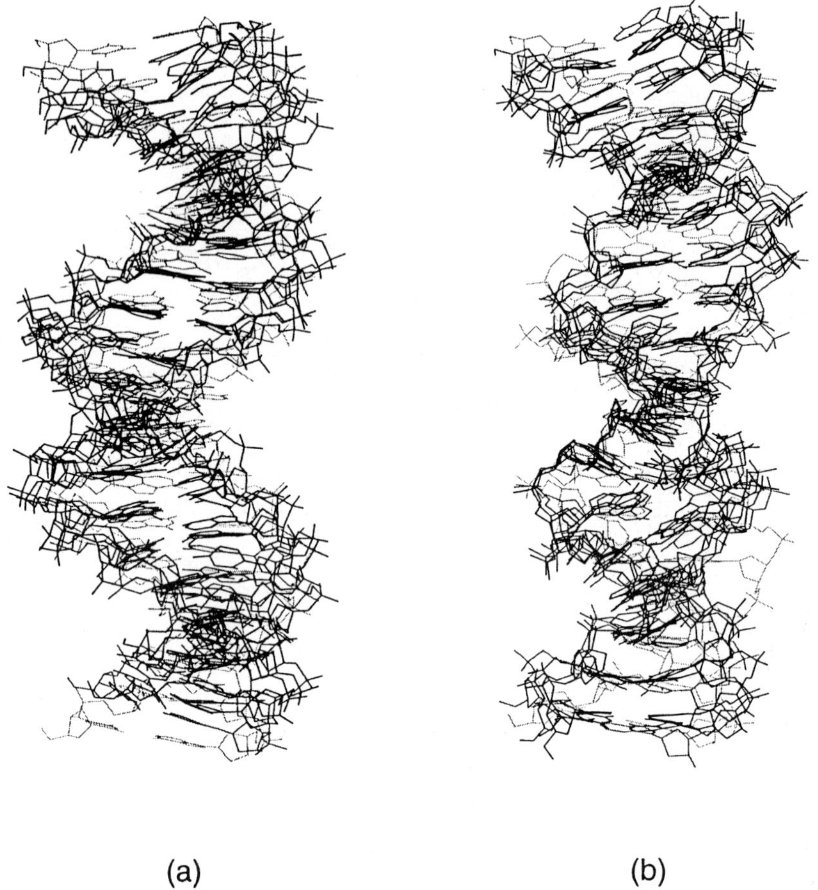

(a)   (b)

Fig. 6.   Comparison of MD structures from the simulation of the free DNA (a) and DNA in complex
(b). See legend of Figure 5 for additional details.

the molecular motions are obtained at a level inaccessible to any other method,
theoretical or experimental. However, due to the empirical nature of the underlying molecular force field, the methods require extensive independent validations
before the results obtained can be considered reliable and accurate. The results
to date are encouraging but suggest a significantly new dynamical behavior in the
ns regime. This may be a consequence of the particular flexibility of the DNA
double helix, but more likely due to relaxation processes occuring over a longer
time scale. Further investigations are clearly necessary to delineate the capabilities
and limitations of MD methodology applied to this class of problems.

Fig. 7. Overlay of the stick figure of the DNA and α-carbon drawing of the protein from the simulation of the complex. The crystal structure and four snapshots from the MD trajectory, taken at equally spaced intervals, are shown. The gray scaling varies from light to dark as the simulation progresses.

Acknowledgements

This project is supported by NIH Grants #GM 37909 from the National Institutes of General Medical Sciences, and RR 07885 from the Division of Research Resources. K. J. McConnell is supported by an NIH Molecular Biophysics Training Grant #GM 08271. Cray C90 and YMP computer facilities for this project were made available to us by the Pittsburgh Supercomputer Center and the NCI Frederick Cancer Research and Development Center of NIH.

References

1. W. F. van Gunsteren and H. J. C. Berendsen: *Angew. Chem. Int. Ed. Engl.* **29**, 992 (1990).
2. W. F. van Gunsteren and H. J. C. Berendsen: in *Molecular Dynamics and Proteins Structure*, J. Hermans, Ed., Polycrystal Book Service, Western Springs, IL, p. 85 (1985).
3. D. L. Beveridge, S. Swaminathan, G. Ravishanker, J. M. Withka, J. Srinivasan, C. Prevost, S. Louise-May, D. R. Langley, F. M. DiCapua, and P. H. Bolton: in *Water and Biological Molecules*, E. Westhof, Ed., The Macmillan Press, Ltd, London, p. 165 (1993).
4. D. L. Beveridge and G. Ravishanker: *Curr. Opin. Struct. Biol.* **4**, 246 (1994).
5. D. L. Beveridge, K. J. McConnell, R. Nirmala, M. A. Young, S. Vijayakumar, and G. Ravishanker: in *Structure and Reactivity in Aqueous Solution*, D. A. Smith, Ed., ACS Symposium Series, in press.
6. K. J. McConnell, R. Nirmala, M. A. Young, G. Ravishanker, and D. L. Beveridge: *J. Am. Chem. Soc.* **116**, 4461 (1994).
7. J. A. McClarin, C. A. Frederick, B.-C. Wang, P. Greene, H. W. Boyer, J. Grabel, and J. M. Rosenberg: *Science* **234**, 1526 (1986).
8. W. F. van Gunsteren and H. J. C. Berendsen: GROMOS86, University of Groningen, Groningen, The Netherlands.
9. H. J. C. Berendsen, J. P. M. Postma, W. F. van Gunsteren, and J. Hermans: in *Intermolecular Forces*, B. Pullman, Ed., D. Reidel, Dordrecht, p. 331 (1981).
10. W. F. van Gunsteren: *Curr. Opin. Struct. Biol.* **3**, 277 (1993).
11. J. Hermans, H. J. C. Berendsen, W. F. van Gunsteren, and J. P. M. Postma: *Biopolymers* **23**, 1513 (1984).
12. H. J. C. Berendsen, W. F. van Gunsteren, H. R. J. Swinderman and R. G. Geurtsen: *Ann. N.Y. Acad. Sci.* **482**, 269 (1986).
13. R. M. Brunne, W. F. van Gunsteren, R. Bruschweiler, and R. R. Ernst: *J. Am. Chem. Soc.* **115**, 4764 (1993).
14. S. Swaminathan, G. Ravishanker, and D. L. Beveridge: *J. Am. Chem. Soc.* **113**, 5027 (1991).
15. W. E. Harte Jr., S. Swaminathan, and D. L. Beveridge: *Proteins: Structure, Function and Energetics* **13**, 175 (1992).
16. S. Vijayakumar, S. Visveshwara, G. Ravishanker, and D. L. Beveridge: *Biophysical J.* **65**, 2304 (1993).
17. C. S. Tung, S. C. Harvey, and A. J. McCammon: *Biopolymers* **23**, 2173 (1984).
18. I. R. Gould and P. A. Kollman: *J. Am. Chem. Soc.* **116**, 2493 (1994).
19. P. Auffinger, R. Nirmala, G. Ravishanker, and D. L. Beveridge: in preparation.
20. G. S. Manning: *Quart. Rev. Biophys.* **11**, 179 (1978).
21. G. Ravishanker and S. Swaminathan: Wesdyn 2.0, Wesleyan University, Middletown, CT.
22. G. Ravishanker and D. L. Beveridge: MDTOOLCHEST, *J. Comp. Chem.*, in preparation.
23. H. R. Drew, R. M. Wing, T. Takano, C. Broka, S. Tanaka, K. Itakura, and R. E. Dickerson: *Proc. Natl. Acad. Sci. USA* **78**, 2179 (1981).
24. S. Kumar, P. A. Kollman, and J. M. Rosenberg: *J. Biomol. Struct. Dyn.* in press.
25. M. A. Young, N. R., J. Srinivasan, K. J. McConnell, G. Ravishanker, and D. L. Beveridge: in *Structural Biology: State of the Art: Proceedings of the Eighth Conversation*, E. R. H. Sarma and M. H. Sarma, Eds., Adenine Press, Albany, N.Y., Vol. 2, p. 197 (1994).
26. H. M. Berman, W. K. Olson, D. L. Beveridge, J. Westbrook, A. Gelbin, T. Demeny, S.-H. Hsieh, A. R. Srinivasan, and B. Schneider: *Biophys. J.* **63**, 751 (1992).
27. R. E. Dickerson, D. S. Goodsell, M. L. Kopka, and P. E. Pjura: *J. Biomol. Struct. Dyn.* **5**, 557 (1987).
28. Z. Shakked: *Curr. Opin. Struct. Biol.* **1**, 446 (1991).
29. A. R. Srinivasan: *Biophys. Chem.* **43**, 279 (1992).
30. H. R. Drew and R. E. Dickerson: *J. Mol. Biol.* **151**, 535 (1981).
31. P. S. Subramanian and D. L. Beveridge: *J. Biomol. Struct. Dyn.* **6**, 1093 (1989.
32. P. S. Subramanian, G. Ravishanker, and D. L. Beveridge: *Proc. Natl. Acad. Sci. USA* **85**, 1836 (1988).
33. C. Branden and J. Tooze: *Introduction to Protein Structure*, Garland Publishing, Inc., New York and London (1991).

34. T. A. Steitz: *Quart. Rev. Biophys.* **23**, 205 (1990).
35. M. Ptashne: *A Genetic Switch*, Cell Press/Blackwell Scientific Publications, Cambridge, MA. (1986).
36. S. R. Jordan and C. O. Pabo: *Science* **242**, 893 (1988).
37. N. D. Clarke, L. J. Beamer, H. R. Goldberg, C. Berkower, and C. O. Pabo: *Science* **254**, 267 (1991).
38. C. O. Pabo and M. Lewis: *Nature* **298**, 443 (1982).
39. G. J. Thomas Jr., B. Prescott, and J. M. Benevides: *Biochemistry* **25**, 6768 (1986).

Practical Tools for Molecular Modeling of Complex Carbohydrates and Their Interactions with Proteins

SERGE PÉREZ,[1] CHRISTOPHE MEYER,[1] and ANNE IMBERTY[2]
[1]*Ingeniérie Moléculaire, INRA, BP 527, 44026 Nantes Cédex 03, France*
[2]*Laboratoire de Synthèse Organique-CNRS, Faculté des Sciences, 2 Rue de la Houssinière, 44072 Nantes Cédex 03, France*

Abstract. Computer modeling has become a valuable component of studies of carbohydrate three-dimensional structures and their relationship to function and properties. In this paper we examine the methods required for conformational modeling of carbohydrates, and we present a series of tools that have been developed to this end. These tools can be integrated into three-dimensional real-time molecular modeling software. A data base of pre-optimized carbohydrate fragments has been established to be used further in the construction of much more complex molecules. In addition we describe some possible uses of a data base of three dimensional structures of the disaccharide fragments present in the glycan moiety of *N*-glycoprotein. A molecular mechanical force field appropriate for the conformational analysis of oligosaccharides has been derived by the addition of new parameters to the Tripos force field and is compatible with protein simulations. The new parametrization has been assessed in three stages of increasing complexity: computations of potential energy surfaces, conformational refinement of relevant oligosaccharides, modeling at the atomic level of a protein/carbohydrate complex.

Key words. Carbohydrate, lectin, modeling, interaction, force field, database.

Introduction

In many instances, the knowledge limited to the primary structure of complex carbohydrates and polysaccharides is not sufficient to understand and explain their function and specificity. For example, the three-dimensional structure of *N*-glycans determines their interactions with other molecules or macromolecules and is important for their function and biological activity [1–3]. Moreover, it has been shown that some of these glycans have a great flexibility which plays an important role in some recognition phenomena [4–7]. It is therefore necessary to determine the ensemble of possible conformations and not only the one having the lowest potential energy.

The major spectroscopic tools for determining three-dimensional structure are X-ray diffraction and nuclear magnetic resonance (NMR) spectroscopy. The amount of available crystallographic data on glycoproteins, glycopeptides and corresponding oligosaccharides, or protein-carbohydrates complexes is very limited. In principle, the solution conformations of oligosaccharides can be derived from NMR studies in a manner similar to that for nucleic acids and proteins. However, due to the flexibility of oligosaccharides, the above approach cannot be successfully applied to carbohydrate molecules, and NMR studies must be generally performed in conjunction with conformational analysis using an appropriate methodology [8–10]. Molecular modeling can also be used as the only tool to determine three-dimensional structure.

A. Pullman et al. (eds.), Modelling of Biomolecular Structures and Mechanisms, 425–454.
© 1995 *Kluwer Academic Publishers. Printed in the Netherlands.*

Advances in molecular modeling of complex carbohydrates and glycoconjugates have been more recent than parallel developments in the nucleic acid and protein fields. For example, the carbohydrate field has been lacking structural data bases even though the implementations of these are required especially with the ever increasing avalanche of new structural data. The Carbohydrate Structure Data Base has recently been implemented [11] which includes items which are pertinent to primary structure. Unfortunately, this data base will not initially contain information about the three-dimensional features of carbohydrates. As for energy calculations, problems in molecular modeling studies of carbohydrates are manifold: the flexibility problem, extensive hydrogen bonding and the occurrence of stereoelectronic effects, such as the anomeric and exo-anomeric effect [12]. Many efforts have been devoted to circumvent these problems, and molecular mechanical parametrizations for carbohydrates have been developed [13–15]. Alternative potential energy parametrizations are being developed that use the potential energy functional forms of the major protein molecular mechanics codes [16, 17], and that can be combined with protein and solvent molecule force fields. Most of these fields constitute the basis of molecular simulation programs which are in turn interfaced with three-dimensional, real-time molecular modeling software that can selectively display and manipulate molecules and their components. Such software accelerates the analysis and simulation of chemical and biological materials.

The present paper is an attempt to approach the three-dimensional elucidation of complex carbohydrates in isolation or in interaction with other molecular partners. First, the presentation will focus on several aspects of increasing complexity, in connection with the different level of carbohydrate flexibility: flexibility of the monosaccharide skeleton, flexibility of the rotational exocyclic groups, i.e. CH_2OH and OH, and flexibility at the glycosidic linkage (Figure 1). In a effort to integrate the latest developments in the conformational analysis of carbohydrates within computer-aided molecular design software, we will describe a force-field capable of predicting both stereochemical features of carbohydrates and protein-carbohydrate interactions.

Monosaccharide Standard Fragment Library

A prerequisite for building up and understanding the structures of the larger oligosaccharides is a detailed description of the structures of the monosaccharide components. For these building blocks, the use of experimental geometry determined by X-ray or neutron crystallography seems to be an obvious solution. Such an approach requires the availability of the pertinent model compounds, which should be persuaded to crystallize. Numerous crystallographic studies are needed to allow generalizations to be made. These structures may show considerable variation arising from such forces as hydrogen bonding and crystal packing and from configurational variation at the glycosidic linkage. Alternatively, an average geometry may be used as pioneered by Arnott and Scott [18], and later extended by Sheldrick and Akrigg [19].

Rather than collecting and comparing structures that correspond to the multiple minima, it may be more appropriate to depict how the energy changes during variations of some of the structural parameters. The conformational energies of

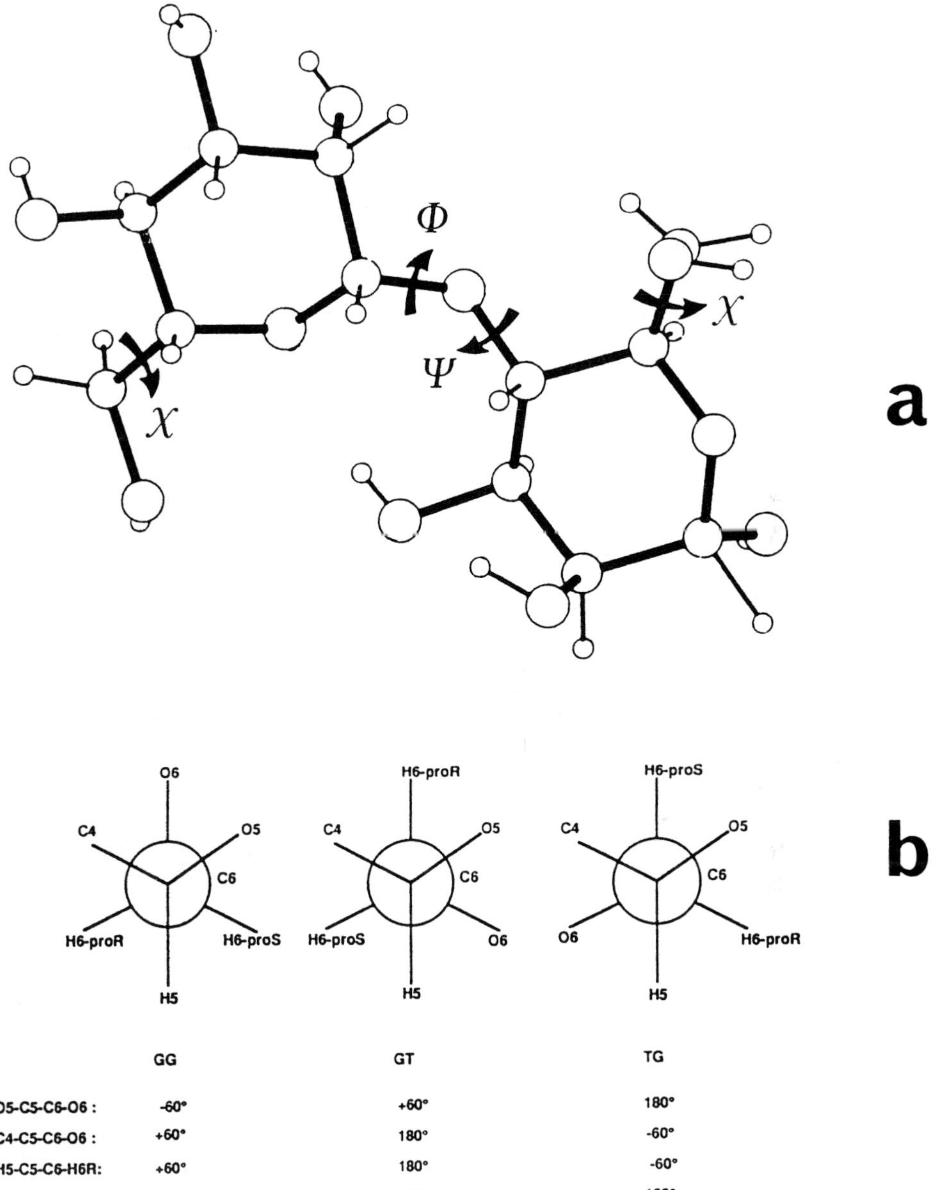

Fig. 1.   Schematic representation of a disaccharide unit along with some definitions of torsional angles. Φ (O-5_C-1_O-1_C′-X) and Ψ (C-1_O-1_C′-X_C′-X + 1) describe the glycosidic linkage, and χ the orientation of the hydroxymethyl group about C-5_C-6 bond. (b) Newman projections showing the low energy conformations of χ, along with the identifications of hydrogen atoms, for the three non-eclipsed conformations.

carbohydrates are calculable, in principle, from either classical quantum-mechanical procedures or by so-called empirical or molecular mechanics calculations. The empirical energy functions used in molecular mechanics calculations consist of terms representing various contributions to the total energy of the molecule [20]. Molecular mechanics is empirical in that it is developed by fitting equations and parameters to experimental data. Provided that these equations and parameters are applied to unknowns that are similar to the structures used to parametrize the program, the approach thus interpolates known data, offering a measure of confidence in the results.

One of the most elaborate molecular mechanics programs, used in many fields of organic chemistry, is Allinger's MM2 program [21] now extended to the MM3 version [22]. The total energy takes into account the stretching, bending, stretch-bending, torsional, dipolar contributions and van der Waals interactions. Anomeric effect is also parametrized. The MM3 program has been shown to reproduce correctly monosaccharide geometries. It also adequately reproduces the dependence of ring geometry on the conformation of the glycosidic linkage as observed in the solid state [23].

Optimized geometries and conformations for common monosaccharides have been derived, and the data are organized in a standardized, database format [24]. Pyranoid sugars are the commonest monosaccharides units in polysaccharides, glycoproteins and glycolipids The 34 pyranoid monosaccharides listed in Table I, constitute more than 95% of the monosaccharides occurring in complex carbohydrates, including N- and O-linked glycans. Monosaccharides having the idose configuration have been omitted because they display conformational variability [25]. As for the sulphated monosaccharides they will be included whenever enough experimental data become available. The optimizations were all performed on the α- and β-glycosides. All the monosaccharides investigated have lateral groups, i.e. CH_2OH and OH which introduce conformational flexibility. These variations cannot be addressed in the present data base. The minimization was therefore performed from one stable orientation as the starting point. We ascertained that the initial substituent orientations had no significant effect on the geometry after full minimization.

The data base is a collection of individual files that contain the coordinate entries, identified by the IUPAC name of the monosaccharide. Graphic representations and geometrical information are also provided. Figure 2 exhibits these information using the example of β-D-galactopyranose. The coordinate entries constitute the central information store. Other files may be created that correspond to such established format as the Cambridge Data Base or Protein Data Base. These formats are readable by most of the molecular modeling packages. This allows the possibility of creating a carbohydrate fragment library to be used further in the construction of much more complex molecules.

Data Bank for Conformation of Disaccharides

The primary conformational determinants in oligosaccharides are the preferred conformations about the glycosidic torsional angles (Figure 1). The orientation of

TABLE I

Monosaccharides included in the database (Pérez and Delage, 1991)

3,6-Anhydro-α-D-galactopyranose
3,6-Dideoxy-α-D-*xylo*-hexopyranose
α-D-Galactopyranose
6-*O*-Methyl-α-D-galactopyranose
α-D-Galactopyranuronic acid
6-*O*-Methyl-α-D-galactopyranosiduronate
2-Anino-2-deoxy-α-D-galactopyranose
2-Acetamido-2-deoxy-α-D-galactopyranose
α-D-Glucopyranuronic acid
α-D-Glucopyranose
α-D-Mannopyranose
α-D-Mannopyranuronic acid
N-Acetyl-α-D-neuraminic acid
α-L-Xylopyranose
3,6-Anhydro-α-L-galactopyranose
α-L-Arabinopyranose
α-L-Fucopyranose
α-L-Galactopyranose
α-L-Gulopyranuronic acid
α-L-Rhamnopyranose
β-D-Arabinopyranose
2-Amino-2-deoxy-β-D-galactopyranose
β-D-Galactopyranose
6-*O*-Acetyl-β-D-galactopyranose
6-*O*-Methyl-β-D-galactopyranosiduronate
β-D-Glucopyranuronic acid
2-Amino-2-deoxy-β-D-glucopyranose
β-D-Glucopyranose
6-*O*-Acetyl-β-D-glucopyranose
2-Acetamido-2-deoxy-β-D-glucopyranose
β-D-Mannuronic acid
β-D-Mannopyranose
β-D-Xylopyranose
β-L-Arabinopyranose

two linked carbohydrate residues is given by torsional angles: Φ and Ψ. Throughout this paper, Φ and Ψ torsion angles are defined as

$$\Phi = \Theta(\text{O-5\_C-1\_O-1\_C'-X})$$
$$\Psi = \Theta(\text{C-1\_O-1\_C'-X\_C'-X} + 1).$$

In the case of (1-6) linkage the second and the third torsional angles are usually referred to as

$$\Psi = \Theta(\text{C-1\_O-1\_C'-6\_C'-5})$$
$$\omega = \Theta(\text{O-1\_C'-6\_C-5'\_O'-5}).$$

The low energy orientations of ω are well documented [26].

β -D-Galactopyranose   (β -D-Galp)

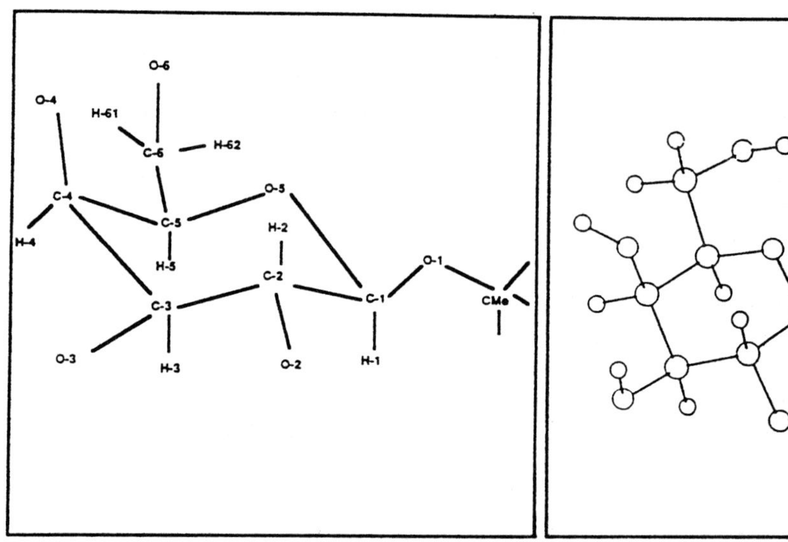


```
C-1        0.0000     0.0000     0.0000      C-1 - C-2 = 1.529      C-3 - C-4 - C-5 = 109.4
C-2       -0.4922    -0.7497     1.2382      C-2 - C-3 = 1.538      C-3 - C-4 - O-4 = 109.1
C-3       -2.0282    -0.7133     1.3044      C-3 - C-4 = 1.539      C-5 - C-4 - O-4 = 109.0
C-4       -2.5193     0.7449     1.2523      C-4 - C-5 = 1.540      C-4 - C-5 - C-6 = 112.9
C-5       -1.9314     1.4387     0.0099      C-5 - C-6 = 1.536      C-4 - C-5 - O-5 = 110.2
C-6       -2.2845     2.9319    -0.0643      C-1 - O-1 = 1.388      C-6 - C-5 - O-5 = 108.4
                                             C-1 - O-5 = 1.404      C-5 - C-6 - O-6 = 108.3
O-1        1.3877     0.0000     0.0000      C-2 - O-2 = 1.409      C-1 - O-5 - C-5 = 116.3
O-2       -0.0304    -2.0781     1.1450      C-3 - O-3 = 1.413
O-3       -2.5103    -1.3188     2.4862      C-4 - O-4 = 1.414
O-4       -2.0761     1.4241     2.4103      C-5 - O-5 = 1.432
O-5       -0.5054     1.3096    -0.0000      C-6 - O-6 = 1.411
O-6       -1.7409     3.4576    -1.2555
                                             C-H = 1.11            O-5 - C-1 - C-2 - C-3 =  55.5
H-1       -0.3365    -0.4797    -0.9337      O-H = 0.97            C-1 - C-2 - C-3 - C-4 = -55.0
H-2       -0.0355    -0.2865     2.1449                           C-2 - C-3 - C-4 - C-5 =  54.3
H-3       -2.4564    -1.2810     0.4421      O-1 - C-1 - C-2 = 108.8   C-3 - C-4 - C-5 - O-5 = -54.2
H-4       -3.6360     0.7825     1.2367      O-1 - C-1 - O-5 = 111.1   C-4 - C-5 - O-5 - C-1 =  58.8
H-5       -2.2751     0.9251    -0.9220      C-2 - C-1 - O-5 = 110.0   C-5 - O-5 - C-1 - C-2 = -59.1
H-61(R)   -1.8435     3.4982     0.7890      C-1 - C-2 - C-3 = 110.2
H-62(S)   -3.3896     3.0861    -0.0879      C-1 - C-2 - O-2 = 107.7   C-5 - O-5 - C-1 - O-1 =-179.6
                                             C-3 - C-2 - O-2 = 110.6   O-1 - C-1 - C-2 - O-2 = -61.9
HO-2       0.9401    -2.0603     1.1266      C-2 - C-3 - C-4 = 109.9   O-2 - C-2 - C-3 - O-3 =  65.1
HO-3      -2.0606    -0.9111     3.2424      C-2 - C-3 - O-3 = 111.5   O-3 - C-3 - C-4 - O-4 =  57.6
HO-4      -2.7806     2.0025     2.7403      C-4 - C-3 - O-3 = 109.0   O-4 - C-4 - C-5 - C-6 = -56.4
HO-6      -0.7756     3.3677    -1.1997                                O-5 - C-5 - C-6 - O-6 =  59.6
                                                                      C-4 - C-5 - C-6 - O-6 =-178.0
HO-1ᵃ      1.6585     0.3838     0.8487


CMeᵇ       1.9198     0.7097    -1.0824
HCMe1ᵇ     3.0287     0.6241    -0.9885
HCMe2ᵇ     1.6021     1.7746    -0.9921
HCMe3ᵇ     1.5666     0.2282    -2.0246
```

Fig. 2.   Example of one card image of the data-bank illustrated for β-D-Galactopyranose residue.

RIGID VERSUS RELAXED POTENTIAL ENERGY MAPS

When the ring geometry, and the glycosidic valence angles are fixed at their equilibrium values, the several conformations attainable for the disaccharide unit are those occurring from rotations about the glycosidic torsion angles. Therefore, when modeling oligosaccharides, computations of individual potential energy surfaces corresponding to each disaccharide unit is a prerequisite. They can adequately be computed in the (Φ, Ψ) or (Φ, Ψ, ω) space using different schemes in which the residues are allowed or not to adjust internally at each increment. If the internal parameters, i.e. bond lengths, valence angles, and all the torsional angles, are allowed to change in response to influences from the other residue, the study is called a flexible-residue analysis. If not, then the study is called a rigid-residue analysis.

For the rigid map calculations, the conformational energy is evaluated by including the partitioned contributions arising from van der Waals interactions [27], torsional potential including exo-anomeric effect [28], and inter residue hydrogen bonding [29] using the PFOS computer program [30]. This program uses a united representation for the hydroxyl groups, this takes care of the multiple minima problem associated with these lateral groups.

As with the conformational analysis of monosaccharide units, it has been found that the MM2CARB parametrization [15] gives a consistent agreement with experimental data for disaccharides. MM2CARB has then be used for the calculations of relaxed potential energy surfaces of various disaccharides such as maltose [31], cellobiose [32] and mannobiose linked $\alpha(1\text{-}3)$ [33]. For the later disaccharide, the NMR observables such NOEs and Tl calculated from the relaxed map agree with the experimental ones. At the present time, MM3 is commonly used for disaccharide conformational studies [34, 35]. In terms of computer resources, there is a difference of approximately three orders of magnitude between the time required to calculate a rigid-residue potential energy surface and a relaxed one for a disaccharide. There are several disadvantages to the rigid-residue method, such as the dependence of the map on the choice of starting model and the overestimating of the energy barriers between minima.

In several cases, we have examined 'rigid' versus 'relaxed' potential energy surfaces. Results for two disaccharides, Man $\alpha(1\text{-}3)$ Man and Man $\alpha(1\text{-}2)$ Man are given in Figures 3 and 4, respectively. The conformations observed in the solid state are shown on the "relaxed" map. For Man $\alpha(1\text{-}3)$ Man, the conformation with $(\Phi = 60°, \Psi = 97°)$ is the one observed in a crystalline trisaccharide [36], whereas the conformation with $(\Phi = 69°, \Psi = 124°)$ and $(\Phi = 90°, \Psi = -74°)$ correspond respectively to a trisaccharide and an octasaccharide complexed with a lectin [37, 38]. In the case of Man $\alpha(1\text{-}2)$ Man the conformation at $\Phi = 64°$ and $\Psi = 136°$ is observed in the crystalline disaccharide [39]. At a first glance, introducing flexibility in the sugar geometry seems to modify the potential energy surface. In fact it can be observed that the locations of the low energy regions remain identical in the 'rigid' and in the 'relaxed' maps studied here. The only difference is that the 'relaxed' potential energy surface map is enlarged compared to the 'rigid' one. The relaxation relieves an important number of small steric conflicts, allowing more conformational states to be reached. The more noticeable expansion

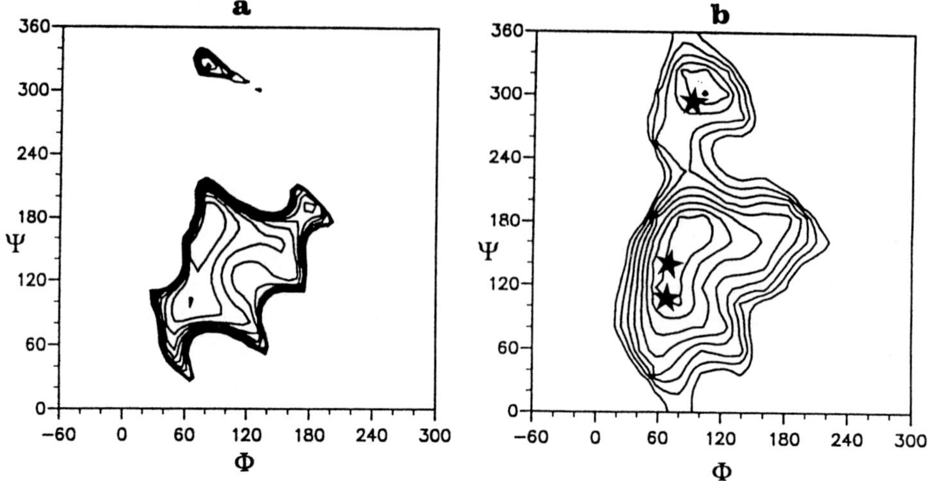

Fig. 3.   Potential energy surfaces of Man α(1-3) Man. (a) Rigid potential energy surface calculated with the PFOS program [15]. The hydrogen bond potential was not included in the calculation. Each residue has a *gauche–gauche* orientation of its hydroxymethyl group. The iso-energy contours are drawn with interpolation of 1 kcal/mol above the minimum. (b) 'Adiabatic relaxed' potential surface created from nine individual relaxed surfaces, each corresponding to one combination of orientation for the hydroxymethyl groups. The conformations displayed by this disaccharide fragment in the solid state are shown on the map.

occurs along the Ψ axis. In the case of Man α(1-3) Man, this expansion creates a pathway between the main low energy domain and the conformations which are located in small islands in the 'rigid' potential energy surface. Whenever the objective of a conformational analysis of disaccharide lies in the location of the several low energy conformers along with an estimate of the energy differences, the use of the 'rigid' residue approximation seems justified.

A DATA BANK OF THREE-DIMENSIONAL STRUCTURES OF DISACCHARIDES

An extensive survey of the primary structures which have been reported for *N*-linked glycans led to the identification of 22 different disaccharide fragments. They are listed in Table II, where each disaccharide is given an identification from A to V. Following already proposed classifications, these fragments were arranged in three families: the oligo-mannose type (identification A to E), the *N*-acetyllacto-saminic type (identification F to O) and the cap disaccharides and other complex types (identification P to V).

The body of the data base [40,41] contains by all the potential energy surfaces, along with the description of the three-dimensional features of all the low energy conformation which are located on these maps. All the potential energy surfaces were calculated using the PFOS program for the possible orientations of the hydroxymethyl groups. These orientations do not affect the overall shapes of the potential energy surface. The only differences seen in these other orientations are very small variations in the Φ, Ψ and energy values for few local minima. In the

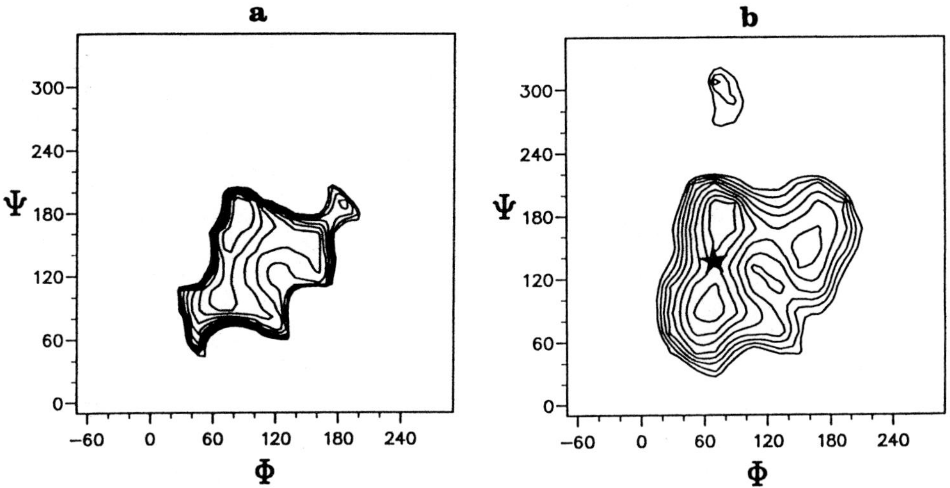

Fig. 4. Potential energy surfaces of Man α(1-2) Man. (a) Rigid potential energy surface calculated with the PFOS program [15]. The hydrogen bond potential was not included in the calculation. Each residue has a *gauche–gauche* orientation of its hydroxymethyl group. The iso-energy contours are drawn with interpolation of 1 kcal/mol above the minimum. (b) 'Adiabatic relaxed' potential surface created from four individual relaxed surfaces, each corresponding to one combination of orientation for the hydroxymethyl groups. The procedure used to produce these maps where all the internal coordinates of the disaccharide are minimized through molecular mechanics calculations follows that described in [33]. The conformation displayed by this disaccharide fragment in the solid state is shown on the map.

case of 1-6 configurations, the stable orientations of ω at the glycosidic linkage were considered, since in this case, the potential energy surfaces are different.

All the Φ, Ψ maps were calculated with and without the contributions of the hydrogen bonds; it was shown that the inclusion of this contribution does not alter the shape of the energy surfaces. Furthermore, no change in the locations of the low energy minima was detected; only the magnitude of the energy of some of them varied. It has been systematically checked that no new minima appear. Calculations of the potential energy surface without the hydrogen bond contribution seem to be appropriate to model solution behaviour of oligosaccharide in water. It may be advocated that an aqueous surrounding medium will favor hydrogen bond between the molecule and the solvent instead of the formation of intramolecular hydrogen bonds. This statement may not be always true, since, in particular case intramolecular hydrogen bonds may be formed as a further stabilization of an already existing low energy conformation. This is frequently the case for crystalline compounds. Therefore, in order to present the most exhaustive description, all the possible hydrogen bonds which can occur for the different low energy conformations have been searched.

A typical illustration of the content of the data base is given in the case of NeuAc α(2-3) Gal (Figure 5 and Table III) along with the conformation observed in the crystalline complex between wheat germ agglutinin and a trisaccharide [42]. As with the other examples, there is a fairly high number of stable conformers

TABLE II
Disaccharide fragments of *N*-linked glycans included in the
data base (Imberty *et al.*, 1990, 1991)

GlcNAc β(1-4) GlcNAc	A	
Man β(1-4) GlcNAc	B	
Man α(1-3) Man	C	Oligomannose type
Man α(1-2) Man	D	
Man a(1-6) Man	E	
GlcNAc β(1-2) Man	F	
GlcNAc β(1-4) Man	G	
GlcNAc β(1-6) Man	H	
Gal β(1-4) GlcNAc	I	
GlcNAc β(1-3) Gal	J	Complex type
Fuc α(1-6) GlcNAc	K	
Fuc α(1-3) GlcNAc	L	
Xyl β(1-2) Man	M	
GlcNAc β(1-6) Gal	N	
Gal β(1-3) GlcNAc	O	
NeuAc α(2-3) Gal	P	
NeuAc α(2-6) Gal	Q	
NeuAc α(2-6) GlcNAc	R	
NeuAc α(2-8) NeuAc	S	Cap disaccharides
Gal α(1-3) Gal	T	
Fuc α(1-2) Gal	U	
Fuc α(1-6) Gal	V	

for each disaccharide. This clearly indicates that depending upon conditions, conformational transitions are expected to occur.

POSSIBLE APPLICATIONS OF THE DATA BANK

The data bank of disaccharide conformations is helpful for various research areas in the field of three-dimensional structures of *N*-glycan. Obviously the data bank provides the basic set of information for either molecular dynamics calculations, or Monte Carlo simulations. It has been used to investigate some possible conformations of the oligo-mannose type glycan Man6-GlcNAc2 [40]. Not unexpectedly, it was found that combining the different low energy conformers would give 756 000 possible starting models, among which many would correspond to stable arrangements, other than those previously described as 'T', 'Y', 'bird' and 'broken wing' shapes [1]. Some of these possible conformations of the Man6-GlcNAc2 oligosaccharide are displayed in Figure 6.

As a tool for ongoing research work, protein crystallographers, when involved in the refinement of a glycoprotein structure may find it interesting to check whether the model they propose for the glycan conformation is compatible with the stereochemical properties of carbohydrates. A typical application is the analysis of the conformation of the biantennary *N*-acetyllactosamine octasaccharide in the crystal structure of its complex with the *Lathyrus ochrus* lectin (LOL) which has been solved at 2.3 A resolution [38]. Figure 7 shows the glycan in the crystalline

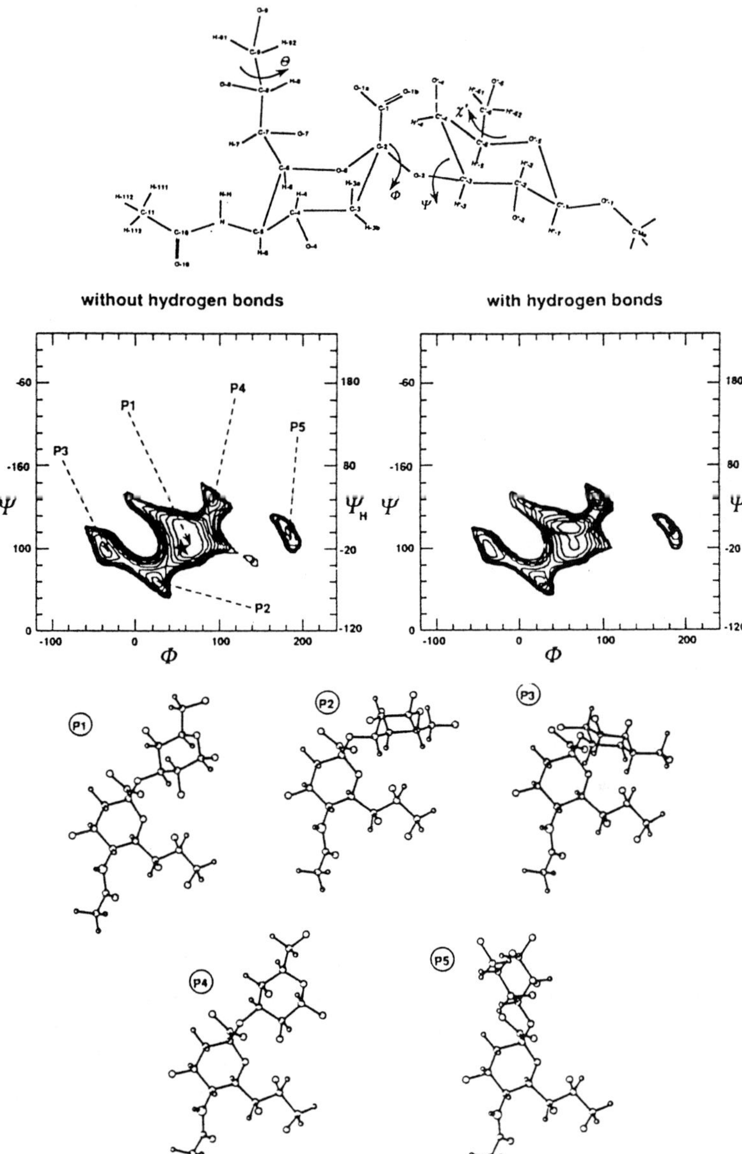

Fig. 5.   Iso-energy map of the NeuAc α(2-3) Gal disaccharide as a function of the Φ and Ψ torsion angles. Iso-energy contours are drawn by extrapolation of 1 kcal/mol with respect to the absolute minimum. All the energy minima P_1 to P_5 have been located on the surface. The naming of the axes at the top and bottom are different since they correspond to the crystallographic and NMR nomenclature, respectively, for torsional angles. The same conventions are used for the left and right axes. The conformation of the reported crystal conformation [42] is shown on the map by the sign '*'.

TABLE III
Illustration of the geometrical features of the low energy conformers of NeuAc α(2-3) Gal

Name (Φ, Ψ)	(Φ, Ψ_H)	(Θ, χ')	$(\Theta_{H,\chi'H})$	Relative energy	Possible hydrogen bond
P1					
(65, 105)	(65, −15)	(−60, 60)	(180, −60)	0.06	O-6···O-2'
(65, 105)	(65, −15)	(60, 60)	(−60, −60)	0.04	O-6···O-2'
(65, 105)	(65, −15)	(180, 60)	(60, −60)	0.04	O-6···O-2'
(65, 105)	(65, −15)	(−60, 180)	(180, 60)	0.02	O-6···O-2'
(65, 105)	(65, −15)	(60, 180)	(−60, 60)	0.00	O-6···O-2'
(65, 105)	(65, −15)	(180, 180)	(60, 60)	0.01	O-6···O-2'
P2					
(25, 60)	(25, −60)	(−60, 60)	(180, −60)	1.57	O-1···O-4'
(25, 60)	(25, −60)	(60, 60)	(−60, −60)	1.46	O-1···O-4'
(25, 60)	(25, −60)	(180, 60)	(60, 60)	1.54	O-1···O-4'
(25, 60)	(25, −60)	(−60, 180)	(180, 60)	1.24	O-1···O-4' O-8···O-6'
(25, 60)	(25, −60)	(60, 180)	(−60, 60)	1.10	O-1···O-4' O-9···O-6'
(25, 60)	(25, −60)	(180, 180)	(60, 60)	1.22	O-1···O-4' O-8···O-6'
P3					
(−35, 95)	(−35, −25)	(−60, 60)	(180, −60)	2.52	−
(−35, 95)	(−35, −25)	(60, 60)	(−60, −60)	2.49	−
(−35, 95)	(−35, −25)	(180, 60)	(60, −60)	2.48	−
(−35, 95)	(−35, −25)	(−60, 180)	(180, 60)	2.34	−
(−35, 95)	(−35, −25)	(60, 180)	(−60, 60)	2.30	−
(−35, 95)	(−35, −25)	(180, 180)	(60, 60)	2.28	−
P4					
(90, 165)	(90, 45)	(−60, 60)	(180, −60)	2.62	O-1···O-2' O-6···O-2' O-8···O-2'
(90, 165)	(90, 45)	(60, 60)	(−60, −60)	2.56	id
(90, 165)	(90, 45)	(180, 60)	(60, −60)	2.61	id
(90, 165)	(90, 45)	(−60, 180)	(180, 60)	2.61	id
(90, 165)	(90, 45)	(60, 180)	(−60, 60)	2.55	id
(90, 165)	(90, 45)	(180, 180)	(60, 60)	2.60	id
P5					
(−175, 115)	(−175, −5)	(−60, 60)	(180, −60)	4.77	O-1···O-2'
(180, 125)	(180, 5)	(60, 60)	(−60, −60)	4.78	O-1···O-2'
(180, 125)	(180, 5)	(180, 60)	(60, −60)	4.78	O-1···O-2'
(−175, 115)	(−175, −5)	(−60, 180)	(180, 60)	4.74	O-1···O-2'
(180, 125)	(180, 5)	(60, 180)	(−60, 60)	4.76	O-1···O-2'
(−175, 115)	(−175, −5)	(180, 180)	(60, 60)	4.74	O-1···O-2'

conformation. The conformations of the oligosaccharide linkages in the crystal are depicted in the corresponding maps in Figure 8. A similar type of study was performed on the Man α(1-3) Man β(1-4) GlcNac trisaccharide complexed with LOL [37]. The conformations observed in that case have been indicated in the corresponding maps, along with those taken by the trisaccharide in its crystal

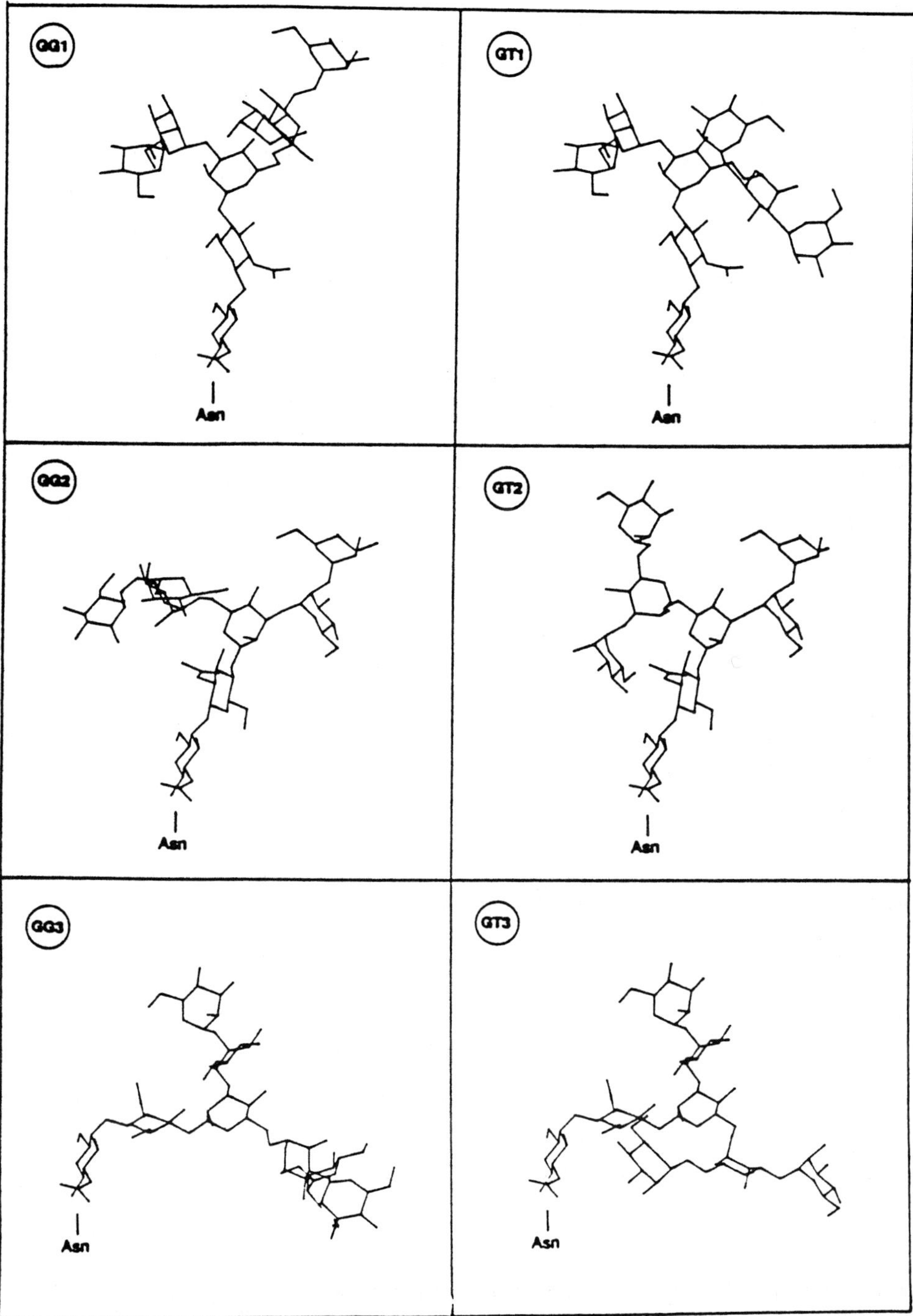

Fig. 6. Three-dimensional representations of several low energy conformations of the $Man_6GlcNac_2$ oligosaccharide built from the calculated low energy minima of the constituting disaccharide fragments.

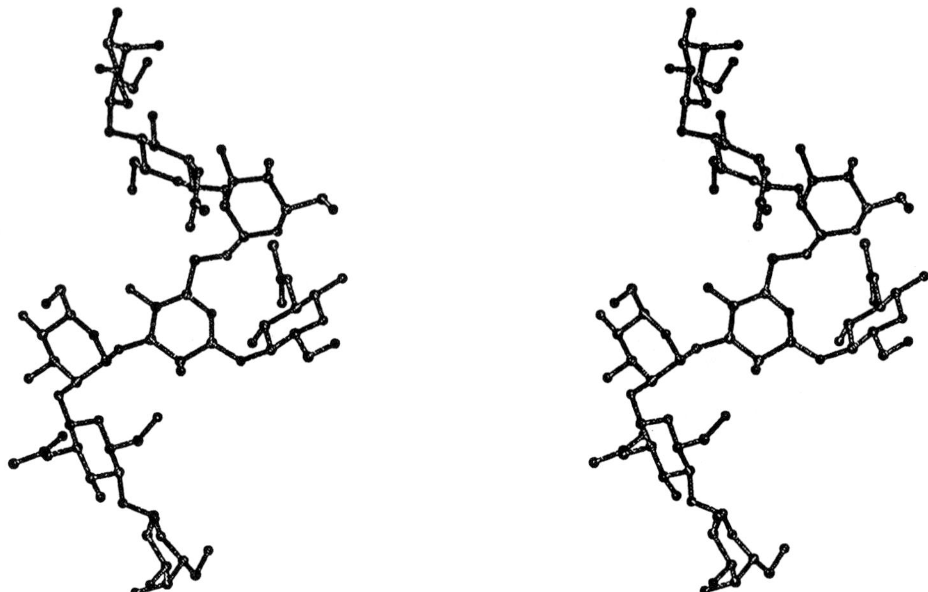

Fig. 7.   Stereoscopic representation of the biantennary octasaccharide as observed in the crystal of the LOL/sugar complex [38].

structure [36]. While the observed conformations are always at predicted minima, they are not always at the minimum predicted to have the lowest energy, specially for the octasaccharide. In particular, the Man α(1-3) Man and the Man β(1-4) GlcNac linkages are trapped in conformations far from the main low energy wells. Previously, it was suggested that these remote wells could be occupied [29]. In the (1-6) linkage, Φ is about 120° from the lowest energy conformation. The GlcNAc β(1-2) Man linkage of the 3 arm and the Gal β(1-4) GlcNAc linkage of the 6 arm would be stabilized even less by the exo-anomeric effect. These are unambiguous illustrations of the flexibility of these oligosaccharides.

Building and Refining Complex Oligosaccharides

Owing to the size and relative rigidity of the intervening monosaccharide units, the rotations at a particular glycosidic linkage could be virtually independent of the nearest neighbor torsion angles. This is the case of linear polysaccharides for which spatial representation are usually described by specifying values for all torsion angles (Φ_i, Ψ_i, ω_i) taken from consecutive dimeric fragments. However, in the case of more complex carbohydrates, the whole structure may not behave as if the glycosidic linkage were independent. Interactions involving non-neighboring residues can occur, either in the case of α(1-2) linkage, or in the case of branching points (Figure 9) [43]. It has to be emphasized that the occurrence of interactions between different entities, i.e. other branches of the glycan, surface of the glyco-protein, neighboring molecules including solvent molecules will only reduce the

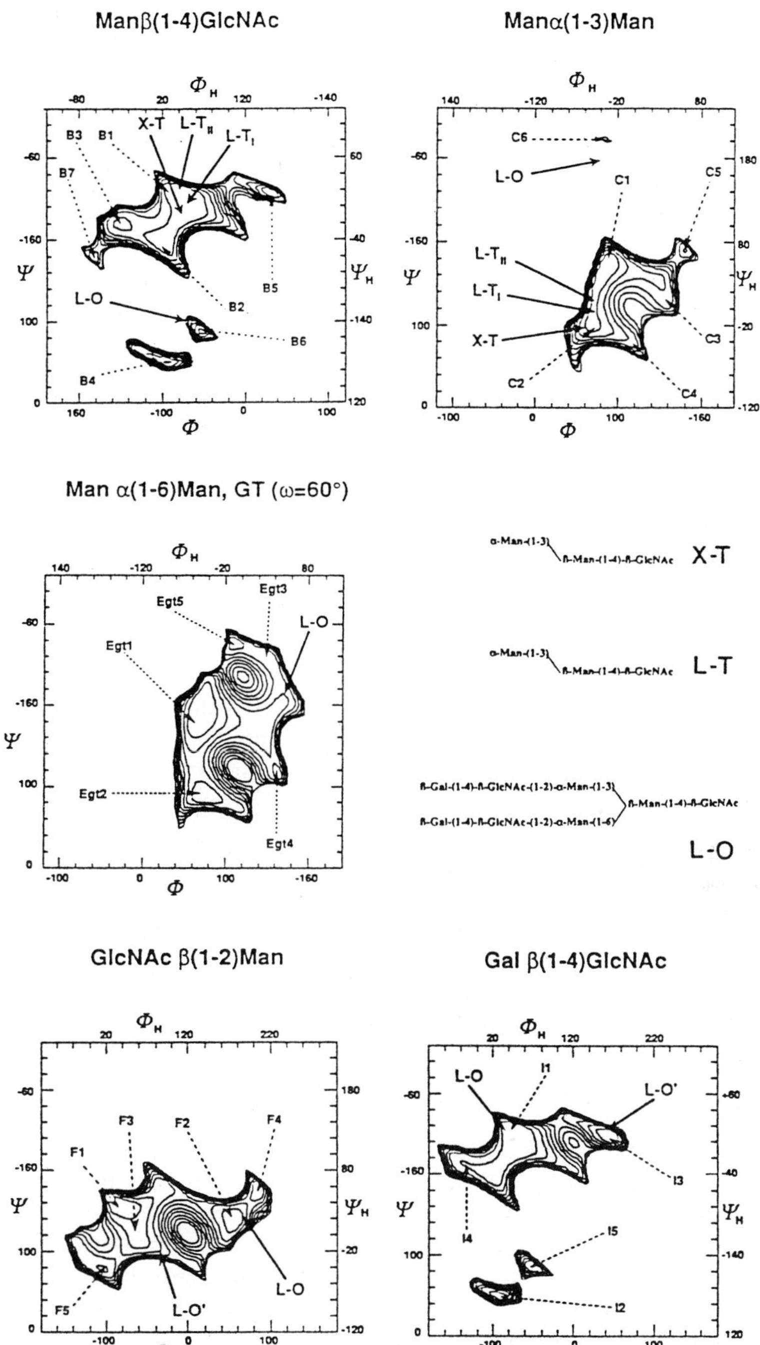

Fig. 8. Iso-energy maps of five disaccharides as a function of Φ and Ψ. The maps are extracted from the data base for disaccharides [40, 41] where all the minima are located on the surfaces and numbered by increasing energy. The conformations observed in three crystal structures of oligosaccharides have been reported on the corresponding maps.

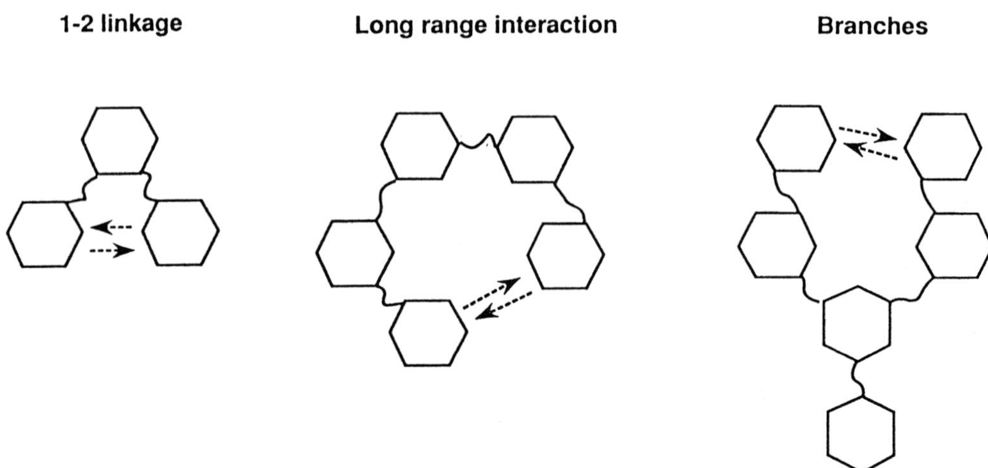

Fig. 9. Examples where the assumption of the independence of consecutive glycosidic torsion angles is not valid. (a) the $\alpha(1\text{-}2)$ case. (b) long range interactions in linear polysaccharides. (c) bi- or multi-antennary structures.

conformational space available for the oligosaccharide as predicted from the conformational analysis of the disaccharide fragments.

Building realistic possible starting conformations from an assembly of the low energy conformers is straightforward using the data base of disaccharides fragments. Then remains the question of minimizing the whole glycan structure. If the starting models do not exhibit steric conflicts between non-adjacent residues, the optimization will only adjust the linkages and ring conformations to the optimum energy, and the linkage will remain in the region of the starting low energy minima. In the case where steric conflicts occur, the optimization will give different results depending upon the severity of the van der Waals surface interpenetrations, and upon the position of the atoms involved. In some cases, the steric conflicts will be relieved only by small adjustments of some side groups, with no concomitant change of the linkage conformations. Alternatively, in the case of severe steric conflicts involving numerous atoms, the whole geometry and conformation of the oligosaccharide will be altered and new minima will be reached for some linkage conformations.

The choice of an appropriate force-field capable of dealing with large carbohydrate structures, typically several hundreds of atoms, eventually interacting with a protein surface or a protein recognition site, is a complicated matter. A few attempts to develop energy parameters, to account for both proteins and carbohydrates, have been made [44, 45]. However, these in-house developed computer programs, evidently lack the power which is provided by molecular graphics packages, where handling, manipulation and visualization of complicated molecular structures is easily performed. Recently, a parametrization for the study of proteins and carbohydrates has been proposed by Homans [17] based on the AMBER force field to be integrated in DISCOVER (Biosym Technologies Inc.). Among others, SYBYL [46] is a widely used molecular graphics package which uses the Tripos force-field; the addition of parameters adequate for handling carbohydrate molecules is highly desirable.

A FORCE-FIELD FOR CARBOHYDRATES TO BE USED IN THE SYSYL SOFTWARE

The Tripos 5.2 force-field, originally developed by White [47] is based on a series of classical molecular mechanics type energy function:

$$E_{tot} = E_{str} + E_{bend} + E_{oop} + E_{tors} + E_{vdw} + E_{ele}$$

including harmonic terms for bond stretching, angle bending, and out-of-plane bending with force constants k^{d_i} and k^{Θ_i}, a torsional term with force constant V^{ω_i}, periodicity n, Lennard–Jones-type nonbonded interactions, and a Coulombic term between partial charges q_i and q_j with a dielectric constant e.

$$E_{str} = \Sigma \tfrac{1}{2} k^{d_i}(d_i - d_i^0)^2 \qquad \text{bond stretching energy term}$$
$$E_{bend} = \Sigma \tfrac{1}{2} k^{\Theta_i}(\Theta_i - \Theta_i^0)^2 \qquad \text{angle bending energy term}$$
$$E_{oop} = \Sigma \tfrac{1}{2} k^{d_i} d_i^2 \qquad \text{out of plane bending energy term}$$
$$E_{tor} = \Sigma \tfrac{1}{2} V^{\omega_i}(1 + S_i \cos(|n_i| \cdot \omega_i)) \qquad \text{torsional energy term}$$
$$E_{vdw} = \Sigma \Sigma E_{ij}(1.0/a_{ij}^{12} - 2.0/a_{ij}^6) \qquad \text{van der Waals energy term } E_{ij}$$

the geometric mean of K_i and K_j, and $a_{ij} = r_{ij}/(R_i + R_j)$

$$E_{ele} = \Sigma \Sigma (q_i q_j / e r_{ij}) \qquad \text{electrostatic energy term}$$

The meaning of the terms in the above equations along with the Tripos energy parameters have been reported recently [48]. The electrostatic term can be switched on and off. When included, charges may be evaluated through several methods. Hydrogen bonding energy is taken into account in the following way: the van der Waals interactions between donor and acceptor, and hydrogen atoms linked to the donor are omitted. Favorable interactions between these atoms are computed when the electrostatic term is taken into account.

Carbohydrate geometries and conformations cannot be properly optimized using the conventional energy parameters of the Tripos 5.2 force-field. This is mainly due to stereoelectronic effects associated with the hemiacetal sequence C—O—C—O—C, and gathered under the name of anomeric and exo-anomeric effect. These effects originate from lone pair orbital back-bonding interactions (see review by Tvaroska and Bleha [49]). It was therefore advisable to derive new parameters for conformational analysis (rather than accurate reproduction of vibrational modes) and we adopted the following procedure [50]. In order to avoid interference with the parameters already in place, new atomic terms were created specific for carbohydrates. Equilibrium values for bond lengths and bond angles for carbohydrates were adapted from a survey of crystal structures [51]. Table IV describes the new atomic types which have been included. During the last year, the previously published parameters [50] have been improved and the new ones are described in Table IV.

The atomic charges were derived from the semi-empirical program MNDO [52]. This program provides charges having magnitude consistent with the ones calculated by the Pullman method [53] used to evaluate atomic charges in the case of proteins within the SYBYL package. Calculations were performed on several

TABLE IV

Atom types used for the description of carbohydrate along with the associated van der Waals parameters. The atoms types indicated by SYBYL are the standard atom types. The other ones (PIM) are specific of carbohydrates. Bond stretching parameters used for carbohydrates. Angle bending parameters used for carbohydrates. Torsional parameters used for carbohydrates. Only the parameters involving carbohydrate atom types are listed here. The parameters corrected from previous version [50] have been indicated by an #

Atom types and van der Waals parameters

Atom type	description	r (Å)	K (kcal/mol)	Origin	
C.C	carbohydrate carbon $sp3$	1.7	0.107	{PIM}	
C.A	α anomeric carbon	1.7	0.107	{PIM}	
C.B	β anomeric carbon	1.7	0.107	{PIM}	
C.3	carbon $sp3$	1.7	0.107	{SYBYL}	
C.2	carbon $sp2$	1.7	0.107	{SYBYL}	
O.A	α anomeric oxygen	1.52	0.116	{PIM}	
O.B	β anomeric oxygen	1.52	0.116	{PIM}	
O.R	sugar ring oxygen	1.52	0.116	{PIM}	
O.3	oxygen $sp3$	1.52	0.116	{SYBYL}	
O.2	oxygen $sp2$	1.52	0.116	{SYBYL}	
H.C	carbohydrate hydr. (on C atom)	1.3	0.042	{PIM}	#
H	hydrogen	1.5	0.042	{SYBYL}	
N.am	nitrogen amide	1.55	0.095	{SYBYL}	

Bond stretching parameters used for carbohydrate molecules

Atom 1	Atom 2	d^0 (Å) equilibrium length	k^d (kcal/mol/Å^2) force constant
C.C	C.C	1.523	633.6
C.C	C.A	1.523	633.6
C.C	C.B	1.523	633.6
C.C	C.2	1.522	639.0
C.C	O.A	1.420	618.9
C.C	O.B	1.420	618.9
C.C	O.R	1.427	618.9
C.C	O.3	1.411	618.9
C.C	H.C	1.100	662.4
C.C	N.am	1.449	677.6
C.A	C.2	1.522	639.0
C.A	O.A	1.395	618.9
C.A	O.R	1.410	618.9
C.A	H.C	1.100	662.4
C.B	C.2	1.522	639.0
C.B	O.B	1.376	618.9
C.B	O.R	1.425	618.9
C.B	H.C	1.100	662.4
O.A	H	0.972	1007.5
O.B	H	0.972	1007.5

TABLE IV.   Continued

Angle bending parameters used for carbohydrate molecules

Angle between			Q^0 (°) equilibrium angle	k^Q (kcal/mol/°^{2}) force constant	
Atom 1	Atom 2	Atom 3			
C.C	C.C	C.C	110.7	0.024	
C.A	C.C	C.C	110.7	0.024	
C.B	C.C	C.C	110.7	0.024	
C.2	C.C	C.C	110.7	0.024	
O.A	C.C	C.C	110.1	0.022	
O.B	C.C	C.C	110.1	0.022	
O.R	C.C	C.C	109.4	0.022	
O.3	C.C	C.C	110.1	0.022	
H.C	C.C	C.C	108.72	0.016	
N.am	C.C	C.C	109.7	0.018	
O.A	C.C	C.A	110.1	0.022	
O.B	C.C	C.A	110.1	0.022	
O.3	C.C	C.A	110.1	0.022	
H.C	C.C	C.A	108.72	0.016	
O.A	C.C	C.B	110.1	0.022	
O.B	C.C	C.B	110.1	0.022	
O.3	C.C	C.B	110.1	0.022	
H.C	C.C	C.B	108.72	0.016	
N.am	C.C	C.B	109.7	0.018	
H.C	C.C	C.2	109.5	0.016	
H.C	C.C	O.A	109.89	0.016	
H.C	C.C	O.B	109.89	0.016	
H.C	C.C	O.R	107.24	0.016	
H.C	C.C	O.3	109.89	0.016	
H.C	C.C	H.C	107.85	0.024	
H.C	C.C	N.am	109.5	0.020	
C.2	C.A	C.C	110.7	0.018	
O.A	C.A	C.C	110.1	0.022	
O.R	C.A	C.C	109.5	0.022	
H.C	C.A	C.C	108.72	0.016	
O.A	C.A	C.2	110.1	0.022	
O.R	C.A	C.2	109.5	0.022	
O.R	C.A	O.A	112.0	0.020	#
H.C	C.A	O.A	109.89	0.016	
H.C	C.A	O.R	107.24	0.016	
C.2	C.B	C.C	110.7	0.018	
O.B	C.B	C.C	110.1	0.022	
O.R	C.B	C.C	109.4	0.022	
H.C	C.B	C.C	108.72	0.016	
O.B	C.B	C.2	110.1	0.022	
O.R	C.B	C.2	109.4	0.022	
O.R	C.B	O.B	107.4	0.020	
H.C	C.B	O.B	109.89	0.016	
H.C	C.B	O.R	107.24	0.016	
O.2	C.2	C.C	120.4	0.026	
N.am	C.2	C.C	116.6	0.020	
C.A	O.A	C.C	113.8	0.044	#
H	O.A	C.A	109.35	0.020	
C.B	O.B	C.C	114.0	0.044	#
H	O.B	C.B	109.35	0.020	
C.A	O.R	C.C	113.8	0.044	
C.B	O.R	C.C	111.9	0.044	
H	O.3	C.C	109.35	0.020	
C.2	N.am	C.C	121.9	0.044	
H	N.am	C.C	118.4	0.020	

TABLE IV.   Continued

Torsional parameters used for carbohydrate molecules

Torsion angle between				n periodicity	V^{ω} (kcal/mol) force constant	
Atom 1	Atom 2	Atom 3	Atom 4			
*	C.C	C.C	*	3	0.5	
*	C.C	C.A	*	3	0.5	
*	C.C	C.B	*	3	0.5	
*	C.C	C.2	*	3	0.126	
*	C.C	O.A	*	3	0.5	
*	C.C	O.B	*	3	0.5	
*	C.C	O.R	*	3	0.5	
*	C.C	O.3	*	3	0.2	
*	C.C	N.am	*	3	0.2	
O.R	C.A	O.A	C.C	−1	1.9	#
C.C	C.A	O.A	C.C	3	1.9	#
H.C	C.A	O.A	C.C	−1	4.6	#
*	C.A	O.A	H	3	0.2	
*	C.A	O.R	*	3	0.5	
O.R	C.B	O.B	C.C	3	2.6	#
C.C	C.B	O.B	C.C	1	3.4	#
H.C	C.B	O.B	C.C	−1	0.4	#
*	C.B	O.B	H	3	0.2	
*	C.B	O.R	*	3	0.5	

monosaccharides and disaccharides in the α and β configuration and the charges
were averaged for each atom type. For disaccharides and for substituted monosaccharides, charges are derived by small adjustments in order to maintain neutrality
(Table V).

Torsional energy function must be able to reproduce the experimental and
theoretical data which indicate that the minimum energy value of Φ is in the
vicinity of 60° for α-D-glycosides and −60° in the case of β-D-glycosides. In the
absence of usable experimental data related to torsional rotation about glycosidic
linkage, we parametrized the torsional potential for Φ on the basis of a series of
ab initio calculations on 2-methoxy-tetrahydropyran (2-MTP) performed using 6-
311 ++ G* *//MP/31G* basis set [54]. Initial geometries of 2-MTP was constructed
in an appropriate configuration to model the α- and β-configuration of glycosides.
A grid search was performed at 10° step about the Φ torsional angle. At each
step, the geometry was fully optimized and the energy was evaluated with the
torsional terms corresponding to Φ set equal to zero, all other energy terms being
the ones listed in Table IV. The energy curve obtained was subtracted from the
ab initio energy curve and the difference between the two curves was fitted by
linear combination of trigonometric terms. The resulting correction functions for
the Φ angles are described by the following equations:

$$E\alpha = 0.95(1 - \cos \Phi) + 0.95(1 + \cos 3\Phi) + 2.3(1 - \cos(\Phi - 120))$$
$$E\beta = 1.3(1 - \cos \Phi) + 1.7(1 + \cos 3\Phi) + 0.2(1 - \cos(\Phi + 120))$$

The final energy curve, along with the one calculated by quantum chemistry

TABLE V

Atomic point charges used for hexopyranoside. When linking monosaccharides, the neutrality of the disaccharide is obtained by adding a charge of 0.04 e to the formal charge of the linked carbon C'-X (for a C-1_ O-1_C'-X linkage)

Atom name	PIM atom type	Charge
C-1	C.A or C.B	0.26
C-2	C.C	0.10
C-3	C.C	0.10
C-4	C.C	0.10
C-5	C.C	0.10
C-6	C.C	0.16
O-1	O.A or O.B	−0.34
O-2	O.3	−0.32
O-3	O.3	−0.32
O-4	O.3	−0.32
O-5	O.R	−0.34
O-6	O.3	−0.32
H-1	H.C	0.03
H-2	H.C	0.03
H-3	H.C	0.03
H-4	H.C	0.03
H-5	H.C	0.03
H-61	H.C	0.03
H-62	H.C	0.03
HO-1	H	0.17
HO-2	H	0.19
HO-3	H	0.19
HO-4	H	0.19
HO-6	H	0.19

For the *N*-acetyl group of GlcNAc residue:

N	N.am	−0.40
C7	C.2	0.35
C8	C.C	0.10
O7	O.2	−0.38
H81	H.C	0.00
H82	H.C	0.00
H83	H.C	0.00
HN	H	0.20

method are displayed in Figures 10a and 10b for α and β configuration, respectively.

VALIDATION OF THE PARAMETRIZATION FOR CARBOHYDRATE

The performance of present parametrization has been assessed in three stages of increasing complexity: computations of potential energy surfaces, refinement of

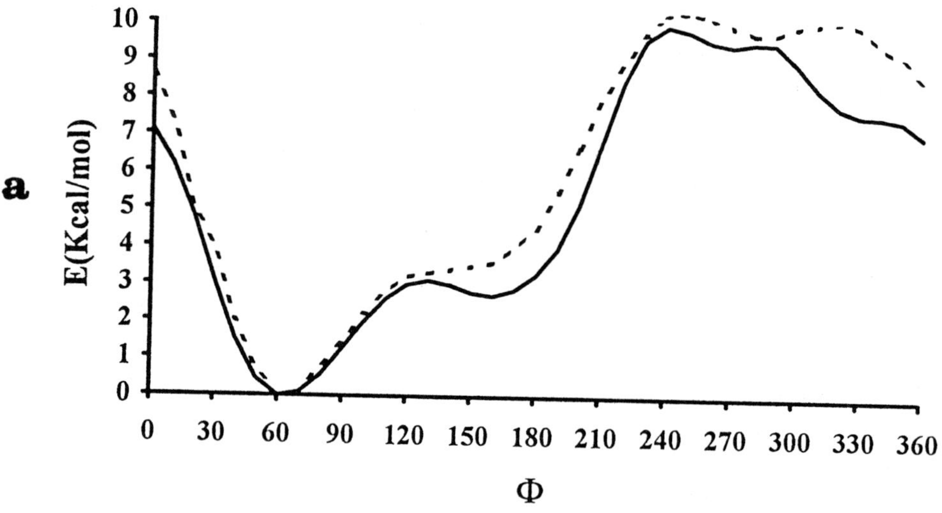

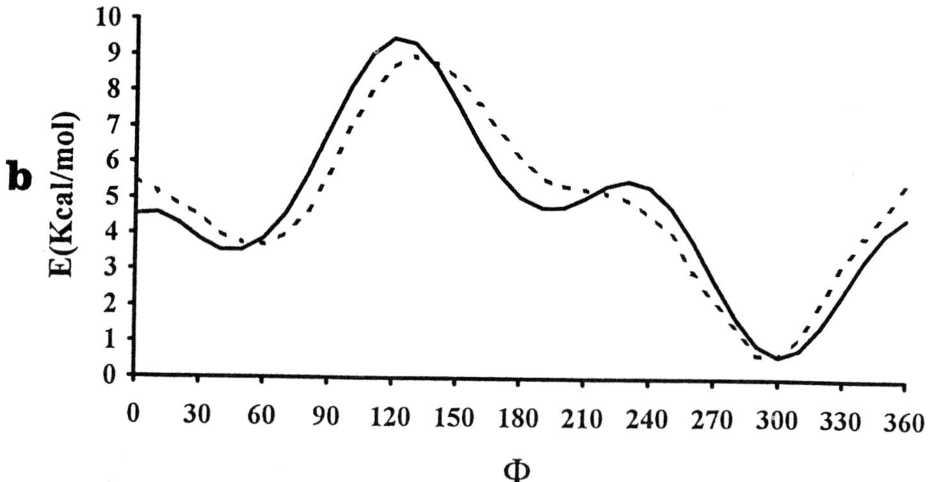

Fig. 10.   Potential energy curves of 2-Methoxytetrahydropyran in the α configuration (a) and the β configuration (b) as a function of the Φ angle. The energy calculated using carbohydrate parametrization of the Tripos force-field (------) can be compared with the energy calculated through *ab initio* method (——) [54].

the conformation of relevant oligosaccharides; computation and refinement of a protein-carbohydrate complex.

Disaccharide Potential Energy Maps

Potential energy surfaces were computed for four disaccharides of biological importance for which conformational analysis is well documented and established, i.e. maltose, lactosamine, Man α(1-3) Man and Man α(1-6) Man. In each case, the starting geometry of the disaccharide was constructed from the carbohydrate standard fragment library [24]. The computations were performed with the rigid residue approximation, in vacuo, using the appropriate atom types and the parameters in Table IV. No electrostatic contribution was taken into account and the hydroxyl hydrogen atoms were not considered. This was performed using the SEARCH option of Sybyl using increments of 5° in Φ and Ψ, over the whole angular range. The four potential energy maps are represented in the form of iso-contour plot in Figure 11. For the four disaccharides the maps correspond closely to the rigid maps which have already been published or shown in this paper. Not only are the external contours reproduced but the location of the different low energy minima both in the main low energy well and in the isolated domains are also reproduced.

Using the updated version of the carbohydrate parametrization, it is also possible to calculate relaxed maps of disaccharide, by using the GRID SEARCH option of Sybyl. A relaxed map of the GlcNAc β(1-2) Man disaccharide has been calculated. The map displayed in Figure 12 is an adiabiatic one since it corresponds to the projection of four maps calculated with different orientations of the secondary hydroxyl groups. The GlcNAc β(1-2) Man relaxed map calculated with the new parameters is very similar to the one which has been obtained recently using the MM3 program [55]. Only the relaxed map is shown here since it appears that in this particular case, the rigid approach is quite inappropriate. This can be explained by the close proximity between the glycosidic linkage and the N-acetyl group. Small adjustments in the geometry of this pendent group have to be taken in account for computing the conformational behavior of the glycosidic linkage.

Oligosaccharide Geometry Optimization

The refinement of the conformation of relevant oligosaccharides for which detailed information is available from crystal structure determinations provides a suitable test, albeit not exhaustive, of the validity of the parametrization The solid-state conformation of the trisaccharide Man α(1-3) Man β(1-4) GlcNac is well-established through X-ray crystallography [36]. The coordinates of the atoms taken from the crystal structure were submitted to a complete energy minimization. This was performed using the force-field for carbohydrates taking into account charges, with a dielectric constant of 4. The agreement between the observed and calculated structures is good, as shown by a root mean square error (rms) of 0.21 A for non-hydrogen atoms (Figure 13). The differences occur mainly at the hydrogen atoms for which X-ray crystallography does not provide accurate locations. Not surprisingly, most of the hydroxyl hydrogen atoms remain in the vicinity of their starting orientation. There is also a slight opening of the bridge angles compared to those in the crystal structure. Such an opening may be due to the relief of the crystal packing forces on the molecule. The obtained agreement indicates that there are no major flaws in the parametrization.

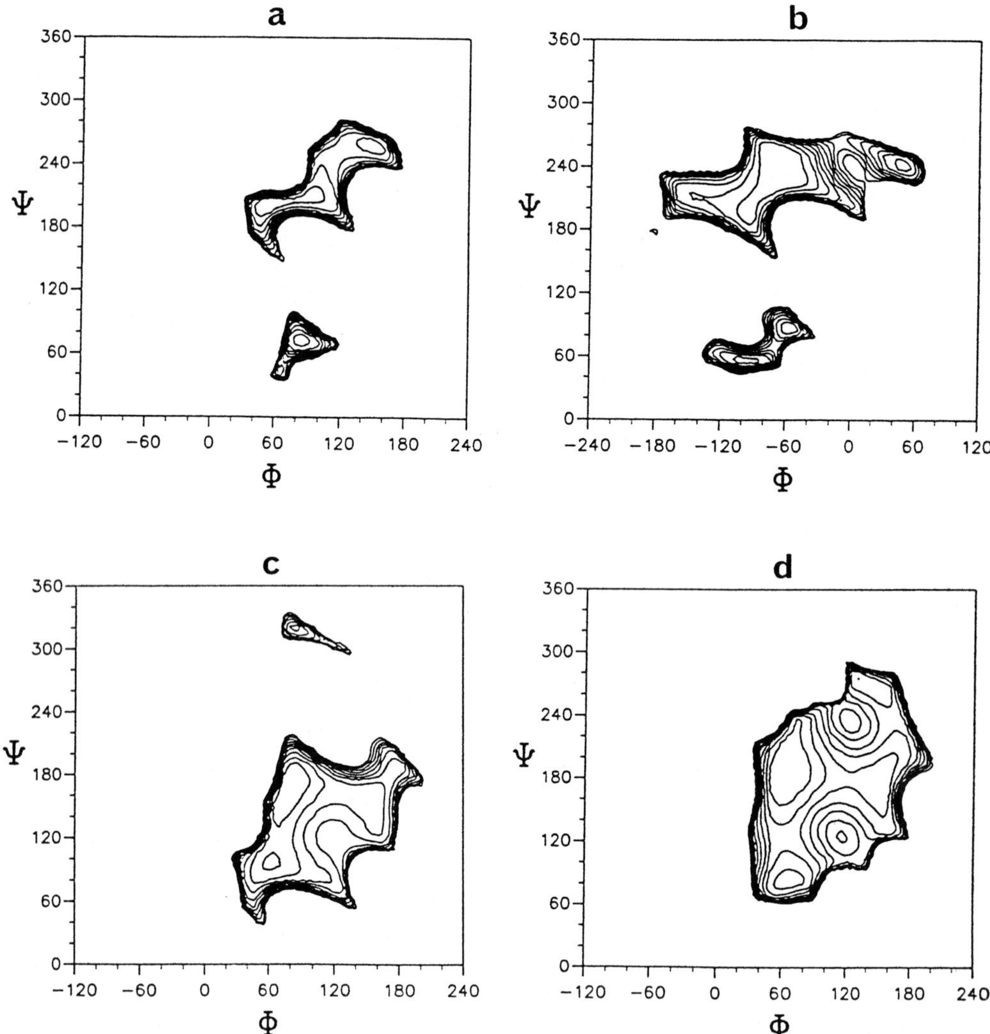

Fig. 11. Potential energy surfaces calculated with the carbohydrate parametrization of the Tripos force-field using the SEARCH option of SYBYL. The iso-energy contours are drawn with interpolation of 1 kcal/mol above the minimum. (a) maltose, (b) lactosamine, (c) Man α(1–3) Man, (d) Man α(1-6) Man (with ω in the GT orientation).

Modeling Protein/Carbohydrate Interactions

The structure of the native Concanavalin A (ConA) has been reported from X-ray crystallographic studies at 1.75 Å resolution [56]. A low resolution crystalline structure of the complex ConA/methyl α-mannopyranoside, allowed the identification of the binding site [57]. In order to address the problem of computer simulation of protein/carbohydrate interactions, we have developed a somehow general procedure. This docking procedure, the so-called 'crankshaft' method,

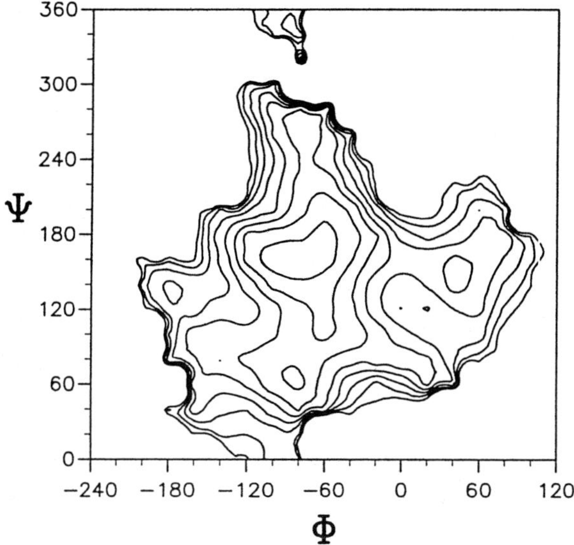

Fig. 12.   Relaxed energy map of the GlcNAc β(1-2) Man disaccharide calculated with the carbohydrate parametrization of the Tripos force-field. The relaxed map is an adiabatic representation of 4 maps calculated for different orientations of the secondary alcohols. Both hydroxymethyl groups have been kept in a GT orientation.

allows for a systematic exploration of the orientations and positions of a ligand into a cavity [50]. Once those relative locations and orientations of the ligand with respect to the protein site are known, further refinements are needed which require a thorough energy minimization of the protein-carbohydrate complex. This was performed using simultaneously our parametrization for carbohydrates and the standard Tripos parametrization for the proteins.

In the present case of ConA interacting with methyl α-D-mannopyranoside, we found that eight possible complexes should be considered. They all displayed binding energy at least 15 kcal/mol lower than the energy corresponding to the case where the protein and the carbohydrate are non interacting. The fact that methyl α-D-mannopyranoside can reach the binding site of ConA in more than one binding orientation has already been stressed [58]. The eight docking solutions have been ordered as a function of the energy and the two lowest energy orientations are displayed in Figure 14. Further validation of the carbohydrate parametrization was provided by the recent crystal structure of the complex of ConA with methyl α-D-mannopyranoside at 2.9 Å resolution [59]. Comparison of the hydrogen bonding schemes listed in Table VI indicates that the observed orientation corresponds to the calculated solution having the lowest energy. The only differences arise from the fact that in the present simulation no water molecule was taken into account. In the simulation, the O-2 oxygen atom is hydrogen bounded to Leu99.N instead of a water molecule.

A more difficult goal is the modeling of interaction phenomena involving flexible ligand. The important role of entropy barrier in protein/oligosaccharide interactions has been demonstrated [6]. The parameters presented here have been used

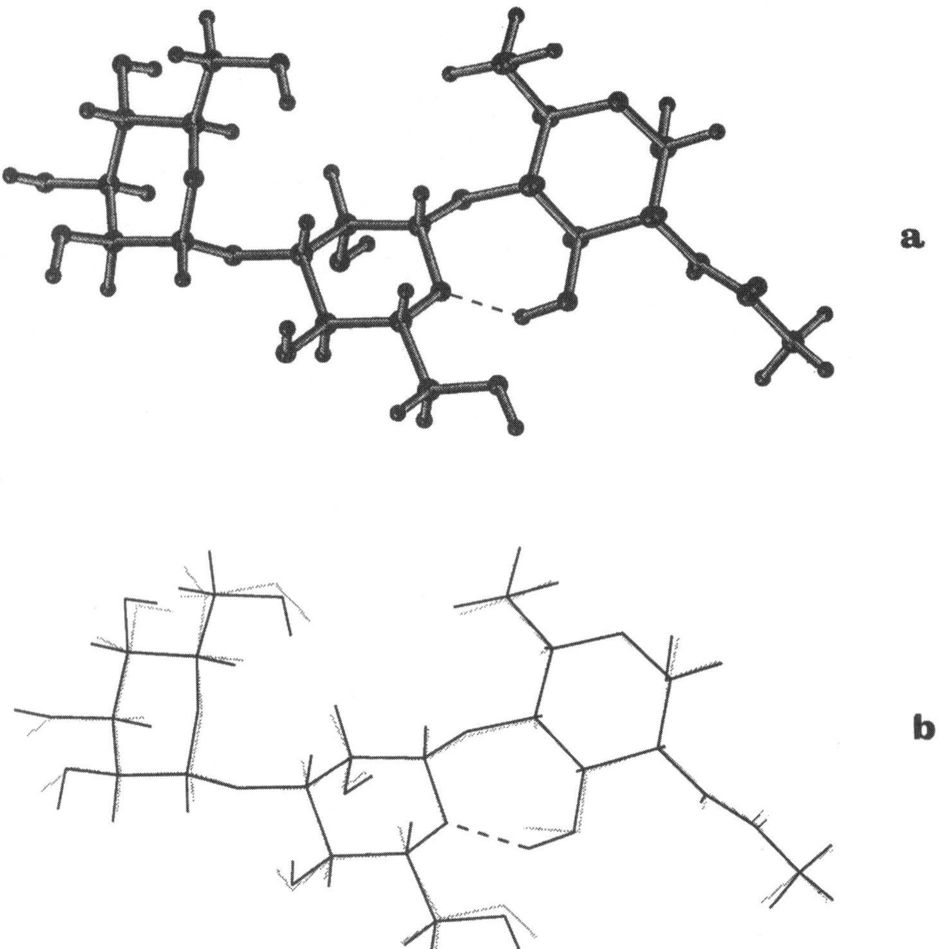

Fig. 13. Three dimensional representation of the trisaccharide Man α(1-3) Man β(1-4) GlcNac. (a) after complete energy minimization using the carbohydrate parametrization of the Tripos force-field. (b) shows the superposition of the crystal structure conformation (thin lines) and the energy minimized conformation (solid lines). Hydrogen bonds are shown as dotted lines.

for the systematic conformational search of di- and trisaccharides interacting with either *Lathyrus ochrus* lectin or ConA [60]. Such approach allowed not only to predict the conformations which are the most favorable for the binding but also to estimate the residual conformational flexibility of the ligand when it is bounded. Thermodynamics calculations from such simulations are now at their beginning and open new perspectives for designing ligands displaying very high affinity for biological receptors.

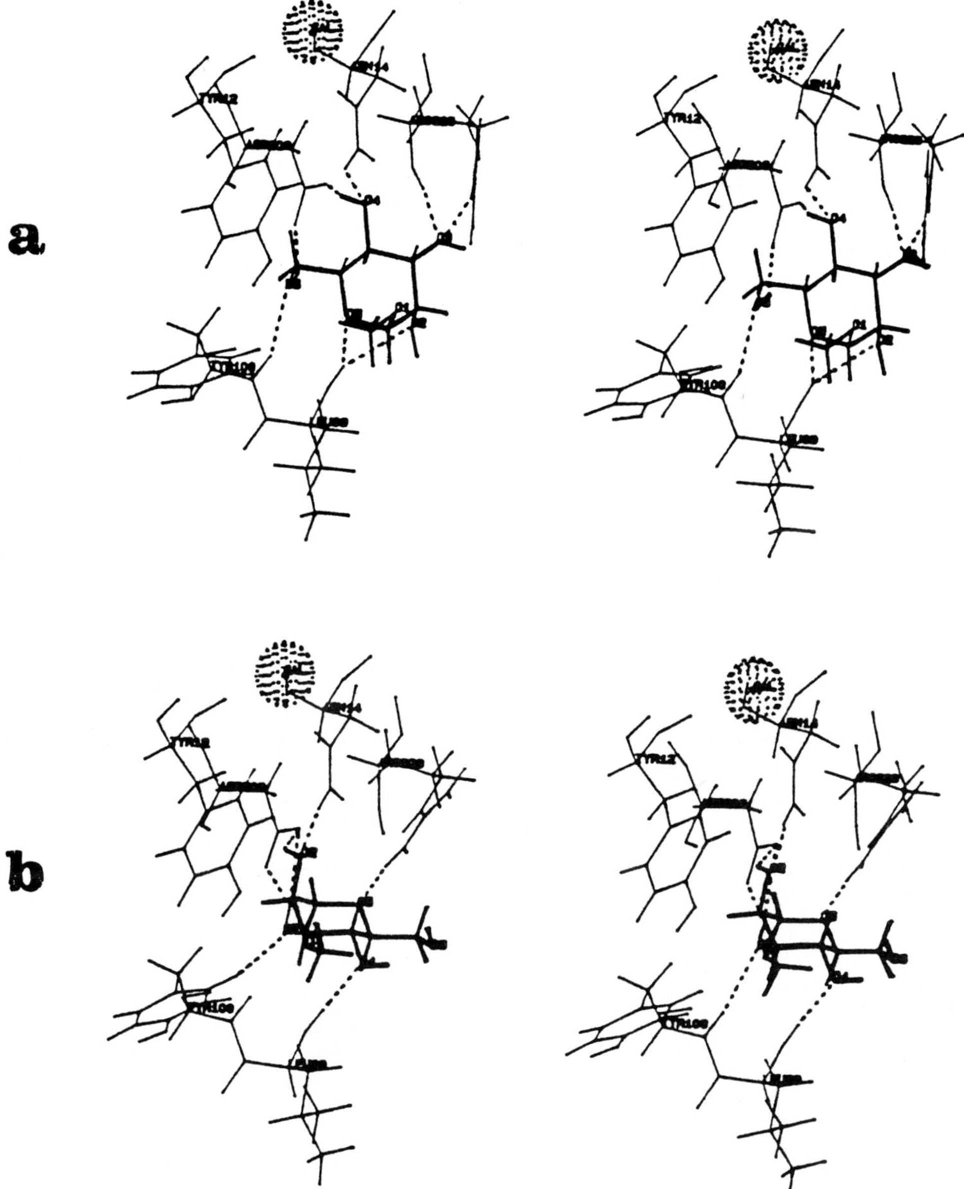

Fig. 14.   Stereoscopic representation of the two best simulated orientations of methyl α-D-mannopyranoside in the binding site of Concanavalin A. For the sake of clarity, only the eight amino acids directly involved in the binding site have been shown. Hydrogen bonds are shown as broken lines.

TABLE VI
Hydrogen bond network between conA and mannose. Comparison between the crystal structure
(Derewenda *et al.*, 1989) and the predicted lowest energy docking

Crystalline complex			Modeled complex		
Donor	Acceptor	D-A dist. (Å)	Donor	Acceptor	D-A dist. (Å)
O-4	Asp208.OD2	2.84	O-4	Asp208.OD2	2.53
O-6	Asp208.OD1	2.72	O-6	Asp208.OD1	3.45
O-2	water	3.13			
			Leu99.N	O.2	3.37
Arg228.N	O-3	3.15	Arg228.N	O-3	2.96
Asn14.ND2	O-4	3.25	Asn14.ND2	O-4	3.06
Leu99.N	O-5	3.02	Leu99.N	O-5	3.27
Leu99.N	O-6	2.92			
Tyr100.N	O-6	3.25	Tyr100.N	O-6	3.05

Conclusions

In this paper, we have examined some of the methods required for accurate molecular modeling of oligosaccharides and complex carbohydrates, in the light of some difficulties inherent to that type of molecules. Two different three-dimensional data bases have been presented which facilitate the construction and/or the analysis of complex carbohydrate architectures. It has been clearly shown that carbohydrate flexibility, especially at the glycosidic linkages has to be taken properly into account.

An additional aspect of the work has been the extension of the Tripos molecular mechanical force field to oligosaccharides. The new parametrization has been assessed in three stages of increasing complexity; it has been shown to be relevant for the computer simulation of protein-carbohydrate interactions.

Thanks to the coupling of such a force field with three-dimensional real-time molecular modeling software, the speed of analysis and simulation of complex carbohydrates either in isolation or in interaction with other macromolecules is dramatically increased. Besides, it is suggested that the availability of such tools to the scientific community should favour reproducibility of the results. Since molecular mechanics and dynamics are becoming important methods to study mono- oligo- and polysaccharides as well as protein–carbohydrate complexes, we hope that the present contribution will permit investigators who are new in the field of oligosaccharide modelling to practice molecular mechanics of carbohydrates with discipline and confidence.

Acknowledgements

The authors are much indebted to Dr. Cambillau (Marseille, France) for giving us access to the coordinates of the *Lathyrus ochrus* lectin complexed with different oligosaccharides. We thank also Dr. Wright (Richmond, USA) for providing the atomic coordinates of *N*-acetylneuraminyl lactose.

References

1. J. Montreuil: *Adv. Carbohydr. Chem. Biochem.* **37**, 157 (1980).
2. T. W. Rademacher, R. B. Parekh, and R. A. Dwek: *Ann. Rev. Biochem.* **57**, 785 (1988).
3. J. P.Carver and D. A. Cumming: *Pure Appl. Chem.* **11**, 1465 (1987).
4. J. P. Carver, D. Mandel, S. W. Michnick, A. Imberty, and J. W. Brady: Conformational Analysis of Oligosaccharides: Reconciliation of Theory with Experiments', in A. D. French and J. W. Brady, Eds., *Computer Modeling of Carbohydrate Molecules*, ACS Symposium Series; No. 430, American Chemical Society, Washington DC., p. 266 (1990).
5. A. Imberty, Y. Bourne, C. Cambillau, and P. Rougé: *Adv. Biophys. Chem*, **3**, 71 (1993).
6. J. P. Carver: *Pure Appl. Chem.* **65**, 763 (1993).
7. S. Pérez, A. Imberty, and J. P. Carver: *Adv. Computational Biol.* **1**, 147 (1994).
8. D. A. Cumming and J. P. Carver: *Biochemistry* **26**, 6664 (1987).
9. C. Hervé du Penhoat, A. Imberty, N. Roques, V. Michon, J. Mentech, G. Descotes, and S. Pérez: *J. Am. Chem. Soc.* **113**, 3720 (1991).
10. C. Meyer, S. Pérez, C. Hervé du Penhoat, and V. Michon: *J. Am. Chem. Soc.* **115**, 10300 (1993).
11. S. Doubet, K. Bock, D. Smith, A. Darvill, and P. Albersheim: *TIBS* **14**, 475 (1989).
12. R. U. Lemieux, K. Bock, L. T. Delbaere, S. Koto and V. S. R. Rao: *Can. J. Chem.* **58**, 631 (1980).
13. S. R. Niketic and K. Rasmussen: *Lecture Notes in Chemistry*. Vol. 3, Springer-Verlag, Berlin (1977).
14. G. A. Jeffrey and R. J. Taylor: *J. Comp. Chem.* **1**, 99 (1980).
15. I. Tvaroska and S. Pérez: *Carbohydr. Res.* **149**, 389 (1986).
16. S. N. Ha, A. Giammona, M. Field, and J. W. Brady: *Carbohydr. Res.* **180**, 207 (1988).
17. S. W. Homans: *Biochemistry* **29**, 9110 (1990).
18. S. Arnott and S. E. Scott: *J. Chem. Soc. Perkin Trans.* **2**, 324 (1972).
19. S. Sheldrick and D. Akrigg: *Acta Crystallogr.* **B36**, 1615 (1980).
20. U. Burkert and N. L. Allinger: *Molecular Mechanics*. Am. Chem. Soc. Monograph No. 177, American Chemical Society, Washington DC (1982).
21. N. L. Allinger: *J. Am. Chem. Soc.* **99**, 8127 (1977).
22. N. L. Allinger, Y. H. Yuh, and J. H. Lii: *J. Am. Chem. Soc.* **11**, 8551 (1989).
23. A. D. French, R. S. Rowland, and N. L. Allinger: 'Modeling of Glucopyranose: The Flexible Monomer of Amylose', in A. D. French and R. W. Brady, Eds., *Computer Modeling of Carbohydrate Molecules*. ACS Symposium Series. No. 430, American Chemical Society, Washington DC., p. 120 (1990).
24. S. Pérez and M. M. Delage: *Carbohydr. Res.* **212**, 253 (1991).
25. M. Ragazzi, D. F. Ferro, and A. Provalosi: *J. Comp. Chem.* **7**, 105 (1987).
26. R. H. Marchessault and S. Pérez: *Biopolymers* **18**, 2369 (1979).
27. R. A. Scott and H. A. Scheraga: *J. Chem. Phys.* **42**, 2209 (1966).
28. I. Tvaroska: *Carbohydr. Res.* **125**, 155 (1984).
29. S. Pérez and C. Vergelati: *Polymer Bull.* **17**, 141 (1987).
30. S. Pérez: D. Sc. thesis, Université de Grenoble (1978).
31. V. H. Tran, A Buléon, A. Imberty, and S. Pérez: *Biopolymers* **28**, 679 (1988).
32. A. D. French, V. H. Tran, S. Pérez: 'Conformational Study of a Disaccharide (Cellobiose) with the Molecular Mechanics Program (MM2)', in A. D. French and R. W. Brady, Eds., *Computer Modeling of Carbohydrate Molecules*, ACS Symposium Series. No. 430, American Chemical Society, Washington DC., p. 191 (1990).
33. A. Imberty, V. Tran and S. Pérez: *J. Comp. Chem.* **11**, 205 (1989).
34. S. Pérez: *Curr. Opinion Structural Biology* **3**, 675 (1993).
35. A. D. French and M. K. Dowd: *J. Mol. Struct.* (*Theochem*) **286**, 183 (1993).
36. V. Warin, F. Baert, R. Fouret, G. Strecker, G. Spik, B. Fournet, and J. Montreuil: *Carbohydr.Res.* **76**, 11 (1979).
37. Y. Bourne, P. Rougé and C. Cambillau: *J. Biol. Chem.* **265**, 18161 (1990).
38. Y. Bourne, P. Rougé, and C. Cambillau: *J. Biol. Chem.* **267**, 197 (1992).
39. T. Srikrishnan, M. S. Chowdhary, and K. L. Matta: *Carbohydr. Res.* **186**, 161 (1989).
40. A. Imberty, S. Gerber, V. Tran, and S. Pérez, *Glycoconj. J.* **7**, 27 (1990).

41. A. Imberty, M. M. Delage, Y. Bourne, C. Cambillau, and S. Pérez: *Glycoconj. J.* **8**, 456 (1991).
42. C. Wright: *J. Mol. Biol.* **215**, 635 (1990).
43. D. A. Brant and M. D. Christ: 'Realistic Conformational Modeling of Carbohydrates: Applications and Limitations in the Context of Carbohydrate-High Polymers', in A. D. French and R. W. Brady, Eds., *Computer Modeling of Carbohydrate Molecules*, ACS Symposium Series No. 430, American Chemical Society, Washington DC., p. 42 (1990).
44. V. S. R. Rao, M. Biswas, C. Mukhopadhyay, and P. V. Balaji: *J. Mol. Struct.* **194**, 203 (1989).
45. R. Stuikeprill and B. Meyer: *Eur. J. Biochem.* **24**, 903 (1990).
46. SYBYL, Tripos Associates, 1669 S. Hanley Road, Suite 303, St Louis, MO, 63144, USA.
47. D. N. J. White: *J. Chem. Soc. Perkin* **2**, 43 (1975).
48. M. Clark, R. D. Cramer III, and N. van Opdenbosch: *J. Comp. Chem.* **10**, 982 (1989).
49. I. Tvaroska and T. Bleha: *Adv. Carbohydr. Chem. Biochem.* **47**, 45 (1989).
50. A. Imberty, K. D. Hardman, J. P. Carver, and S. Pérez: *Glycobiology* **1**, 456 (1991).
51. G. A. Jeffrey, J. A. Pople, J. S. Binkley, and S. Vishveshwara: *J. Am. Chem. Soc.* **100**, 373 (1978).
52. M. J. S. Dewar and W. Thiel: *J. Am. Chem. Soc.* **99**, 4899 (1977).
53. H. Berthod and A. Pullman: *J. Chem. Phys.* **62**, 942 (1965).
54. I. Tvaroska and J. P. Carver: *J. Phys. Chem.* **98**, 9477 (1994).
55. J. Koca, S. Pérez, and A. Imberty: *J. Comp. Chem.* (in press).
56. K. D. Hardman, R. C. Agarwal, and M. J. Freiser: *J. Biol. Chem.* **157**, 69 (1982).
57. K. D. Hardman and C. F. Ainsworth: *Biochemistry* **15**, 1120 (1976).
58. Y. C. Sekharudu and V. S. R. Rao: *Int. J. Biol. Macromol.* **6**, 337 (1984).
59. Z. Derewenda, J Yariv, J. R. Helliwell, A. J. Kalb, E. J. Dodson, M. Z. Papiz, T. Wan, and J. Campbell: *EMBO J.* **8**, 2189 (1989).
60. A. Imberty and S. Pérez: *Glycobiology* **4**, 351 (1994).

3-D Structure of Acetylcholinesterase and Its Complexes with Anticholinesterase Agents

J. L. SUSSMAN,[1] M. HAREL,[1] M. RAVES,[1] D. QUINN,[2] and I. SILMAN[3]

[1]*Department of Structural Biology, The Weizmann Institute of Science, Rehovot 76100, Israel*
[2]*Department of Chemistry, University of Iowa, Iowa City, Iowa 52242-1294 U.S.A.*
[3]*Department of Neurobiology, The Weizmann Institute of Science, Rehovot 76100, Israel*

Abstract. The principal biological role of acetylcholinesterase (AChE) is termination of impulse transmission at cholinergic synapses by rapid hydrolysis of the neurotransmitter acetylcholine (ACh). Based on our recent X-ray crystallographic structure determination of AChE from *Torpedo californica* we can see, at atomic resolution, a protein binding pocket for the neurotransmitter ACh. We found that the active site consists of a catalytic triad (Ser^{200}-His^{440}-Glu^{327}) which lies close to the bottom of a deep and narrow gorge, that is lined with the rings of 14 aromatic amino acid residues. Despite the complexity of this array of aromatic rings, we suggested, on the basis of modelling which involved docking of the ACh molecule in an all-trans conformation, that the quaternary group of the choline moiety makes close contact with the indole ring of Trp^{84}. In order to study experimentally this interaction, in detail, we soaked into crystals of AChE a series of different inhibitors, including the competitive inhibitor edrophonium (EDR) and the transition-state analog (*N*,*N*,*N*-trimethylammonio)-trifluoroacetophenone (TFK), and determined their 3-D structures.

Key words. Acetylcholinesterase, anticholinesterase, aromatic rings.

Introduction

The principal biological role of acetylcholinesterase (AChE, acetylcholine hydrolase, EC 3.1.1.7) is to terminate signal transmission at cholinergic synapses by rapid hydrolysis of the neurotransmitter, acetylcholine (ACh) [1,2]. In keeping with this requirement, AChE possesses a remarkably high specific activity, especially for a serine hydrolase [3], functioning at a rate approaching that of a diffusion-controlled reaction [4]. Early kinetic studies indicated that the active site of AChE consists of two subsites, the 'esteratic' and 'anionic' subsites, corresponding to the catalytic machinery and the choline-binding pocket, respectively [5]. A second, 'peripheral', anionic site exists, so named because it appears to be distant from the active site [6]. The recent elucidation of the three-dimensional structure of *Torpedo* AChE [7] has served to confirm these earlier studies, and has shown that AChE contains a catalytic triad similar to that present in other serine hydrolases [8]. Unexpectedly, it has also revealed that this triad is located near the bottom of a deep and narrow cavity, *ca.* 20 Å deep, which has been named the 'active-site gorge'. The cavity is lined by the rings of fourteen aromatic residues which are conserved in the AChE sequences published so far [9]. Much of the subsequent research on structure-function relationships in AChE has been concerned with the functional significance of the gorge and with the role of the aromatic rings which account for more than 50% of its surface area [10]. Thus, structural evidence [7,10], as well as evidence obtained by modelling [11], by

A. Pullman et al. (eds.), *Modelling of Biomolecular Structures and Mechanisms*, 455–460.

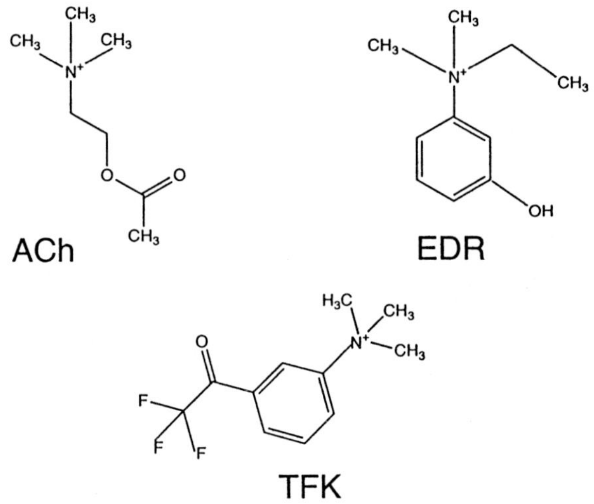

Fig. 1. Chemical structures of the neurotransmitter acetylcholine (ACh); a competitive inhibitor of acetylcholinesterase, i.e., edrophonium (EDR); and a transition-state analog (*N*,*N*,*N*-trimethylammonio)trifluoroacetophenone (TFK).

chemical modification [12–14], and by site-directed mutagenesis [11,15–18], all point to important roles for certain of these conserved aromatic residues in both the esteratic and 'anionic' subsites of the active site, and in the 'peripheral' anionic site.

In the following we will present structural evidence, derived from X-ray data sets collected for complexes of AChE with a competitive inhibitor and with a transition-state analog, which substantiates the involvement of aromatic residues in the binding of the quaternary group of ACh in the so-called 'anionic' subsite of the active site, and provides experimental corroboration for a plausible orientation of ACh within the active-site suggested by computer-docking.

Results and Discussion

In our initial report concerning the three-dimensional structure of *Torpedo* AChE [7], we suggested, on the basis of manual docking, that the primary interaction of the quaternary group of the substrate, ACh (Figure 1a), was with the indole ring of the conserved tryptophan residue, Trp[84]. Thus, Figure 2 shows that if ACh is docked in an all-trans configuration, with the C^α carbon of its acyl group positioned to make a tetrahedral bond with O^γ of Ser[200], and the acyl oxygen pointing towards a putative oxyanion hole, composed of the main-chain nitrogens of Gly[118], Gly[119] and Ala[201], the quaternary group is within van der Waals distance (*ca.* 3.5 Å) of Trp[84]. This assignment was at odds with the accepted notion that the 'anionic' binding site for the positive quaternary group contains a cluster of negative charges [19]. However, support for such an assignment was provided by both affinity labeling of Trp[84] in the *Torpedo* enzyme [12] and by a study which used both model host-guest systems and theoretical considerations to argue that aromatic

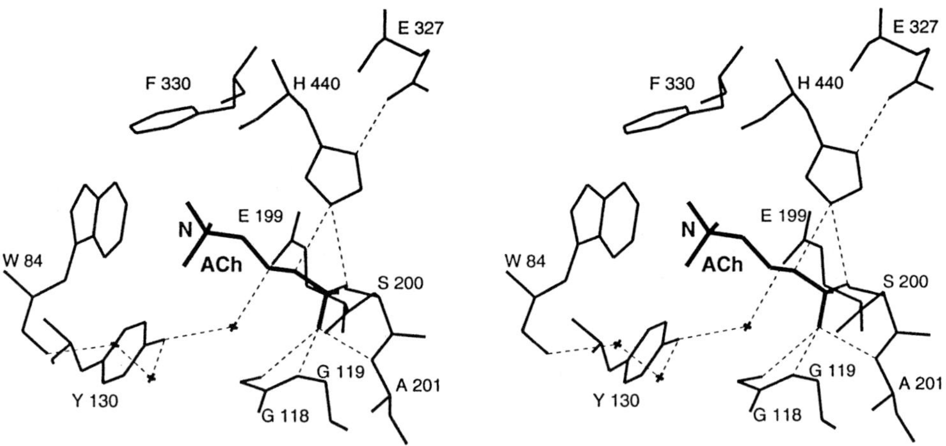

Fig. 2.   Stereo view of a model of ACh docked in the active site of AChE in the vicinity of the active site [7].

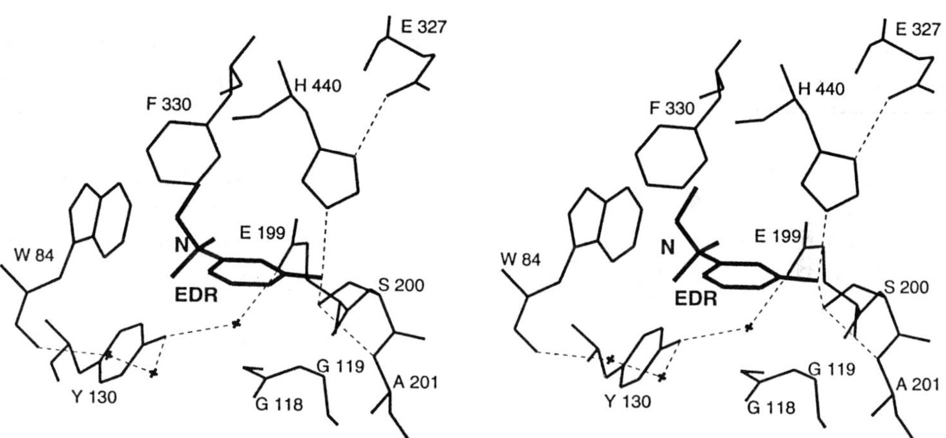

Fig. 3.   Stereo view of the crystal structure of the AChE : edrophonium complex in the vicinity of the active site [14].

rings might play an important role in receptors for quaternary ligands, via an interaction of their π electrons with the positive charge [20]. Nevertheless, we felt that it was important to confirm this assignment at the X-ray level by studying appropriate AChE-ligand complexes.

Edrophonium (EDR, Figure 1b) is a powerful competitive inhibitor of AChE [21,22], used clinically in the diagnosis of myasthenia gravis [23]. Due to its quaternary character, it does not penetrate cell membranes or the blood-brain barrier; it thus acts primarily at peripheral sites such as the muscle endplate [23]. In the EDR–AChE complex (Figure 3), the quaternary group of the ligand nestles

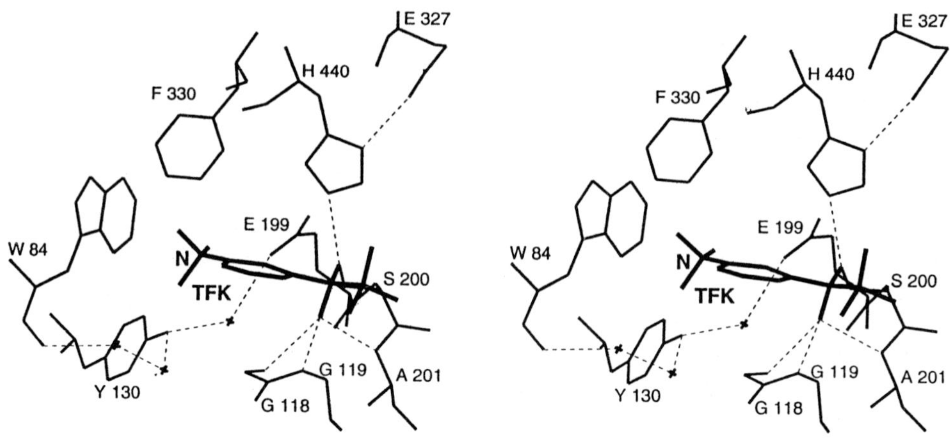

Fig. 4.   Stereo view of the crystal structure of AChE:TFK complex in the vicinity of the active site.

adjacent to the indole of Trp[84], in a position equivalent to that assigned to the quaternary group of ACh in the docking procedure [7]. In the affinity labeling study mentioned above [12], this tryptophan was covalently labeled by the aziridinium ion, which is similar in structure to EDR, and EDR protected against the labeling by aziridinium. Our data thus demonstrate good correspondence between the crystal structure and that in solution. The *m*-hydroxyl group is positioned between $His^{440}N^{\epsilon 2}$ and $S^{200}O^{\gamma}$, thus making hydrogen bonds of 3.0 and 3.5 Å, respectively, to two of the three members of the catalytic triad. There is also a 3.5-Å hydrogen bond to $Gly^{119}N$, which is part of the oxyanion hole. This provides a structural basis for the observation that such *meta*-substituted anilinium ions are much more potent inhibitors of AChE than the homologous ligand which lacks the hydroxyl group [21].

Transition-state analogs are designed to provide a stable adduct which closely resembles the transition state which is formed transiently during substrate hydrolysis [24]. A number of studies have described the design, synthesis and functional characterization of transition-state analogs for AChE [see, for example [25,26]. One such compound, *m*-(*N,N,N*-trimethylammonio)trifluoroacetophenone (TFK, Figure 1c), is a potent time-dependent inhibitor of both *Electrophorus* and *Torpedo* AChE [27,28]. In aqueous solution, TFK is an equilibrium mixture of the free ketone and of the corresponding ketone hydrate, with only the free ketone serving as an inhibitor of the enzyme. The second-order rate constant for inhibitor binding K_{on}, is 6–$7 \times 10^9 \, M^{-1} \, s^{-1}$, and probably monitors a diffusion-controlled process. In contrast, dissociation is very slow, with $k_{off} = 10^{-5} \, s^{-1}$ for *Electrophorus* AChE and $10^{-4} \, s^{-1}$ for *T. californica* AChE. The corresponding inhibition constants, K_i, based on the concentration of the free ketone are 1.3 and 15 fM, respectively [28].

In the TFK–AChE complex (Figure 4), the quaternary group lies virtually in the same position as in the EDR–AChE complex, interacting with the aromatic groups of Trp[84] and Phe[330]. The latter group, which assumes different orientations in various AChE–ligand complexes studied [14], so as to interact favourably with the bound ligand, adopts virtually the same orientation in the TFK–AChE complex as in the EDR–AChE complex. The keto function, which is in the '*m-*' position, relative to the quaternary group, appears well positioned to form a hemiacetal with $Ser^{200}O^{\gamma}$, with its carbon atom also being in hydrogen-bonding distance of $His^{440}N^{\epsilon 2}$. The overall orientation thus seems to be very similar to that of ACh modelled in its all-*trans* conformation [7]. Moreover, the oxygen atom of TFK is within hydrogen-bonding distance of the main-chain nitrogens of Gly^{118}, Gly^{119} and Ala^{201}, which we had suggested might serve as the 'oxyanion' hole for the carbonyl of ACh [7]. Although the three fluorine atoms on the proximal carbon presumably serve to enhance the stability of the transition-state analog due to their high electronegativity, close inspection of the structure of the TFK–AChE complex reveals that all three also appear to make specific interactions with the protein, near the aromatic residues Phe^{288} and Phe^{290}.

The structure presented in Figure 4 shows that the TFK–AChE complex indeed displays the structural characteristics to be expected of a transition-state complex. Detailed inspection of the structure should provide valuable insights into the structural features of the active site which are largely responsible for the catalytic power of this potent hydrolase.

Acknowledgements

We thank Lilly Toker for preparation of the AChE. This project was supported by the U.S. Army Medical Research and Development Command under Contract DAMD17-89-C9063, the Association Franco-Israelienne pour la Recherche Scientifique et Technologique (AFIRST), the Minerva Foundation, Munich, Germany and the Kimmelman Center for Biomolecular Structure and Assembly, Rehovot.

References

1. E. A. Barnard: 'Neuromuscular Transmission – Enzymatic Destruction of Acetylcholine', in J. I. Hubbard, Ed., *The Peripheral Nervous System*, Plenum, New York, p. 201 (1974).
2. B. Katz: *Nerve. Muscle and Synapse*, McGraw-Hill, New York (1966).
3. D. M. Quinn: *Chem. Rev.* **87**, 955 (1987).
4. M. Bazelyansky, C. Robey, and J. F. Kirsch: *Biochemistry* **25**, 125 (1986).
5. D. Nachmansohn and I. B. Wilson: *Adv. Enzymol.* **12**, 259 (1951).
6. P. Taylor and S. Lappi: *Biochemistry* **14**, 1989 (1975).
7. J. L. Sussman, M. Harel, F. Frolow, C. Oefner, A. Goldman, L. Toker, and I. Silman: *Science* **253**, 872 (1991).
8. T. A. Steitz and R. G. Shulman: *Ann. Rev. Biophys. Bioeng.* **11**, 419 (1982).
9. M. K. Gentry and B. P. Doctor: 'Alignment of Amino Acid Sequences of Acetylcholinesterases and Butyrylcholinesterases', in J. Massoulié, F. Bacou, E. Barnard, A. Chatonnet, B.P. Doctor and D. M. Quinn, Eds., *Cholinesterases: Structure, Function, Mechanism, Genetics and Cell Biology*, American Chemical Society, Washington, DC, 394 (1991).
10. P. H. Axelsen, M. Harel, I. Silman, and J. L. Sussman: *Prot. Sci.* **3**, 188 (1994).
11. M. Harel, J. L. Sussman, E. Krejci, S. Bon, P. Chanal, J. Massoulié, and I. Silman: *Proc. Natl. Acad. Sci. USA* **89**, 10827 (1992).
12. C. Weise, H.-J. Kreienkamp, R. Raba, A. Pedak, A. Aaviksaar, and F. Hucho: *EMBO J.* **9**, 3885 (1990).
13. I. Schalk, L. Ehret-Sabatier, F. Bouet, M. Goeldner, and C. Hirth: 'Structural Analysis of

Acetylcholinesterase Ammonium Binding Sites', in A. Shafferman and B. Velan, Eds., *Multidisciplinary Approaches to Cholinesterase Functions*, Plenum Press, New York, p. 117 (1992).
14. M. Harel, I. Schalk, L. Ehret-Sabatier, F. Bouet, M. Goeldner, C. Hirth, P. Axelsen, I. Silman, and J. L. Sussman: *Proc. Natl. Acad. Sci. USA* **90**, 9031 (1993).
15. A. Shafferman, B. Velan, A. Ordentlich, C. Kronman, H. Grosfeld, M. Leitner, Y. Flashner, S. Cohen, D. Barak, and N. Ariel: 'Acetylcholinesterase Catalysis – Protein Engineering Studies', in A. Shafferman and B. Velan, Eds., *Multidisciplinary Approaches to Cholinesterase Functions*, Plenum Press, New York, p. 165 (1992).
16. A. Ordentlich, D. Barak, C. Kronman, Y. Flashner, M. Leitner, Y. Segall, N. Ariel, S. Cohen, B. Velan, and A. Shafferman: *J. Biol. Chem.* **268**, 17083 (1993).
17. D. C. Vellom, Z. Radic, Y. Li, N. A. Pickering, S. Camp, and P. Taylor: *Biochemistry* **32**, 12 (1993).
18. Z. Radic, N. A. Pickering, D. C. Vellom, S. Camp, and P. Taylor: *Biochemistry* **32**, 12074 (1993).
19. H.-J. Nolte, T. L. Rosenberry, and E. Neumann: *Biochemistry* **19**, 3705 (1980).
20. D. A. Dougherty and D. A. Stauffer: *Science* **250**, 1558 (1990).
21. I. B. Wilson and C. Quan: *Arch. Biochem. Biophys.* **73**, 131 (1958).
22. F. Hobbiger: *Brit. J. Pharmacol.* **7**, 223 (1952).
23. P. Taylor: 'Anticholinesterase Agents', in A. G. Gilman, A. S. Nies, T. W. Rall, and P. Taylor, Eds., *The Pharmacological Basis of Therapeutics. 5th edition*, MacMillan, New York, 131 (1990).
24. R. Wolfenden: *Ann. Rev. Biophys. Bioeng.* **5**, 271 (1976).
25. A. Dafforn, M. Anderson, D. Ash, J. Campagna, E. Daniel, R. Horwood, P. Kerr, G. Rych, and F. Zappitelli: *Biochim. Biophys. Acta* **484**, 375 (1977).
26. M. H. Gelb, J. P. Svaren, and R. H. Abeles: *Biochemistry* **24**, 1813 (1985).
27. U. Brodbeck, K. Schweikert, R. Gentinetta, and M. Rottenberg: *Biochim. Biophys. Acta* **567**, 357 (1979).
28. H. K. Nair, K. Lee, and D. M. Quinn: *J. Am. Chem. Soc.* **115**, 9939 (1993).

Sequence Markers of Segmented Protein Structure

EUGENE KOLKER and EDWARD N. TRIFONOV
Department of Structural Biology, The Weizmann Institute of Science, Rehovot 76100, Israel

Abstract. We have recently shown that ≈20% of proteins are apparently made of standard size units of ≈123 amino acids for eukaryotes and ≈152 amino acids for prokaryotes. The observed phenomenon may reflect some ancient evolution events, e.g., the shuffling of DNA segments of a certain size. If that were the case, the original protein segments could have had not only standard sizes but some sequence structure as well, indicating a preference of some amino acids for certain positions within the segments. In particular, and first of all, the former initiation triplets (methionines) would be expected to preferentially appear at the borders between the fused segments. In this study we confirmed the underlying protein segment organization by Fourier analysis of the distributions calculated from up-to-date sequence collection, and analyzed positional distributions of various amino acids along the segments. The expected methionine recurrences are indeed found. Weak positional preferences of some other amino acids are detected and the preliminary scheme of the prototype segment structure is outlined.

Key words. Proteins, sequence length, segmented structure, amino acid (aa) distribution, methionine, recombination

Introduction

Sixty-five years ago Svedberg [1] suggested that proteins in general might perhaps be made of standard-sized units. For many years this was basically only an idea, neither proved nor disproved by several studies on protein lengths [2–5]. Recent extensive analysis of protein sequence length distributions [6] revealed that the expected regularity in the protein organization does indeed exist, although in a hidden form. We have shown that an appreciable fraction of proteins make a ladder of standard sizes 123 and 152 aa in eukaryotes and prokaryotes, respectively, and this phenomenon is of a universal nature rather than a result of any domination by one or another species or protein group. Among other possible reasons that may cause the observed regularity, molecular recombination was suggested [6,7]. One could imagine that earlier in genome evolution the standard size genes – DNA circles – were shuffled by their insertions and excisions. Such recombination events would more likely happen at the borders between fused genes to keep the older parts of the new protein intact. In this case, the initiation triplets coding for methionine (Met) residues would be located right at the borders between these standard length sequences [7]. (We call these unit size sequences 'sequence segments' or segments'.) This crucial prediction was found to be the case: we did observe the periodic methionine recurrences at the segment borders in eukaryotic sequences [8].

The goal of this paper is twofold: to verify our previous estimations of the standard segment sizes by using the up-to-date collection of protein sequences, and to analyze positional occurrences of various aa along the sequence segments. The protein sequences were taken from the last release of the Swiss-Prot data bank [9], and were subjected to an extensive data cleaning, in order to get

A. Pullman et al. (eds.), *Modelling of Biomolecular Structures and Mechanisms*, 461–471.
© 1995 *Kluwer Academic Publishers. Printed in the Netherlands.*

unbiased sequence sets. The study of both protein length distributions and aa positional distributions included their Fourier analyses.

The standard sizes of the segments and periodicity of the Met distribution derived earlier are confirmed, and some features of possible prototype primary structure of the protein segments are outlined.

Sequences and Methods

Eukaryotic protein sequences were obtained from the Swiss-Prot data bank [9]. To construct representative sequence collection (more precisely, collection of independent polypeptide chains) for analysis of protein lengths, we applied the following cleaning procedure [6]. First, all irrelevant entries were removed, such as fragments, unidentified and open reading frame sequences, duplicated and almost duplicated entries. Mitochondrial and chloroplast genome sequences were also discarded. This reduced the data to 13133 eukaryotic sequences with lengths ⩽600 aa (Swiss-Prot data bank release 27), which corresponds to ≈50% of its original size.

For analysis of primary structure of the sequence segments an additional cleaning is necessary to avoid any possible domination of some sequence types. In contrast to the above cleaning procedure, in this case the sequences matter more than their lengths. We used [8]: original dipeptide composition technique [10], and n-peptide (here $n = 7$) sequence comparison [11]. All sequences with ⩾50% of similarity were discarded. Then, only sequences of two-, three-, and four-segment sizes (245 ± 15, 370 ± 15, and 490 ± 15 aa, correspondingly) were selected. This resulted in a set of 509 nonredundant sequences of the above lengths (Swiss-Prot data bank release 24).

The smoothed distributions for protein lengths and for aa occurrences along the sequences were obtained by calculating running averages with a 40-aa window. Standard deviations (SD) were calculated as square roots of the corresponding sample means.

To evaluate the most typical periods both in protein lengths and in oscillations of aa occurrences, we applied Fourier transforms, based on the following equations:

$$A(j) = \frac{\sum_{i=1}^{n} H(i) \cos(2\pi i/j)}{\sum_{i=1}^{n} \cos^2(2\pi i/j)}; \qquad B(j) = \frac{\sum_{i=1}^{n} H(i) \sin(2\pi i/j)}{\sum_{i=1}^{n} \sin^2(2\pi i/j)};$$

$$C(j) = \sqrt{A^2(j) + B^2(j)}$$

where $H(i)$ is oscillating part of the corresponding distribution; n = total number of points, in aa residues; i = variable size (distance) in aa residues; j = a given period value; $A(j)$, $B(j)$, and $C(j)$ = amplitudes of cosine and sine components and of their sum for a given period, respectively. Oscillating components of the corresponding distributions were calculated by subtracting from the actual distributions their smoothed versions, with the window equal to a given period. Cos-Fourier transforms were calculated similarly, by taking component $A(j)$ only.

TABLE I

Sequence segment sizes for eucaryotic species

Species	No. of sequences	Calculated average unit length, aa
Human (*Homo sapiens*)	1819	121 ± 6
Mouse (*Mus musculus*)	1163	121 ± 8
Rat (*Rattus norvegicus*)	1029	125 ± 8
Yeast (*S. cerevisiae*)	848	141 ± 9
Bovine (*Bos tarus*)	469	125 ± 12*
Chicken (*Gallus gallus*)	347	128 ± 14
Fruit flu (*D. melanogaster*)	344	123 ± 14
Pig (*Sus scrofa*)	322	121 ± 14
African clawed frog (*X. laevis*)	234	112 ± 17
Rabbit (*Oryctolagus cuniculus*)	231	125 ± 17
Maize (*Zea mays*)	143	122 ± 21
Rice (*Oryza sativa*)	126	124 ± 23
>100 other species	6404**	124 ± 4
Total	13133**	123 ± 3

* Highest maximum in the 100- to 150-aa range.
** Like-named proteins of different species are represented by only one copy.

Results and Discussion

UNDERLYING POLYPEPTIDE CHAIN LENGTH PERIODICITY

Eukaryotic sequence length distribution derived from the new release of the Swiss-Prot protein data bank has four peaks (not shown) at the lengths approximately equal to standard unit size and multiples of it, as in previous work [6]. Using the cos-Fourier transform we confirmed the most preferred eukaryotic protein size, that is 123 ± 3 aa (6.5 SD for the 40-aa window). As it was also found earlier, the observed protein-size periodicity is universal, i.e. it is not caused by domination of one or few organisms or protein families. Proteins of most (if not all) eukaryotic species display such regularity (see Table I). Compared to previous data [6], the calculated average unit lengths for all species became somewhat closer to the common standard size. About 20% of proteins of the purified sequence collection are of the standard size or multiples thereof.

The unit size in prokaryotic proteins is 152 ± 4 aa [6]. Interestingly, the average segment length in yeasts (141 ± 9 aa, Table I) appears to be different from the proteins of other eukaryotes, being rather halfway between eukaryotic and prokaryotic standard sizes. It may have some evolutionary implications.

Worth noting, that in such conservative proteins as enzymes the proportion of the standard length sequences is about twice higher. Figures 1 and 2 present the length distribution for the eukaryotic enzymes and corresponding cos-Fourier spectrum, respectively. Total number of eukaryotic enzymes taken for the analysis is 3840, that is ≈30% of an entire set. The cos-Fourier transform of the periodic pattern in the Figure 1 gives the value of the period: 125 ± 5 aa (8.3 SD). In the spectral data along with major maximum at the standard size, few smaller maxima are observed. The second largest is the 37-aa peak. This is close to the typical

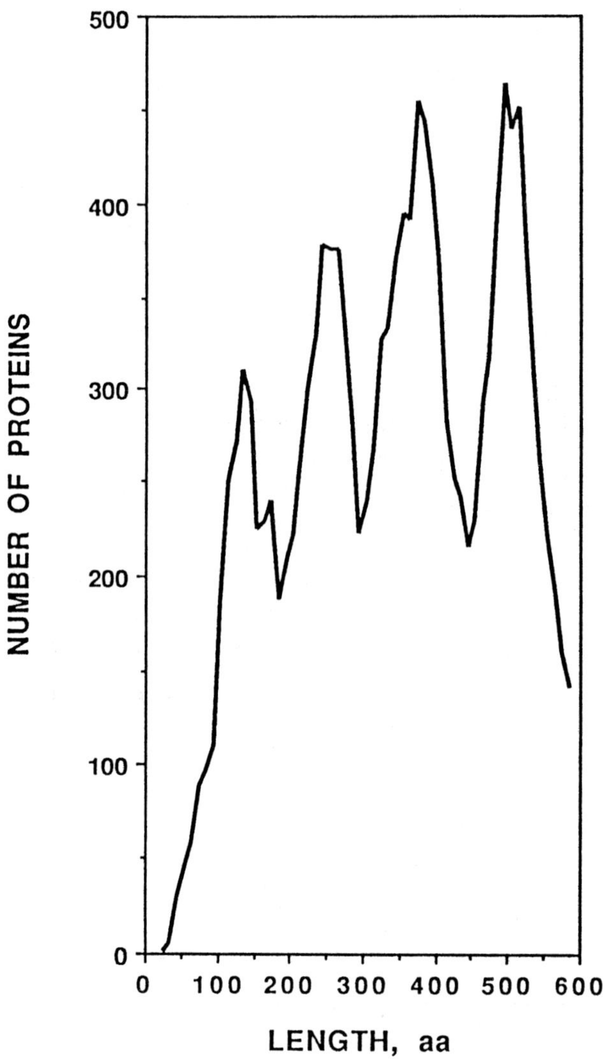

Fig. 1.   Length distribution of polypeptide chains of eukaryotic enzymes. The smooth curve corresponds to the running averages with a 40-aa window.

exon size, which is ≈40 aa [12,13]. The amplitude of the peak, however, is insufficient for conclusions to be drawn on the exon-size regularity.

Several possible reasons for the observed underlying organization in protein lengths have been discussed earlier [6–8]. The size regularity in general could be due to typical functional or optimal folding domain sizes, due to exons or nucleosome sizes, or could be result of several pressures both of divergent and convergent nature. The stronger periodicity of the protein sizes for the conservative sequences indicates that the regular protein organization is of an ancient origin.

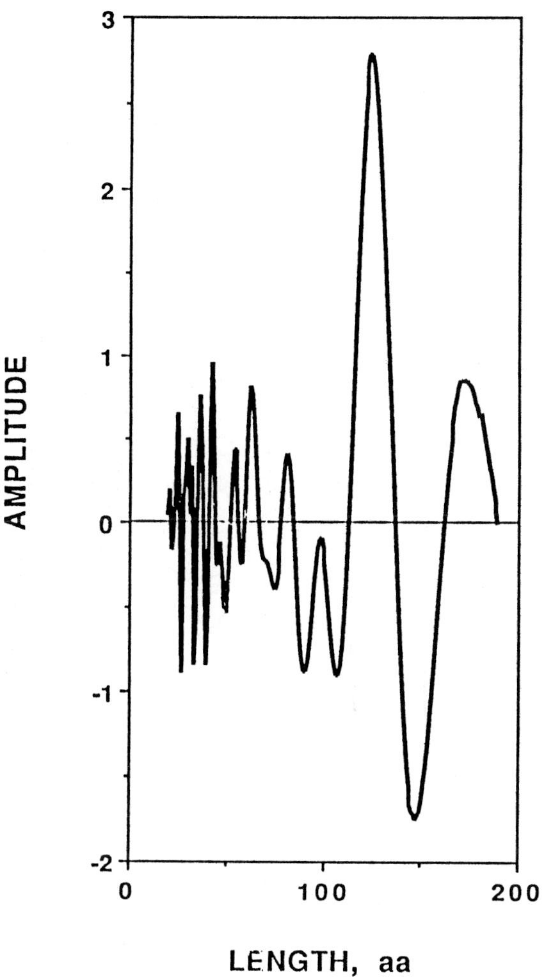

Fig. 2.   Cos-Fourier spectrum of the length distribution of eucaryotic enzymes.

RECOMBINATION MECHANISM OF PROTEIN SEGMENTATION

It was proposed [6,7] that at some early stages in protein evolution a frequent gene fusion had been taking place. The standard size (circular) DNA segments with encoded 100–160 aa proteins were fused and shuffled by recombination events, that is insertions and deletions of these segments. Their lengths (300–500 bp) correspond to the optimal size region for DNA circularization [14]. To be planar, i.e. without any torsional constraints, and to keep reading frames intact, these standard size circles have to have their exact contour lengths divisible both by helical repeat of DNA and by 3 bp [6,7,15]. These two conditions are satisfied for only limited number of possible DNA circles with the contour lengths and DNA helical repeats within the observed regions (300–500 bp; 10.50–10.60 bp per

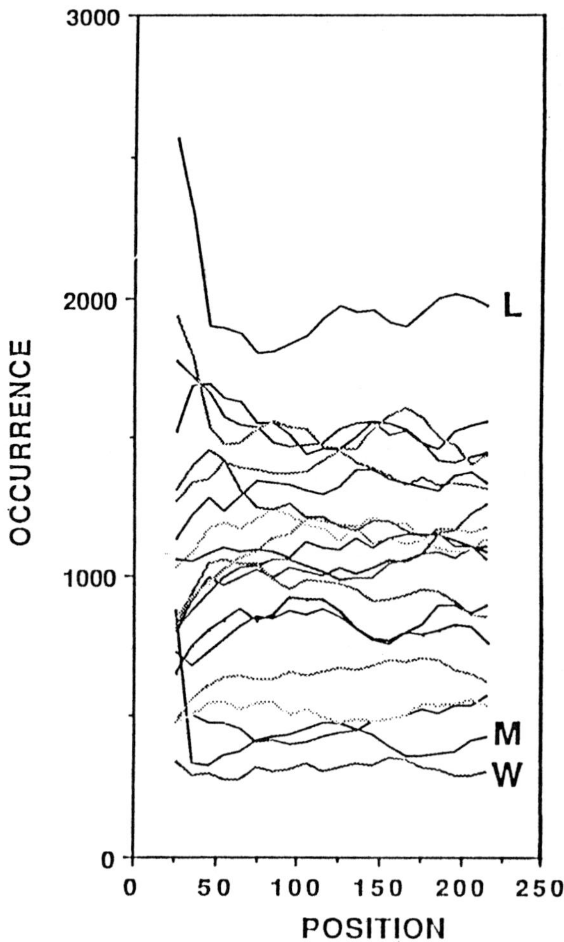

Fig. 3.   Distriutions of all aa occurrences along eukaryotic sequences. The curves correspond to complete sequence set, with chain lengths close to 245, 370 and 490 aa. The two-, three-, and four-segment sets contain 171, 188 and 150 sequences, respectively. The smooth curves correspond to running averages with a 40-aa window.

turn). (For more details about interrelationships between the standard protein sizes, the elementary protein-coding circles and the DNA helical repeats see [7,15].)

The recombination events could have preferentially happened at the very ends of the fused coding segments, to incorporate the precursor genes into new protein product while keeping all the fused genes intact. Consequently, one can expect more frequent appearance of the initiation triplets (and Met, respectively) at the borders between the internal sequence segments [7]. The positional preference of Met could be still detectable in modern proteins, despite numerous subsequent mutational changes. In contrast to the unique initiation triplet, the termination

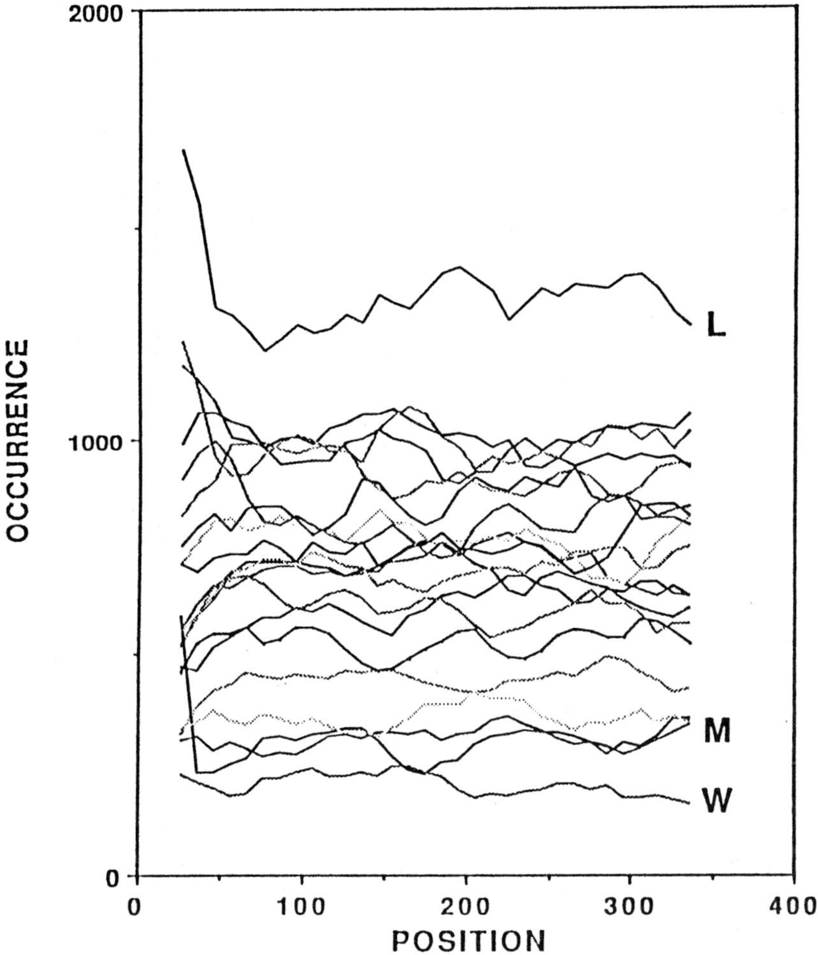

Fig. 4.   Distributions of all aa occurrences along eukaryotic sequences. The curves correspond to
three- and four-segment long sequences. The smoothed distributions are shown.

triplets had to mutate into one of at least 18 single change derivatives to be
acquired into new recombination product. The positional preference of these 18
triplets to the border regions would be very weak, perhaps not detectable in such
a small sequence ensemble. To test the hypothesis concerning the positional
preferences of Met and perhaps of some other aa, we analyzed distributions of all
aa along the sequences.

POSITIONAL DISTRIBUTIONS OF AMINO ACIDS AND SPECTRAL CHARACTERISTICS

Data on occurrences of all aa at distances up to the lengths of two standard sizes
as well as three segment sequences are shown in Figures 3 and 4, respectively.
All aa from the most frequent Leu down to the very rare Trp show rather small

TABLE II

Spectral characteristics of positional distributions of amino acid within the sequence segment*

Amino Acid	Apparent Period (residues)	Average occurrence**	Amplitude	SD***
Arg	100 ± 10	19.980	1.787	2.4
Phe	120 ± 12	14.708	1.517	2.3
Cys	108 ± 10	7.352	1.047	2.3
Ser	101 ± 10	24.574	1.884	2.3
Thr	146 ± 12	18.315	1.718	2.2
Ala	150 ± 10	24.853	1.989	2.2
His	135 ± 13	8.793	1.017	1.9
Gly	147 ± 11	25.687	1.744	1.9
Val	124 ± 13	23.524	1.571	1.9

* Preferred positions of aa within the segment are not indicated (see Figure 6).

** This value is an average occurrence within corresponding region for Fourier analysis.

*** SD are calculated from the total occurrences within the 40-aa windows.

positional variations around their means, except Met and only Met. This is one of the rarest aa, but the amplitude of its oscillation is one of the highest. Met indeed shows maxima at the standard sizes, as predicted (see also [8]).

To evaluate the amplitudes and periods of possible oscillating components in the aa occurrences, the Fourier transforms were applied to the corresponding distributions. The cos-Fourier analysis also showed that the only amino acid that displays significant periodicity with expected period is Met. The period observed (119 ± 6 aa, 3.8 SD) is indistinguishable from the expected 123-aa size. Other aa do not show any statistically significant preference to the segment border region. There could be however some preference of certain amino acids to internal positions within the 120-aa long sequence segments. To detect these a complete Fourier analysis was performed. For each aa the largest amplitude Fourier component was determined in the 100- to 150-aa range of periods. Table II presents the spectral characteristics for the aa with the oscillating components of highest SD, based on the three-segment size sequence set (Figure 4). Only for this set the positional data are available for at least two full periods, which is necessary for such calculations. None of the observed oscillations is statistically solid, largely because of small sample size and perhaps of actual absence of the signal.

POSSIBLE PRIMARY STRUCTURE OF THE SEGMENTS

The special role and position of the Met residues is demonstrated by Figure 5. Along with location of Met at the start position in the Met occurrences for the sequences of at least two standard lengths only a single maximum is present at ≈120 aa (Figure 5A). For three and more segment size sequences the same peak at ≈120 aa is observed, as well as the second peak at ≈240 aa, which corresponds to a second-to-third segment border (Figure 5B). In the distribution of four-segment sequences the third weak maximum apparently corresponds to the border between third and fourth segments (Figure 5C). Thus, both the positional distributions and the cos-Fourier calculations demonstrate the Met recurrences with the standard 120-aa period.

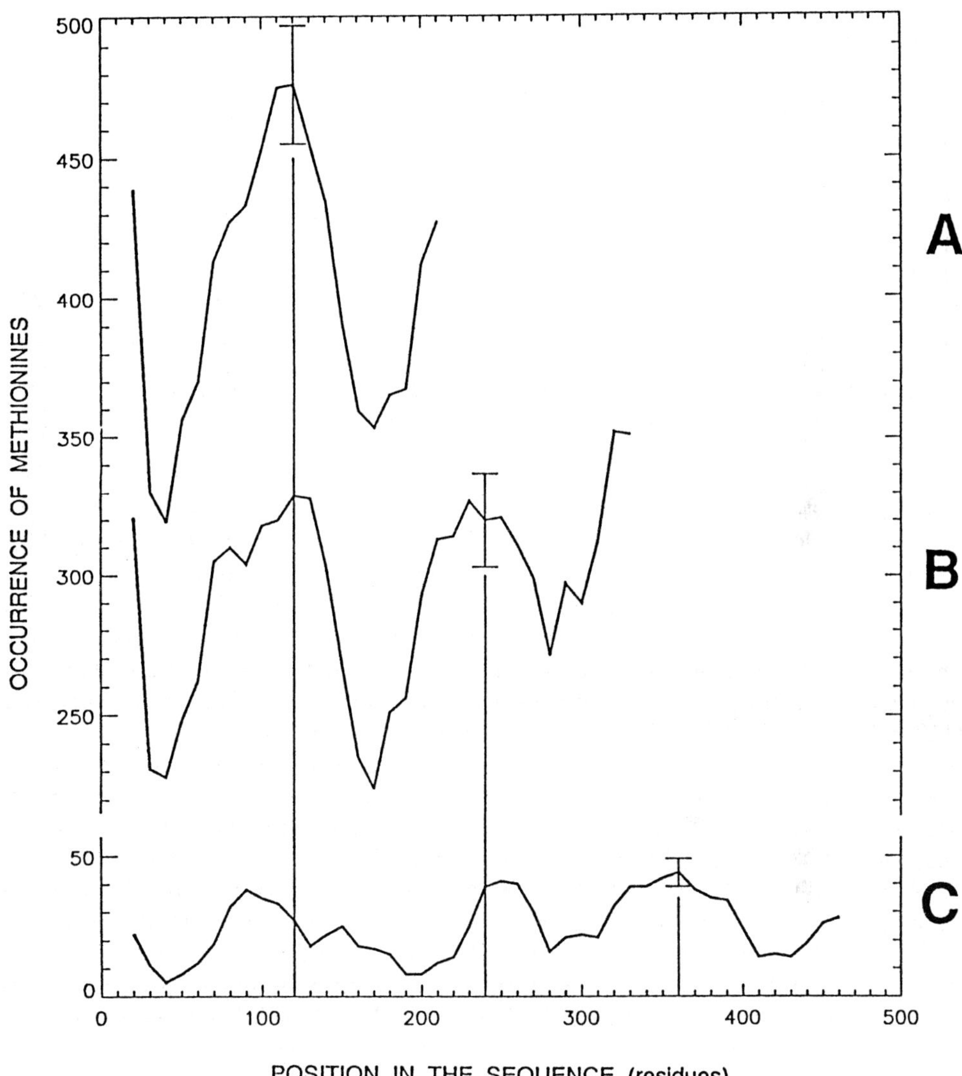

Fig. 5. Distribution of methionines along eukaryotic polypeptide chains. The curves correspond to three sequence sets: complete set, with chain lengths close to 245, 370 and 490 aa (two, three and four segments) (A), three- and four-segment long sequences (B), and four-segment long sequences (C). The smoothed distributions are shown. Error bars were calculated as square roots of corresponding mean values of the methionine occurrences.

Taking the four aa (Phe, Cys, His, and Val) with estimated oscillating components closest to the standard period (see Table II) as potential 'candidate' features of the prototype pnmary structure of the protein segments, one could propose the following tentative scheme of the sequence segment structure (Figure 6). Only

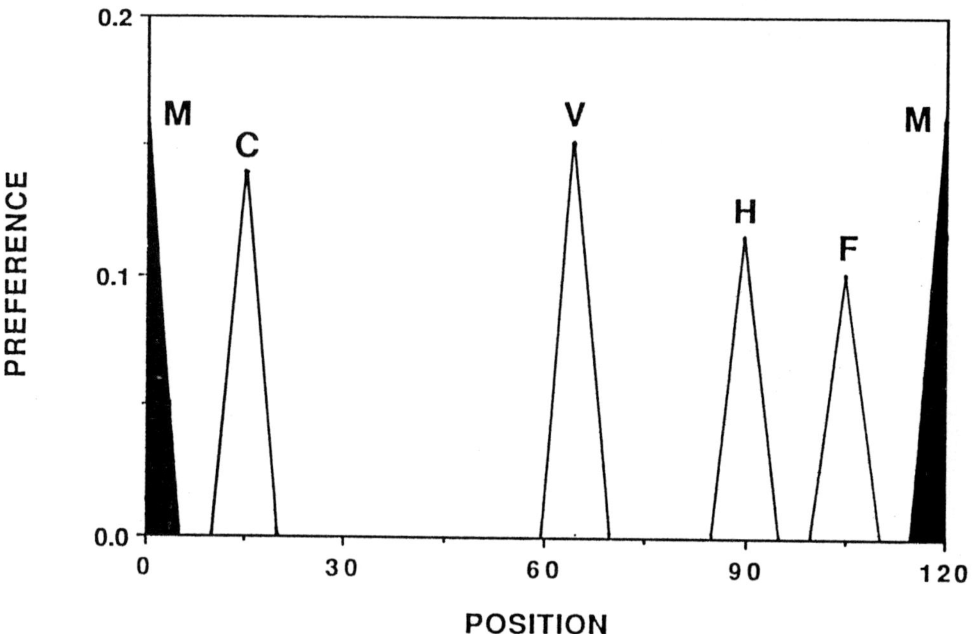

Fig. 6. Scheme of a prototype primary structure of the protein sequence segment. Preferential positions of most likely component residues are indicated together with the relative magnitudes of their preferences.

further analysis on larger sequence ensemble would add more details to this scheme and verify the tentative positional preferences.

In conclusion, the ladder of the standard sizes in protein lengths and periodic recurrence of Met demonstrate the segmented structure of protein sequences. Stronger length preferences in the conservative proteins indicate that the sequence segments are of the recombinational origin. The positional preferences of some aa may serve as a tentative outline of possible prototype protein segment structure.

Acknowledgments

We are grateful to summer students O. Klinkenberg and T. Andersson for support. This work was partly funded by The Genome Committee of The Weizmann Institute of Science.

References

1. T. Svedberg: *Nature* **123**, 871 (1929).
2. T. Svedberg: *Nature* **139**, 1051 (1937).
3. E. D. Kieh and J. T. Holland: *Nature* **226**, 544 (1970).
4. S. S. Sommer and J. E. Cohen: *J. Mol. Evol.* **15**, 37 (1980).
5. M. A. Savageau: *Proc. Natl. Acad. Sci. USA* **83**, 1198 (1986).
6. A. L. Berman, E. Kolker, and E. N. Trifonov: *Proc. Natl. Acad. Sci. USA* **91**, 4044 (1994).

7. E. N. Trifonov: *J. Mol. Evol.* **38**, 543 (1994).

8. E. Kolker and E. N. Trifonov: *Proc. Natl. Acad. Sci. USA* (1995), in press.

9. A. Bairoch and B. Boeckmann: *Nucleic Acids Res.* **21**, 3093 (1993).

10. S. Pietrokovski, J. Hirshon, and E. N. Trifonov: *J. Biomol. Struct. Dyn.* **7**, 1251 (1990).

11. V. Brendel: *Math. Comput. Modelling* **16**, 37 (1992).

12. J. D. Hawkins: *Nucleic Acids Res.* **16**, 9893 (1988).

13. R. L. Dorit, L. Schoenbach, and W. Gilbert: *Science* **250**, 1377 (1990).

14. D. Shore, J. Langowski, and R. L. Baldwin: *Proc. Natl. Acad. Sci. USA* **78**, 4833 (1981).

15. E. N. Trifonov: this volume.

Hidden Segmentation of Protein Sequences: Structural Connection with DNA

E. N. TRIFONOV
Department of Structural Biology, The Weizmann Institute of Science, Rehovot 76100, Israel

Abstract. The chain lengths of about 20% of all proteins appear as multiples of 123 aa residues for eukaryotes and 152 aa residues for prokaryotes. This is interpreted as reflection of a recombination process early in evolution when many of the genes, presumably, existed in form of separate DNA rings of close to standard sizes, about 369 and 456 bp for early eukaryotes and prokaryotes, respectively. This theory finds confirmation in the analysis of theoretically possible torsionally non-constrained (planar) rings, within the experimentally observed ranges of the overall ring sizes (330–500 bp) and DNA helical repeats (10.50–10.60 bp/turn). The condition of planarity is satisfied only if the contour lengths of the rings correspond to integral numbers of the DNA helical periods while continuity of the reading frames after recombination of the rings imposes another condition: that of divisibility by 3. Of 24 theoretically possible rings within the indicated regions of the ring sizes and helical repeats of DNA two rings very accurately correspond to observed sizes of the protein sequence segments and to mean value of experimental estimates of the DNA helical periodicity. Other possible rings, contemporaries of the early eukaryotes and prokaryotes, may be still reflected in some of modern protein sequences, in their lengths or in the DNA helical repeat values for corresponding protein-coding sequences.

Key words. Protein segments, DNA rings, evolution.

Introduction

The path of the polypeptide chain in a folded protein can be viewed as a topologically linear array of local foldings, structural domains, perhaps of several inclusively hierarchical types. Such organization is believed to be one (if not the only one) way to ensure rapid and stable folding of long chains [1,2]. The structural domains may have certain typical sizes [2,3] as well as some amino-acid sequence preferences, such that the protein sequence can be also viewed as a combination of respective sequence segments joined together in one chain. Careful analysis of protein chain lengths revealed, indeed, that eukaryotic proteins have well detectable preference to the dominating size, close to mean value 123 ± 3 aa residues, and multiples of this size, while prokaryotic proteins are characterized by a unit size 152 ± 4 residues [4]. This underlying order in protein sequence organization can be interpreted as reflection of some early stage of protein evolution, when larger protein molecules were formed as a result of combinatorial fusion of small unit-size genes, DNA circles with contour lengths around 369 ± 9 and 456 ± 12 base pairs [5]. This range of DNA lengths is known to be optimal for DNA ring closure [6]. Both the recombination and closure of such small circles are most efficient when there is no torsional constraint in the circles [6,7]. This condition is fulfilled only when the contour length of the DNA ring corresponds to integral number of DNA helical repeats which also results in the planarity of the ring (no superhelicity). Together with condition of divisibility by 3, for conservation of the

A. Pullman et al. (eds.), Modelling of Biomolecular Structures and Mechanisms, 473–479.
© 1995 *Kluwer Academic Publishers. Printed in the Netherlands.*

TABLE I
Helical repeat of free DNA

Source	bp/turn
1. Strauss *et al.*, 1981 [9]	10.55 (.10)
2. Strauss *et al.*, 1981 [9]	10.60 (.10)
3. Strauss *et al.*, 1981 [9]	10.60 (.10)
4. Strauss *et al.*, 1981 [9]	10.50 (.10)
5. Peck and Wang, 1981 [10]	10.60 (.10)
6. Shore and Baldwin, 1983 [11]	10.45 (.02)
7. Horowitz and Wang, 1984 [12]	10.54 (.02)
8. Diekmann and Wang, 1985 [13]	10.58 (.02)
9. Gottesfeld, 1987 [14]	10.54 (.15)
10. Goulet *et al.*, 1987 [15]	10.54 (.02)
11. Goulet *et al.*, 1987 [15]	10.59 (.02)
12. Goulet *et al.*, 1987 [15]	10.54 (.02)
13. Goulet *et al.*, 1987 [15]	10.55 (.02)
14. Goulet *et al.*, 1987 [15]	10.58 (.02)
Weighted average	10.55 (.01)

Experimental estimates of the mixed-sequence natural DNA helical repeat are presented in chronological order. The error bars are indicated in parentheses.

reading frame after the insertion of the flat ring, this leads to a very sharp estimate of the helical repeat of the early DNA, 10.54 bp per turn [5]. This value is indistinguishable from the experimental mean for the mixed-sequence natural extant DNA, thus quantitatively supporting the recombinational theory. One important prediction of this theory is expected recurrence of methionine residues at the sites where the ancient genes are assumed to have been fused. Indeed, the methionines of eukaryotic proteins are found to appear more frequently at the sites located approximately 120, 240 and 360 residues from the start residue [8].

In this paper the known measurements of the DNA helical periodicity are reviewed, to compare with the calculated parameters of a full spectrum of possible flat rings, in broader range of DNA helical repeats, satisfying the above divisibility conditions. Of over 20 such rings two (of contour lengths 369 bp and 453 bp) are found to be an exact match to the observed sizes, 369 ± 9 and 456 ± 12 bp for eukaryotes and prokaryotes, respectively. The theory, thus, gives correct values for the observed ring sizes as well.

Helical Periodicity of Natural Mixed-Sequence DNA

Solution measurements of the DNA helical periodicity, known from literature, are presented in Table I. Data for 14 different DNA fragments, excluding those of strongly biased compositions, are weight averaged, giving highest weight to lowest dispersion estimates. Two related techniques had been used for these measurements, based on the sensitivity of the gel electrophoresis of circular DNA to its writhe modulated by inserts of various lengths and sequences. The measurements cover the range 10.50–10.60 bp/turn and their weighted average is 10.55 ± 0.01 bp/turn. This is to compare with the nearest theoretical estimate for

TABLE II
Helical repeat of DNA in the nucleosome

Source	Technique	bp/turn
1. Prunell *et al.*, 1979 [21]	DNaseI	10.35 (.05)
2. Lutter, 1981 [22]	DNaseII	10.40 (.05)
3. Prunell, 1983 [23]	ExoIII	10.45 (.06)*
4. Cockell *et al.*, 1983 [24]	MNase	10.44 (.05)*
5. Satchwell *et al.*, 1986 [25]	Sequence periodicity	10.28 (.10)**
6. Gale *et al.*, 1987 [26]	Pyrimidine dimers	10.30 (.10)
7. Franchet-Benzit *et al.*, 1993 [27]	Beta-rays	10.39 (.16)
8. Ioshikhes *et al.*, 1994 [28]	Sequence periodicity	10.35 (.20)
	Weighted average	10.39 (.02)

Independent experimental estimates are sorted in chronological order. The error bars are indicated in parentheses.
* Calculated from Table 1 of the source.
** Calculated from Table 2 of the source.

the planar rings of DNA, 10.54 bp/turn [5]. The mean value of natural DNA helical periodicity, thus, exactly corresponds to the calculated pitch of the early DNA. It is still conserved perhaps due to long period of selection in favor of the torsionally non-constrained rings at early stages of genome evolution. At later times after formation of comparatively large genomes the above selection pressure presumably decreased, probably, close to none. This should have resulted in broadening of the original presumably narrow distribution of the values of the helical repeat, and some shift toward higher values, closer to the helical repeat for a random sequence, 10.61 bp/turn [16]. This possible trend, however, is not detectable within the experimental uncertainty of the measured periodicity.

Helical Periodicity of Free Nucleosomal DNA

An independent series of measurements has been made for evaluation of the helical periodicity of DNA wrapped around histone octamers in the nucleosome core particles. This winding does not involve any additional torsion, apart from the one caused by the superhelical writhe of the winding [17]. In this case the helical repeat h_c of the DNA wound on imaginary cylindrical surface of the histone octamer should be smaller by 0.15 ± 0.015 bp/turn [18,19] as calculated by using Cricks formula [20] for superhelical DNA trajectories. With this correction the helical periodicity of the free nucleosomal DNA can be evaluated from the measurements of the helical repeat of DNA in the nucleosome.

These measurements are summarized in Table II. The corresponding weighted mean value is 10.39 ± 0.02 bp/turn, practically unchanged since a weight averaging derived 11 years ago [29]. Introducing the above correction one derives the value for the helical repeat of the free nucleosomal DNA: 10.54 ± 0.03 bp/turn, indistinguishable from the experimental mean value given in the previous section.

Yet another estimate can be derived from the available nucleosomal DNA sequences [30] by summing up corresponding twist angles for all dinucleotide

TABLE III
Parameters of torsionally unconstrained (planar) DNA rings

	10.50 series			10.54 series			10.58 series		
	bp	aa	bp/turn	bp	aa	bp/turn	bp	aa	bp/turn
1.	483	161	10.500	495	165	10.532	486	162	10.565
2.	462	154	10.500	474	158	10.533	465	155	10.568
3.	441	147	10.500	**453**	**151**	**10.535**	444	148	10.571
4.	420	140	10.500	432	144	10.537	423	141	10.575
5.	399	133	10.500	411	137	10.538	402	134	10.579
6.	378	126	10.500	390	130	10.541	381	127	10.583
7.	357	119	10.500	**369**	**123**	**10.543**	360	120	10.588
8.	336	112	10.500	348	116	10.545	339	113	10.594

The contour lengths indicated are divisible by 3 and by DNA helical repeat indicated. Each ring corresponds to integral number of the helical repeats (from 32 to 47, not indicated). The rings corresponding to actually observed protein sequence segments in eukaryotes and prokaryotes are highlighted in bold face.

steps [16]. This calculation for 143 nucleosomal sequences gives the mean value 10.56 ± 0.01 bp/turn for the helical repeat of free nucleosomal DNA (E. Shpigelman, personal communication), the same as two other estimates above. Combining all three estimates (10.55 ± 0.01, 10.54 ± 0.03 and 10.56 ± 0.01 bp/turn) results in the average 10.55 ± 0.01 bp/turn which thus represents the helical repeat for (predominantly eukaryotic) mixed-sequence extant DNA. Its good correspondence with the basic DNA helical repeat of the flat ring [5] suggests that the mean value of the DNA helical periodicity indeed is still maintained unchanged from the earliest days of the genome evolution.

Other Possible Rings

The helical periodicity of any given DNA fragment is not necessarily very close to the above constant. Actually observed DNA pitch values cover the range 10.50 to 10.60 bp/turn. One cannot rule out that at the hypothetical DNA ring stage of genome evolution the rings of DNA with other values of the helical repeat existed, beyond the sharp range 10.54–10.56 bp/turn for the mean value of the observed helical repeats. In Table III the contour lengths and helical repeats for all theoretically possible flat DNA rings are listed (flat ring constants) covering the broad range of repeats within 10.50 to 10.60 bp/turn. The contour lengths of the rings are taken in the range 330 to 500 bp, which corresponds to the optimal DNA ring closure sizes [6] and covers the protein-coding DNA segment sizes as reflected in the length distributions of eukaryotic and prokaryotic proteins [4], 110–165 aa residues (330–495 bp). The total of 24 rings in the indicated ranges of contour lengths and helical repeats clearly subdivide in three groups, with the helical repeats 10.50 and around 10.54 and 10.58 bp/turn. It is better seen in Figure 1 where the dots correspond to the flat rings of Table III. Two particular rings (369 bp, 10.543 bp/turn and 453 bp, 10.535 bp/turn, larger dots) closely correspond to the observed protein segment sizes (4), 123 ± 3 aa and 152 ± 4 aa for eukaryotes

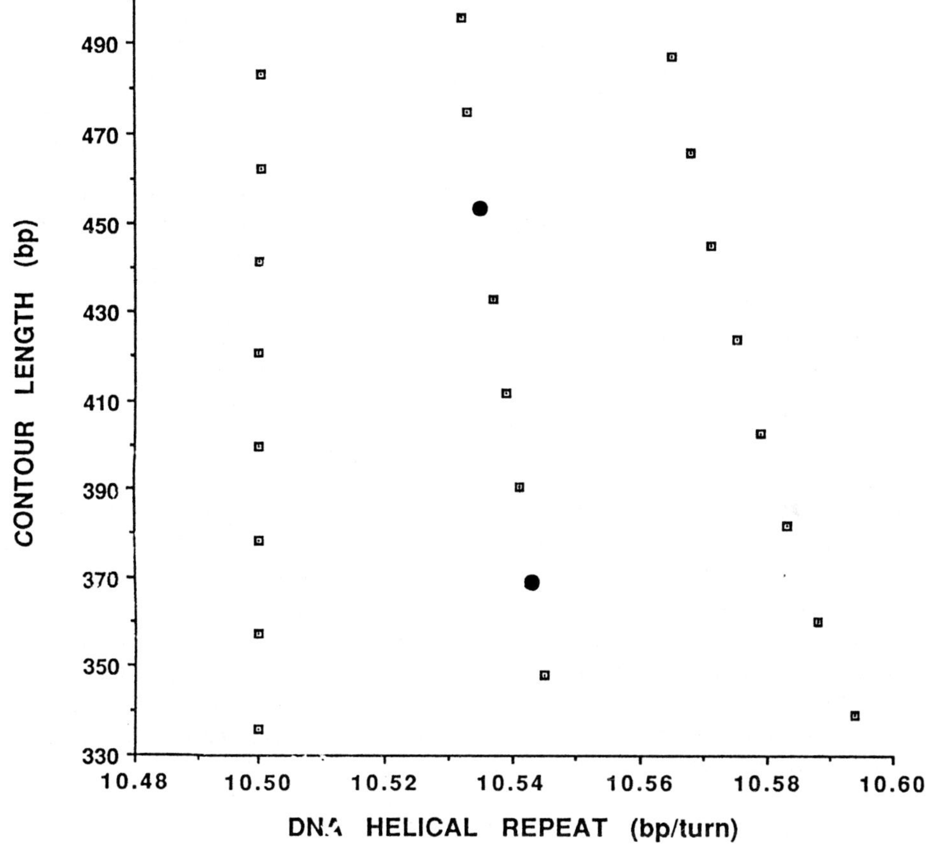

Fig. 1. Structural parameters of theoretically possible torsionally non-constrained (planar) DNA rings, within the ranges of helical repeats and contour lengths indicated. The rings are represented by dots. Two rings, for prokaryotes (top) and eukaryotes, are marked by larger dots.

and prokaryotes, respectively, and to the experimental mean value of DNA helical repeat, 10.55 ± 0.01 bp/turn. Parameters of none of 22 other rings are as close to and as consistent with these experimental values.

An immediate speculation is suggested by the calculations presented in Table III and in Figure 1: apart from the large separate families of eukaryotic and prokaryotic genes there could be in the distant past some other gene groups (DNA ring types) perhaps extinct today, with different parameters of the DNA rings, as in Table III, or even further beyond the likely ranges of the contour lengths and the helical repeats. Since the hypothetical gene groups fall in three series according to their helical repeat values (Table III) one may expect that modern protein-coding sequences would still have their respective helical repeat values clustered around the original theoretical means for the three series.

Selection Pressures for the Ring Sizes

The closeness to the optimal ring closure DNA size easily explains the range of
the observed values for the early eukaryotic and prokaryotic DNA rings. The ring
closure optimum is rather broad [6], and both sizes are well within the optimal
range. The fact that the sizes are different, 369 bp and 451 bp rings for eukaryotes
and prokaryotes respectively, is rather puzzling. In particular, the question arises
of why the corresponding protein segments of the prokaryotic and eukaryotic sizes
are not mixed (by DNA ring fusion), in which case, of course, some average value
would be observed, common for eukaryotes and prokaryotes. One possibility is
that some specific recombination signals in the sequences could have existed which
were significantly different to preclude interkingdom recombination events. This,
however, would not maintain the contour lengths unchanged. There should have
been a selection pressure towards conservation of this very length. Thus, one has
to look for purely structural reasons for the conservation. The identity of contour
lengths could have been a precondition for the recombination of two elementary
circles. For example, the circles may form a side-by-side stack secured, perhaps,
by some proteins. The stability of such complex would be expected to depend on
the match of the partners lengthwise. This possible pressure of ring-to-ring stacking
would be active only at the very initial stages of the gene fusion. For the longer
fusion products to be also a part of the length conservation scheme somewhat
more complicated recombination structures have to be considered. For example,
the complexes could form between the elementary rings and the inserts of the
same unit size, looped outside – side-by-side loop-ring stacks. Such a structure
would still exert the length selection pressure, since only in this case could all the
ends of the elementary segments involved be put in register, in the exact juxtaposition necessary for proper recombination. Similarly, side-by-side loop-loop stacks
could have been formed between larger molecules involved in the recombination
exchange. In this case the pressure towards the length conservation would be
exerted as well.

This early recombination scheme and the unit sizes could still be reflected not
only in the protein lengths [4] and in periodical distribution of methionines [8],
but perhaps in contour lengths of modern DNA mobile elements as well, if the
latter were considered to be descendants of the early DNA rings. Indeed, this
expectation is at least partially confirmed in the apparent 350–400 bp periodicity
of sizes of the extrachromosomal closed circular DNA [31], as compared with
369 bp contour lengths of eukaryotic flat rings, as above. The typical lengths of
prokaryotic insertion sequences, on the other hand, are close to 1300 bases [32]
which corresponds to three elementary prokaryotic units. A thorough analysis of
the size distributions of various classes of the DNA mobile elements (work in
progress) will show whether this correspondence is of a general nature.

References

1. D. B. Wetlaufer: *Proc. Natl. Acad. Sci. USA* **70**, 697 (1973).
2. K. A. Dill: *Biochemistry* **24**, 1501 (1985).
3. D. S. Goodsell and A. J. Olson: *Trends Biochem. Sci.* **18**, 65 (1993).
4. A. L. Berman, E. Kolker, and E. N. Trifonov: *Proc. Natl. Acad. Sci. USA* **91**, 4044 (1994).

5. E. N. Trifonov: *J. Molec. Evol.* **38**, 543 (1994).
6. D. Shore and R. L. Baldwin: *J. Mol. Biol.* **170**, 957 (1983).
7. R. Hoess, A. Wierzbicki, and K. Abremski: *Gene* **40**, 325 (1985).
8. E. Kolker and E. N. Trifonov: *Proc. Natl. Acad. Sci. USA*, in press.
9. F. Strauss, C. Gaillard, and A. Prunell: *Eur. J. Bioch.* **118**, 215 (1981).
10. L. J. Peck and J. C. Wang: *Nature* **292**, 375 (1981).
11. D. Shore and R. L. Baldwin: *J. Mol. Biol.* **170**, 983 (1983).
12. D. S. Horowitz and J. C. Wang: *J. Mol. Biol.* **173**, 75 (1984).
13. S. Diekmann and J. C. Wang: *J. Mol. Biol.* **186**, 1 (1985).
14. J. M. Gottesfeld: *Mol. Cell Biol.* **7**, 1612 (1987).
15. I. Goulet, Y. Zivanovic, and A. Prunell: *Nuci. Acids Res.* **15**, 2803 (1987).
16. W. Kabsch, C. Sander, and E. N. Trifonov: *Nucl. Acids Res.* **10**, 1097 (1982).
17. Y. Zivanovic, I. Goulet, B. Revet, M. Le Bret, and A. Prunell: *J. Mol. Biol.* **200**, 267 (1988).
18. L. E. Ulanovsky and E. N. Trifonov: *Cell Biophysics* **5**, 281 (1983).
19. M. Le Bret: *J. Mol. Biol.* **200**, 285 (1988).
20. F. H. C. Crick: *Proc. Natl. Acad. Sci. USA* **73**, 2639 (1976).
21. A. Prunell, R. Kornberg, L. Lutter, A. Klug, M. Levitt, and F. Crick: *Science* **204**, 855 (1979).
22. L. Lutter: *Nucl. Acids Res.* **9**, 4251 (1981).
23. A. Prunell: *Biochemistry* **22**, 4887 (1983).
24. M. Cockell, D. Rhodes, and A. Klug: *J. Mol. Biol.* **170**, 423 (1983).
25. S. C. Satchwell, H. R. Drew, and A. A. Travers: *J. Mol. Biol.* **191**, 659 (1986).
26. J. M. Gale, K. A. Nissen, and M. J. Smerdon: *Proc. Natl. Acad. Sci. USA* **84**, 6644 (1987).
27. J. Franchet-Benzit, M. Spotheim-Maurizot, R. Sabattier, B. Blazy-Baudras, and M. Charlier: *Biochemistry* **32**, 2104 (1993).
28. I. Ioshikhes, A. Bolshoy, M. Borodovsky, and E. N. Trifonov: in preparation.
29. E. N. Trifonov: 'Nucleosomal DNA Structure', in B. Pullman and J. Jortner (Eds.), *Nucleic Acids: The Vectors of Life*, D. Reidel Publishing Company, Dordrecht–Boston–Lancaster, p. 373 (1983).
30. I. Ioshikhes and E. N. Trifonov: *Nucl. Acids Res.* **21**, 4857 (1993).
31. J. W. Gaubatz: *Mutation Res.* **237**, 271 (1990).
32. D. J. Galas and M. Chandler: 'Bacterial Insertion Sequences', in D. Berg and M. M. Howe (Eds.), *Mobile DNA*, ASM, Washington, p. 109 (1989).